T0275711

LONDON MATHEMATICAL SOCIETY LECTURE NOTE SERIES

Managing Editor: Professor J.W.S. Cassels, Department of Pure Mathematics and Mathematical Statistics, University of Cambridge, 16 Mill Lane, Cambridge CB2 1SB, England

The titles below are available from booksellers, or, in case of difficulty, from Cambridge University Press.

34 Representation theory of Lie groups, M.F. ATIYAH *et al*
46 p-adic analysis: a short course on recent work, N. KOBLITZ
50 Commutator calculus and groups of homotopy classes, H.J. BAUES
59 Applicable differential geometry, M. CRAMPIN & F.A.E. PIRANI
66 Several complex variables and complex manifolds II, M.J. FIELD
69 Representation theory, I.M. GELFAND *et al*
76 Spectral theory of linear differential operators and comparison algebras, H.O. CORDES
77 Isolated singular points on complete intersections, E.J.N. LOOIJENGA
83 Homogeneous structures on Riemannian manifolds, F. TRICERRI & L. VANHECKE
86 Topological topics, I.M. JAMES (ed)
87 Surveys in set theory, A.R.D. MATHIAS (ed)
88 FPF ring theory, C. FAITH & S. PAGE
89 An F-space sampler, N.J. KALTON, N.T. PECK & J.W. ROBERTS
90 Polytopes and symmetry, S.A. ROBERTSON
92 Representation of rings over skew fields, A.H. SCHOFIELD
93 Aspects of topology, I.M. JAMES & E.H. KRONHEIMER (eds)
94 Representations of general linear groups, G.D. JAMES
95 Low-dimensional topology 1982, R.A. FENN (ed)
96 Diophantine equations over function fields, R.C. MASON
97 Varieties of constructive mathematics, D.S. BRIDGES & F. RICHMAN
98 Localization in Noetherian rings, A.V. JATEGAONKAR
99 Methods of differential geometry in algebraic topology, M. KAROUBI & C. LERUSTE
100 Stopping time techniques for analysts and probabilists, L. EGGHE
104 Elliptic structures on 3-manifolds, C.B. THOMAS
105 A local spectral theory for closed operators, I. ERDELYI & WANG SHENGWANG
107 Compactification of Siegel moduli schemes, C-L. CHAI
108 Some topics in graph theory, H.P. YAP
109 Diophantine analysis, J. LOXTON & A. VAN DER POORTEN (eds)
110 An introduction to surreal numbers, H. GONSHOR
113 Lectures on the asymptotic theory of ideals, D. REES
114 Lectures on Bochner-Riesz means, K.M. DAVIS & Y-C. CHANG
115 An introduction to independence for analysts, H.G. DALES & W.H. WOODIN
116 Representations of algebras, P.J. WEBB (ed)
118 Skew linear groups, M. SHIRVANI & B. WEHRFRITZ
119 Triangulated categories in the representation theory of finite-dimensional algebras, D. HAPPEL
121 Proceedings of *Groups - St Andrews 1985*, E. ROBERTSON & C. CAMPBELL (eds)
122 Non-classical continuum mechanics, R.J. KNOPS & A.A. LACEY (eds)
125 Commutator theory for congruence modular varieties, R. FREESE & R. MCKENZIE
126 Van der Corput's method of exponential sums, S.W. GRAHAM & G. KOLESNIK
128 Descriptive set theory and the structure of sets of uniqueness, A.S. KECHRIS & A. LOUVEAU
129 The subgroup structure of the finite classical groups, P.B. KLEIDMAN & M.W. LIEBECK
130 Model theory and modules, M. PREST
131 Algebraic, extremal & metric combinatorics, M-M. DEZA, P. FRANKL & I.G. ROSENBERG (eds)
132 Whitehead groups of finite groups, ROBERT OLIVER
133 Linear algebraic monoids, MOHAN S. PUTCHA
134 Number theory and dynamical systems, M. DODSON & J. VICKERS (eds)
135 Operator algebras and applications, 1, D. EVANS & M. TAKESAKI (eds)
136 Operator algebras and applications, 2, D. EVANS & M. TAKESAKI (eds)
137 Analysis at Urbana, I, E. BERKSON, T. PECK, & J. UHL (eds)
138 Analysis at Urbana, II, E. BERKSON, T. PECK, & J. UHL (eds)
139 Advances in homotopy theory, S. SALAMON, B. STEER & W. SUTHERLAND (eds)
140 Geometric aspects of Banach spaces, E.M. PEINADOR & A. RODES (eds)
141 Surveys in combinatorics 1989, J. SIEMONS (ed)
144 Introduction to uniform spaces, I.M. JAMES
145 Homological questions in local algebra, JAN R. STROOKER
146 Cohen-Macaulay modules over Cohen-Macaulay rings, Y. YOSHINO
147 Continuous and discrete modules, S.H. MOHAMED & B.J. MÜLLER
148 Helices and vector bundles, A.N. RUDAKOV *et al*
149 Solitons, nonlinear evolution equations and inverse scattering, M. ABLOWITZ & P. CLARKSON
150 Geometry of low-dimensional manifolds 1, S. DONALDSON & C.B. THOMAS (eds)
151 Geometry of low-dimensional manifolds 2, S. DONALDSON & C.B. THOMAS (eds)
152 Oligomorphic permutation groups, P. CAMERON

153 L-functions and arithmetic, J. COATES & M.J. TAYLOR (eds)
154 Number theory and cryptography, J. LOXTON (ed
155 Classification theories of polarized varieties, TAKAO FUJITA
156 Twistors in mathematics and physics, T.N. BAILEY & R.J. BASTON (eds)
157 Analytic pro-*p* groups, J.D. DIXON, M.P.F. DU SAUTOY, A. MANN & D. SEGAL
158 Geometry of Banach spaces, P.F.X. MÜLLER & W. SCHACHERMAYER (eds)
159 Groups St Andrews 1989 volume 1, C.M. CAMPBELL & E.F. ROBERTSON (eds)
160 Groups St Andrews 1989 volume 2, C.M. CAMPBELL & E.F. ROBERTSON (eds)
161 Lectures on block theory, BURKHARD KÜLSHAMMER
162 Harmonic analysis and representation theory, A. FIGA-TALAMANCA & C. NEBBIA
163 Topics in varieties of group representations, S.M. VOVSI
164 Quasi-symmetric designs, M.S. SHRIKANDE & S.S. SANE
165 Groups, combinatorics & geometry, M.W. LIEBECK & J. SAXL (eds)
166 Surveys in combinatorics, 1991, A.D. KEEDWELL (ed)
167 Stochastic analysis, M.T. BARLOW & N.H. BINGHAM (eds)
168 Representations of algebras, H. TACHIKAWA & S. BRENNER (eds)
169 Boolean function complexity, M.S. PATERSON (ed)
170 Manifolds with singularities and the Adams-Novikov spectral sequence, B. BOTVINNIK
171 Squares, A.R. RAJWADE
172 Algebraic varieties, GEORGE R. KEMPF
173 Discrete groups and geometry, W.J. HARVEY & C. MACLACHLAN (eds)
174 Lectures on mechanics, J.E. MARSDEN
175 Adams memorial symposium on algebraic topology 1, N. RAY & G. WALKER (eds)
176 Adams memorial symposium on algebraic topology 2, N. RAY & G. WALKER (eds)
177 Applications of categories in computer science, M. FOURMAN, P. JOHNSTONE, & A. PITTS (eds)
178 Lower K- and L-theory, A. RANICKI
179 Complex projective geometry, G. ELLINGSRUD *et al*
180 Lectures on ergodic theory and Pesin theory on compact manifolds, M. POLLICOTT
181 Geometric group theory I, G.A. NIBLO & M.A. ROLLER (eds)
182 Geometric group theory II, G.A. NIBLO & M.A. ROLLER (eds)
183 Shintani zeta functions, A. YUKIE
184 Arithmetical functions, W. SCHWARZ & J. SPILKER
185 Representations of solvable groups, O. MANZ & T.R. WOLF
186 Complexity: knots, colourings and counting, D.J.A. WELSH
187 Surveys in combinatorics, 1993, K. WALKER (ed)
188 Local analysis for the odd order theorem, H. BENDER & G. GLAUBERMAN
189 Locally presentable and accessible categories, J. ADAMEK & J. ROSICKY
190 Polynomial invariants of finite groups, D.J. BENSON
191 Finite geometry and combinatorics, F. DE CLERCK *et al*
192 Symplectic geometry, D. SALAMON (ed)
193 Computer algebra and differential equations, E. TOURNIER (ed)
194 Independent random variables and rearrangement invariant spaces, M. BRAVERMAN
195 Arithmetic of blowup algebras, WOLMER VASCONCELOS
196 Microlocal analysis for differential operators, A. GRIGIS & J. SJÖSTRAND
197 Two-dimensional homotopy and combinatorial group theory, C. HOG-ANGELONI, W. METZLER & A.J. SIERADSKI (eds)
198 The algebraic characterization of geometric 4-manifolds, J.A. HILLMAN
199 Invariant potential theory in the unit ball of $\mathbf{C}^n$, MANFRED STOLL
200 The Grothendieck theory of dessins d'enfant, L. SCHNEPS (ed)
201 Singularities, JEAN-PAUL BRASSELET (ed)
202 The technique of pseudodifferential operators, H.O. CORDES
203 Hochschild cohomology of von Neumann algebras, A. SINCLAIR & R. SMITH
204 Combinatorial and geometric group theory, A.J. DUNCAN, N.D. GILBERT & J. HOWIE (eds)
205 Ergodic theory and its connections with harmonic analysis, K. PETERSEN & I. SALAMA (eds)
206 An introduction to noncommutative differential geometry and its physical applications, J. MADORE
207 Groups of Lie type and their geometries, W.M. KANTOR & L. DI MARTINO (eds)
208 Vector bundles in algebraic geometry, N.J. HITCHIN, P. NEWSTEAD & W.M. OXBURY (eds)
209 Arithmetic of diagonal hypersurfaces over finite fields, F.Q. GOUVÊA & N. YUI
210 Hilbert C*-modules, E.C. LANCE
211 Groups 93 Galway / St Andrews I, C.M. CAMPBELL *et al*
212 Groups 93 Galway / St Andrews II, C.M. CAMPBELL *et al*
214 Generalised Euler-Jacobi inversion formula and asymptotics beyond all orders, V. KOWALENKO, N.E. FRANKEL, M.L. GLASSER & T. TAUCHER
215 Number theory, S. DAVID (ed)
216 Stochastic partial differential equations, A. ETHERIDGE (ed)
217 Quadratic forms with applications to algebraic geometry and topology, A. PFISTER
218 Surveys in combinatorics, 1995, PETER ROWLINSON (ed)
220 Algebraic set theory, A. JOYAL & I. MOERDIJK
221 Harmonic approximation, S.J. GARDINER
222 Advances in linear logic, J.-Y. GIRARD, Y. LAFONT & L. REGNIER (eds)
223 Analytic semigroups and semilinear initial boundary value problems, KAZUAKI TAIRA
226 Novikov conjectures, index theorems and rigidity I, S. FERRY, A. RANICKI & J. ROSENBERG (eds)
227 Novikov conjectures, index theorems and rigidity II, S. FERRY, A. RANICKI & J. ROSENBERG (eds)

London Mathematical Society Lecture Note Series. 226

Novikov Conjectures, Index Theorems and Rigidity

Volume 1

Oberwolfach 1993

Edited by

Steven C. Ferry
State University of New York, Binghampton

Andrew Ranicki
University of Edinburgh

Jonathan Rosenberg
University of Maryland

CAMBRIDGE UNIVERSITY PRESS
Cambridge, New York, Melbourne, Madrid, Cape Town, Singapore, São Paulo

Cambridge University Press
The Edinburgh Building, Cambridge CB2 8RU, UK

Published in the United States of America by Cambridge University Press, New York

www.cambridge.org
Information on this title: www.cambridge.org/9780521497961

First published 1995

A catalogue record for this publication is available from the British Library

ISBN 978-0-521-49796-1 paperback

Transferred to digital printing 2008

To Sergei P. Novikov
with respect and admiration

Contents

Contents, Volume 1

Contents, Volume 2

Preface

These volumes grew out of the conference which we organized at the Mathematisches Forschungsinstitut Oberwolfach in September, 1993, on the subject of "Novikov conjectures, index theorems and rigidity." The aim of the meeting was to examine the Novikov conjecture, one of the central problems of the topology of manifolds, along with the vast assortment of refinements, generalizations, and analogues of the conjecture which have proliferated over the last 25 years. There were 38 participants, coming from Australia, Canada, France, Germany, Great Britain, Hong Kong, Poland, Russia, Switzerland, and the United States, with interests in topology, analysis, and geometry. What made the meeting unusual were both its interdisciplinary scope and the lively and constructive interaction of experts from very different fields of mathematics. The success of the meeting led us to try to capture its spirit in print, and these two volumes are the result.

It was not our intention to produce the usual sort of conference proceedings volume consisting of research announcements by the participants. There are enough such tomes gathering dust on library shelves. Instead, we have hoped to capture a snapshot of the status of work on the Novikov conjecture and related topics, now that the subject is about 25 years old. We have also tried to produce volumes which will be helpful to beginners in the area (especially graduate students), and also to those working in some aspect of the subject who want to understand the connection between what they are doing and what is going on in other fields. Accordingly, we have included here :

(a) a fairly detailed historical survey of the Novikov conjecture, including an annotated reprint of the original statement (both in the original Russian and in English translation), and a reasonably complete bibliography of the subsequent developments;

(b) the texts of hitherto unpublished classic papers by Milnor, Browder, and Kasparov relevant to the Novikov conjecture, which are known to the experts but hard for the uninitiated to locate;

(c) several papers (Ferry, Ferry-Weinberger, Ranicki, Rosenberg) which, while they present some new work, also attempt to survey aspects of the subject; and

(d) research papers which reflect the wide range of current techniques used to attack the Novikov conjecture: geometry, analysis, topology, algebra, ...

All the research papers have been refereed.

We hope that the reader will find the two volumes worthwhile, not merely as a technical reference tool, but also as stimulating reading to be browsed through at leisure.

We should like to thank the Director and staff of the Mathematisches Institut Oberwolfach for their expert logistical help, for their financial support, and for the marvellous working environment that made possible the 1993 conference that got this project started. Thanks are due as well to all the participants at the meeting, to the contributors to the two volumes, to the referees of the research papers, to the contributors to the problem list and bibliography, and to Roger Astley and David Tranah of Cambridge University Press. We thank the National Science Foundation of the U. S. for its support under grants DMS 90-03746, DMS-93-05758 and DMS 92-25063, the European Union for its support via the K-theory Initiative under Science Plan SCI–CT91–0756, as well as the Centenary Fund of the Edinburgh Mathematical Society.

Steve Ferry
(Binghamton, NY)
Andrew Ranicki
(Edinburgh, Scotland)
Jonathan Rosenberg
(College Park, MD)

June, 1995

Conference Programme

Oberwolfach 1993

Novikov Conjectures, Index Theorems and Rigidity

Monday, 6th September

9.30–10.30	A. S. Mishchenko	Analytic torsion over C^*-algebras
11.00–12.00	J. Roe	Coarse geometry and index theory
16.00–17.00	G. Kasparov	Groups acting on 'bolic' spaces and the Novikov conjecture
17.15–18.15	S. Weinberger	Coarse geometry and the Novikov conjecture

Informal evening session

20.00–20.30	A. Ranicki	Algebraic Novikov for analysts
20.45–21.15	J. Rosenberg	Analytic Novikov for topologists

Tuesday, 7th September

9.30–10.30	E. K. Pedersen	Controlled algebra and the Novikov conjecture
11.00–12.00	J. Eichhorn	Rigidity of the index on open manifolds
16.00–16.30	V. Mathai	Homotopy invariance of spectral flow and η-invariants
16.40–17.10	J. Kaminker	Duality for dynamical system-C^*-algebras and the Novikov conjecture
17.20–17.50	T. Koźniewski	On the Nil and UNil groups
18.00–18.30	S. Prassidis	K-theory rigidity of virtually nilpotent groups

Informal evening session

20.00–	S. Ferry (chair)	Various proofs of Novikov's theorem on the topological invariance of the rational Pontrjagin classes

Wednesday, 8th September

9.30–10.00	O. Attie	The Borel conjecture for manifolds with bounded geometry
10.15–10.45	E. Troitsky	The homotopy invariance of some higher signatures
11.00–11.30	P. Julg	The KK-ring and Baum-Connes conjecture for complex hyperbolic groups
11.45–12.15	W. Lück	L^2-Betti numbers and the Novikov-Shubin invariants
14.00–15.00	M. Gromov	Reflections on the Novikov conjecture
16.30–17.15	S. Hurder	Exotic index theory and the Novikov conjecture
17.30–18.15	B. Williams	A proof of a conjecture of Lott

Thursday, 9th September

9.30–10.30	M. Puschnigg	Cyclic homology and the Novikov conjecture
11.00–11.30	U. Bunke	Glueing problems for indices and the η-invariant
11.45–12.15	M. Yan	The role of signature in periodicity

Informal evening session
20.00– **Problem session**

Friday, 10th September

9.30–10.00	R. Jung	Elliptic homology and the Novikov conjecture
10.15–10.45	C. Stark	Approximate finiteness properties
11.00–12.00	J. Cuntz	On excision in cyclic homology
14.00–14.30	F. X. Connolly	Ends of G-manifolds and stratified spaces
14.40–15.10	J. Miller	Signature operators and surgery groups

Participant List

Oberwolfach 1993

Novikov Conjectures, Index Theorems and Rigidity

Note: Affiliations are as of the time of the conference. Addresses are as current as possible as of June, 1995.

Organizers:

Ferry, Steve, SUNY at Binghamton
mail: Dept. of Mathematical Sciences, SUNY at Binghamton, Binghamton, NY 13901, U.S.A.
email: steve@math.binghamton.edu

Ranicki, Andrew, University of Edinburgh
mail: Dept. of Mathematics and Statistics, James Clerk Maxwell Bldg., University of Edinburgh, Edinburgh EH9 3JZ, Scotland, U.K.
email: a.ranicki@edinburgh.ac.uk

Rosenberg, Jonathan, University of Maryland
mail: Dept. of Mathematics, University of Maryland, College Park, MD 20742, U.S.A.
email: jmr@math.umd.edu

Other participants:

Attie, Oliver, McMaster University
mail: Dept. of Mathematics, University of California at Los Angeles, Los Angeles, CA 90024, U.S.A.
email: oattie@math.ucla.edu

Block, Jonathan, University of Pennsylvania
mail: Dept. of Mathematics, University of Pennsylvania, Philadelphia, PA 19104, U.S.A.
email: blockj@archimedes.math.upenn.edu

Bunke, Ulrich, Humboldt University
mail: Dept. of Mathematics, Humboldt Universität, Unter den Linden 6, 10099 Berlin, Germany.
email: bunke@mathematik.hu-berlin.de

Connolly, Frank, University of Notre Dame
mail: Dept. of Mathematics, University of Notre Dame, Notre Dame, IN 46556, U.S.A.
email: francis.x.connolly.1@math.nd.edu

Cuntz, Joachim, Heidelberg University
mail: Mathematisches Institut, Universität Heidelberg, Im Neuenheimer Feld 288, 69120 Heidelberg, Germany.
email: `cuntz@math.uni-heidelberg.de`

Davis, James, Indiana University
mail: Dept. of Mathematics, Indiana University, Bloomington, IN 47405, U.S.A.
email: `jfdavis@ucs.indiana.edu`

tom Dieck, Tammo, Göttingen University
mail: Mathematisches Institut, Universität Göttingen, Bunsenstraße 3–5, 37073 Göttingen, Germany.
email: `tammo@cfgauss.uni-math.gwdg.de`

Eichhorn, Jürgen, Greifswald University
mail: Fachbereich Mathematik, Universität Greifswald, Jahnstraße 15a, 17487 Greifswald, Germany.
email: `eichhorn@math-inf.uni-greifswald.dbp.de`

Gromov, Misha, I.H.E.S.
mail: Institut des Hautes Études Scientifiques, 35 route de Chartres, 91440 Bures-sur-Yvette, France.
email: `gromov@ihes.fr; gromov@math.umd.edu`

Hughes, Bruce, Vanderbilt University
mail: Dept. of Mathematics, Vanderbilt University, Nashville, TN 37240, U.S.A.
email: `hughescb@athena.cas.vanderbilt.edu`

Hurder, Steve, University of Illinois at Chicago
mail: Dept. of Mathematics (m/c 249), University of Illinois at Chicago, 851 S. Morgan St., Chicago, IL 60607, U.S.A.
email: `hurder@boss.math.uic.edu`

Julg, Pierre, Louis Pasteur University
mail: Institut de Recherche Mathématique Avancée, Université Louis Pasteur, 7 rue René Descartes, 67084 Strasbourg Cedex, France.
email: `julg@math.u-strsbg.fr`

Jung, Rainer, University of Mainz
mail: Fachbereich Mathematik, Johannes Gutenberg Universität Mainz, 55099 Mainz, Germany.
email: `jung@topologie.mathematik.uni-mainz.de`

Kaminker, Jerry, Indiana/Purdue University at Indianapolis
mail: Dept. of Mathematics, Indiana Univ./Purdue Univ. at Indianapolis, 402 N. Blackford Street, Indianapolis, IN 46202, U.S.A.
email: `kaminker@math.iupui.edu`

Kasparov, Gennadi, University of Marseille–Luminy
mail: Département Mathématique–Informatique, Université de Marseille–Luminy, 13288 Marseille Cedex 9, France.
email: `kasparov@lumimath.univ-mrs.fr`

Kassel, Christian, Louis Pasteur University
mail: Institut de Recherche Mathématique Avancée, Université Louis Pasteur, 7 rue René Descartes, 67084 Strasbourg Cedex, France.
email: `kassel@math.u-strasbg.fr`

Koźniewski, Tadeusz, University of Warsaw
mail: Inst. of Mathematics, Uniwersytet Warszawski, 00–901 Warszawa, Poland.
email: `tkozn@mimuw.edu.pl`

Lesch, Matthias, Augsburg University
mail: Mathematisches Institut, Universität Augsburg, Universitätsstraße 14, 86135 Augsburg, Germany.
email: `lesch@uni-augsburg.de`

Lück, Wolfgang, Mainz University
mail: Fachbereich Mathematik, Johannes Gutenberg Universität Mainz, 55099 Mainz, Germany.
email: `lueck@topologie.mathematik.uni-mainz.de`

Luke, Glenys, Oxford University
mail: St. Hugh's College, Oxford OX2 6LE, England, U.K.
email: `glenys.luke@st_hugh.ox.ac.uk`

Mathai, Varghese, University of Adelaide
mail: Dept. of Pure Mathematics, University of Adelaide, Adelaide, SA 5005, Australia.
email: `vmathai@maths.adelaide.edu.au`

Mavra, Boris, Oxford University
mail: Mathematical Institute, Oxford University, 24–29 St. Giles', Oxford OX1 3LB, England, U.K.
email: `mavra@maths.ox.ac.uk`

Miller, John, Indiana/Purdue University at Indianapolis
mail: Dept. of Mathematics, Indiana Univ./Purdue Univ. at Indianapolis, 402 N. Blackford Street, Indianapolis, IN 46202, U.S.A.
email: `jmiller@math.iupui.edu`

Mishchenko, Alexander, Moscow State University
mail: Dept. of Mathematics and Mechanics, Moscow State University, Moscow 119899, Russia.
email: `asmish@mech.math.msu.su`

Pedersen, Erik, SUNY at Binghamton
mail: Dept. of Mathematical Sciences, SUNY at Binghamton, Binghamton, NY 13901, U.S.A.
email: `erik@math.binghamton.edu`

Prassidis, Stratos, Vanderbilt University
mail: Dept. of Mathematics, Vanderbilt University, Nashville, TN 37240, U.S.A.
email: `prassie@athena.cas.vanderbilt.edu`

Puschnigg, Michael, Heidelberg University
mail: Mathematisches Institut, Universität Heidelberg, Im Neuenheimer Feld 288, 69120 Heidelberg, Germany.
email: `puschnig@vogon.mathi.uni-heidelberg.de`

Roe, John, Oxford University
mail: Jesus College, Oxford OX1 3DW, England, U.K.
email: `jroe@spinoza.jesus.ox.ac.uk`

Rothenberg, Mel, University of Chicago
mail: Dept. of Mathematics, University of Chicago, Chicago IL 60637, U.S.A.
email: `mel@math.uchicago.edu`

Stark, Chris, University of Florida
mail: Dept. of Mathematics, University of Florida, P. O. Box 11800, Gainesville, FL 32611, U.S.A.
email: `cws@math.ufl.edu`

Troitsky, Evgenii, Moscow State University
mail: Dept. of Mathematics and Mechanics, Moscow State University, Moscow 119899, Russia.
email: `troitsky@mech.math.msu.su`

Valette, Alain, University of Neuchâtel
mail: Institut de Mathématiques, Université de Neuchâtel, Chantemerle 20, CH-2007 Neuchâtel, Switzerland.
email: `valette@maths.unine.ch`

Weinberger, Shmuel, University of Chicago
mail: Dept. of Mathematics, University of Pennsylvania, Philadelphia, PA 19104, U.S.A.
email: `shmuel@archimedes.math.upenn.edu`

Williams, Bruce, University of Notre Dame
mail: Dept. of Mathematics, University of Notre Dame, Notre Dame, IN 46556, U.S.A.
email: `bruce@bruce.math.nd.edu`

Yan, Min, Hong Kong University of Science and Technology
mail: Dept. of Mathematics, Hong Kong University of Science and Technology, Clear Water Bay, Kowloon, Hong Kong.
email: `mamyan@usthk.ust.hk`

A History and Survey of the Novikov Conjecture

Steven C. Ferry, Andrew Ranicki, and Jonathan Rosenberg

Contents

S. C. Ferry was partially supported by NSF Grant # DMS-93-05758.
J. Rosenberg was partially supported by NSF Grant # DMS-92-25063.

1. Precursors of the Novikov Conjecture

Characteristic classes

The Novikov Conjecture has to do with the question of the relationship of the characteristic classes of manifolds to the underlying bordism and homotopy theory. For smooth manifolds, the characteristic classes are by definition the characteristic classes of the tangent (or normal) bundle, so basic to this question is another more fundamental one: how much of a vector bundle is determined by its underlying spherical fibration? The Stiefel-Whitney classes of vector bundles are invariants of the underlying spherical fibration, and so the Stiefel-Whitney numbers of manifolds are homotopy invariants. Furthermore, they determine unoriented bordism. The Pontrjagin classes of vector bundles are not invariants of the underlying spherical fibration, and the Pontrjagin numbers of manifolds are not homotopy invariants. However, together with the Stiefel-Whitney numbers, they do determine oriented bordism. The essential connection between characteristic numbers and bordism was established by Thom [Th1] in the early 1950's.

Geometric rigidity

As we shall see later, the Novikov Conjecture is also closely linked to problems about rigidity of aspherical manifolds. As everyone learns in a first course in geometric topology, closed 2-manifolds are determined up to homeomorphism by their fundamental groups. In higher dimensions, of course, nothing like this is true in general, but one can still ask if *aspherical* closed manifolds (closed manifolds having contractible universal cover) are determined up to homeomorphism by their fundamental groups. That this should be the case is the *Borel Conjecture* formulated by Armand Borel in the 50's (according to Hsiang in [Hs3]), and communicated to various people in the 60's. In dimension 2, restricting attention to aspherical manifolds is little loss of generality, since S^2 and $\mathbb{RP}^2$ are the only closed 2-manifolds which are *not* aspherical. The Mostow Rigidity Theorem was the most dramatic early evidence for the Borel Conjecture, proving the conjecture for closed manifolds which are locally symmetric spaces.

The Hirzebruch signature theorem

The actual history of the Novikov Conjecture starts with the Hirzebruch signature theorem [Hir], which expresses the signature of an oriented closed $4k$-dimensional manifold M in terms of characteristic classes:

$$\text{signature}(M) \;=\; \langle \mathcal{L}(M),\, [M]\rangle \in \mathbb{Z}\ .$$

Here, $\mathcal{L}(M) \in H^{4*}(M;\mathbb{Q})$ is the $\mathcal{L}$-class of M, a certain formal power series in the Pontrjagin classes $p_*(M) \in H^{4*}(M)$ with rational coefficients. The formula is surprising in that the left hand side is an integer which only depends on the structure of the cohomology ring of M, whereas the right hand side is a sum of rational numbers which are defined (at least *a priori*) in terms of the differentiable structure. The (inhomogeneous) class $\mathcal{L}(M)$ determines all of the rational Pontrjagin classes of M, but only the component of $\mathcal{L}(M)$ in the dimension of M is homotopy invariant – in fact, the other components are not even bordism invariants. Milnor [Miln1] used the signature theorem to verify that the homotopy spheres he constructed do indeed have exotic differentiable structures.

The converse of the signature theorem (Browder, Novikov)

Following the development of Thom's bordism theory, Milnor [Miln2] proved that two manifolds are bordant if and only if they are related by a finite sequence of surgeries. This was the beginning of the use of surgery as a fundamental tool in differential topology. Soon afterwards, Kervaire and Milnor [KerM] used surgery to classify exotic spheres in dimensions ≥ 7. In 1962, Browder ([Br1],[Br4]) and Novikov [Nov1], working independently, applied the same technique to manifolds with more complicated homology. They used surgery theory to establish a converse to the Hirzebruch signature theorem in dimensions ≥ 5: if X is a simply-connected $4k$-dimensional Poincaré space, such that the signature of X is the evaluation on the fundamental class $[X] \in H_{4k}(X)$ of $\mathcal{L}(-\nu)$ for some vector bundle ν with spherical Thom class, then X is homotopy equivalent to a smooth closed manifold M with stable normal bundle pulled back from ν. A consequence of this is that for simply-connected $4k$-dimensional manifolds in high dimensions, the top degree term of the $\mathcal{L}$-class is essentially the *only* homotopy-invariant rational characteristic class. Novikov ([Nov1], [Nov2], [Nov3]) extended these ideas to the study of the *uniqueness* properties of manifold structures within a homotopy type. Sullivan [Sul] then combined the Browder-Novikov surgery theory with homotopy theory to reformulate the surgery classification of manifolds in terms of the *surgery exact sequence* of pointed sets, which for a $4k$-dimensional simply-connected manifold M ($k > 1$) has the form:

$$0 \to \mathcal{S}(M) \xrightarrow{\theta} [M, G/O] \xrightarrow{A} \mathbb{Z}$$

with $\mathcal{S}(M)$ the *structure set* of M, consisting of the equivalence classes of pairs (N, f) with N a closed manifold and $f : N \to M$ a homotopy equivalence. Two such pairs (N, f), (N', f') are equivalent if there exists a diffeomorphism $g : N \to N'$ with a homotopy $f'g \simeq f : N \to M$. Here G/O is the homotopy fiber of the forgetful map J from the classifying space BO for stable vector bundles to the classifying space BG for stable spherical fibrations, and the map θ sends an element (N, f) of the structure set to the difference between the stable normal bundle of M and the push-forward under f of the stable normal bundle of N. (Both are lifts of the same underlying spherical fibration, the Spivak normal fibration [Spv].) The map A sends an element of $[M, G/O]$, represented by a vector bundle η over M with a fiber homotopy trivialization, to

$$\langle \mathcal{L}(\tau_M \oplus \eta) - \mathcal{L}(\tau_M), [M] \rangle,$$

where τ_M is the tangent bundle of M.

Topological invariance of the rational Pontrjagin classes (Novikov)

Around 1957, Thom [Th2] and Rokhlin and Shvarts [RokS], working independently, proved that the *rational* Pontrjagin classes of PL manifolds are combinatorial invariants. As we have just explained, the work of Browder showed that the Pontrjagin classes are very far from being homotopy invariants of closed differentiable manifolds. Nevertheless, Novikov in 1966 ([Nov4], [Nov5], [Nov6]) was able to prove a most remarkable fact: the rational Pontrjagin classes are *topological* invariants. An essential feature of the proof was the use of non-simply-connected compact manifolds with free abelian fundamental group (e.g., tori), and of their non-compact universal covers.

Non-simply-connected surgery theory (Novikov, Wall)

While the basic methods used in Browder-Novikov surgery theory make sense without assuming simple connectivity, it was soon realized that formulating the correct results in the non-simply connected case is not so easy. For one thing, correctly understanding Poincaré duality in this context requires using homology with local coefficients. In fact, the correct algebraic approach required developing a theory of quadratic forms defined over an arbitrary ring with involution, the prototype being the integral group ring $\mathbb{Z}[\pi]$ of the fundamental group π. This algebra was developed in the even-dimensional case by Novikov ([Nov8], [Nov10]) and Wall, working independently, and in the odd-dimensional case by Wall [Wall1]. Using this algebra, Wall [Wall2] developed a non-simply connected version of the surgery exact sequence for a closed n-dimensional manifold M with $n \geq 5$:

$$\cdots \to L_{n+1}(\mathbb{Z}[\pi_1(M)]) \to \mathcal{S}(M) \xrightarrow{\theta} [M, G/O] \xrightarrow{A} L_n(\mathbb{Z}[\pi_1(M)]) \ .$$

The L-groups are Witt groups of $(-)^k$-quadratic forms on finitely generated free modules over the group ring for even $n = 2k$, and stable automorphism groups of such forms for odd $n = 2k + 1$. The L-groups are periodic in n, with period 4. While $\mathcal{S}(M)$ is only a pointed set, not a group, it has an affine structure: $L_{n+1}(\mathbb{Z}[\pi_1(M)])$ acts on $\mathcal{S}(M)$, and two elements with the same image under θ lie in the same orbit. Rationally, the map $\theta : \mathcal{S}(M) \to [M, G/O]$ sends a homotopy equivalence $f : N \to M$ to the difference $f_*(\mathcal{L}(N)) - \mathcal{L}(M)$. Here the push-forward map f_* can be defined as g^*, where g is a homotopy inverse to f.

Shortly after the work of Wall, new advances by Kirby and Siebenmann made it possible to carry surgery theory over from the category of differentiable manifolds to the category of topological manifolds [KirS]. One again obtained a surgery exact sequence of the same form as before, but with

G/O replaced by G/TOP. This theory made it possible to reinterpret Novikov's theorem on the topological invariance of rational Pontrjagin classes as the fact that the forgetful map $G/TOP \to G/O$ induces an isomorphism on rational homology. The classifying spaces G/O and G/TOP both have rational cohomology rings which are formal power series algebras in the Pontrjagin classes.

Higher signatures

Let Γ be a discrete group. A rational cohomology class $x \in H^*(B\Gamma; \mathbb{Q})$ may be interpreted as a characteristic class for manifolds with fundamental group Γ. If $\Gamma = \pi_1(M)$ for a manifold M, obstruction theory implies that one can always find a map $u : M \to B\Gamma$ which induces an isomorphism on π_1. For oriented M the class x defines a (rational) characteristic number, called a *higher signature*:

$$\text{signature}_x(M, u) \;=\; \langle \mathcal{L}(M) \cup u^*(x), [M] \rangle \in \mathbb{Q} .$$

This characteristic number is said to be *homotopy invariant* if for all orientation-preserving homotopy equivalences $f : N \to M$ of closed oriented manifolds and all maps $u : M \to B\Gamma$,

$$\text{signature}_x(M, u) = \text{signature}_x(N, u \circ f) \in \mathbb{Q} .$$

It is now possible to determine when this is the case. Because of the L-groups in the surgery sequence, there can be far more homotopy-invariant characteristic classes than in the simply connected case. Let $\theta^*_{(M,u)}$ be the map sending $x \in H^*(B\Gamma; \mathbb{Q})$ to the functional on structure sets sending

$$N \xrightarrow{f} M \xrightarrow{u} B\Gamma$$

to

$$\text{signature}_x(M, u) - \text{signature}_x(N, u \circ f) .$$

By definition, the homotopy-invariant higher signatures are exactly those signature_x's for which x is in the kernel of $\theta^*_{(M,u)}$, for all M and u. The surgery exact sequence shows that these are precisely the x's in the image of a certain map

$$A^* : \text{Hom}\,(L_*(\mathbb{Z}[\Gamma]), \mathbb{Q}) \to H^*(B\Gamma; \mathbb{Q}) .$$

The Novikov Conjecture is that *every* higher signature is homotopy-invariant, or equivalently that A^* is onto, for every discrete group Γ.

Discovery of special cases of the Novikov Conjecture (Rokhlin, Novikov)

Novikov's use of manifolds with free abelian fundamental group in the proof of the topological invariance of rational Pontrjagin classes led him to the study of homotopy invariance properties of other characteristic classes as well. In particular, he studied the mod-p Pontrjagin classes of homotopy lens spaces ([Nov7], [Nov9]), and the higher signatures of general non-simply connected manifolds. Novikov himself discovered that the higher signature (in this case there is essentially only one) of a manifold with infinite cyclic fundamental group is a homotopy invariant [Nov7], and Rokhlin [Rokh] studied the case of $\Gamma = \mathbb{Z}\times\mathbb{Z}$. These examples led Novikov to the formulation of the general conjecture.

2. The Original Statement of the Novikov Conjecture

The statement that is now usually known as the Novikov Conjecture first appears in complete form in §11 of S. P. Novikov's monumental paper [Nov10]. A slightly different formulation was given in Novikov's talk at the International Congress in Nice in 1970 [Nov8]. More preliminary versions had appeared in the lectures of Novikov for the de Rham Festschrift [Nov9] and the Moscow International Congress [Nov7]. Since the name "Novikov Conjecture" these days seems to mean quite different things to different people, in the interests of historical accuracy, we quote here the complete text of Novikov's original (Izvestia) formulation, both in the original Russian and in an English translation. As we shall see shortly, Novikov's original formulation already includes the three main approaches to the conjecture: the analytic, the topological, and the algebraic. Here is first the original Russian (with a few misprints corrected) and then a translation (our own correction of the printed translation in [Nov10]). The footnote indexed * is Novikov's; numbered footnotes in the English version are ours.

О нерешенных задачах

1. Здесь мы обсудим первоначально следующий общий вопрос: что такое "общая неодносвязная Формула Хирцебруха"?

На этот вопрос можно ответить таким образом: должен сущест-вовать некоторый гомоморфизм "обобщенных сигнатур"

$$\sigma_k : U_1^n(A) \to H_{n-4k}(\pi_1; \mathbb{Q})$$

такой, что для любого n-мерного замкнутого ориентированного многообразия M^n с фундаментальной группой π_1 и естественного отображения $f : M^n \to K(\pi_1, 1)$ скалярное произведение

$$\langle L_k(M^n), Df^*(x)\rangle$$

гомотопически инвариантно при всех $x \in H^*(\pi_1; \mathbb{Q})$, и DL_k как линейная форма на $H^*(\pi_1)$—или элемент $H_*(\pi_1; \mathbb{Q})$—принадлежит образцу σ_k. Мы явно построили такие гомоморфизмы для одной абелевой группы—они оказались здесь даже изоморфизмами над $\mathbb{Q}$ (неэффективно это было известно в топологии—см. [HsS], [Sh1], [Wall1]).

Конечно, эта задача может быть поставлена и для конечных модулей p—по крайней мере для больших p сравнительно с n.

Заметим, что ряд соображений подсказывает, что, например, для фундаментальных групп "солв" и "ниль"-многообразий такого рода гомоморфизм существует и является эпиморфизмом над $\mathbb{Q}$, так что допустимые классы циклов—это не только пересечение циклов коразмерности 1. Здесь можно ввести "некоммутативное расширение" кольца A—прибавление z, z^{-1} без коммутирования с A—обобщить теорню операторов типа Басса. Однако общего вопроса это не проясняет. Разумеется, более прост вопрос об "относительных Формулах Хирцебруха". Отметим, что существенно более сложным является вопрос о внутреннем вычислении скалярных произведений L_k с циклами вида $Df^*(x)$ даже для абелевых π—он не решен уже для $\pi = \mathbb{Z} \times \mathbb{Z}$ (см. [Nov6], [Nov7], [Rokh]).*

2. Посмотрим, во что переходит вопрос о "неодносвязной формуле Хирцебруха" и построении гомоморфизмов "обобщенных сигнатур"

$$\sigma : U_1^*(A) \to H^*(\pi; \mathbb{Q})$$

при замене групповых колец A кольцами функций $A = C(X)$.

Если заменить $H_*(\pi; \mathbb{Q})$ на $H^*(X)$, то мы ириходим к кадаче об абстрактно алгебраическом построении характера Черна

$$\mathrm{Ch} : U^*(A) = K^*(X) \to H^*(X) \ .$$

При этом надо исходить из каого-то чисто кольцевого алгебраического формализма в построении $H^*(X)$—от кольца $C(X)$.

---

*А. С. Мищенко нашёл своеобразный аналог классической сигнатуры—многообразию гомотопически инвариантным способом сопоставляется элемент из $U^*(\pi_1)\otimes\mathbb{Z}[\frac{1}{2}]$, что определяет гомоморфизм теории бордизмов $\Omega_*^{SO}(\pi_1) \to U^*(\pi_1)\otimes\mathbb{Z}[\frac{1}{2}]$ в эрмитову K-теорию, связанный, вероятно, с L-родом.

[An English Version:] Unsolved Problems

1. Here we consider first of all the following question: what should be the "general non-simply connected Hirzebruch formula"?

The question can be answered as follows: there should exist a certain "generalized signature"[1] homomorphism

$$\sigma_k : U_1^n(A) \to H_{n-4k}(\pi_1; \mathbb{Q}) \quad {}^{2}$$

such that for any n-dimensional closed oriented manifold M^n with fundamental group π_1 and for the natural map[3] $f : M^n \to K(\pi_1, 1)$, the scalar product $\langle L_k(M^n), Df^*(x)\rangle$ [4] is homotopy-invariant for any $x \in H^*(\pi_1; \mathbb{Q})$; and DL_k as a linear form on $H^*(\pi_1)$—or regarded as an element of $H_*(\pi_1, \mathbb{Q})$ [5]—belongs to the image of σ_k. We have explicitly constructed such homomorphisms for one class of abelian groups (viz., free abelian groups)—they turn out to be isomorphisms over $\mathbb{Q}$ (this was known non-effectively from results in topology—cf. [HsS], [Sh1], [Wall1]).

Of course, this problem can be posed for a finite modulus p, at least for p large compared with n.[6]

Let us note that a number of considerations suggest that, for example, for the fundamental groups of "solv-" and "nil-" manifolds, such a homomorphism exists and is an epimorphism over $\mathbb{Q}$, such that the allowable homology classes are not just the intersections of cycles of codimension 1. Here we can introduce a "non-commutative extension" of the ring A, by adjoining z and z^{-1} without assuming that they commute with A, to gen-

[1] In modern language, perhaps "higher signature" would be more appropriate.

[2] Here $A = \mathbb{Z}[\pi_1]$ is the group ring, and $U_1^n(A)$ is a certain variant of the Wall group $L_n(A)$; the exact decoration on the surgery group is unimportant since we are ignoring torsion here anyway. The homomorphisms σ_* are exactly what one needs to have a rational splitting of the L-theory assembly map.

[3] the classifying map for the universal cover of M

[4] Here L_k is the component of the total Hirzebruch L-class in degree $4k$, and D denotes the Poincaré dual, or $\cap$-product with the fundamental class $[M]$ determined by the orientation.

[5] meaning $f_*(DL_k(M^n))$

[6] It seems that here Novikov is referring back to a problem discussed in §3 of his paper [Nov9], concerning topological and homotopy invariance of "mod-p" Pontrjagin classes. While the mod-p Pontrjagin classes are in general not even homeomorphism invariants, Corollary C in [Nov9, §3] asserts that for any integer $n \geq 2$, the tangential homotopy type of lens spaces obtained as quotients of S^{2n-1} by linear representations of $\mathbb{Z}/p$ on $\mathbb{C}^n$ (free away from the origin) is a topological invariant, provided that p is sufficiently large compared with n.

eralize the theory of operators of Bass type.[7] However, this does not clarify the general question. It goes without saying that the question of a "relative Hirzebruch formula" is simpler.[8] Let us note that the question of the intrinsic calculation of scalar products of L_k with cycles of the form $Df^*(x)$ is essentially more complicated even for abelian π—it has not been solved even for $\pi = \mathbb{Z} \times \mathbb{Z}$ (see [Nov6], [Nov7], [Rokh]).*[9]

2. Let us see what the question about a "general non-simply connected Hirzebruch formula" and the construction of "generalized signature" homomorphisms

$$\sigma_k : U_1^*(A) \to H_*(\pi; \mathbb{Q})$$

becomes when we replace the group ring A by a ring of functions $A = C(X)$.[10]

If we replace $H_*(\pi; \mathbb{Q})$ by $H^*(X)$ then we arrive at a problem about the abstract algebraic construction of the Chern character

$$\mathrm{Ch} : U^*(A) = K^*(X) \to H^*(X) .$$

For this it is necessary to start from some purely ring-theoretic formalism for constructing $H^*(X)$ from the ring $C(X)$.[11]

---

[7]Without saying so, Novikov is sketching here an inductive method of proving the Novikov Conjecture for poly-$\mathbb{Z}$ groups in a purely algebraic way. For the free abelian case, one needs an analogue in L-theory of the Bass-Heller-Swan decomposition of the K-theory of a Laurent polynomial ring $A[z, z^{-1}]$ ([BHS]). This was first provided in work of Shaneson [Sh2]. A similar method will work for poly-$\mathbb{Z}$ groups but it is necessary to work with twisted Laurent rings or crossed products

$$A \rtimes_\alpha \mathbb{Z} = A_\alpha[z, z^{-1}] := \langle A, z, z^{-1} \mid zaz^{-1} = \alpha(a), \quad a \in A \rangle$$

and to prove a "Bass type" theorem for those. Such theorems for twisted Laurent rings were later provided by Farrell and Hsiang ([FarHs2], [FarHs3]) for algebraic K-theory, by Cappell [Cap2] and Ranicki [Ran3] for algebraic L-theory, and by Pimsner-Voiculescu [PimV] for the K-theory of C^*-algebras. Specific applications to the Novikov Conjecture were provided in [FarHs4], [FarHs5] and in [Ros2], [Ros4].

[8]Novikov has pointed out to us that he was referring here to (relative) invariants of degree-one normal maps, which are easier to define than (absolute) invariants for closed manifolds.

*A. S. Mishchenko has found an analogue of the classical signature—a homotopy-invariant element of $U^*(\pi_1) \otimes \mathbb{Z}[\frac{1}{2}]$ associated to a manifold, which defines a homomorphism from bordism theory $\Omega_*^{SO}(\pi_1) \to U^*(\pi_1) \otimes \mathbb{Z}[\frac{1}{2}]$ to hermitian K-theory, related, apparently, to the L-genus.

[9]This is the symmetric signature of Mishchenko [Mis1] and Ranicki ([Ran4], [Ran5]).

[10]Here Novikov is anticipating what later became a major industry, of studying the Novikov Conjecture in the context of C^*-algebras rather than group rings. A ring of the form $C(X)$ is exactly the most general commutative (complex) C^*-algebra.

[11]This is of course exactly what Connes has done with the introduction of cyclic

3. Work related to the Novikov Conjecture: The First 12 Years or So

Statements of the Novikov and Borel Conjectures

Let Γ be a discrete group.

Novikov Conjecture for Γ. *The higher signatures determined by Γ are all homotopy invariant, i.e. for every rational cohomology class $x \in H^*(B\Gamma; \mathbb{Q})$, for every orientation-preserving homotopy equivalence $f : N \to M$ of closed oriented manifolds and for every map $u : M \to B\Gamma$*

$$signature_x(M, u) = signature_x(N, u \circ f) \in \mathbb{Q} .$$

Borel Conjecture for Γ. *Every homotopy equivalence $f : N \to M$ of closed aspherical manifolds with $\pi_1(M) = \Gamma$ is homotopic to a homeomorphism. More generally, if $f : (N, \partial N) \to (M, \partial M)$ is a homotopy equivalence of compact manifolds with boundary such that M is aspherical, $\pi_1(M) = \Gamma$ and $\partial f : \partial N \to \partial M$ is a homeomorphism, then f is homotopic rel boundary to a homeomorphism.*

The first part of the Borel Conjecture only applies to discrete groups Γ such that the classifying space $B\Gamma$ is realized by a closed aspherical manifold M with $\pi_1(M) = \Gamma$, $\pi_i(M) = 0$ for $i \geq 2$. The more general part applies to any discrete group Γ such that $B\Gamma$ is realized by a finite aspherical polyhedron K, since then any regular neighbourhood of K in a high-dimensional Euclidean space is a compact manifold with boundary $(M, \partial M)$ such that $M \simeq K \simeq B\Gamma$ is aspherical. Such Γ are finitely presented, but in a later section we shall also formulate a version of the Borel Conjecture for *non-compact* manifolds, which applies to Γ which need not be finitely generated.

The Novikov and Borel Conjectures are only interesting for *infinite* groups Γ.

The *h-cobordism* version of the Borel Conjecture has the same hypothesis, but it is only required that the homotopy equivalence be h-cobordant to a homeomorphism. There is also an *s-cobordism* version of the Borel Conjecture in which it is required that the homotopy equivalence be simple: by the s-cobordism theorem for dimensions ≥ 6 there exists a homotopy to a

homology, though one complication that Novikov seems not to have anticipated is the need to make a good choice of a dense subalgebra $\mathcal{A}$ of the C^*-algebra $A = C(X)$, which on the one hand has the property that the inclusion $\mathcal{A} \hookrightarrow A$ induces an isomorphism on (topological) K-theory, and on the other hand gives the correct cyclic homology groups.

homeomorphism if and only if there exists an s-cobordism to a homeomorphism. The h- and s-cobordism versions of the Borel Conjecture only differ from the actual Borel Conjecture in Whitehead torsion considerations. In particular, if $\mathrm{Wh}(\Gamma) = 0$ the three versions of the conjecture coincide.

Surgery theory shows that the Borel Conjecture for Γ implies the Novikov Conjecture for Γ, and that in fact the Borel Conjecture is an integral version of the Novikov Conjecture.

Also at about the same that Novikov's *Izvestia* paper appeared in print, Wall's monumental book [Wall2] appeared, giving for the first time a complete published account of the theory of non-simply connected surgery. The appendices to this book, written later than the main body of the text, contain Wall's slight reformulation of the Novikov Conjecture. Using Mishchenko's work on the symmetric signature (which is described in the next section) Wall made the first study of what is now :

Integral Novikov Conjecture for Γ. *The assembly map in quadratic L-theory*

$$A_\Gamma : H_*(B\Gamma; \mathbb{L}_\bullet(\mathbb{Z})) \to L_*(\mathbb{Z}[\Gamma])$$

is an isomorphism for a torsion-free group Γ.

See the section below on surgery spectra for an account of the quadratic L-theory assembly map.

For a group Γ which is the fundamental group of an aspherical manifold $M \simeq B\Gamma$ and is such that the Whitehead group of Γ vanishes the Integral Novikov Conjecture is in fact equivalent to the Borel Conjecture in dimensions ≥ 5.

Mishchenko and the symmetric signature

As Novikov indicated in a footnote (marked above with an asterisk) to his *Izvestia* paper, a useful technical tool, the *symmetric signature*, was developed by Mishchenko ([Mis1], [Mis2]) shortly after Novikov was led to the first version of his conjecture. Mishchenko worked not with quadratic forms over the integral group ring $\mathbb{Z}[\Gamma]$ of the fundamental group Γ, but rather with symmetric forms over the rational group ring $\mathbb{Q}[\Gamma]$ (though for rings containing $\frac{1}{2}$ there is no essential difference between quadratic and symmetric forms), and more generally with chain complexes C over an arbitrary ring with involution A, with a symmetric Poincaré duality $C^{n-*} \simeq C$. In more modern language, Mishchenko had in effect introduced the *symmetric L-groups* $L^n(A)$, as the cobordism groups of n-dimensional symmetric Poincaré complexes over A. The symmetric signature of an n-dimensional Poincaré duality space M with $\pi_1(M) = \Gamma$ is the cobordism class $\sigma^*(M) \in L^n(\mathbb{Z}[\Gamma])$ of the chain complex $C(\widetilde{M})$ of the universal cover $\widetilde{M}$. This is a homotopy invariant of M, which for $n \equiv 0 (\bmod 4)$, $\Gamma = \{1\}$, is just the ordinary signature. The symmetrization maps $1 + T : L_*(A) \to L^*(A)$ from the Wall quadratic L-groups $L_*(A)$ are isomorphisms modulo 2-primary torsion, for any ring with involution A. The symmetrization of the surgery obstruction $\sigma_*(f,b) \in L_n(\mathbb{Z}[\Gamma])$ of an n-dimensional normal map $(f,b) : N \to M$ is the difference of the symmetric signatures

$$(1+T)\sigma_*(f,b) = \sigma^*(N) - \sigma^*(M) \in L^n(\mathbb{Z}[\Gamma]) .$$

Mishchenko and Soloviev ([Mis5], [MisS1]) used sheaves of symmetric Poincaré complexes to define assembly maps[12]

$$A : H_*(M; \mathbb{L}^\bullet(\mathbb{Z})) \to L^*(\mathbb{Z}[\Gamma]) .$$

Here, $\mathbb{L}^\bullet(\mathbb{Z})$ is the spectrum of the symmetric L-theory of $\mathbb{Z}$, and

$$H_*(M; \mathbb{L}^\bullet(\mathbb{Z})) \otimes \mathbb{Q} \cong \sum_{k=0}^{\infty} H_{*-4k}(M; \mathbb{Q}) .$$

The surgery obstruction of a normal map $(f,b) : N \to M$ of closed n-dimensional manifolds is determined modulo 2-primary torsion by an element $[f,b]^\bullet \in H_n(M; \mathbb{L}^\bullet(\mathbb{Z}))$[13]. If $u : M \to B\Gamma$ is the classifying map for the universal cover of M, the assembly map A for M factors as

$$A : H_n(M; \mathbb{L}^\bullet(\mathbb{Z})) \xrightarrow{u_*} H_n(B\Gamma; \mathbb{L}^\bullet(\mathbb{Z})) \xrightarrow{A_\Gamma} L^n(\mathbb{Z}[\Gamma])$$

with A_Γ the assembly map for the classifying space $B\Gamma$, and

$$A[f,b]^\bullet = A_\Gamma u_*[f,b]^\bullet = (1+T)\sigma_*(f,b) = \sigma^*(N) - \sigma^*(M) \in L^n(\mathbb{Z}[\Gamma])$$

is the difference of the symmetric signatures. The torsion-free part

$$\begin{aligned} u_*[f,b]^\bullet \otimes 1 &\in H_n(B\Gamma; \mathbb{L}^\bullet(\mathbb{Z})) \otimes \mathbb{Q} \\ &= H_{n-4*}(B\Gamma; \mathbb{Q}) = \mathrm{Hom}_{\mathbb{Q}}(H^{n-4*}(B\Gamma; \mathbb{Q}), \mathbb{Q}) \end{aligned}$$

---

[12] This construction of A required M to be a manifold, which is not necessary in the construction of A due to Ranicki [Ran9]; cf. the section below on surgery spectra.

[13] The symmetric L-theory homology class $[f,b]^\bullet$ does not depend on the bundle map b, being the difference $[f,b]^\bullet = f_*[N]_{\mathbb{L}} - [M]_{\mathbb{L}}$ of absolute invariants with $A[M]_{\mathbb{L}} = \sigma^*(M)$, $A[N]_{\mathbb{L}} = \sigma^*(N)$. See Ranicki ([Ran9],[Ran10]) for the symmetric L-theory orientation of manifolds. The actual surgery obstruction is the quadratic L-theory assembly $\sigma_*(f,b) = A[f,b]_\bullet \in L_n(\mathbb{Z}[\Gamma])$ of a quadratic L-theory homology class $[f,b]_\bullet \in H_n(M; \mathbb{L}_\bullet)$ with $(1+T)[f,b]_\bullet = [f,b]^\bullet$, which is not in general the difference of absolute invariants.

determines and is determined by the differences of the higher signatures

$$\text{signature}_x(N, u \circ f) - \text{signature}_x(M, u) \in \mathbb{Q}, \qquad x \in H^{n-4*}(B\Gamma; \mathbb{Q}) .$$

If $f : N \to M$ is a homotopy equivalence of closed manifolds, then $\sigma^*(N) = \sigma^*(M)$ and

$$u_*[f, b]^\bullet \in \ker\big(A_\Gamma : H_n(B\Gamma; \mathbb{L}^\bullet(\mathbb{Z})) \to L^n(\mathbb{Z}[\Gamma])\big) .$$

Thus if A_Γ is a rational injection then the higher signatures of M and N are equal and the Novikov Conjecture on the homotopy invariance of the higher signatures holds for Γ. In fact, the following is true :

Proposition. *The Novikov Conjecture holds for a group Γ if and only if the assembly map in symmetric L-theory*

$$A_\Gamma : H_*(B\Gamma; \mathbb{L}^\bullet(\mathbb{Z})) \to L^*(\mathbb{Z}[\Gamma])$$

is a rational split injection.

Many proofs of the Novikov Conjecture use special properties of some class of groups Γ to construct (rational) splittings $L^*(\mathbb{Z}[\Gamma]) \to H_*(B\Gamma; \mathbb{L}^\bullet(\mathbb{Z}))$ of A_Γ.

Lusztig and the analytic approach

In his thesis, published in 1972 [Lus], Lusztig made a major contribution to the theory of the Novikov Conjecture by being the first one to use *analysis*, more specifically, *index theory*, to attack the conjecture. Lusztig's paper was in fact the prototype for what was ultimately to be the largest body of literature related to the conjecture. The basic idea of Lusztig's work was to relate the higher signatures of a manifold to *a priori* homotopy invariants coming from the de Rham complex with local coefficients in a flat vector bundle. In the case of an oriented closed manifold M^{2m} with free abelian fundamental group $\mathbb{Z}^k$, the flat line bundles over M are parametrized by a torus T^k, and a choice of a Riemannian metric on M gives rise to a signature operator $D = d + d^*$ which can be "twisted" by any of these line bundles. Twisting by a line bundle does not change the index of D, which is just signature(M), but viewing all the twists simultaneously gives a *family* of elliptic operators on M parametrized by T^k. Lusztig showed that the index of this family, in the sense of the Atiyah-Singer Index Theorem for Families, is on the one hand a homotopy invariant, but on the other hand related to the higher signatures. He was thus able to give an analytic proof of the Novikov Conjecture for manifolds with free abelian fundamental group. His methods also gave partial results for other fundamental groups with "lots" of finite-dimensional representations (for which one can again twist the signature operator by a family of flat bundles).

Splitting theorems for polynomial extensions

Surgery on codimension 1 submanifolds has been an important feature of the study of non-simply-connected manifolds in general, and the Novikov Conjecture in particular. Browder ([Br2], [Br3]) used surgery on codimension 1 submanifolds to deal with the homotopy properties of simply-connected open manifolds, and non-simply-connected closed manifolds with $\pi_1 = \mathbb{Z}$. Novikov used an iteration of codimension 1 surgeries to prove the topological invariance of the rational Pontrjagin classes. On the algebraic side, codimension 1 surgery corresponds to the algebraic K- and L-theory properties of polynomial rings and their generalizations, starting with $\mathbb{Z}[\mathbb{Z}] = \mathbb{Z}[z, z^{-1}]$. We have already seen (in footnote 7 above) that Novikov in his *Izvestia* paper recognized the significance for his conjecture of the "fundamental theorem of algebraic K-theory" proved by Bass, Heller and Swan [BHS] and Bass [Bass]: for any ring A

$$K_1(A[z,\, z^{-1}]) \cong K_1(A) \oplus K_0(A) \oplus \widetilde{\mathrm{Nil}}_0(A) \oplus \widetilde{\mathrm{Nil}}_0(A)$$

with $\widetilde{\mathrm{Nil}}_0(A)$ the nilpotent class group. Farrell-Hsiang ([FarHs1], [FarHs3]) gave a geometric interpretation of the fundamental theorem in terms of splitting homotopy equivalences of manifolds $N \to M \times S^1$ along the codimension 1 submanifold $M \times \{*\} \subset M \times S^1$, with

$$A = \mathbb{Z}[\pi_1(M)]\ ,\ A[z, z^{-1}] = \mathbb{Z}[\pi_1(M \times S^1)] = \mathbb{Z}[\pi_1(M) \times \mathbb{Z}]\ .$$

Shaneson ([Sh1], [Sh2]) used this codimension 1 splitting theorem to give a geometric proof of the analogous L-theory splitting theorem

$$L^s_n(\mathbb{Z}[\Gamma \times \mathbb{Z}]) \cong L^s_n(\mathbb{Z}[\Gamma]) \oplus L^h_{n-1}(\mathbb{Z}[\Gamma])$$

for any finitely presented group Γ. Novikov [Nov10] gave an algebraic proof of the L-theory splitting theorem modulo 2-torsion:

$$L_n(A[z, z^{-1}]) \otimes \mathbb{Z}[1/2] \cong (L_n(A) \oplus L_{n-1}(A)) \otimes \mathbb{Z}[1/2]$$

for any ring with involution A with $1/2 \in A$, with the involution extended by $\bar{z} = z^{-1}$. This splitting was used to give the algebraic proof of the Novikov Conjecture for free abelian groups in [Nov10]. Farrell and Hsiang

[FarHs3] gave the corresponding geometric proof. The 2-torsion restrictions were removed by Ranicki ([Ran1], [Ran2]), and the splitting theorems

$$L^s_n(A[z,z^{-1}]) \cong L^s_n(A) \oplus L^h_{n-1}(A)$$
$$L^h_n(A[z,z^{-1}]) \cong L^h_n(A) \oplus L^p_{n-1}(A)$$

were obtained algebraically for any ring with involution A, with $L^s_*(A)$ (resp. $L^h_*(A)$, $L^p_*(A)$) the simple (resp. free, projective) L-groups. The simple L-groups $L^s_*(A)$ are the original surgery obstruction groups of Wall [Wall2]; there is only a 2-primary torsion difference between $L^s_*(A)$, $L^h_*(A)$ and $L^p_*(A)$. The lower K-groups $K_{-i}(A)$ and the lower NK-groups $NK_{-i}(A)$ were defined by Bass [Bass] for any ring A, to fit into splittings

$$K_{-i+1}(A[z,z^{-1}]) \cong K_{-i+1}(A) \oplus K_{-i}(A) \oplus NK_{-i+1}(A) \oplus NK_{-i+1}(A)$$

for all $i \geq 0$, with $NK_1(A) = \widetilde{\mathrm{Nil}}_0(A)$. The analogous lower L-groups $L^{\langle -i \rangle}(A)$ were defined in [Ran2] for any ring with involution, to fit into splittings

$$L^{\langle -i+1 \rangle}_n(A[z,z^{-1}]) \cong L^{\langle -i+1 \rangle}_n(A) \oplus L^{\langle -i \rangle}_{n-1}(A)$$

for all $i \geq 0$, with $L^{\langle 0 \rangle}_*(A) = L^p_*(A)$. The forgetful maps $L^{\langle -i+1 \rangle}_n(A) \to L^{\langle -i \rangle}_n(A)$ are isomorphisms modulo 2-primary torsion, with the relative terms the Tate $\mathbb{Z}_2$-cohomology of the duality involution on $K_{-i}(A)$.

Cappell and codimension 1 splitting theorems

With the development of surgery theory for non-simply connected manifolds, machinery was finally in place that could be used to determine when a homotopy equivalence $f : M' \to M$ of closed manifolds "splits" with respect to submanifold P of M, in other words, when it can be deformed so as to restrict to a homotopy equivalence $P' \to P$. The general splitting obstruction theory was worked out by Wall [Wall2, §12]. As noted in the previous section, the case where P is of codimension 1 in M (i.e. a hypersurface) is of particular importance for Novikov Conjecture. Suppose P is a separating hypersurface in M, so that M is the union of two codimension-zero compact submanifolds, M_+ and M_-, each with boundary P. Then assuming that f splits, we get a comparable decomposition of M' as the union of two submanifolds M'_+ and M'_-, each with boundary P'. By Van Kampen's Theorem, $\pi_1(M)$ splits as an amalgamated free product:

$$\pi_1(M) = \pi_1(M_+) *_{\pi_1(P)} \pi_1(M_-)\,.$$

So the question of whether or not f preserves higher signatures can be reduced to questions about the restrictions of f to the various pieces of M', and about higher signatures for the groups $\pi_1(M_+)$, $\pi_1(P)$, $\pi_1(M_-)$. If these groups are simpler than $\pi_1(M)$, there is some hope to use this strategy to give an inductive proof of the Novikov Conjecture for a large class of groups that can be built up from amalgamated free products. As we saw above, Novikov was certainly aware that such a strategy might be useful, especially for studying polycyclic groups, but Cappell ([Cap2], [Cap3]) was the one to finally work out the applications to the Novikov Conjecture using both this case of amalgamated free products (corresponding geometrically to separating hypersurfaces) and the case of HNN extensions (corresponding geometrically to non-separating hypersurfaces, generalizing the polynomial extensions considered in the previous section). Cappell's theory was the first successful attempt to inductively verify the Novikov Conjecture for a large class of well-behaved fundamental groups. An unexpected subtlety which Cappell had to overcome was the "UNil obstruction" to splitting, involving the L-theoretic analogues of the nilpotent class group $\widetilde{\mathrm{Nil}}_0$; since this involves only 2-torsion it has little impact on the higher signatures, but it does play an important role in any attempts to correctly formulate an integral Novikov Conjecture.

Mishchenko and Fredholm representations

Meanwhile, in the wake of Lusztig's thesis, others hoped to use index theory to attack the Novikov Conjecture for large numbers of fundamental groups. But Lusztig's methods required having families of flat vector bundles, which may not be available for non-commutative groups. Mishchenko ([Mis3], [Mis4]) suggested an important idea for overcoming this difficulty, namely the use of *Fredholm representations* of the fundamental group. A Fredholm representation ρ of a group Γ on a ($\mathbb{Z}/2$-graded) Hilbert space $\mathcal{H}$ is a pair (ρ^0, ρ^1) of unitary representations of Γ on $\mathcal{H}^{(0)}$ and on $\mathcal{H}^{(1)}$, respectively, together with a Fredholm operator $T : \mathcal{H}^{(0)} \to \mathcal{H}^{(1)}$ which intertwines the two representations modulo compact operators. One should think of ρ as being the formal difference $\rho^0 - \rho^1$, which one can think of as being approximately finite-dimensional, even though ρ^0 and ρ^1 are themselves infinite-dimensional, so that the case where T is a precise intertwiner is uninteresting. A general discrete group π always has lots of Fredholm representations, even though it may have very few finite-dimensional representations. We can think of these as parameterizing certain generalized flat vector bundles over manifolds with Γ as fundamental group. Mishchenko's idea was to prove the appropriate index theorem for the signature operator with coefficients in a Fredholm representation of the fundamental group, then to substitute these for genuine flat vector bundles in Lusztig's machinery.

Another of Mishchenko's major contributions was to notice that this program works especially well in the presence of a "non-positive curvature assumption" in a model for $B\Gamma$, when for example $B\Gamma$ is a compact manifold whose universal cover $E\Gamma$ is a locally symmetric space of non-compact type. Then the "outward-pointing vector field" on $E\Gamma$ "asymptotically commutes" with covering translations by Γ, and thus gives rise to what Connes later called the "dual Dirac" operator. By using the machinery of Fredholm representations and the dual Dirac (or its analogue in the case of Bruhat-Tits buildings), Mishchenko and his co-workers ([Mis3], [Mis4], [Mis5], [Mis6], [MisS2], [MisS3]) were able to verify the Novikov Conjecture for a number of geometrically interesting fundamental groups.

Farrell-Hsiang and the geometric topology approach

At about the same time, Farrell and Hsiang embarked on a program to systematically attack not only the Novikov Conjecture but also the Borel Conjecture for classes of groups of geometric interest, using methods of geometric topology. Farrell and Hsiang began [FarHs3] by proving both conjectures for free abelian fundamental groups by purely topological methods, using splitting machinery growing out of Farrell's thesis work [Far] on when a manifold fibers over a circle. Then they went on to study the conjectures for flat manifolds (Bieberbach groups) ([FarHs4], [FarHs6], [FarHs7]), non-positively curved manifolds ([FarHs8], [FarHs10]), and almost flat manifolds (infra-nilpotent groups) [FarHs11]. Of special interest in their work was a new idea which they applied to the study of the Novikov Conjecture for Bieberbach groups [FarHs7]: the application of "controlled" topology. The rough idea of how Farrell and Hsiang applied this, reformulated in terms of a fundamental theorem of Chapman and Ferry [ChapF],[14] is the following. The Chapman-Ferry "α-approximation theorem" says that given a closed manifold M^n, with $n > 4$, there is a constant $\varepsilon > 0$ such that a homotopy equivalence $f : M' \to M$ of closed manifolds is homotopic to a homeomorphism provided that there is a homotopy inverse $g : M \to M'$ to f such that the homotopies from fg and gf to the identity maps don't move points more than a distance ε (as measured in M). Suppose one has a homotopy equivalence $f : M' \to M$, and suppose the fundamental group of M is such that there exist coverings $M \to M$ of arbitrarily large degree, which stretch distances by an arbitrarily large amount. Then lifting f by such coverings, one can get new homotopy equivalences $M' \to M$ which are "controlled"

[14] Farrell and Hsiang actually quoted earlier papers of Chapman and of Ferry that use some of the same ideas.

as much as one likes, so that eventually these are homotopic to homeomorphisms. This is a major step in proving the Borel Conjecture. For further discussion of this and other geometric approaches to rigidity, see the subsections on "Farrell-Hsiang, Ferry-Weinberger and tangentiality" and on "The Farrell-Jones Program."

Kasparov and operator-theoretic K-homology

Still another analytic attack on the Novikov Conjecture, motivated both by Lusztig's thesis and by ideas of Atiyah and Singer concerning possible reformulations of index theory, was begun by Gennadi Kasparov in the 1970's. The idea behind this program was to give a good analytic model for the homology theory dual to K-theory, so that elliptic operators on manifolds would naturally give rise to K-homology classes. Then the index of an elliptic operator is computed merely by taking the image, in K-homology of a point, of the corresponding K-homology class. By analyzing the K-homology class of the signature operator on a non-simply connected manifold, one could hope to redo what Lusztig had done, but in a more powerful setting. Kasparov's earliest results in this direction, as well as the first announcements of his results on the Novikov Conjecture, appeared in [Kas1] and [Kas2], although the power of his methods did not become clear until the development of the "KK calculus" in [Kas3]. (For more informal expositions, see also [Black] and [Fack1].) While Mishchenko's Fredholm representations were basically equivalent to K-homology classes in the Kasparov sense, Kasparov's "intersection product" in KK gave more powerful technical tools for overcoming a drawback of Mishchenko's method pointed out in [HsR]. In his famous "Conspectus" [Kas4], Kasparov for the first time was able to sketch a complete analytic proof of a result not yet provable by purely topological methods: that the Novikov Conjecture (in fact, even an integral version, after localizing away from the prime 2) holds for groups Γ which are fundamental groups of complete Riemannian manifolds of non-positive curvature, or which can be realized as discrete subgroups of connected Lie groups.

Surgery spectra and assembly (Quinn)

The global approach to surgery theory initiated by Sullivan and Wall was carried forward by Quinn, and has proved useful in attacking the Borel and Novikov Conjectures using various mixtures of geometry and algebra. Given an n-dimensional Poincaré duality space M, let $\mathcal{S}^{TOP}(M)$ be the *topological manifold structure set* of M, consisting of the equivalence classes of pairs (N, f) with N a closed n-dimensional topological manifold and $f : N \to M$ a homotopy equivalence. Two such pairs (N, f), (N', f') are equivalent if there exists an h-cobordism $(W;N,N')$ with a homotopy equivalence $(g;f,f'):(W;N,N')\to M\times(I;\{0\},\{1\})$. The Browder-Novikov-Sullivan-Wall theory provides an obstruction theory for deciding if $\mathcal{S}^{TOP}(M)$ is non-empty, and for manifold M there is an exact sequence of pointed sets

$$\cdots\to L_{n+1}(\mathbb{Z}[\Gamma])\to\mathcal{S}^{TOP}(M)\xrightarrow{\theta}[M,G/TOP]\xrightarrow{A}L_n(\mathbb{Z}[\Gamma])$$

with $\Gamma=\pi_1(M)$, and G/TOP the homotopy fiber of the forgetful map J from the classifying space $BTOP$ for stable topological bundles to the classifying space BG for stable spherical fibrations. For a manifold M the map $\theta:\mathcal{S}^{TOP}(M)\to[M,G/TOP]$ sends an element (N,f) of the structure set to the difference between the stable normal bundle of M and the push-forward under f of the stable normal bundle of N. The map $A:[M,G/TOP]\to L_n(\mathbb{Z}[\Gamma])$ sends an element of $[M,G/TOP]$, represented by a topological bundle η over M with a fiber homotopy trivialization, to the surgery obstruction $\sigma_*(f,b)\in L_n(\mathbb{Z}[\Gamma])$ of any normal map $(f,b):N\to M$ with $b:\nu_N\to\eta$. The interpretation of G/TOP as a surgery classifying space came from the work of Casson and Sullivan on the manifold Hauptvermutung, which grew out of Novikov's proof of the topological invariance of the rational Pontrjagin classes ([Ran10], [Ran11]).

Quinn ([Q1], [Q2], [Q3]) constructed for each space X a spectrum $\mathbb{L}_\bullet(X)$ consisting of normal maps with a reference map to X, such that the homotopy groups are the surgery obstruction groups

$$\pi_*(\mathbb{L}_\bullet(X)) = L_*(\mathbb{Z}[\pi_1(X)])\ ,$$

and with a homotopy equivalence

$$\mathbb{L}_\bullet(\mathrm{pt.})_0 \simeq L_0(\mathbb{Z})\times G/TOP\ .$$

The surgery obstruction map A for an n-dimensional topological manifold X was interpreted as the abelian group morphism

$$A:[X,G/TOP]\subseteq[X,L_0(\mathbb{Z})\times G/TOP]=H_n(X;\mathbb{L}_\bullet(\mathrm{pt.}))\xrightarrow{A}L_*(\mathbb{Z}[\pi_1(X)])$$

defined by the *geometric surgery assembly map*. Let $\mathbb{H}_\bullet(X;\mathbb{L}_\bullet(\mathrm{pt.}))$ be the spectrum with homotopy groups the generalized homology groups with $\mathbb{L}_\bullet(\mathrm{pt.})$-coefficients

$$\pi_*(\mathbb{H}_\bullet(X;\mathbb{L}_\bullet(\mathrm{pt.})))=H_*(X;\mathbb{L}_\bullet(\mathrm{pt.}))\ .$$

The surgery assembly map is induced by a map of spectra $A:\mathbb{H}_\bullet(X;\mathbb{L}_\bullet(\mathrm{pt.}))\to\mathbb{L}_\bullet(X)$. For the surgery exact sequence it is necessary to work with the

1-connective simply-connected surgery spectrum $\mathbb{L}_\bullet = \mathbb{L}_\bullet\langle 1\rangle(\text{pt.})$ such that $(\mathbb{L}_\bullet)_0 \simeq G/TOP$. The cofibre $\mathbb{S}_\bullet(X)$ of the spectrum-level 1-connective surgery assembly map fits into a (co)fibration sequence

$$\mathbb{H}_\bullet(X;\mathbb{L}_\bullet) \xrightarrow{A} \mathbb{L}_\bullet(X) \to \mathbb{S}_\bullet(X) ,$$

and the homotopy groups $\pi_*(\mathbb{S}_\bullet(X)) = \mathbb{S}_*(X)$ fit into a long exact sequence of abelian groups

$$\cdots \to \mathbb{S}_{n+1}(X) \to H_n(X;\mathbb{L}_\bullet) \xrightarrow{A} L_n(\mathbb{Z}[\pi_1(X)]) \to \mathbb{S}_n(X) \to \dots .$$

There is such a sequence both for the free L-groups $L_* = L^h_*$ and for the simple L-groups $L_* = L^s_*$. If M is a closed n-dimensional TOP manifold then

$$[M, G/TOP] \cong H^0(M;\mathbb{L}_\bullet) \cong H_n(M;\mathbb{L}_\bullet)$$

and the structure set

$$\mathcal{S}^{TOP}(M) = \mathbb{S}_{n+1}(M)$$

have abelian group structures, as indeed do all the rel ∂ structure sets

$$\mathcal{S}^{TOP}(M \times D^k \operatorname{rel} \partial) = \mathbb{S}_{n+k+1}(M) \ (k \geq 0) .$$

Nicas [Ni1] used the abelian group structures to prove induction theorems for the structure set. If $(M^n, \partial M)$ is an n-dimensional manifold with boundary and $\pi_1(M) = \Gamma$, then Siebenmann Periodicity [KirS][15] shows that there is a monomorphism $\mathcal{S}^{TOP}(M^n) \to \mathcal{S}^{TOP}(M^n \times D^4 \text{ rel } \partial)$, $n \geq 6$, which is an isomorphism for $\partial M \neq \emptyset$.

4. Work related to the Novikov Conjecture: The Last 12 Years or So, I: Homotopy Theory and Algebra

Algebraic surgery theory (Ranicki)

It was already suggested by Wall [Wall2] that a development of chain complexes with Poincaré duality would be the appropriate formulation for the 'whole setup' of surgery. The symmetric Poincaré complex theory of Mishchenko [Mis2] was extended by Ranicki ([Ran4], [Ran5], [Ran 7]) to a comprehensive theory of chain complexes with Poincaré duality, including the quadratic structures required for the Wall surgery obstruction. The surgery obstruction of an n-dimensional normal map $(f, b) : M \to X$ was

[15] See [Ni1] and [CaW1] for a correction.

expressed as a quadratic Poincaré cobordism class $\sigma_*(f,b) \in L_n(\mathbb{Z}[\pi_1(X)])$ of the quadratic Poincaré duality on the algebraic mapping cone $C(f^!)$ of the *Umkehr* $\mathbb{Z}[\pi_1(X)]$-module chain map

$$f^! : C(\widetilde{X}) \simeq C(\widetilde{X})^{n-*} \xrightarrow{f^*} C(\widetilde{M})^{n-*} \simeq C(\widetilde{M}) .$$

The main application of the theory to the Novikov Conjecture is by way of the algebraic surgery assembly map, as follows.

Ranicki ([Ran6], [Ran7], [Ran8], [Ran9], [LeRa]) used quadratic Poincaré complexes to define, for any ring with involution R, an algebraic surgery spectrum $\mathbb{L}_\bullet(R)$ such that $\pi_*(\mathbb{L}_\bullet(R)) = L_*(R)$, and an *algebraic surgery assembly map*

$$A \ : \ H_*(X;\mathbb{L}_\bullet(\mathbb{Z})) \ \to \ L_*(\mathbb{Z}[\pi_1(X)])$$

for any space X. These are algebraic versions of Quinn's geometric constructions, particularly the surgery exact sequence. Taylor and Williams [TW] determined the homotopy types of the algebraic L-spectra, generalizing Sullivan's determination of the homotopy type of G/TOP.

Proposition. *The Novikov Conjecture holds for a group Γ if and only if the algebraic surgery assembly map $A_\Gamma : H_*(B\Gamma;\mathbb{L}_\bullet(\mathbb{Z})) \to L_*(\mathbb{Z}[\Gamma])$ is a rational split injection.*

Proposition. *The h-cobordism Borel Conjecture holds for a group Γ if and only if the algebraic surgery assembly map $A_\Gamma : H_*(B\Gamma;\mathbb{L}_\bullet(\mathbb{Z})) \to L^h_*(\mathbb{Z}[\Gamma])$ is an isomorphism. Similarly for the s-cobordism Borel Conjecture with L^s_*.*

See [Ran10] for a somewhat more detailed account. The algebraic surgery assembly map is a special case of the universal assembly construction of Weiss and Williams ([WW3], [WW4]). Study of the assembly map for polynomial extensions and amalgamated free products ([MilgR], [Ran8], [Ran10]) has been used to prove some special cases of the (integral) Novikov Conjecture, extending the method of Cappell.

The homotopy-limit problem, descent (Carlsson)

The homotopy theoretic approach to the Novikov Conjecture is based on experience with analogous problems arising with finite groups. We first recall G. B. Segal's equivariant stable theory. For any finite group G, Segal constructs a G-space $Q^G(S^0)$, whose fixed point set is described as

$$\prod\nolimits_{K \subseteq G} Q(BW_G(K)_+)$$

where Q denotes $\Omega^\infty\Sigma^\infty$, $W_G(K)$ denotes the "Weyl group" $N_G(K)/K$, and "+" denotes disjoint basepoint added. The factor corresponding to the trivial subgroup is thus $Q(BG_+)$. This factor can be seen to be the image of a certain transfer map, which is formally similar to the assembly in its definition. The affirmative solution of Segal's Burnside ring conjecture now shows that in the case of a finite p-group, this factor includes as a factor in the homotopy fixed set of the G-action on $Q^G(S^0)$ after p-adic completion. See [Car1] or [Car2] for a more thorough discussion. Similarly, it follows from results of Atiyah [At1] that the assembly map for K-theory of the complex group ring of a finite p-group is split injective, indeed that it is an equivalence, after p-adic completion. Here it is crucial that we consider the periodic complex K-theory. In this case, the inverse to the assembly map is given by the natural map to the homotopy fixed set of the action of G on BU.

Both these constructions suggest that if one wants to study the assembly map more generally, one should attempt to construct a splitting map using the homotopy fixed set of an action of the group Γ in question on the K-theory of the coefficient ring. In the finite group cases mentioned above, the existence of the corresponding G-action and homotopy fixed set was self-evident, arising from the actions of the G by conjugation on symmetric groups or complex matrix groups. In the case of infinite groups of geometric interest, the obvious actions on rings of infinite matrices yields nothing, since the K-theory of infinite matrix rings is trivial by an appropriate use of the "Eilenberg swindle". It turns out, though, that the Pedersen-Weibel bounded K-theory [PeW] gives the right model. For any torsion-free group Γ, it is possible to construct a bounded K-theory spectrum K^Γ on which Γ acts with fixed point set equal to the algebraic K-theory of the group ring. Furthermore, it is possible to construct an equivariant assembly map α from the "locally finite homology" of the universal cover of $B\Gamma$ to K^Γ, whose induced map on fixed-point sets is the usual assembly. When one can prove that the map α is an equivalence of spectra, standard facts about homotopy fixed-point sets allow one to conclude that the usual assembly map is onto a wedge factor. See [Car3] for an expository discussion of this. Details are done in [Car5]. There, the bounded K-theory is computed for G/K, where G is a connected Lie group and K is its maximal compact subgroup. This allows one to prove the integral form of the Novikov Conjecture for cocompact torsion-free discrete subgroups of Lie groups. Similar methods work for the case of cocompact torsion-free discrete subgroups of p-adic Lie groups, as in the Princeton thesis of P. Mostad.

The Carlsson-Pedersen approach

An alternative is to observe that one perhaps does not need to compute the bounded K-theory explicitly, but only to produce a Γ-spectrum to which the bounded K-theory maps equivariantly, and so that the composite of this map with the map α above is an equivalence. This is the approach taken in [CarP], where it is assumed that the universal cover of the classifying space can be compactified to a contractible space with Γ-action, with certain hypotheses on the action on the boundary. The method of Carlsson and Pedersen combines this equivariant homotopy theory with the categorical approach of Pedersen and Weibel [PeW], the continuously controlled categories of Anderson, Connolly, Ferry and Pedersen [AnCFP], and the lower L-theory of Ranicki [Ran10]. (See the section below on bounded and controlled topology for more information.) In particular, Carlsson and Pedersen proved the integral form of the Novikov Conjecture for, e.g., Gromov's word-hyperbolic groups. They are currently in the process of extending this work to the case of groups which are not torsion-free, using analogues of the Baum-Connes ideas discussed below.

Controlled, continuously controlled and bounded topology

Controlled topology gives geometric methods for approximating homotopy equivalences by homeomorphisms [ChapF]. As we shall see in the sections below on the work of Farrell-Jones and Ferry-Weinberger, these methods can be directly applied to proofs of the Novikov and Borel Conjectures. The controlled algebra of Quinn [Q3] uses a mixture of algebra and topology for recognizing certain types of spectra to be generalized homology spectra. In [Q4], Quinn gave applications to the algebraic K-theory of polycyclic groups. For certain groups Γ it is possible to show that there is enough codimension 1 transversality to prove that the surgery spectrum $\mathbb{L}(\mathbb{Z}[\Gamma])$ is a generalized homology spectrum, verifying the Conjectures by showing that the assembly map $A_\Gamma : H_*(B\Gamma; \mathbb{L}_\bullet(\mathbb{Z})) \to L_*(\mathbb{Z}[\Gamma])$ is an isomorphism. Yamasaki ([Ya1], [Ya2]) applied these methods to the case when Γ is a crystallographic group. Other results, both positive and negative, on topological rigidity statements for crystallographic groups may be found in the work of Connolly and Koźniewski ([CyK1]–[CyK3]). Bounded topology also gives methods for recognizing generalized homology spectra, using the categorical methods initiated by Pedersen and Weibel [PW]. The most effective results on the Novikov Conjecture obtained algebraically use the continuously controlled category of Anderson, Connolly, Ferry and Pedersen [AnCFK] — see the section above on the work of Carlsson and Pedersen. For more details on how bounded and continuously controlled topology are related to the Novikov Conjecture, see the papers of Pedersen [Pe2] and

Ferry-Weinberger [FW2].

K-theoretic analogues of the Novikov and Borel Conjectures

Proposition. *If the Borel Conjecture holds for Γ then the Whitehead group of Γ vanishes,* $\mathrm{Wh}(\Gamma) = 0$.

Proof. Let K be a finite aspherical polyhedron with $\pi_1(K) = \Gamma$. Let M be a regular neighborhood of K in some high-dimensional euclidean space. If $\tau \in \mathrm{Wh}(\Gamma)$, then we can build an h-cobordism rel boundary, (W, M, M') so that $\tau(W, M) = \tau$. Let $f : (W, M, M') \to (M \times [0,1], M \times \{0\}, M \times \{1\})$ be a homotopy equivalence with $f|M \times \{0\} = id$ and $f|W - M \times \{0,1\}$ a homeomorphism. Since $f|M'$ is a homotopy equivalence, the Borel Conjecture implies that $f|M'$ is homotopic to a homeomorphism rel boundary. By the homotopy extension theorem, we can assume that $f|M'$ is a homeomorphism. Applying the conjecture again, f is homotopic to a homeomorphism rel ∂W. Since Whitehead torsion is a topological invariant and $\tau(M \times [0,1], M \times \{0\}) = 0$, $\tau = 0$.

Proposition. *The Borel Conjecture holds for a group Γ if and only if the algebraic surgery assembly map $A_\Gamma : H_*(B\Gamma; \mathbb{L}_\bullet(\mathbb{Z})) \to L_*(\mathbb{Z}[\Gamma])$ is an isomorphism and* $\mathrm{Wh}(\Gamma) = 0$, *with $L_* = L_*^h = L_*^s$..*

Thus if the Borel Conjecture holds for Γ then $\mathbb{S}_*(B\Gamma) = 0$, and if $(M, \partial M)$ is an n-dimensional manifold with boundary such that $\pi_1(M) = \Gamma$ and M is aspherical then

$$\mathcal{S}^{TOP}(M\,\mathrm{rel}\,\partial) = \mathbb{S}_{n+1}(M) = \mathbb{S}_{n+1}(B\Gamma) = 0 \ .$$

Remark. The "fundamental theorem of K-theory" of Bass-Heller-Swan ([BHS], [Bass]) for the Whitehead group of a product $\Gamma \times \mathbb{Z}$ is

$$\mathrm{Wh}(\Gamma \times \mathbb{Z}) \ = \ \mathrm{Wh}(\Gamma) \oplus \widetilde{K}_0(\mathbb{Z}[\Gamma]) \oplus \widetilde{\mathrm{Nil}}_0(\mathbb{Z}[\Gamma]) \oplus \widetilde{\mathrm{Nil}}_0(\mathbb{Z}[\Gamma]) \ .$$

Thus $\widetilde{K}_0(\mathbb{Z}[\Gamma])$ and $\widetilde{\mathrm{Nil}}_0(\mathbb{Z}[\Gamma])$ are direct summands in $\mathrm{Wh}(\Gamma \times \mathbb{Z})$, and these must vanish as well if the conjecture holds for $\Gamma \times \mathbb{Z}$. Similarly, the conjecture for products of Γ with free abelian groups implies that all the negative K-groups of $\mathbb{Z}[\Gamma]$ must vanish.

Thus, rigidity for aspherical manifolds with boundary requires that the Whitehead groups and projective class groups of the fundamental groups of these manifolds should vanish. Consequently, a number of authors have proven vanishing theorems for Whitehead groups of fundamental groups of aspherical manifolds and polyhedra, beginning with the proof in [Hi] that $\mathrm{Wh}(\mathbb{Z}) = 0$ and in [BHS] that $\mathrm{Wh}(\mathbb{Z}^n) = 0$. Other efforts along these lines include [FarHs2], [FarHs9], [FJ3], [FJ12], [Ni2], [Ni3], [Q4] and [Wald1]. A notable recent effort is the paper [Hu] in which Hu proves that $\mathrm{Wh}(\Gamma) = 0$ when Γ is the fundamental group of a finite nonpositively curved ($\equiv$ CAT(0)) polyhedron. The paper uses a Gromov hyperbolization trick to reduce the problem to the case of a nonpositively curved PL manifold. This case is then handled by an extension of the methods of [FJ13]. Hu also proves that the Whitehead group of Γ vanishes for any Γ which is isomorphic to a torsion-free cocompact discrete subgroup of $SL_n(\mathbb{Q}_p)$, where $\mathbb{Q}_p$ is the field of p-adic numbers.

There is an assembly map in algebraic K-theory, first introduced by Loday [Lod], which is analogous to the one for L-theory:

$$A : H_*(B\Gamma; \mathbb{K}(\mathbb{Z})) \to K_*(\mathbb{Z}[\Gamma]) .$$

(For further information on how to understand this map, see [Wald1], §15.) In his address to the 1983 ICM, W.-C. Hsiang, [Hs2], proposed four conjectures as K-theory analogues of the Novikov and Borel Conjectures. The last of these refers to Loday's assembly map.

Conjecture 1. *Let Γ be a finitely presented group. Then $K_{-i}(\mathbb{Z}[\Gamma]) = 0$ for $i \geq 2$. At least, $K_{-i}(\mathbb{Z}[\Gamma]) = 0$ for $i \gg 0$.*

Conjecture 2. *Let Γ be the fundamental group of a closed $K(\Gamma,1)$-manifold. Then* $\mathrm{Wh}(\Gamma) = \widetilde{K}_0(\mathbb{Z}[\Gamma]) = K_{-i}(\mathbb{Z}[\Gamma]) = 0, \ (i \geq 1)$.

Conjecture 3. *Let Γ be a torsion-free group such that $B\Gamma$ has the homotopy type of a finite CW-complex. Then* $\mathrm{Wh}(\Gamma) = \widetilde{K}_0(\mathbb{Z}[\Gamma]) = K_{-i}(\mathbb{Z}[\Gamma]) = 0, \ (i \geq 1)$.

Conjecture 4. *If Γ is a torsion-free group such that $B\Gamma$ is of the homotopy type of a finite complex, then*

$$A \otimes id : H_*(B\Gamma; \mathbb{K}(\mathbb{Z})) \otimes \mathbb{Q} \to K_*(\mathbb{Z}[\Gamma]) \otimes \mathbb{Q}$$

is an isomorphism.

The first conjecture is somewhat tangential to the concerns of this survey and will not be discussed here, though it is true for finite groups by work of Carter [Carter]. The second is the K-theory part of the Borel Conjecture for closed aspherical manifolds. The third is a generalization of the second to the case of finite aspherical polyhedra. As we have seen, this must hold if the Borel Conjecture mentioned above is true for compact aspherical manifolds. The fourth conjecture, however, is a true analogue of the classical Novikov Conjecture, though it is only stated for a restricted class of groups. A still closer analogue of the classical Novikov Conjecture would be:

Conjecture 5. *For any group Γ, the rational K-theory assembly map*

$$A \otimes id : H_*(B\Gamma; \mathbb{K}(\mathbb{Z})) \otimes \mathbb{Q} \to K_*(\mathbb{Z}[\Gamma]) \otimes \mathbb{Q}$$

is injective.

The integral version of Conjecture 4 is:

Algebraic K-theory Isomorphism Conjecture. *If Γ is the fundamental group of a finite aspherical polyhedron, then the assembly map*

$$A : H_i(B\Gamma; \mathbb{K}(\mathbb{Z})) \to K_i(\mathbb{Z}[\Gamma])$$

is an isomorphism for all $i \geq 0$.

For $i = 0$ this is just the conjecture that $\widetilde{K}_0(\mathbb{Z}[\Gamma]) = 0$, and for $i = 1$ that $\mathrm{Wh}(\Gamma) = 0$.

Results on the K-theory Isomorphism Conjecture for discrete subgroups of Lie groups and for groups satisfying non-negative curvature assumptions may be found in [FJ12] and [Car5]. For various technical reasons, it turns out that in approaching Conjecture 5, it is best to introduce Waldhausen's *algebraic K-theory of spaces* $A(X)$ [Wald2], a theory which has geometrical interpretations in terms of pseudo-isotopies of manifolds [Wald3]. There are analogues of the above conjectures for A-theory as well, results about which may be found in the papers by Carlsson, Carlsson-Pedersen, and Farrell-Jones already cited.

In a technical *tour de force* ([BöHM1], [BöHM2]), Bökstedt, Hsiang, and Madsen have proved Conjecture 5 for a class of groups including all groups Γ such that $H_i(B\Gamma)$ is finitely generated in each dimension. This, of course, includes all groups of type FP^∞ — groups Γ such that $B\Gamma$ has finite n-skeleton for all n. Their argument is homotopy-theoretic and relies on both a topological version of cyclic homology theory and use of Waldhausen's A-theory.

5. Work related to the Novikov Conjecture: The Last 12 Years or So, II: Geometric Topology

Farrell-Hsiang, Ferry-Weinberger and tangentiality

In [FarHs8], Farrell and Hsiang gave a proof of the integral Novikov Conjecture for closed Riemannian manifolds of nonpositive curvature. Their approach was to use Quinn's geometric description of the L-spectrum, [Q1], to construct an explicit splitting of the assembly map. An interesting feature of their construction is that it uses the nonpositive curvature assumption to produce a suitably nice compactification of the universal cover. The existence of this compactification is the only aspect of nonpositive curvature used in the proof.

In [FW1], Ferry and Weinberger extended the Farrell-Hsiang argument to include the noncompact case. This recovers Kasparov's theorem and extends it to include the prime 2. We shall sketch a proof of the Novikov Conjecture for closed Riemannian manifolds of nonpositive curvature which blends elements of the Farrell-Hsiang approach with elements of the argument from [FW1]. One pleasant aspect of this argument is that it shows that if $f : N \to M$ is a homotopy equivalence from a closed n-manifold N to a closed nonpositively curved n-manifold M, then f is covered by an *unstable* equivalence of tangent bundles. This unstable equivalence was used in [FW1] to show that the A-theory assembly map (see the section on K-theory above)

$$W_+ \wedge A(*) \to A(W)$$

also splits for W a complete Riemannian manifold of nonpositive curvature.

Let M^n be a closed Riemannian manifold with nonpositive curvature. To prove that the assembly map

$$A : H_n(M; \mathbb{L}_\bullet) \to L_n(\mathbb{Z}[\pi_1(M)])$$

is a monomorphism, it suffices to show that the map $\mathcal{S}^{TOP}(M) \to H_n(M; \mathbb{L}_\bullet)$ is zero. Thus, if $f : N \to M$ is a homotopy equivalence, we need to show that f is normally cobordant to the identity. That is, we must show that there is a cobordism (W, N, M) and a map $F : W \to M \times I$ so that $F|_N = f$, $F|_M = id$, and so that F is covered by a map of stable normal bundles.

Let $\tilde{f} : \widetilde{N} \to \widetilde{M}$ be the induced map on universal covers and form the diagram:

$$\begin{array}{ccc} \widetilde{N} \times_\Gamma \widetilde{N} & \xrightarrow{\tilde{f} \times_\Gamma \tilde{f}} & \widetilde{M} \times_\Gamma \widetilde{M} \\ {\scriptstyle proj_1} \big\downarrow & & \big\downarrow {\scriptstyle proj_1} \\ N & \xrightarrow{f} & M, \end{array}$$

where $\Gamma = \pi_1(M) = \pi_1(N)$ acts diagonally on the product spaces. The fiber-preserving map $\tilde{f} \times_\Gamma \tilde{f}$ restricts to $\tilde{f}$ on the fibers of the bundle projections $proj_1$.

Since M is nonpositively curved, $\widetilde{M}$ has a natural compactification to a disk, $\overline{M} \cong D^n$. The action of Γ on $\widetilde{M}$ extends to $\overline{M}$ and we obtain a disk bundle $\widetilde{M} \times_\Gamma \overline{M} \to M$. This compactification induces a similar compactification of $\widetilde{N}$ by adding an $(n-1)$-sphere at infinity. By the Černavskii-Seebeck theorem (see [Fer1]), the compactified fibers $\overline{N}$ are disks and an argument using local contractibility of the homeomorphism group shows that $\widetilde{N} \times_\Gamma \overline{N} \to N$ is a disk bundle. The induced map of boundaries is a homeomorphism on each fiber, so fiberwise application of the Alexander coning trick shows that $\tilde{f} \times_\Gamma \tilde{f}$ is homotopic to a fiber-preserving homeomorphism. Making this homotopy transverse to the zero section $s(m) = [\tilde{m}, \tilde{m}]$, where $m \in M$ and $\tilde{m}$ is any lift of m to $\widetilde{M}$, gives the desired normal cobordism. The twisted products $\widetilde{M} \times_\Gamma \widetilde{M}$ and $\widetilde{N} \times_\Gamma \widetilde{N}$ contain copies of the tangent microbundle so, by Kister's theorem [Kis], they are isomorphic to the tangent microbundles of M and N and unstable tangentiality follows. For details, see [FW1].

We shall now show that the above assembly map is split. As we have seen in the section on surgery spectra (Quinn) the surgery exact sequence is the long exact homotopy sequence of a (co)fibration of spectra:

$$\mathbb{H}_\bullet(M; \mathbb{L}_\bullet) \xrightarrow{A} \mathbb{L}_\bullet(\mathbb{Z}[\pi_1(M)]) \to \mathbb{S}_\bullet(M) .$$

A spectrum-level version of the vanishing result above shows that the map $\mathbb{S}_\bullet(M) \to \Sigma\mathbb{H}_\bullet(M; \mathbb{L}_\bullet)$ is nullhomotopic. Since

$$\mathbb{L}_\bullet(\mathbb{Z}[\pi_1(M)]) \to \mathbb{S}_\bullet(M) \to \Sigma\mathbb{H}_\bullet(M; \mathbb{L}_\bullet)$$

is also a (co)fibration sequence, the nullhomotopy allows us to lift a map homotopic to the identity map $\mathbb{S}_\bullet(M) \to \mathbb{S}_\bullet(M)$ to $\mathbb{L}_\bullet(\mathbb{Z}[\pi_1(M)])$, splitting the homotopy sequence of the (co)fibration. It follows that the assembly map is split injective.

We begin our discussion of the extension to complete Riemannian manifolds of nonpositive curvature by stating a noncompact Borel Conjecture, which applies to discrete groups Γ which need not be finitely generated.

Noncompact Borel Conjecture for Γ. *Let $(M, \partial M)$ be an open n-dimensional noncompact manifold with boundary such that M is aspherical and $\pi_1(M) = \Gamma$. Let $f : N^n \to M^n$ be a proper homotopy equivalence which is a homeomorphism on the union of ∂N with a neighborhood of infinity. Then f is homotopic to a homeomorphism relative to the boundary and relative to a (possibly smaller) neighborhood of infinity.*

To see how this is related to the Borel Conjecture consider the surgery exact sequence

$$\cdots \to \mathcal{S}^{TOP}(M \text{rel } \partial M \cup \{\textit{nbhd. of infinity}\}) \\ \to [M, \partial M \cup \{\textit{nbhd. of infinity}\}; G/TOP, *] \to L_n(\mathbb{Z}[\pi_1(M)]) \ .$$

The middle term is the cohomology of M rel ∂M with compact supports and coefficients in G/TOP, so that

$$[M, \partial M \cup \{\textit{nbhd. of infinity}\}; G/TOP, *] \\ \cong H^0_c(M, \partial M; \mathbb{L}_\bullet) \cong H_n(M; \mathbb{L}_\bullet) \cong H_n(B\Gamma; \mathbb{L}_\bullet) \ .$$

As in the compact case considered in the section on "K-theory analogues" we have :

Proposition. *The Noncompact Borel Conjecture holds for a group Γ if and only if the algebraic surgery assembly map $A_\Gamma : H_*(B\Gamma; \mathbb{L}_\bullet(\mathbb{Z})) \to L_*(\mathbb{Z}[\Gamma])$ is an isomorphism and* $\mathrm{Wh}(\Gamma) = 0$.

Here is an analogous Novikov-type result for complete Riemannian manifolds of nonpositive curvature.

Theorem ([FW1]). *Let W be a complete Riemannian manifold of nonpositive curvature and dimension ≥ 4. Suppose that $f : W \to W'$ is a proper homotopy equivalence and a homeomorphism on the complement of some compact set. Then f is canonically covered by an isomorphism of unstable tangent bundles. The isomorphism produced agrees with the isomorphism given by f outside of a, perhaps larger, compact set.*

Instead of compactifying and coning, the proof uses the Chapman-Ferry α-approximation theorem, a rescaling argument, and local contractibility of the homeomorphism group to produce a tangent bundle isomorphism covering f. The advantage is that this construction only sees a finite neighborhood of the zero-section, so the bundle isomorphism automatically becomes the given isomorphism near infinity.

The Farrell-Jones program

In this section, we shall outline the work of Farrell-Jones on topological rigidity and the Borel Conjecture. This work was a direct continuation of the work of Farrell-Hsiang discussed above. By the Proposition in the section on "K-theory analogues," the Borel Conjecture holds for a discrete group Γ if and only if $\mathrm{Wh}(\Gamma) = 0$ and the algebraic surgery assembly map $A : H_*(B\Gamma; \mathbb{L}_\bullet) \to L_*(\mathbb{Z}[\Gamma])$ is an isomorphism.

In [FarHs7], Farrell and Hsiang proved the Borel Conjecture in case $B\Gamma = M^n$, $n > 4$, is a flat Riemannian manifold with odd order holonomy group. The argument is an interesting combination of geometry and algebra. By an argument of Epstein-Shub, the manifold M supports expanding endomorphisms. It follows that given $\epsilon > 0$ and a homotopy equivalence $f : N \to M$, a map $e : M \to M$ can be chosen so that the pullback f^e of f over e is an ϵ-controlled homotopy equivalence over (the upper copy of) M in the sense of [ChF] :

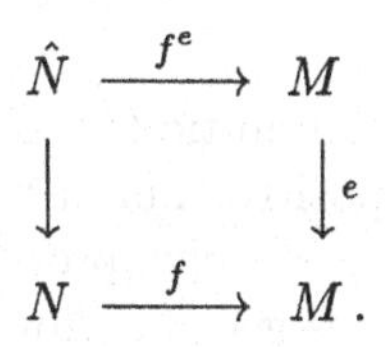

$$\begin{array}{ccc} \hat{N} & \xrightarrow{f^e} & M \\ \downarrow & & \downarrow e \\ N & \xrightarrow{f} & M . \end{array}$$

It follows from the "α-Approximation" theorem of Chapman-Ferry quoted above that for small ϵ, f^e is homotopic to a homeomorphism. Farrell and Hsiang then use Frobenius induction to analyze the surgery exact sequence and show that the structure given by the original f is trivial. The idea is that passage to finite covers corresponds to an algebraic transfer map and that if enough transfers of a structure are trivial, then the original structure is trivial, as well.

In [FJ5], Farrell and Jones proved the conjecture for closed hyperbolic manifolds. As in the work of Farrell-Hsiang, the idea is first to use differential geometry and a transfer argument to show that the transfer of an obstruction dies, and then to use an algebraic argument to deduce that the original obstruction is also zero.

In this case, the relevant geometry is the geometry of the geodesic flow on the unit sphere bundle of a hyperbolic manifold M^n. This flow is Anosov, which means that the sphere bundle admits a pair of transverse foliations such that the flow is expanding along one foliation and contracting along the other. Farrell and Jones show that it is possible to lift a homotopy equivalence $f : N \to M$ in such a way that the tracks of the lifted homotopies are pushed close to flow lines by the geodesic flow. This is the *asymptotic transfer*. Farrell and Jones use this construction to generalize results of Chapman and Ferry [ChF] and Quinn [Q3] to obtain a foliated control theorem. A full discussion of their results would occupy too much space for this survey, but we include a few precise statements to give the flavor of their work.

Definition. *A path γ in the unit sphere bundle SM is said to be (β, ϵ)-controlled if there is a second path ϕ in SM such that*

(1) *The image of ϕ is contained in an arc of length β inside a flow line of the geodesic flow.*
(2) *$d(\gamma(t), \phi(t)) < \epsilon$ for all $t \in [0, 1]$.*

An h-cobordism (W, SM) is said to be (β, ϵ)-controlled if the tracks of the strong deformation retractions are (β, ϵ)-controlled when pushed into SM.

Theorem. *Given a closed hyperbolic manifold M with dim $M > 2$ and a positive real number β, there exists a number $\epsilon > 0$ such that the following is true. Every (β, ϵ)-controlled h-cobordism (W, SM) is a product.*

The key property of the asymptotic transfer is that given positive real numbers β and ϵ there is a positive number t_0 so that if α is a smooth path in M of length $< \beta$, then the asymptotic lift $\bar{\alpha}$ of α to SM becomes $(\sqrt(2)\beta, \epsilon)$-controlled if we let the geodesic flow act on $\bar{\alpha}$ for any time $t \geq t_0$.

Together, these results show that if (W, M) is an h-cobordism on M, then its transfer to the unit sphere bundle is a trivial h-cobordism. This is a key tool in their proof of topological rigidity for closed hyperbolic manifolds. We should emphasize that by itself this does not prove very much. Even though the fundamental groups of M and SM are isomorphic, the transfer map on the Whitehead groups is multiplication by 2 for n odd and multiplication by 0 for n even. Part of the solution is to modify this transfer to get a transfer to a disk bundle. Here, the transfer is multiplication by 1 and the analogous foliated control theorem shows that the Whitehead group of M vanishes. Further effort is required to achieve topological rigidity. See [FJ3], [FJ5], [FJ7] for details.

In [FJ6], Farrell and Jones announced a proof of topological rigidity for closed nonpositively curved manifolds. The proof of this result is similar in outline, but much more complicated in execution. They begin by showing in [FJ10] that the Whitehead groups of the fundamental groups of these manifolds vanish. Given this, it suffices to show that if $f : N \to M$ is a homotopy equivalence then $f \times id : M \times S^1 \to N \times S^1$ is homotopic to a homeomorphism. Instead of transferring to the unit sphere bundle of M, they then transfer to a certain bundle over $M \times S^1$ whose fiber is a stratified space with three strata. The asymptotic transfer, which does nothing in the nonpositively curved case, is replaced by a *focal transfer*. See [FJ13] for details. The work of Farrell and Jones has also given significant results on *integral* versions of the Novikov Conjecture for algebraic K-theory [FJ12]; cf. the section on K-theory above.

Problems about group actions (Weinberger *et al.*)

Weinberger ([W3]–[W6]) has pointed out a number of interesting connections between the Novikov Conjecture and group actions. Aside from being interesting in itself, this work, together with the structure of the proofs of various versions of the Novikov Conjecture, has led to a study of "equivariant" versions of the Novikov Conjecture for manifolds equipped with group actions, a subject which is still under active development.

In general, equivariant surgery theory is considerably more complicated than ordinary surgery theory, both for geometric reasons (a manifold M equipped with an action of, say, a compact group G is stratified according to the various orbit types, and the relations between the various strata can be quite complicated) and for algebraic ones (in general, G does not act on the fundamental group of M, but only on the fundamental groupoid; also, the algebra required to keep track of all the strata can be quite complicated). Weinberger in [W3]–[W5] studied the case where the algebra is as simple as possible, namely the case of "homologically trivial" actions, and found (this should not be so surprising, after all) that when M is not simply connected, the theory of such actions is closely linked to *ordinary* non-simply connected surgery. As a result, he was able to prove the following:

Theorem [W4]. *Let G be a non-trivial finite group. Then the Novikov Conjecture is equivalent to the statement that the higher signatures vanish for any connected oriented manifold M cobordant (by a bordism preserving the fundamental group) to a manifold N admitting a free homologically trivial G-action (i.e, to a manifold N admitting a free G-action, such that $\pi_1(N/G) \cong \pi_1(N) \times G$, and such that G operates trivially on the homology of N with local coefficients).*

Theorem [W5]. *Let p be a prime, $G = \mathbb{Z}/(p)$. Suppose G acts on a connected oriented manifold M with non-empty connected fixed set F, with trivial action on $\Gamma = \pi_1(M)$ (computed at a basepoint in F) and on the homology of M with local coefficients. (Also assume G preserves the orientation if $p = 2$.) By the G-signature theorem (the equivariant version of the Hirzebruch signature theorem), there is a rational characteristic class $\mathcal{D}$ of the equivariant normal bundle ν of F, such that*

$$\text{signature}(M) = \langle \mathcal{L}(F) \cup \mathcal{D}(\nu),\ [F] \rangle \in \mathbb{Z}.$$

Then the Novikov Conjecture is equivalent to the statement that, in this context, the higher signatures of M agree with the higher signatures of F twisted by $\mathcal{D}(\nu)$, i.e., that if $u : M \to B\Gamma$ is the classifying map for the universal cover of M, then

$$\textit{signature}_x(M, u) = \langle \mathcal{L}(F) \cup \mathcal{D}(\nu) \cup u^*(x), [F] \rangle$$

for each cohomology class $x \in H^*(B\Gamma; \mathbb{Q})$.

Subsequent work by Weinberger, discussed in detail in his book [W9], deals with applications and analogues of the Novikov Conjecture in much more complicated situations involving stratified spaces or manifolds with group actions. We shall discuss just one aspect of this here, the concept of the *equivariant Novikov Conjecture*, which is considered in [FRW], [RW1]–[RW3], and [Gong]. Using Kasparov's notion of the K-homology class $[D_M]$ of the signature operator D_M on a closed oriented manifold M, and following [RW2], we may explain this as follows. Suppose $f : N \to M$ is an orientation-preserving homotopy equivalence of closed manifolds. The ordinary Novikov Conjecture says that if X is an aspherical space and $u : M \to X$,[16] then

$$u_*([D_M]) = (u \circ f)_*([D_N]),$$

at least rationally. Now suppose that a compact group G acts on M and N and that the map f is also G-equivariant (though not necessarily an equivariant homotopy equivalence). A G-space X is called *equivariantly aspherical* if for every subgroup H of G (including the trivial subgroup!), every connected component of the fixed set X^H is aspherical. For example, it follows from the Cartan-Hadamard Theorem (see [RW2, Proposition 1.5]) that if X is a complete Riemannian manifold of nonpositive curvature, and if G acts on X by isometries, then X is equivariantly aspherical. The equivariant Novikov Conjecture asserts that for suitable equivariantly aspherical spaces X, where $u : M \to X$ is a G-map and $f : N \to M$ is as above, then again $u_*([D_M]) = (u \circ f)_*([D_N])$, the equality holding in the equivariant K-homology $K_*^G(X)$. While this is definitely false in some cases, it seems plausible when X satisfies some equivariant finiteness conditions, and various cases are proved in the references cited above.

---

[16] Any map will do here. Often u is the classifying map for the universal cover, but this need not be so.

6. Work related to the Novikov Conjecture: The Last 12 Years or So, III: Elliptic Operators and Operator Algebras

In the last decade and a half, the analytic approach to the Novikov Conjecture, first introduced by Lusztig, has led to an explosion of research on various problems concerning elliptic operators and operator algebras. While we can only hint at some of these developments, we shall at least try to give the reader some impression of the various directions in which the subject is moving.

Further development of Kasparov KK-theory

To understand any of the further analytic developments, one needs to examine some of the ideas behind the work of Mishchenko and Kasparov, and in particular, why the K-theory of group C^*-algebras plays such a critical role in analytic approaches to the Novikov Conjecture. We shall be very brief; for further details, see the survey [Ros6] elsewhere in this volume.

The following is the basic idea of the analytic approach. Consider, say, a closed manifold M with fundamental group Γ, equipped with the classifying map $f : M \to B\Gamma$ for the universal cover. Without great loss of generality, suppose the dimension n of M is even, $n = 2k$. We compare a certain analytic invariant of M, which one can call the *analytic higher signature*, with an *a priori* homotopy invariant, the Mishchenko symmetric signature. The former is the generalized index of a certain (generalized) elliptic operator; it plays the role of the index of a family of twisted signature operators in Lusztig's proof. Recall that the index of a family of operators parameterized by a compact space Y is a certain formal difference of vector bundles over Y, in other words an element of the Grothendieck group of vector bundles, $K^0(Y)$. By the Serre-Swan Theorem, vector bundles over Y correspond precisely (via passage to the space of continuous sections) to finitely generated projective modules over the ring $C(Y)$ of continuous functions on Y, so that $K^0(Y)$ may be identified with $K_0(C(Y))$. In Lusztig's case, Γ was free abelian, Y is the Pontrjagin dual $\widehat{\Gamma}$, and by Fourier analysis, $C(Y)$ may be identified with a C^*-algebra completion of the group ring $\mathbb{C}[\Gamma]$. In a similar fashion, in the general case the analytic higher signature is a formal difference of finitely generated projective modules over the *completed group ring* $C^*(\Gamma)$, and thus takes its values in the K-group $K_0(C^*(\Gamma))$. On the other hand, the Mishchenko symmetric signature lives in the Wall group $L_{2k}(C^*(\Gamma)) = L_0(C^*(\Gamma))$. (In this case the symmetric and quadratic L-groups coincide and are 2-periodic, since we are taking $C^*(\Gamma)$ to be an algebra over $\mathbb{C}$.)

The key idea that makes it possible to compare the two invariants is an observation originally going back to Gelfand and Mishchenko [GM]: for complex C^*-algebras, unlike general rings with involution, the functors L_0 and K_0 are naturally isomorphic. (This is due to the Spectral Theorem for self-adjoint operators on a Hilbert space, which implies that any non-singular hermitian form on a C^*-algebra can be split as a direct sum of a positive-definite form and a negative-definite form.) One can then show ([Kas4], [Kas8], [KM2]) that under this isomorphism, the analytic higher signature coincides with the Mishchenko symmetric signature, and is thus homotopy-invariant. On the other hand, one can prove an index theorem, which implies that the analytic higher signature is the image under a certain *analytic assembly map*

$$A : K_*(B\Gamma) \to K_*(C^*(\Gamma))$$

of $f_*([D_M])$, where $[D_M]$ is the class of the Atiyah-Singer signature operator on M in Kasparov's analytic K-homology group. Thus if A is injective, $f_*([D_M])$, which under the Chern character corresponds (up to some powers of 2) to f_* of the total L-class, is a homotopy invariant of M. In other words, the higher signatures of M are homotopy invariants.

Thus, just as in the algebraic approaches to the Novikov Conjecture, the conjecture boils down to the injectivity of a certain assembly map A. Within the last decade or so, various strategies for proving this injectivity have been simplified and strengthened. The approach of Kasparov and his co-workers has basically been to construct a splitting s to the assembly map using a "generalized elliptic operator" (some variant of "dual Dirac"), and to prove that $s \circ A = \text{id}$ using the KK-calculus. In fact, this has usually been done by showing that A and s come from *equivariant* KK-classes on the universal cover $E\Gamma$ of $B\Gamma$, and then applying the equivariant KK-calculus ([Kas4], [Kas7]). This program has by now been extended to quite a number of situations: discrete subgroups of Lie groups [Kas7], groups acting on buildings ([JuV], [KS1], [KS2]), hyperbolic or even "bolic" groups ([HilS], [KS3]), and so on.

The cyclic homology approach (Connes *et al.*)

One of the most important new developments has been the introduction by Connes (from the point of view of geometry and analysis, [Con1]–[Con7]) and by Loday-Quillen [LodQ] and Tsygan [Tsy] (from the point of view of algebra and topology) of *cyclic homology and cohomology* HC_* and HC^*, homology and cohomology theories for algebras which, when specialized to the algebra $C^\infty(M)$ of smooth functions on a manifold, recover de Rham homology and cohomology theory, based on differential forms and currents. The fundamental perspective here is that cyclic homology should be viewed as a "linearized" version of K-theory, and bears the same relationship to the algebraic K-theory of general algebras that ordinary (co)homology bears to topological K-theory. Motivated by this relationship, one can construct a natural transformation, the *Chern character*, from K-theory to cyclic homology.

The idea of applying cyclic homology to the Novikov Conjecture is based on the hope of finding a completion $\mathcal{A}(\Gamma)$ of the group ring $\mathbb{C}[\Gamma]$ so that $\mathbb{C}[\Gamma] \subseteq \mathcal{A}(\Gamma) \subseteq C^*(\Gamma)$ [17] and the following two properties hold:

(1) $\mathcal{A}(\Gamma)$ is "big enough" so that the inclusion $\mathcal{A}(\Gamma) \hookrightarrow C^*(\Gamma)$ induces an isomorphism on K_0, and
(2) $\mathcal{A}(\Gamma)$ is "small enough" so that the inclusion $\mathbb{C}[\Gamma] \hookrightarrow \mathcal{A}(\Gamma)$ induces an injective map on cyclic homology.

Now there is an analogue of the assembly map A for cyclic homology, and in the case of the group ring $\mathbb{C}[\Gamma]$, it is quite easy to show this assembly map is a split injection (for any group Γ) [Bur]. On the other hand, as we mentioned above, the usual Novikov Conjecture is a consequence of injectivity of the assembly map for K-theory of the group C^*-algebra, because of the special relationship between L-theory and K-theory for C^*-algebras. (Knowledge of K_0 is all one needs to handle even-dimensional manifolds, and odd-dimensional manifolds can be handled by crossing with a circle and replacing Γ by $\Gamma \times \mathbb{Z}$.) The idea of the cyclic homology approach is therefore to construct a commutative diagram

$$\begin{array}{ccc}
K_*(B\Gamma) & \xrightarrow{A} & K_*(C^*(\Gamma)) \\
\Vert & & \uparrow \\
K_*(B\Gamma) & \xrightarrow{A} & K_*(\mathcal{A}(\Gamma)) \\
{\scriptstyle \mathrm{Ch}}\downarrow & & \downarrow{\scriptstyle \mathrm{Ch}} \\
H_*(B\Gamma, \mathbb{C}) & \xrightarrow{A} & HC_*(\mathcal{A}(\Gamma)) \\
\Vert & & \uparrow \\
H_*(B\Gamma, \mathbb{C}) & \xrightarrow{A} & HC_*(\mathbb{C}[\Gamma]) ,
\end{array}$$

where the Chern character Ch induces an isomorphism $K_*(B\Gamma) \otimes_{\mathbb{Z}} \mathbb{C} \xrightarrow{\cong}$

[17]The experts will realize that sometimes one needs to use the *reduced* group C^*-algebra here.

$H_*(B\Gamma, \mathbb{C})$, and thus to deduce from (1) and (2) that the top assembly map is rationally injective.

In some cases, this program has been carried out successfully. As notable successes of the program, we mention in particular [CM1]–[CM2], [CGM2], [Jo1]–[Jo2], [Ji1]–[Ji3].

K-theory of group C^*-algebras: the Connes-Kasparov and Baum-Connes Conjectures

As we indicated above, the work of Mishchenko and Kasparov showed that the Novikov Conjecture for a group Γ is a consequence of something stronger (now often called the "Strong Novikov Conjecture"), the injectivity of the assembly map $K_*(B\Gamma) \xrightarrow{A} K_*(C^*(\Gamma))$. Thus the study of the Novikov Conjecture naturally leads to the study of the K-theory of group C^*-algebras. This study has led in turn to a number of related conjectures, which have been verified in many cases.

To begin with, the classifying space BG and the group C^*-algebra $C^*(G)$ are defined not only for discrete groups, but also for all locally compact groups G, and circumstantial evidence [Ros1] suggests a close connection between the K-theory of the classifying space and of the C^*-algebra for arbitrary Lie groups.

Secondly, it is useful to try to examine what sort of *surjectivity* one should expect for the assembly map. For non-amenable discrete groups Γ, the "full" C^*-algebra $C^*(\Gamma)$ turns out to be "too big" (in the sense that even for nice torsion-free groups, one cannot expect surjectivity), so it seems the appropriate object of study is the "reduced" C^*-algebra $C^*_r(\Gamma)$. This is a quotient of $C^*(\Gamma)$ for which one also has a functorial assembly map A. By extrapolating from carefully studied examples, Baum and Connes ([BC3], [BC4], [BCH]) arrived at the conjecture that a certain modified assembly map

$$A' : K^\Gamma_*(\mathcal{E}\Gamma) \to K_*(C^*_r(\Gamma))$$

should be an isomorphism for *any* discrete group. Here $\mathcal{E}\Gamma$ is the universal *proper* Γ-space, just as $B\Gamma$ is the quotient of the universal *free* Γ-space $E\Gamma$ by the Γ-action. When Γ is torsion-free, Γ acts freely on $\mathcal{E}\Gamma = E\Gamma$, so the equivariant K-group on the left becomes just $K_*(B\Gamma)$, and $A' = A$. Hence the Baum-Connes Conjecture predicts that for torsion-free groups, the assembly map of Mishchenko and Kasparov is an isomorphism if one uses the reduced C^*-algebra. In general, A factors through A', and the natural map

$$K_*(B\Gamma) \cong K^\Gamma_*(E\Gamma) \to K^\Gamma_*(\mathcal{E}\Gamma)$$

is at least rationally injective, so the Baum-Connes Conjecture implies the Novikov Conjecture.

Various extensions and strengthenings of the Baum-Connes Conjecture have been proposed. For example, Baum and Connes [BCH] also suggested that if Γ is replaced by any countably connected Lie group G, acting on a C^*-algebra$A$, then there should be a canonical assembly isomorphism from $K_*^G(\mathcal{E}G;\, A)$ to $K_*(A \rtimes_r G)$. Here $A \rtimes_r G$, also sometimes written $C_r^*(G,\, A)$, is the reduced crossed product C^*-algebra. The previous conjecture is just the special case where $A = \mathbb{C}$ and G is discrete. When G is a connected Lie group with maximal compact subgroup K, then $\mathcal{E}G = G/K$, so if we take $A = \mathbb{C}$ again, the conjecture reduces to the assertion that the "Dirac induction" map

$$\mu : R(K) = K_*^K(pt) \cong K_*^G(G/K) \to K_*(C_r^*(G))$$

(here $R(G)$ is the representation ring of K) should be an isomorphism. This special case of the generalized Baum-Connes Conjecture is usually known as the *Connes-Kasparov Conjecture*, and has been proved for connected linear reductive Lie groups by Wassermann [Was]. Similar results are known for some p-adic Lie groups (e.g., [Pl3], [BHP], [BCH]). For arbitrary closed subgroups G of amenable connected Lie groups, of $SO(n,\, 1)$, or of $SU(n,\, 1)$, the generalized Baum-Connes Conjecture follows from still stronger results of Kasparov *et al.* ([Kas4], [Kas5], [Kas7], [JuK]).

Parallels with positive scalar curvature (Gromov-Lawson, Rosenberg *et al.*)

Around 1980, Gromov and Lawson ([GL1], [GL2]) began to notice an interesting parallel between the Novikov Conjecture and a problem in Riemannian geometry, that of determining what smooth manifolds admit Riemannian metrics of positive scalar curvature. They conjectured in particular that a closed aspherical manifold could not admit such a metric, and that a general spin manifold (except perhaps in dimensions 3 and 4) should admit such a metric exactly when certain "higher index invariants" vanish. Originally, the parallel with the Novikov Conjecture was just phenomenological: both problems were related to index theory on non-simply connected manifolds, and in both cases the best results were for aspherical manifolds homotopy-equivalent to $K(\Gamma,\, 1)$'s satisfying some sort of "non-positive curvature" condition.

In a series of papers [Ros2]–[Ros5], Rosenberg showed that this coincidence was not accidental, and that in fact both problems are closely related to the Strong Novikov Conjecture. The formal similarity between the two problems is explained via the index theory of Mishchenko and Fomenko [MisF], applied to the canonical flat bundle $\mathcal{V} = \widetilde{M} \times_\Gamma C^*(\Gamma)$ over a manifold M, whose fibers are rank-one free modules over the group C^*-algebra $C^*(\Gamma)$ of the fundamental group. The main differences between the two problems are that:

(1) the Novikov Conjecture is related to the signature operator, whereas the positive scalar curvature problem (on a spin manifold M) is related to the Dirac operator $\not{D}_M$ via the *Lichnerowicz identity* $\not{D}_M^2 = \nabla^*\nabla + \frac{s}{4}$, where s is the scalar curvature;

(2) the Novikov Conjecture and the Borel Conjecture are most directly related to the assembly map for L-theory, whereas the positive scalar curvature problem is related to the assembly map for KO-theory. In particular, the two theories behave very differently at the prime 2 [Ros6].

Rosenberg gave [Ros5] a reformulation of the Gromov-Lawson Conjecture with some hope of being true for arbitrary spin manifolds, and he and Stolz (whose techniques have proven to be crucial for the problem) have now verified a "stable" form of the Gromov-Lawson Conjecture for a wide variety of fundamental groups, including in particular all finite groups [RS1]–[RS3].

Analogues for foliations

Around 1980, A. Connes [Con1] pointed out that one can attach operator algebras to foliations, and use them to develop index theory for operators associated with foliated manifolds: at first, operators which are "elliptic along the leaves," such as the Dirac and signature operators of the leaves [CS], and later, certain transversally elliptic operators (i.e., elliptic operators on the "leaf space") [Con2]. From these beginnings it was only a small step to the study of an assembly map

$$A : K_*(B(M, \mathcal{F})) \to K_*(C_r^*(M, \mathcal{F}))$$

[BC1] for foliated manifolds $(M, \mathcal{F})$ and to analogues for foliations of the Novikov, Borel, and Baum-Connes Conjectures. The first formulation of a Novikov Conjecture for foliations was given by Baum and Connes in [BC2]. The conjecture states that if $(M, \mathcal{F})$, $(M', \mathcal{F}')$ are foliated closed manifolds, with M and M' compact and oriented and with $\mathcal{F}$ and $\mathcal{F}'$ orientable, then if $f : M' \to M$ is an orientation-preserving *leafwise homotopy equivalence* (i.e., leaf-preserving map which is a homotopy equivalence in a leaf-preserving way) and if $u : M \to B\pi$ is the canonical map to the classifying space of the fundamental groupoid of $(M, \mathcal{F})$ along the leaves, then for any $x \in H^*(B\pi; \mathbb{Q})$, the "higher signatures"

$$\langle \mathcal{L}(M) \cup u^*(x), [M] \rangle \qquad \text{and} \qquad \langle \mathcal{L}(M') \cup (u \circ f)^*(x), [M'] \rangle$$

should be equal. As pointed out by Baum and Connes, when M and M' each consist of a single leaf, this is the usual Novikov Conjecture, and when M and M' are each foliated by points (i.e., $\mathcal{F}$ and $\mathcal{F}'$ are zero-dimensional), this reduces to Novikov's theorem on the homeomorphism invariance of rational Pontrjagin classes. Using a variant of Kasparov's methods, Baum and Connes verified their Conjecture when there is a Riemannian metric on M for which the sectional curvatures of the leaves are all non-positive. Other cases of the Novikov Conjecture for foliations and of the analogues of the Baum-Connes and Gromov-Lawson Conjectures for foliations have been settled in [Na2], [To], [Mac], [Tak1]–[Tak3], and [Hu4].

Flat and almost flat bundles revisited (Connes-Gromov-Moscovici, Gromov *et al.*)

A major theme in analytic work on the Novikov Conjecture, which already appeared in the pioneering work of Lusztig, has been the use of flat and almost flat bundles. These have also been used by Gromov [Gr6] within the last year to give a very slick proof of Novikov's theorem on the topological invariance of rational Pontrjagin classes. Since this is a little easier than the work on the Novikov Conjecture, we mention it first. As is well known, Novikov's theorem is essentially equivalent (see [Ran10, §2] for a few more details) to the statement that if M^{4k+r} is a closed oriented manifold and N^{4k} is the inverse image of a regular value of a map $M \to S^r$, then the signature of N is a homeomorphism invariant of M. Since homeomorphisms restrict on open subsets to proper homotopy equivalences, it is then enough to show that signature(N) only depends on the proper homotopy type of some open tubular neighborhood U. One can reduce to the case where $r = 2m+1$ is odd, and then embed in S^r a product $(\Sigma^2)^m \times \mathbb{R}$, where Σ^2 is a closed Riemann surface of genus > 1. (That such an embedding is possible follows from the fact that Σ is stably parallelizable.) Gromov then uses the fact that Σ admits a flat rank-2 real vector bundle X with structure group $SL(2, \mathbb{R})$ and non-trivial Euler class. Since $SL(2, \mathbb{R}) \cong Sp(\mathbb{R}^2)$, this bundle comes with a natural symplectic structure, which gives rise (because of anti-symmetry of the cup product for odd cohomology classes) to a non-degenerate *symmetric* bilinear form on the cohomology with local coefficients, $H^1(\Sigma; X)$. One can compute that the signature σ of this form turns out to be non-zero. Putting m copies of X together and pulling back to U, Gromov obtains a twisted signature invariant for U that can be shown (using the original ideas of Novikov) to be a proper homotopy invariant, but which differs from signature(N) only by a non-zero constant factor (σ^m), and so Novikov's theorem follows. See [Ran10, §4] for the surgery-theoretic interpretations of Novikov's and Gromov's proofs of the topological invariance of the rational Pontrjagin classes.

The application of flat bundles to the Novikov Conjecture similarly originates from the rather simple observation that if M^{4k} is a closed oriented manifold and X is a flat vector bundle over M with structure group $O(m)$ or $U(m)$, then one can define a symmetric or hermitian pairing on the cohomology with local coefficients, $H^{2k}(M;X)$, and thus a "twisted signature" $\text{signature}_X(M)$. It is clear that this signature is an oriented homotopy invariant. It is also not hard to show that $\text{signature}_X(M)$ coincides with the higher signature $\text{signature}_x(M, u)$, where (since X is flat) the Chern character x of X is the pull-back under some $u : M \to B\Gamma$ of a cohomology class $x \in H^*(B\Gamma; \mathbb{Q})$. The only difficulty is that since flat vector bundles with structure group $O(m)$ or $U(m)$ have trivial rational characteristic classes, all one gets from this argument is the homotopy invariance of the usual signature.

Nevertheless, there are several ways of getting around this difficulty to use this argument to prove results on the Novikov Conjecture. One idea (in effect the idea of the Mishchenko and Kasparov methods) is to replace the ordinary vector bundle X by a flat bundle with *infinite-dimensional* fibers (which are finitely generated projective modules over $C^*(\Gamma)$). Another possibility, explored by Lusztig [Lus] and Gromov [Gr6], is to use *indefinite* orthogonal or unitary groups in place of $O(m)$ or $U(m)$. But an alternative is to use ordinary finite-dimensional bundles, but which are only "approximately" flat, yet which come from representations of the fundamental group. It is this approach which is adopted in [CGM1] and [HilS], and also discussed in [Gr6]. Here the idea is roughly that since the twisted signature (i.e., the index of the signature operator with coefficients in a bundle) is a "discrete" invariant, "small" amounts of non-flatness do not affect that argument that this is a homotopy invariant.

Index theory on non-compact manifolds (Roe *et al.*)

The original ideas of Novikov for proving topological invariance of rational Pontrjagin classes made essential use of non-compact manifolds, so it is not surprising that analysis on such manifolds is starting to play a bigger and bigger role in recent work on the Novikov Conjecture. While it would be impossible to survey here everything that has been done using analysis on non-compact manifolds that is related to the Novikov Conjecture, we shall mention a few key themes, especially as found in the work of J. Roe and his coworkers. Much of the impetus for this work came from an important paper of Atiyah [At2], which showed how the theory of operator algebras could be used to study index theory on the universal covers of compact manifolds. To illustrate how the theory works, we shall begin by discussing some results on the Gromov-Lawson Conjecture (see the section above on positive scalar curvature). The index theory of the Dirac operator $\not{D}_M$ shows via the Lichnerowicz identity that if M is a closed spin manifold with $\hat{A}$-genus

$$\hat{A}(M) = \langle \hat{\mathcal{A}}(M), M \rangle \neq 0$$

(here $\hat{\mathcal{A}}(M)$ is a certain formal power series in the rational Pontrjagin classes of M), then M does not admit a metric of positive scalar curvature. (Positive scalar curvature would imply $\not{D}_M$ is invertible by the Lichnerowicz identity, whereas the index of $\not{D}_M$ is $\hat{A}(M) \neq 0$ by the Atiyah-Singer Theorem.) If M has fundamental group Γ and universal cover $\widetilde{M}$, then one could similarly argue using Atiyah's index theorem [At2] that $\widetilde{M}$ does not admit a Γ-invariant metric of positive scalar curvature. However, this argument does not yet exploit the extra "flexibility" of non-compact manifolds as compared with closed manifolds, since most metrics on $\widetilde{M}$ are not Γ-invariant. A more powerful result, proved by Roe [Roe1] using a more robust version of index theory on non-compact manifolds, is that under these circumstances, $\widetilde{M}$ does not admit any metric, *quasi-isometric* to a Γ-invariant metric, whose scalar curvature function is uniformly positive off a compact set.

Once the technology for proving this result was in place, it became possible to prove that certain aspherical manifolds M (such as tori, which have vanishing $\hat{A}$-genus) do not admit Riemannian metrics of positive scalar curvature, by showing that $\widetilde{M}$ does not admit any metric of uniformly positive scalar curvature in the appropriate quasi-isometry class. It is here that "coarse geometry" and "coarse homology" (also known as "exotic homology," though this is less descriptive) come into play. The Atiyah-Singer Theorem on a compact manifold computes the index of an elliptic operator, which is a certain obstruction to its non-invertibility, as a certain characteristic cohomology class (determined by the symbol of the operator) paired against the fundamental class in homology. Roe's index theory in [Roe3] does something similar for elliptic operators on non-compact manifolds, but now one is only interested in the asymptotic behavior of the manifold at infinity, and compact sets can be thrown away. One still finds that there are homological obstructions to the non-invertibility of the operator, but they involve the homology of the manifold "at infinity." This may be made precise either in terms of the concept of a "corona" (introduced in [Hig2]—as a prototypical example, the sphere at infinity S^{n-1} is a corona for the Euclidean space $\mathbb{R}^n$) or in terms of Roe's coarse homology theory [Roe3].

Corresponding to these results on the Gromov-Lawson Conjecture, there are also results on the Novikov Conjecture itself using the machinery of "coarse homology." For more details, we refer the reader to the paper [HigR2] in these proceedings. However, the main idea can be stated briefly as follows. For any "proper metric space" X,[18] one can define a coarse assembly map

$$A^c : K_*^{\ell f}(X) \to K_*(C^*(X)) .$$

Here $K_*^{\ell f}(X)$ is the locally finite K-homology of X, in other words, the reduced Steenrod K-homology of the one-point compactification [Fer2], and $K_*(C^*(X))$ is the topological K-theory of the C^*-algebra constructed in [Roe3] out of "generalized pseudodifferential operators" on X. This assembly map factors through the "coarse K-homology" $KX_*(X)$. In favorable cases, one can prove injectivity of this coarse assembly map A^c with $X = E\Gamma$, and then deduce from a commutative diagram the injectivity of the usual assembly map

$$A : K_*(B\Gamma) \to C_r^*(\Gamma) ,$$

in other words, the "strong Novikov Conjecture" for Γ. (See [HigR2], [Yu3]–[Yu5], [FW2], [PeRW], and [Ros6] for more information.)

References

[Ad] A. Adem, *Characters and K-theory of discrete groups*, Invent. Math. **114** (1993), 489–514.

[AnCFK] D. R. Anderson, F. X. Connolly, S. C. Ferry, and E. K. Pedersen, *Algebraic K-theory with continuous control at infinity*, J. Pure Appl. Alg. **94** (1994), 25–48.

[AnM1] D. R. Anderson and H. J. Munkholm, *Boundedly Controlled Topology*, Lect. Notes in Math., vol. 1323, Springer-Verlag, Berlin, Heidelberg, New York, 1988.

[AnM2] D. R. Anderson and H. J. Munkholm, *Geometric modules and algebraic K-homology theory*, K-Theory **3** (1990), 561–602.

[Ang] N. Anghel, *On the index of Callias-type operators*, Geom. and Funct. Anal. **3** (1993), 431–438.

[At1] M. Atiyah, *Characters and cohomology of finite groups*, Publ. Math. Inst. Hautes Études Sci. **9** (1961), 23–64.

[At2] M. Atiyah, *Elliptic operators, discrete groups and von Neumann algebras*, Colloque analyse et topologie en l'honneur de Henri Cartan, Astérisque, vol. 32–33, Soc. Math. de France, Paris, 1976, pp. 43–72.

[Att1] O. Attie, *Quasi-isometry classification of some manifolds of bounded geometry*, Math. Z. **216** (1994), 501–527.

[Att2] O. Attie, *A surgery theory for manifolds of bounded geometry*, preprint, McMaster Univ. (1994).

[18] In the case of present interest, X is a non-compact manifold, equipped with a complete Riemannian metric—for example, the universal cover $\widetilde{M}$ of a compact manifold M, equipped with the pull-back of a Riemannian metric on M.

[AtBW] O. Attie, J. Block, and S. Weinberger, *Characteristic classes and distortion of diffeomorphisms*, J. Amer. Math. Soc. **5** (1992), 919–921.

[AsV] A. H. Assadi and P. Vogel, *Semifree finite group actions on compact manifolds*, Algebraic and geometric topology (New Brunswick, NJ, 1983), Lect. Notes in Math., vol. 1126, Springer-Verlag, Berlin, Heidelberg, New York, 1985, pp. 1–21.

[Bass] H. Bass, *Algebraic K-Theory*, W. A. Benjamin, New York and Amsterdam, 1968.

[BHS] H. Bass, A. Heller and R. Swan, *The Whitehead group of a polynomial extension*, Publ. Math. Inst. Hautes Études Sci. **22** (1964), 61–79.

[BC1] P. Baum and A. Connes, *Geometric K-theory for Lie groups and foliations*, Preprint, Inst. Hautes Études Sci., 1982.

[BC2] P. Baum and A. Connes, *Leafwise homotopy equivalence and rational Pontrjagin classes*, Foliations: Proceedings, Tokyo, 1983 (I. Tamura, ed.), Advanced Studies in Pure Math., vol. 5, Kinokuniya and North-Holland, Tokyo, Amsterdam, 1985, pp. 1–14.

[BC3] P. Baum and A. Connes, *Chern character for for discrete groups*, A fête of topology, Academic Press, Boston, 1988, pp. 163–232.

[BC4] P. Baum and A. Connes, *K-theory for discrete groups*, Operator algebras and applications, Vol. 1 (D. Evans and M. Takesaki, eds.), London Math. Soc. Lecture Note Ser., vol. 135, Cambridge Univ. Press, Cambridge, 1988, pp. 1–20.

[BCH] P. Baum, A. Connes, and N. Higson, *Classifying space for proper actions and K-theory of group C^*-algebras*, C^*-algebras: 1943–1993, A Fifty-Year Celebration (R. S. Doran, ed.), Contemp. Math., vol. 167, Amer. Math. Soc., Providence, RI, 1994, pp. 241–291.

[BHP] P. Baum, N. Higson, and R. Plymen, *Equivariant homology for $SL(2)$ of a p-adic field*, Index Theory and Operator Algebras (J. Fox and P. Haskell, eds.), Contemp. Math., vol. 148, Amer. Math. Soc., Providence, RI, 1993, pp. 1–18.

[BeM] S. Bentzen and I. Madsen, *Trace maps in algebraic K-theory and the Coates-Wiles homomorphism*, J. reine angew. Math. **411** (1990), 171–195.

[Bi] R. Bieri, *A group with torsion-free 2-divisible homology and Cappell's result on the Novikov Conjecture*, Invent. Math. **33** (1976), 181–184.

[Black] B. Blackadar, *K-Theory for Operator Algebras*, Math. Sciences Research Inst. Publications, vol. 5, Springer-Verlag, New York, Berlin, Heidelberg, London, Paris, Tokyo, 1986.

[BlocW] J. Block and S. Weinberger, *Aperiodic tilings, positive scalar curvature and amenability of spaces*, J. Amer. Math. Soc. **5** (1992), 907–918.

[BöHM1] M. Bökstedt, W. C. Hsiang, and I. Madsen, *The cyclotomic trace and the K-theoretic analogue of Novikov's conjecture*, Proc. Nat. Acad. Sci. U.S.A. **86** (1989), 8607–8609.

[BöHM2] M. Bökstedt, W. C. Hsiang, and I. Madsen, *The cyclotomic trace and algebraic K-theory of spaces*, Invent. Math. **111** (1993), 465–539.

[Br1] W. Browder, *Homotopy type of differentiable manifolds*, Proc. Århus Topology Symp., unpublished mimeographed notes, 1962, reprinted in these Proceedings.

[Br2] W. Browder, *Structures on $M \times \mathbb{R}$*, Proc. Camb. Phil. Soc. **61** (1965), 337–345.

[Br3] W. Browder, *Manifolds with $\pi_1 = \mathbb{Z}$*, Bull. Amer. Math. Soc. **72** (1966), 238–244.

[Br4] W. Browder, *Surgery on Simply-Connected Manifolds*, Ergebnisse der Math. und ihrer Grenzgebiete, vol. 65, 1972.

[BrHs] W. Browder and W. C. Hsiang, *G-actions and the fundamental group*, Invent. Math. **65** (1982), 411–424.

[BryF] J. Bryant, S. Ferry, W. Mio, and S. Weinberger, *Topology of homology manifolds*, Bull. Amer. Math. Soc. **28** (1993), 324–328.

[BryP] J. L. Bryant and M. E. Petty, *Splitting manifolds as $M \times \mathbb{R}$ where M has a k-fold end structure*, Topology Appl. **14** (1982), 87–104.

[Bur] D. Burghelea, *The cyclic cohomology of group rings*, Comment. Math. Helv. **60** (1985), 354–365.

[Cap1] S. E. Cappell, *A splitting theorem for manifolds and surgery groups*, Bull. Amer. Math. Soc. **77** (1971), 281–286.

[Cap2] S. E. Cappell, *A splitting theorem for manifolds*, Invent. Math. **33** (1976), 69–170.

[Cap3] S. E. Cappell, *On homotopy invariance of higher signatures*, Invent. Math. **33** (1976), 171–179.

[CaSW] S. E. Cappell, J. L. Shaneson, and S. Weinberger, *Classes topologiques caractéristiques pour les actions de groupes sur les espaces singuliers*, C. R. Acad. Sci. Paris Sér. I Math. **313** (1991), 293–295.

[CaW1] S. E. Cappell and S. Weinberger, *A geometric interpretation of Siebenmann's periodicity phenomenon*, Geometry and topology (Athens, Ga. 1985), Lecture Notes in Pure and Appl. Math., vol. 105, Dekker, New York, NY, 1987, pp. 47–52.

[CaW2] S. E. Cappell and S. Weinberger, *Classification de certains espaces stratifiés*, C. R. Acad. Sci. Paris Sér. I Math. **313** (1991), 399–401.

[Car1] G. Carlsson, *Equivariant stable homotopy and Segal's Burnside ring conjecture*, Ann. of Math. **120** (1984), 189–224.

[Car2] G. Carlsson, *Segal's Burnside ring conjecture and the homotopy limit problem*, Homotopy Theory (E. Rees and J. D. S. Jones, eds.), London Math. Soc. Lect. Notes, vol. 117, Cambridge Univ. Press, Cambridge, 1987, pp. 6–34.

[Car3] G. Carlsson, *Homotopy fixed points in the algebraic K-theory of certain infinite discrete groups*, Advances in Homotopy Theory (S. Salamon, B. Steer, and W. Sutherland, eds.), London Math. Soc. Lect. Notes, vol. 139, Cambridge Univ. Press, Cambridge, 1990, pp. 5–10.

[Car4] G. Carlsson, *Proper homotopy theory and transfers for infinite groups*, Algebraic Topology and its Applications (G. Carlsson, R. Cohen, W.-C. Hsiang, and J. D. S. Jones, eds.), M. S. R. I. Publications, vol. 27, Springer-Verlag, New York, 1994, pp. 1–14.

[Car5] G. Carlsson, *Bounded K-theory and the assembly map in algebraic K-theory, I*, these Proceedings.

[CarP] G. Carlsson and E. K. Pedersen, *Controlled algebra and the Novikov conjectures for K- and L-theory*, Topology (to appear).

[Carter] D. W. Carter, *Localization in lower algebraic K-theory*, Comm. Algebra **8** (1980), 603–622; *Lower K-theory of finite groups*, Comm. Algebra **8** (1980), 1927–1937.

[ChapF] T. A. Chapman and S. C. Ferry, *Approximating homotopy equivalences by homeomorphisms*, Amer. J. Math. **101** (1979), 583–607.

[ChG] J. Cheeger and M. Gromov, *Bounds on the von Neumann dimension of L^2-cohomology and the Gauss-Bonnet theorem for open manifolds*, J. Diff. Geom. **21** (1985), 1–34.

[ChG] J. Cheeger and M. Gromov, *L^2-cohomology and group cohomology*, Topology

25 (1986), 189–215.

[CoJ] R. L. Cohen and J. D. S. Jones, *Algebraic K-theory of spaces and the Novikov conjecture*, Topology **29** (1990), 317–344.

[Con1] A. Connes, *A survey of foliations and operator algebras*, Operator Algebras and Applications (Kingston, Ont., 1980) (R. V. Kadison, ed.), Proc. Symp. Pure Math., vol. 38, Part 1, Amer. Math. Soc., Providence, 1982, pp. 521–628.

[Con2] A. Connes, *Cyclic cohomology and the transverse fundamental class of a foliation*, Geometric methods in operator algebras (Kyoto, 1983) (H. Araki and E. G. Effros, eds.), Pitman Res. Notes Math., vol. 123, Longman Sci. Tech., Harlow, 1986, pp. 52–144.

[Con3] A. Connes, *Introduction to noncommutative differential geometry*, Arbeitstagung Bonn 1984, Lecture Notes in Math., vol. 1111, Springer-Verlag, Berlin, Heidelberg, New York, 1985, pp. 3–16.

[Con4] A. Connes, *Cyclic cohomology and noncommutative differential geometry*, Proc. International Congress of Mathematicians (Berkeley, Calif., 1986), vol. 1, Amer. Math. Soc., Providence, RI, 1987, pp. 879–889.

[Con5] A. Connes, *Cyclic cohomology and noncommutative differential geometry*, Géometrie différentielle (Paris, 1986), Travaux en Cours, vol. 33, Hermann, Paris, 1988, pp. 33–50.

[Con6] A. Connes, *Géometrie non commutative*, InterEditions, Paris, 1990.

[Con7] A. Connes, *Noncommutative Geometry*, Academic Press, San Diego, 1994.

[CGM1] A. Connes, M. Gromov, and H. Moscovici, *Conjecture de Novikov et fibrés presques plats*, C. R. Acad. Sci. Paris Sér. I Math. **310** (1990), 273–277.

[CGM2] A. Connes, M. Gromov, and H. Moscovici, *Group cohomology with Lipschitz control and higher signatures*, Geom. Funct. Anal. **3** (1993), 1–78.

[CM1] A. Connes and H. Moscovici, *Conjecture de Novikov et groupes hyperboliques*, C. R. Acad. Sci. Paris Sér. I Math. **307** (1988), 475–480.

[CM2] A. Connes and H. Moscovici, *Cyclic cohomology, the Novikov conjecture and hyperbolic groups*, Topology **29** (1990), 345–388.

[CS] A. Connes and G. Skandalis, *The longitudinal index theorem for foliations*, Publ. Res. Inst. Math. Sci. (Kyoto Univ.) **20** (1984), 1139–1183.

[CST] A. Connes, D. Sullivan and N. Teleman, *Quasiconformal mappings, operators on Hilbert space, and local formulæ for characteristic classes*, Topology **33** (1994), 663–681.

[CyK1] F. X. Connolly and T. Koźniewski, *Finiteness properties of classifying spaces of proper Γ-actions*, J. Pure Applied Algebra **41** (1986), 17–36.

[CyK2] F. X. Connolly and T. Koźniewski, *Rigidity and crystallographic groups, I*, Invent. Math. **99** (1990), 25–48.

[CyK3] F. X. Connolly and T. Koźniewski, *Examples of lack of rigidity in crystallographic groups*, Algebraic Topology, Poznań 1989 (S. Jackowski, B. Oliver, and K. Pawałowski, eds.), Lecture Notes in Math., vol. 1474, Springer-Verlag, Berlin, Heidelberg, and New York, 1991, pp. 139–145.

[Cu1] J. Cuntz, *The K-groups for free products of C^*-algebras*, Operator Algebras and Applications (Kingston, Ont, 1980) (R. V. Kadison, ed.), Proc. Symp. Pure Math., vol. 38, Part 1, Amer. Math. Soc., Providence, 1982, pp. 81–84.

[Cu2] J. Cuntz, *K-theory for certain C^*-algebras, II*, J. Operator Theory **5** (1981), 101–108.

[Da] M. W. Davis, *Coxeter groups and aspherical manifolds*, Algebraic topology, Aarhus 1982, Lecture Notes in Math., vol. 1051, Springer-Verlag, Berlin, Heidelberg, New York, 1984, pp. 197–221.

[Di] Do Ngoc Diep, *C^*-complexes de Fredholm, I*, Acta Math. Vietnam. **9** (1984), 121–130; *II*, ibid. **9** (1984), 193–199.

[Do] R. G. Douglas, *C^*-algebra extensions and K-homology*, Ann. of Math. Studies, vol. 95, Princeton University Press, Princeton, NJ, 1980.

[DrFW] A. N. Dranishnikov, S. Ferry, and S. Weinberger, *A large Riemannian manifold which is flexible*, preprint, SUNY at Binghamton.

[Fack1] T. Fack, *K-théorie bivariante de Kasparov*, Séminaire Bourbaki, 1982–83, exposé no. 605, Astérisque **105–106** (1983), 149–166.

[Fack2] T. Fack, *Sur la conjecture de Novikov*, Index theory of elliptic operators, foliations, and operator algebras (J. Kaminker, K. C. Millett and C. Schochet, eds.), Contemp. Math., vol. 70, Amer. Math. Soc., Providence, RI, 1988, pp. 43–102.

[FarL] M. S. Farber and J. P. Levine, *Jumps of the eta-invariant*, with an appendix by S. Weinberger, preprint, Brandeis Univ., 1994.

[Far] F. T. Farrell, *The obstruction to fibering a manifold over a circle*, Indiana Univ. Math. J. **21** (1971), 315–346.

[FarHs1] F. T. Farrell and W. C. Hsiang, *A geometric interpretation of the Künneth formula for algebraic K-theory*, Bull. Amer. Math. Soc. **74** (1968), 548–553.

[FarHs2] F. T. Farrell and W.-C. Hsiang, *A formula for $K_1 R_\alpha[T]$*, Applications of Categorical Algebra (New York, 1968), Proc. Symp. Pure Math., vol. 17, Amer. Math. Soc., Providence, 1970, pp. 192–218.

[FarHs3] F. T. Farrell and W.-C. Hsiang, *Manifolds with $\pi_1 = G \times_\alpha T$*, Amer J. Math. **95** (1973), 813–848.

[FarHs4] F. T. Farrell and W.-C. Hsiang, *Rational L-groups of Bieberbach groups*, Comment. Math. Helv. **52** (1977), 89–109.

[FarHs5] F. T. Farrell and W.-C. Hsiang, *On the rational homotopy groups of the diffeomorphism groups of discs, spheres and aspherical manifolds*, Algebraic and geometric topology, Proc. Sympos. Pure Math., vol. 32 (Part 1), Amer. Math. Soc., Providence, RI, 1978, pp. 325–337.

[FarHs6] F. T. Farrell and W.-C. Hsiang, *Remarks on Novikov's conjecture and the topological-Euclidean space form problem*, Algebraic topology, Aarhus 1978, Lecture Notes in Math., vol. 763, Springer-Verlag, Berlin, Heidelberg, New York, 1979, pp. 635–642.

[FarHs7] F. T. Farrell and W.-C. Hsiang, *The topological-Euclidean space form problem*, Invent. Math. **45** (1978), 181–192.

[FarHs8] F. T. Farrell and W.-C. Hsiang, *On Novikov's conjecture for nonpositively curved manifolds, I*, Ann. of Math. **113** (1981), 199–209.

[FarHs9] F. T. Farrell and W.-C. Hsiang, *The Whitehead group of poly-(finite or cyclic) groups*, J. London Math. Soc. **24** (1981), 308–324.

[FarHs10] F. T. Farrell and W.-C. Hsiang, *The stable topological-hyperbolic space form problems for complete manifolds of finite volume*, Invent. Math. **69** (1982), 155–170.

[FarHs11] F. T. Farrell and W.-C. Hsiang, *Topological characterization of flat and almost flat Riemannian manifolds M^n $(n \neq 3, 4)$*, Amer. J. Math. **105** (1983), 641–672.

[FarHs12] F. T. Farrell and W.-C. Hsiang, *On Novikov's conjecture for cocompact discrete subgroups of a Lie group*, Algebraic topology, Aarhus 1982, Lecture Notes in Math., vol. 1051, Springer-Verlag, Berlin, Heidelberg, New York, 1984, pp. 38–48.

[FJ1] F. T. Farrell and L. E. Jones, *Implications of the geometrization conjecture for the algebraic K-theory of 3-manifolds*, Geometry and topology (Athens, GA, 1985), Lect. Notes in Pure and Appl. Math., vol. 105, Marcel Dekker, New York, 1987, pp. 109–113.

[FJ2] F. T. Farrell and L. E. Jones, *h-cobordisms with foliated control*, Bull. Amer. Math. Soc. **15** (1986), 69–72; *Erratum*, ibid. **16** (1987), 177.

[FJ3] F. T. Farrell and L. E. Jones, *K-theory and dynamics, I*, Ann. of Math. **124** (1986), 531–569; *II*, ibid. **126** (1987), 451–493.

[FJ4] F. T. Farrell and L. E. Jones, *Topological rigidity for hyperbolic manifolds*, Bull. Amer. Math. Soc. **19** (1988), 277–282.

[FJ5] F. T. Farrell and L. E. Jones, *A topological analogue of Mostow's rigidity theorem*, J. Amer. Math. Soc. **2** (1989), 257–370.

[FJ6] F. T. Farrell and L. E. Jones, *Rigidity and other topological aspects of compact nonpositively curved manifolds*, Bull. Amer. Math. Soc. **22** (1990), 59–64.

[FJ7] F. T. Farrell and L. E. Jones, *Classical Aspherical Manifolds*, Conf. Board of the Math. Sciences Regional Conf. Ser. in Math., vol. 75, Amer. Math. Soc., Providence, RI, 1990.

[FJ8] F. T. Farrell and L. E. Jones, *Rigidity in geometry and topology*, Proc. International Congress of Mathematicians (Kyoto, 1990), vol. 1, Math. Soc. Japan, Tokyo, 1991, pp. 653–663.

[FJ9] F. T. Farrell and L. E. Jones, *Computations of stable pseudoisotopy spaces for aspherical manifolds*, Algebraic Topology, Poznań 1989 (S. Jackowski, B. Oliver, and K. Pawałowski, eds.), Lecture Notes in Math., vol. 1474, Springer-Verlag, Berlin, Heidelberg, and New York, 1991, pp. 59–74.

[FJ10] F. T. Farrell and L. E. Jones, *Stable pseudoisotopy spaces of compact non-positively curved manifolds*, J. Diff. Geom. **34** (1991), 769–834.

[FJ11] F. T. Farrell and L. E. Jones, *Foliated control without radius of injectivity restrictions*, Topology **30** (1991), 117–142.

[FJ12] F. T. Farrell and L. E. Jones, *Isomorphism conjectures in algebraic K-theory*, J. Amer. Math. Soc. **6** (1993), 249–297.

[FJ13] F. T. Farrell and L. E. Jones, *Topological rigidity for compact nonpositively curved manifolds*, Differential Geometry (Los Angeles, 1990) (R. E. Greene and S.-T. Yau, eds.), Proc. Symp. Pure Math., vol. 54, Part 3, Amer. Math. Soc., Providence, 1993, pp. 229–274.

[Fer1] S. C. Ferry, *Homotoping ϵ-maps to homeomorphisms*, Amer. J. Math **101** (1979), 567–582.

[Fer2] S. Ferry, *Remarks on Steenrod homology*, these Proceedings.

[FP] S. Ferry and E. K. Pedersen, *Epsilon surgery theory*, these Proceedings.

[FRR] S. Ferry, A. Ranicki, and J. Rosenberg, *A history and survey of the Novikov Conjecture*, these Proceedings.

[FRW] S. Ferry, J. Rosenberg, and S. Weinberger, *Phénomènes de rigidité topologique équivariante*, C. R. Acad. Sci. Paris Sér. I Math. **306** (1988), 777–782.

[FW1] S. Ferry and S. Weinberger, *Curvature, tangentiality, and controlled topology*, Invent. Math. **105** (1991), 401–414.

[FW2] S. Ferry and S. Weinberger, *A coarse approach to the Novikov Conjecture*, these Proceedings.

[FH] J. Fox and P. Haskell, *K-theory and the spectrum of discrete subgroups of Spin(4, 1)*, Operator Algebras and Topology (Craiova, 1989), Pitman Res. Notes Math., vol. 270, Longman Sci. Tech., Harlow, 1992, pp. 30–44.

[GM] I. M. Gelfand and A. S. Mishchenko, *Quadratic forms over commutative group rings, and K-theory*, Funkc. Anal i Priložen. **3** no. 4 (1969), 28–33; English translation, Funct. Anal. and its Appl. **3** (1969), 277–281.

[Gong] D. Gong, *Equivariant Novikov Conjecture for groups acting on Euclidean buildings*, preprint, Univ. of Chicago.

[GonL] D. Gong and K. Liu, *On the rigidity of higher elliptic genera*, preprint, Univ. of Chicago.

[Gor] V. V. Gorbatsevich, *On some cohomology invariants of compact homogeneous manifolds*, Lie groups, their discrete subgroups, and invariant theory, Adv. Soviet Math., vol. 8, Amer. Math. Soc., Providence, RI, 1992, pp. 69–85.

[Gr1] M. Gromov, *Large Riemannian manifolds*, Curvature and Topology of Riemannian Manifolds, Proc., Katata, 1985 (K. Shiohama, T. Sakai and T. Sunada, eds.), Lecture Notes in Math., vol. 1201, 1986, pp. 108–121.

[Gr2] M. Gromov, *Hyperbolic groups*, Essays in Group Theory (S. M. Gersten, ed.), Math. Sciences Research Inst. Publications, vol. 8, Springer-Verlag, New York, Berlin, Heidelberg, London, Paris, Tokyo, 1987, pp. 75–263.

[Gr3] M. Gromov, *Kähler hyperbolicity and L_2-Hodge theory*, J. Diff. Geometry **33** (1991), 263–292.

[Gr4] M. Gromov, *Asymptotic invariants for infinite groups*, Geometric Group Theory (G. A. Niblo and M. A. Roller, eds.), London Math. Soc. Lect. Notes, vol. 182, Cambridge Univ. Press, Cambridge, 1993, pp. 1–295.

[Gr5] M. Gromov, *Geometric reflections on the Novikov Conjecture*, these Proceedings.

[Gr6] M. Gromov, *Positive curvature, macroscopic dimension, spectral gaps and higher signatures*, Functional Analysis on the Eve of the 21st Century (Proc. conf. in honor of I. M. Gelfand's 80th birthday) (S. Gindikin, J. Lepowsky and R. Wilson, eds.), Progress in Math., Birkhäuser, Boston, 1995 (to appear).

[GL1] M. Gromov and H. B. Lawson, Jr., *Spin and scalar curvature in the presence of a fundamental group. I*, Ann. of Math. **111** (1980), 209–230.

[GL2] M. Gromov and H. B. Lawson, Jr., *Positive scalar curvature and the Dirac operator on complete Riemannian manifolds*, Publ. Math. Inst. Hautes Études Sci. **58** (1983), 83–196.

[HaMTW] I. Hambleton, J. Milgram, L. Taylor and B. Williams, *Surgery with finite fundamental group*, Proc. London Math. Soc. **56** (1988), 349–379.

[Has] P. Haskell, *Direct limits in an equivariant K-theory defined by proper cocycles*, Michigan Math. J. **36** (1989), 17–27.

[Hi] G. Higman, *Units in group rings*, Proc. Lond. Math. Soc. **(2) 46** (1940), 231–248.

[Hig1] N. Higson, *A primer on KK-theory*, Operator theory: operator algebras and applications (Durham, NH, 1988), Proc. Sympos. Pure Math., vol. 51, Part 1, Amer. Math. Soc., Providence, RI, 1990, pp. 239–283.

[Hig2] N. Higson, *On the relative K-homology theory of Baum and Douglas*, Unpublished preprint, 1990.

[Hig3] N. Higson, *On the cobordism invariance of the index*, Topology **30** (1991), 439–443.

[HigR1] N. Higson and J. Roe, *A homotopy invariance theorem in coarse cohomology and K-theory*, Trans. Amer. Math. Soc. **345** (1994), 347–365.

[HigR2] N. Higson and J. Roe, *On the coarse Baum-Connes conjecture*, these Proceedings.

[HigRY] N. Higson, J. Roe, and G. Yu, *A coarse Mayer-Vietoris principle*, Math. Proc. Camb. Philos. Soc. **114** (1993), 85–97.

[Hil] M. Hilsum, *Fonctorialité en K-théorie bivariante pour les variétés lipschitziennes*, K-Theory **3** (1989), 401–440.

[HilS] M. Hilsum and G. Skandalis, *Invariance par homotopie de la signature à coefficients dans un fibré presque plat*, J. reine angew. Math. **423** (1992), 73–99.

[Hir] F. Hirzebruch, *The signature theorem: reminiscences and recreation*, Prospects in mathematics (Proc. Sympos., Princeton Univ., Princeton, N.J., 1970), Ann. of Math. Studies, vol. 70, Princeton Univ. Press, Princeton, NJ, 1971, pp. 3–31.

[Ho] J. G. Hollingsworth, *A short list of problems*, Geometric topology (Proc. Georgia Topology Conf., Athens, Ga., 1977), Academic Press, New York and London, 1979, pp. 685–692.

[Hs1] W.-C. Hsiang, *On* π_i Diff M^n, Geometric topology (Proc. Georgia Topology Conf., Athens, Ga., 1977), Academic Press, New York and London, 1979, pp. 351–365.

[Hs2] W.-C. Hsiang, *Geometric applications of algebraic K-theory*, Proc. International Congress of Mathematicians (Warsaw, 1983), Polish Scientific Publishers, Warsaw, 1984, pp. 99–119.

[Hs3] W.-C. Hsiang, *Borel's conjecture, Novikov's conjecture and the K-theoretic analogues*, Algebra, analysis and geometry (Taipei, 1988), World Sci. Publishing, Teaneck, NJ, 1989, pp. 39–58.

[HsR] W.-C. Hsiang and H. D. Rees, *Mishchenko's work on Novikov's conjecture*, Operator algebras and K-theory (San Francisco, 1981) (R. G. Douglas and C. Schochet, eds.), Contemp. Math., vol. 10, Amer. Math. Soc., Providence, RI, 1982, pp. 77–98.

[HsS] W.-C. Hsiang and J. Shaneson, *Fake tori, the annulus conjecture, and the conjectures of Kirby*, Proc. Nat. Acad. Sci. U.S.A. **62** (1969), 687–691.

[Hu] B. Hu, *Whitehead groups of finite polyhedra with nonpositive curvature*, J. Diff. Geom. **38** (1993), 501–517.

[Hu1] S. Hurder, *Eta invariants and the odd index theorem for coverings*, Geometric and topological invariants of elliptic operators (Bowdoin Coll., 1988) (J. Kaminker, ed.), Contemp. Math., vol. 105, Amer. Math. Soc., Providence, RI, 1990, pp. 47–82.

[Hu2] S. Hurder, *Topology of covers and spectral theory of geometric operators*, Index Theory and Operator Algebras (Boulder, CO, 1991) (J. Fox and P. Haskell, eds.), Contemp. Math., vol. 148, Amer. Math. Soc., Providence, RI, 1993, pp. 87–119.

[Hu3] S. Hurder, *Exotic index theory and the Novikov Conjecture*, these Proceedings.

[Hu4] S. Hurder, *Exotic index theory for foliations*, preprint, Univ. of Illinois at Chicago.

[Ji1] R. Ji, *Smooth dense subalgebras of reduced group* C^**-algebras, Schwartz cohomology of groups, and cyclic cohomology*, J. Functional Analysis **107** (1992), 1–33.

[Ji2] R. Ji, *Some applications of cyclic cohomology to the study of group* C^**-algebras*, Index Theory and Operator Algebras (Boulder, CO, 1991) (J. Fox and P. Haskell, eds.), Contemp. Math., vol. 148, Amer. Math. Soc., Providence, RI, 1993, pp. 121–129.

[Ji3] R. Ji, *Nilpotency of Connes' periodicity operator and the idempotent conjectures*, K-Theory (to appear).

[Jo1] P. Jolissaint, *K-theory of reduced C^*-algebras and rapidly decreasing functions on groups*, K-Theory **2** (1989), 723–736.

[Jo2] P. Jolissaint, *Rapidly decreasing functions in reduced C^*-algebras of groups*, Trans. Amer. Math. Soc. **317** (1990), 167–196.

[Ju1] P. Julg, *C^*-algèbres associées à des complexes simpliciaux*, C. R. Acad. Sci. Paris Sér. I Math. **308** (1989), 97–100.

[Ju2] P. Julg, *K-théorie des C^*-algèbres associées à certains groupes hyperboliques*, Dissertation, Univ. Louis Pasteur, Publ. Inst. Recherche Math. Avancée, Univ. Louis Pasteur, Strasbourg, 1991.

[JuK] P. Julg and G. Kasparov, *L'anneau $KK_G(\mathbb{C}, \mathbb{C})$ pour $G = SU(n, 1)$*, C. R. Acad. Sci. Paris Sér. I Math. **313** (1991), 259–264.

[JuV] P. Julg and A. Valette, *Fredholm modules associated to Bruhat-Tits buildings*, Proc. Miniconferences on Harmonic Analysis and Operator Algebras (Canberra, 1987) (M. Cowling, C. Meaney, and W. Moran, eds.), Proc. Centre for Mathematical Analysis, vol. 16, Centre for Mathematical Analysis, Australian National Univ., Canberra, 1987, pp. 143–155.

[KM1] J. Kaminker and J. Miller, *A comment on the Novikov conjecture*, Proc. Amer. Math. Soc. **83** (1981), 656–658.

[KM2] J. Kaminker and J. Miller, *Homotopy invariance of the analytic index of signature operators over C^*-algebras*, J. Operator Theory **14** (1985), 113–127.

[Kas1] G. G. Kasparov, *A generalized index for elliptic operators*, Funkc. Anal i Priložen. **7** no. 3 (1973), 82–83; English translation, Funct. Anal. Appl. **7** (1973), 238–240.

[Kas2] G. G. Kasparov, *Topological invariants of elliptic operators, I: K-homology*, Izv. Akad. Nauk SSSR, Ser. Mat. **39** (1975), 796–838; English translation, Math. USSR–Izv. **9** (1975), 751–792.

[Kas3] G. G. Kasparov, *The operator K-functor and extensions of C^*-algebras*, Izv. Akad. Nauk SSSR, Ser. Mat. **44** (1980), 571–636; English translation, Math. USSR–Izv. **16** (1981), 513–572.

[Kas4] G. G. Kasparov, *K-theory, group C^*-algebras, and higher signatures: Conspectus, I, II*, preprint, Inst. for Chemical Physics of the Soviet Acad. of Sci., Chernogolovka; reprinted with annotations, these Proceedings.

[Kas5] G. G. Kasparov, *Operator K-theory and its applications: elliptic operators, group representations, higher signatures, C^*-extensions*, Proc. International Congress of Mathematicians (Warsaw, 1983), vol. 2, Polish Scientific Publishers, Warsaw, 1984, pp. 987–1000.

[Kas6] G. G. Kasparov, *Operator K-theory and its applications*, Current problems in mathematics. Newest results, Itogi Nauki i Tekhniki, vol. 27, Akad. Nauk SSSR, Vsesoyuz. Inst. Nauchn. i Tekhn. Inform., Moscow, 1985, pp. 3–31.

[Kas7] ———, *Equivariant KK-theory and the Novikov conjecture*, Invent. Math. **91** (1988), 147–201.

[Kas8] ———, *Novikov's conjecture on higher signatures: the operator K-theory approach*, Representation Theory of Groups and Algebras (J. Adams, R. Herb, S. Kudla, J.-S. Li, R. Lipsman, and J. Rosenberg, eds.), Contemporary Math., vol. 145, Amer. Math. Soc., Providence, 1993, pp. 79–100.

[KS1] G. G. Kasparov and G. Skandalis, *Groupes agissant sur des immeubles de Bruhat-Tits, K-théorie operatorielle et conjecture de Novikov*, C. R. Acad. Sci. Paris Sér. I Math. **310** (1990), 171–174.

[KS2] G. G. Kasparov and G. Skandalis, *Groups acting on buildings, operator K-theo-*

ry, and Novikov's conjecture, K-Theory **4**, Special issue in honor of A. Grothendieck (1991), 303–337.

[KS3] G. G. Kasparov and G. Skandalis, *Groupes "boliques" et conjecture de Novikov*, C. R. Acad. Sci. Paris Sér. I Math. **319** (1994), 815–820.

[Kee] J. Keesling, *The one-dimensional Čech cohomology of the Higson corona and its compactification*, preprint, Univ. of Florida (1994).

[KerM] M. Kervaire and J. Milnor, *Groups of homotopy spheres, I*, Ann. of Math. **77** (1963), 504–537.

[KirS] R. Kirby and L. Siebenmann, *Foundational Essays on Topological Manifolds, Smoothings, and Triangulations*, Ann. of Math. Studies, vol. 88, Princeton Univ. Press, Princeton, NJ, 1977.

[Kis] J. Kister, *Microbundles are fiber bundles*, Ann. of Math. **80** (1964), 190–199.

[LaMi] H. B. Lawson, Jr. and M.-L. Michelsohn, *Spin Geometry*, Princeton Math. Ser., vol. 38, Princeton Univ. Press, Princeton, NJ, 1990.

[LeRa] N. Levitt and A. A. Ranicki, *Intrinsic transversality structures*, Pacific J. Math. **129** (1987), 85–144.

[Lod] J.-L. Loday, *K-théorie algébrique et représentations des groupes*, Ann. Sci. École Norm. Sup. (4) **9** (1976), 309–377.

[LodQ] J.-L. Loday and D. Quillen, *Cyclic homology and the Lie algebra homology of matrices*, Comment. Math. Helv. **59** (1984), 565–591.

[Lott] J. Lott, *Superconnections and higher index theory*, Geom. Funct. Anal. **4** (1992), 421–454.

[Lott] J. Lott, *Higher eta-invariants*, K-Theory **6** (1992), 191–233.

[Lus] G. Lusztig, *Novikov's higher signature and families of elliptic operators*, J. Diff. Geom. **7** (1972), 229–256.

[Mac] M. Macho-Stadler, *La conjecture de Baum-Connes pour un feuilletage sans holonomie de codimension un sur une variété fermée*, Publ. Mat. **33** (1989), 445–457.

[Ma] I. Madsen, *Reidemeister torsion, surgery invariants and spherical space forms*, Proc. London Math. Soc. **46** (1983), 193–240.

[Mat1] V. Mathai, *On the homotopy invariance of reduced eta and other signature type invariants*, preprint, Univ. of Adelaide.

[Mat2] V. Mathai, *Spectral flow, eta invariants, and von Neumann algebras*, J. Funct. Anal. **109** (1992), 442–456.

[Milg] R. J. Milgram, *Surgery with finite fundamental group, II: the oozing conjecture*, Pacific J. Math. **151** (1991), 117–150.

[MilgR] R. J. Milgram and A. A. Ranicki, *The L-theory of Laurent polynomial extensions and genus 0 function fields*, J. reine angew. Math. **406** (1990), 121–166.

[Mill] J. Miller, *Signature operators and surgery groups over C^*-algebras*, Preprint, Indiana Univ./Purdue Univ. (1992).

[Miln1] J. Milnor, *On manifolds homeomorphic to the 7-sphere*, Ann. of Math. **64** (1956), 399–405.

[Miln2] J. Milnor, *A procedure for killing the homotopy groups of differentiable manifolds*, Differential Geometry (Tucson, 1960) (C. B. Allendoerfer, ed.), Proc. Symp. Pure Math., vol. 3, 1961, pp. 39–55.

[Mis1] A. S. Mishchenko, *Homotopy invariants of nonsimply connected manifolds, I: rational invariants*, Izv. Akad. Nauk SSSR Ser. Mat. **34** (1970), 501–514; English translation, Math. USSR–Izv. **4** (1970), 506–519.

[Mis2] A. S. Mishchenko, *Homotopy invariants of nonsimply connected manifolds, III:*

Higher signatures, Izv. Akad. Nauk SSSR Ser. Mat. **35** (1971), 1316–1355; Erratum , *ibid.* **36** (1972), 910.

[Mis3] A. S. Mishchenko, *Infinite dimensional representations of discrete groups, and higher signatures*, Izv. Akad. Nauk SSSR Ser. Mat. **38** (1974), 81–106; English translation, Math. USSR–Izv. **8** (1974), 85–111.

[Mis4] A. S. Mishchenko, *On Fredholm representations of discrete groups*, Funkc. Anal i Priložen. **9** no. 2 (1975), 36–41; English translation, Funct. Anal. Appl. **9** (1975), 121–125.

[Mis5] A. S. Mishchenko, *Hermitian K-theory. The theory of characteristic classes and methods of functional analysis*, Uspekhi Mat. Nauk **31** no. 2 (188) (1976), 69–134; English translation, Russian Math. Surv. **31** no. 2 (1972), 71–138.

[Mis6] A. S. Mishchenko, *C^*-algebras and K-theory*, Algebraic Topology, Århus 1978 (J. L. Dupont and I. Madsen, eds.), Lecture Notes in Math., vol. 763, Springer-Verlag, Berlin, Heidelberg, and New York, 1979, pp. 262–274.

[MisF] A. S. Mishchenko and A. T. Fomenko, *The index of elliptic operators over C^*-algebras*, Izv. Akad. Nauk SSSR Ser. Mat. **43** (1979), 831–859; English translation, Math. USSR–Izv. **15** (1980), 87–112.

[MisS1] A. S. Mishchenko and Y. P. Soloviev, *A classifying space for hermitian K-theory*, Trudy Sem. Vect. and Tensor Anal. **18** (1976), 140–168.

[MisS2] A. S. Mishchenko and Yu. P. Soloviev, *Infinite-dimensional representations of fundamental groups, and formulas of Hirzebruch type*, Dokl. Akad. Nauk SSSR **234** (1977), 761–764; English translation, Soviet Math. Dokl. **18** (1977), 767–771.

[MisS3] A. S. Mishchenko and Yu. P. Soloviev, *Representations of Banach algebras and formulas of Hirzebruch type*, Mat. Sb. **111(153)** (1980), 209–226; English translation, Math. USSR–Sb. **39** (1980), 189–205.

[Mos] H. Moscovici, *Cyclic cohomology and invariants of multiply connected manifolds*, Proc. International Congress of Mathematicians (Kyoto, 1990), vol. 1, Math. Soc. Japan, Tokyo, 1991, pp. 675–688.

[Na1] T. Natsume, *The Baum-Connes conjecture, the commutator theorem, and Rieffel projections*, C. R.-Math. Rep. Acad. Sci. Canada **10** (1988), 13–18.

[Na2] T. Natsume, *Topological K-theory for codimension one foliations without holonomy*, Foliations (Tokyo, 1983), Adv. Stud. Pure Math., vol. 5, 1985, pp. 15–27.

[Ne] W. D. Neumann, *Signature-related invariants of manifolds, I: monodromy and γ-invariants*, Topology **18** (1979), 147–160.

[Ni1] A. J. Nicas, *Induction theorems for groups of homotopy manifold structures*, Mem. Amer. Math. Soc. **39** no. 267 (1982).

[Ni2] A. J. Nicas, *On* Wh_3 *of a Bieberbach group*, Math. Proc. Cambridge Philos. Soc. **95** (1984), 55–60.

[Ni3] A. J. Nicas, *On the higher Whitehead groups of a Bieberbach group*, Trans. Amer. Math. Soc. **287** (1985), 853–859.

[Nis1] V. Nistor, *Group cohomology and the cyclic cohomology of crossed products*, Invent. Math. **99** (1990), 411–424.

[Nis2] V. Nistor, *Cyclic cohomology of crossed products by algebraic groups*, Invent. Math. **112** (1993), 615–638.

[Nov1] S. P. Novikov, *Diffeomorphisms of simply connected manifolds*, Dokl. Akad. Nauk SSSR **143** (1962), 1046–1049; English translation, Soviet Math. Dokl. **3** (1962), 540–543.

[Nov2] S. P. Novikov, *Homotopically equivalent smooth manifolds. I*, Izv. Akad. Nauk

SSSR, Ser. Mat. **28** (1964), 365–474; English translation, Amer. Math. Soc. Transl. (2) **48** (1965), 271–396.

[Nov3] S. P. Novikov, *Rational Pontrjagin classes. Homeomorphism and homotopy type of closed manifolds. I*, Izv. Akad. Nauk SSSR, Ser. Mat. **29** (1965), 1373–1388.

[Nov4] S. P. Novikov, *The homotopy and topological invariance of certain rational Pontrjagin classes*, Dokl. Akad. Nauk SSSR **162** (1965), 1248–1251; English translation, Soviet Math. Dokl. **6** (1965), 854–857.

[Nov5] S. P. Novikov, *Topological invariance of rational classes of Pontrjagin*, Dokl. Akad. Nauk SSSR **163** (1965), 298–300; English translation, Soviet Math. Dokl. **6** (1965), 921–923.

[Nov6] S. P. Novikov, *On manifolds with free abelian fundamental group and their application*, Izv. Akad. Nauk SSSR, Ser. Mat. **30** (1966), 207–246; English translation, Amer. Math. Soc. Transl. (2) **71** (1968), 1–42.

[Nov7] S. P. Novikov, *Pontrjagin classes, the fundamental group and some problems of stable algebra*, Proc. International Congress of Mathematicians (Moscow, 1966), Amer. Math. Soc. Transl. (2), vol. 70, Amer. Math. Soc., Providence, RI, 1968, pp. 172–179.

[Nov8] S. P. Novikov, *Analogues hermitiens de la K-théorie*, Actes du Congrès International des Mathématiciens (Nice, 1970), vol. 2, Gauthier-Villars, Paris, pp. 39–45.

[Nov9] S. P. Novikov, *Pontrjagin classes, the fundamental group and some problems of stable algebra*, Essays on Topology and Related Topics (Mémoires dédiés à Georges de Rham) (A. Haefliger and R. Narasimhan, eds.), Springer-Verlag, New York, Heidelberg, and Berlin, 1970, pp. 147–155.

[Nov10] S. P. Novikov, *Algebraic construction and properties of Hermitian analogues of K-theory over rings with involution from the viewpoint of the Hamiltonian formalism. Applications to differential topology and the theory of characteristic classes*, Izv. Akad. Nauk SSSR, Ser. Mat. **34** (1970), 253–288, 475–500; English translation, Math. USSR–Izv. **4** (1970), 257–292, 479–505.

[Ogle1] C. Ogle, *Assembly maps, K-theory, and hyperbolic groups*, K-Theory **6** (1992), 235–265.

[Ogle2] C. Ogle, *Simplicial rapid decay algebras associated to a discrete group*, Preprint, Ohio State Univ., 1995.

[Ogle3] C. Ogle, *Filtrations of simplicial functors, topological K-theory and the strong Novikov conjecture*, Preprint, Ohio State Univ., 1995.

[Pe1] E. K. Pedersen, *On the bounded and thin h-cobordism theorem parametrized by* $\mathbb{R}^k$, Transformation Groups, Poznań 1985 (S. Jackowski and K. Pawałowski, eds.), Lecture Notes in Math., vol. 1217, Springer-Verlag, Berlin, Heidelberg, and New York, 1986, pp. 306–320.

[Pe2] E. K. Pedersen, *Bounded and continuous control*, these Proceedings.

[PeRW] E. K. Pedersen, J. Roe and S. Weinberger, *On the homotopy invariance of the boundedly controlled analytic signature of a manifold over an open cone*, these Proceedings.

[PeW] E. K. Pedersen and C. Weibel, *K-theory homology of spaces*, Algebraic Topology, Proc., 1986 (G. Carlsson R. L. Cohen, H. R. Miller, and D. C. Ravenel, eds.), Lecture Notes in Math., vol. 1370, Springer-Verlag, Berlin, Heidelberg, and New York, 1989, pp. 346–361.

[Pet] M. E. Petty, *Waldhausen's theory of k-fold end structures*, Topology Appl. **14**

(1982), 71–85.

[Pim1] M. Pimsner, *KK-groups of crossed products by groups acting on trees*, Invent. Math. **86** (1986), 603–634.

[Pim2] M. Pimsner, *K-theory for groups acting on trees*, Proc. International Congress of Mathematicians (Kyoto, 1990), vol. 1, Math. Soc. Japan, Tokyo, 1991, pp. 979–986.

[PimV1] M. Pimsner and D. Voiculescu, *Exact sequences for K-groups and Ext-groups for certain cross products of C^*-algebras*, J. Operator Theory **4** (1980), 93–118.

[PimV2] M. Pimsner and D. Voiculescu, *K-groups of reduced crossed products by free groups*, J. Operator Theory **8** (1982), 131–156.

[Pl1] R. Plymen, *K-theory for the reduced C^*-algebra of $SL_2(\mathbb{Q}_p)$*, Operator Algebras and their Connections with Topology and Ergodic Theory (H. Araki, C. C. Moore, Ş. Strătilă and D. Voiculescu, eds.), Lect. Notes in Math., vol. 1132, Springer-Verlag, Berlin, Heidelberg, New York, 1985, pp. 409–420.

[Pl2] R. Plymen, *Reduced C^*-algebra of the p-adic group $GL(n)$*, J. Funct. Anal. **72** (1987), 1–12.

[Pl3] R. Plymen, *Reduced C^*-algebra for reductive p-adic groups*, J. Funct. Anal. **88** (1990), 251–266.

[Q1] F. Quinn, *A geometric formulation of surgery*, Topology of manifolds (Athens, Ga. 1969) (J. C. Cantrell and C. H. Edwards, ed.), Markham Math. Ser., Markham Press, Chicago, 1970, pp. 500–511.

[Q2] F. Quinn, *$B_{(TOP_n)^\sim}$ and the surgery obstruction*, Bull. Amer. Math. Soc. **77** (1971), 596–600.

[Q3] F. Quinn, *Ends of maps, II*, Invent. Math. **68** (1982), 353–424.

[Q4] F. Quinn, *Algebraic K-theory of poly-(finite or cyclic) groups*, Bull. Amer. Math. Soc. **12** (1985), 221–226.

[Q5] F. Quinn, *Applications of topology with control*, Proc. International Congress of Mathematicians (Berkeley, Calif., 1986), vol. 1, Amer. Math. Soc., Providence, RI, 1987, pp. 598–606.

[Q6] F. Quinn, *Homotopically stratified spaces*, J. Amer. Math. Soc. **1** (1988), 441–499.

[Q7] F. Quinn, *Assembly maps in bordism-type theories*, these Proceedings.

[Ran1] A. Ranicki, *Algebraic L-theory I. Foundations*, Proc. London Math. Soc. **27** (1973), 101–125.

[Ran2] A. Ranicki, *Algebraic L-theory II. Laurent extensions*, Proc. London Math. Soc. **27** (1973), 126–158.

[Ran3] A. Ranicki, *Algebraic L-theory III. Twisted Laurent extensions*, Hermitian K-Theory and Geometric Applications (H. Bass, ed.), Lecture Notes in Math., vol. 343, Springer-Verlag, Berlin, Heidelberg, and New York, 1973, pp. 412–463.

[Ran4] A. Ranicki, *The algebraic theory of surgery I. Foundations*, Proc. London Math. Soc. **40** (1980), 87–192.

[Ran5] A. Ranicki, *The algebraic theory of surgery II. Applications to topology*, Proc. London Math. Soc. **40** (1980), 193–283.

[Ran6] A. Ranicki, *The total surgery obstruction*, Algebraic topology, Aarhus 1978, Lecture Notes in Math., vol. 763, Springer-Verlag, Berlin, Heidelberg, New York, 1979, pp. 275–316.

[Ran7] A. Ranicki, *Exact Sequences in the Algebraic Theory of Surgery*, Princeton Math. Notes, vol. 26, Princeton Univ. Press, Princeton, 1981.

[Ran8] A. Ranicki, *Lower K- and L-Theory*, London Math. Soc. Lecture Notes, vol. 178,

Cambridge Univ. Press, Cambridge, 1992.

[Ran9] A. Ranicki, *Algebraic L-Theory and Topological Manifolds*, Cambridge Tracts in Math., vol. 102, Cambridge Univ. Press, Cambridge, 1992.

[Ran10] A. Ranicki, *On the Novikov conjecture*, these Proceedings.

[Ran11] A. Ranicki (editor), *The Hauptvermutung Book, papers by A. J. Casson, D. P. Sullivan, M. A. Armstrong, G. E. Cooke, C. P. Rourke and A. A. Ranicki*, *K-Theory* book series, vol. 1 (to appear).

[Roe1] J. Roe, *An index theorem on open manifolds, I*, J. Diff. Geom. **27** (1988), 87–113; *II*, ibid., 115–136.

[Roe2] J. Roe, *Hyperbolic metric spaces and the exotic cohomology Novikov conjecture*, K-Theory **4** (1990), 501–512; *Erratum*, ibid. **5** (1991), 189.

[Roe3] J. Roe, *Coarse cohomology and index theory on complete Riemannian manifolds*, Mem. Amer. Math. Soc. no. 497 (1993).

[Rokh] V. A. Rokhlin, *Pontrjagin-Hirzebruch class of codimension 2*, Izv. Akad. Nauk SSSR Ser. Mat. **30** (1966), 705–718.

[RokS] V. A. Rokhlin and A. S. Shvarts, *The combinatorial invariance of Pontrjagin classes*, Dokl. Akad. Nauk SSSR **114** (1957), 490–493.

[Ros1] J. Rosenberg, *Group C^*-algebras and topological invariants*, Operator algebras and group representations, Vol. II (Neptun, 1980), Monographs Stud. Math., vol. 18, Pitman, Boston and London, 1984, pp. 95–115.

[Ros2] J. Rosenberg, *C^*-algebras, positive scalar curvature, and the Novikov conjecture*, Publ. Math. Inst. Hautes Études Sci. **58** (1983), 197–212.

[Ros3] J. Rosenberg, *C^*-algebras, positive scalar curvature, and the Novikov conjecture, II*, Geometric methods in operator algebras (Kyoto, 1983) (H. Araki and E. G. Effros, eds.), Pitman Res. Notes Math., vol. 123, Longman Sci. Tech., Harlow, 1986, pp. 341–374.

[Ros4] J. Rosenberg, *C^*-algebras, positive scalar curvature, and the Novikov conjecture, III*, Topology **25** (1986), 319–3362.

[Ros5] J. Rosenberg, *The KO-assembly map and positive scalar curvature*, Algebraic Topology, Poznań 1989 (S. Jackowski, B. Oliver, and K. Pawałowski, eds.), Lecture Notes in Math., vol. 1474, Springer-Verlag, Berlin, Heidelberg, and New York, 1991, pp. 170–182.

[Ros6] J. Rosenberg, *Analytic Novikov for topologists*, these Proceedings.

[RS1] J. Rosenberg and S. Stolz, *Manifolds of positive scalar curvature*, Algebraic Topology and its Applications (G. Carlsson, R. Cohen, W.-C. Hsiang, and J. D. S. Jones, eds.), M. S. R. I. Publications, vol. 27, Springer-Verlag, New York, 1994, pp. 241–267.

[RS2] J. Rosenberg and S. Stolz, *A "stable" version of the Gromov-Lawson conjecture*, The Čech Centennial: Proc. Conf. on Homotopy Theory (M. Cenkl and H. Miller, eds.), Contemp. Math., vol. 181, Amer. Math. Soc., Providence, RI, 1995, pp. 405–418.

[RS3] J. Rosenberg and S. Stolz, *Metrics of Positive Scalar Curvature*, DMV-Seminar, Birkhäuser, Basel, Boston, in preparation.

[RW1] J. Rosenberg and S. Weinberger, *Higher G-indices and applications*, Ann. Sci. École Norm. Sup. **21** (1988), 479–495.

[RW2] J. Rosenberg and S. Weinberger, *An equivariant Novikov conjecture*, with an appendix by J. P. May, K-Theory **4**, Special issue in honor of A. Grothendieck (1990), 29–53.

[RW3] J. Rosenberg and S. Weinberger, *Higher G-signatures for Lipschitz manifolds*,

K-Theory **7** (1993), 101–132.

[Sh1] J. Shaneson, *Wall's surgery obstruction groups for $\mathbb{Z} \times G$, for suitable groups G*, Bull. Amer. Math. Soc. **74** (1968), 467–471.

[Sh2] J. Shaneson, *Wall's surgery obstruction groups for $\mathbb{Z} \times G$*, Ann. of Math. **90** (1969), 296–334.

[Shar] B. L. Sharma, *Topologically invariant characteristic classes*, Topology Appl. **21** (1985), 135–146.

[SharSi] B. L. Sharma and N. Singh, *Topological invariance of integral Pontrjagin classes mod p*, Topology Appl. **63** (1995), 59–67.

[Si] N. Singh, *On topological and homotopy invariance of integral Pontrjagin classes modulo a prime p*, Topology Appl. **38** (1991), 225–235.

[Sk1] G. Skandalis, *Kasparov's bivariant K-theory and applications*, Expositiones Math. **9** (1991), 193–250.

[Sk2] G. Skandalis, *Approche de la conjecture de Novikov par la cohomologie cyclique (d'après A. Connes, M. Gromov et H. Moscovici)*, Séminaire Bourbaki, 1990–91, exposé no. 739, Astérisque **201–203** (1992), 299–320.

[So] Yu. P. Soloviev, *A theorem of Atiyah-Hirzebruch type for infinite groups*, Vestnik Moskov. Univ., Ser. I–Mat. Meh. **30** no. 4 (1975), 26–35; English translation, Moscow Univ. Math. Bull. **30** no. 3/4 (1975), 77–85.

[Spv] M. Spivak, *Spaces satisfying Poincaré duality*, Topology **6** (1967), 77–102.

[Sul] D. Sullivan, *Triangulating homotopy equivalences*, Ph.D. dissertation, Princeton Univ., Princeton, 1965.

[Tak1] H. Takai, *A counterexample of strong Baum-Connes conjectures for foliated manifolds*, Mappings of operator algebras (Philadelphia, PA, 1988), Progress in Math., vol. 84, Birkhäuser, Boston, 1990, pp. 183–197.

[Tak2] H. Takai, *Baum-Connes conjectures and their applications*, Dynamical systems and applications (Kyoto, 1987), World Sci. Adv. Ser. Dyn. Syst., vol. 5, World Sci., Singapore, 1987, pp. 89–116.

[Tak3] H. Takai, *On the Baum-Connes conjecture*, The study of dynamical systems (Kyoto, 1989), World Sci. Adv. Ser. Dyn. Syst., vol. 7, World Sci., Teaneck, NJ, 1989, pp. 149–154.

[TaW] L. Taylor and B. Williams, *Surgery spaces: formulae and structure*, Algebraic topology, Waterloo, 1978, Lecture Notes in Math., vol. 741, Springer-Verlag, Berlin, Heidelberg, New York, 1979, pp. 170–195.

[Th1] R. Thom, *Quelques propriétés globales des variétés différentiables*, Comment. Math. Helv. **28** (1954), 17–86.

[Th2] R. Thom, *Les classes caractéristiques de Pontrjagin des variétés triangulées*, Proc. Symp. Internacional de Topologica Algebraica, Univ. Nacional Autonoma de México, Mexico City, 1958, pp. 54–67.

[Ths1] C. B. Thomas, *Splitting theorems for certain PD^3-groups*, Math. Z. **186** (1984), 201–209.

[Ths2] C. B. Thomas, *3-manifolds and PD(3)-groups*, these Proceedings.

[Ti] U. Tillmann, *K-theory of fine topological algebras, Chern character, and assembly*, K-Theory **6** (1992), 57–86.

[To] A. M. Torpe, *K-theory for the leaf space of foliations by Reeb components*, J. Funct. Anal. **61** (1985), 15–71.

[Tr1] E. V. Troitsky, *The index of the generalized Hirzebruch operator and the homotopically invariant higher signatures*, Selected topics in Algebra, Geometry and Discr. Math., Moscow State Univ., Moscow, 1992, pp. 151–152.

[Tr2] E. V. Troitsky, *The Hirzebruch operator on PL manifolds and the higher signatures*, Uspekhi Mat. Nauk **48** no. 1 (1993), 189–190; English translation, Russian Math. Surv. **48** no. 1 (1993), 197–198.

[Tsy] B. L. Tsygan, *Homology of matrix Lie algebras over rings and Hochschild homology*, Uspekhi Mat. Nauk **38** no. 2 (1983), 217–218; English translation, Russian Math. Surv. **38** no. 2 (1983), 198–199.

[Va] A. Valette, *Le rôle des fibrés de rang fini en théorie de Kasparov équivariante*, Mem. Acad. Roy. Belg., Classes des Sciences (1986).

[Vi] M. F. Vigneras, *Homologie cyclique, principe de Selberg et pseudo-coefficient*, Invent. Math. **116** (1994), 651–676.

[Wald1] F. Waldhausen, *Algebraic K-theory of amalgamated free products*, Ann. of Math. **108** (1978), 135–256.

[Wald2] F. Waldhausen, *Algebraic K-theory of topological spaces, I*, Algebraic and Geometric Topology (Stanford, 1976), Proc. Symp. Pure Math., vol. 32, Amer. Math. Soc., Providence, RI, 1978, pp. 35–60; *II*, Algebraic topology, Aarhus 1978, Lecture Notes in Math., vol. 763, Springer-Verlag, Berlin, Heidelberg, New York, 1979, pp. 356–394.

[Wald3] F. Waldhausen, *Algebraic K-theory of spaces, a manifold approach*, Current Trends in Algebraic Topology, Part 1 (London, Ont., 1981), Canad. Math. Soc. Conf. Proc., vol. 2, Amer. Math. Soc., Providence, RI, 1982, pp. 141–184.

[Wall1] C. T. C. Wall, *Surgery of non-simply-connected manifolds*, Ann. of Math. **84** (1966), 217–276.

[Wall2] C. T. C. Wall, *Surgery on Compact Manifolds*, London Math. Soc. Monographs, vol. 1, Academic Press, London and New York, 1970.

[Was] A. Wassermann, *A proof of the Connes-Kasparov conjecture for connected reductive linear Lie groups*, C. R. Acad. Sci. Paris Sér. I Math. **304** (1987), 559–562.

[W1] S. Weinberger, *The Novikov conjecture and low-dimensional topology*, Comment. Math. Helv. **58** (1983), 355–364.

[W2] S. Weinberger, *Constructions of group actions: a survey of some recent developments*, Group actions on manifolds (Boulder, Colo., 1983) (R. Schultz, ed.), Contemp. Math., vol. 36, Amer. Math. Soc., Providence, RI, 1985, pp. 269–298.

[W3] S. Weinberger, *Group actions and higher signatures*, Proc. Nat. Acad. Sci. U.S.A. **82** (1985), 1297–1298.

[W4] S. Weinberger, *Homologically trivial group actions, II: nonsimply connected manifolds*, Amer. J. Math. **108** (1986), 1259–1275.

[W5] S. Weinberger, *Group actions and higher signatures, II*, Comm. Pure Appl. Math. **40** (1987), 179–187.

[W6] S. Weinberger, *Class numbers, the Novikov Conjecture, and transformation groups*, Topology **27** (1988), 353–365.

[W7] S. Weinberger, *Homotopy invariance of eta invariants*, Proc. Nat. Acad. Sci. U.S.A. **85** (1988), 5362–5363.

[W8] S. Weinberger, *Aspects of the Novikov conjecture*, Geometric and topological invariants of elliptic operators (Bowdoin Coll., 1988) (J. Kaminker, ed.), Contemp. Math., vol. 105, Amer. Math. Soc., Providence, RI, 1990, pp. 281–297.

[W9] S. Weinberger, *The Topological Classification of Stratified Spaces*, Chicago Lectures in Math., Univ. of Chicago Press, Chicago, 1994.

[W10] S. Weinberger, *Nonlocally linear manifolds and orbifolds*, Proc. International Congress of Mathematicians (Zürich, 1994) (to appear).

[WW1] M. Weiss and B. Williams, *Automorphisms of manifolds and algebraic K-theory: I*, K-Theory **1** (1988), 575–626.

[WW2] M. Weiss and B. Williams, *Automorphisms of manifolds and algebraic K-theory: II*, J. Pure and Appl. Algebra **62** (1989), 47–107.

[WW3] M. Weiss and B. Williams, *Automorphisms of manifolds and algebraic K-theory: Finale*, Preprint, 1994, 62pp.

[WW4] M. Weiss and B. Williams, *Assembly*, these Proceedings.

[WW5] M. Weiss and B. Williams, *Pro-excisive functors*, these Proceedings.

[Wu] F. Wu, *The higher index theorem for manifolds with boundary*, J. Funct. Anal. **103** (1992), 160–189.

[Ya1] M. Yamasaki, *L-groups of crystallographic groups*, Invent. Math. **88** (1987), 571–602.

[Ya2] M. Yamasaki, *L-groups of virtually polycyclic groups*, Topology and Appl. **33** (1989), 223–233.

[Ya3] M. Yamasaki, *Maps between orbifolds*, Proc. Amer. Math. Soc. **109** (1990), 223–232; *Erratum*, ibid. **115** (1992), 875.

[Yo] M. Yoshida, *Surgery obstruction of twisted products*, Math. J. Okayama Univ. **24** (1982), 73–97.

[Yu1] G. Yu, *K-theoretic indices of Dirac-type operators on complete manifolds and the Roe algebra*, preprint, Univ. of Colorado.

[Yu2] G. Yu, *Cyclic cohomology and higher indices for noncompact complete manifolds*, J. Funct. Anal. (to appear).

[Yu3] G. Yu, *Baum-Connes conjecture and coarse geometry*, K-Theory (to appear).

[Yu4] G. Yu, *Localization algebra and coarse Baum-Connes conjecture*, preprint, Univ. of Colorado.

[Yu5] G. Yu, *Zero-in-the-spectrum conjecture, positive scalar curvature and asymptotic dimension*, preprint, Univ. of Colorado.

Department of Mathematical Sciences, SUNY at Binghamton, Binghamton, NY 13901, U.S.A.

email: `steve@math.binghamton.edu`

Department of Mathematics and Statistics, James Clerk Maxwell Bldg., University of Edinburgh, Edinburgh EH9 3JZ, Scotland, U.K.

email: `a.ranicki@edinburgh.ac.uk`

Department of Mathematics, University of Maryland, College Park, MD 20742, U.S.A.

email: `jmr@math.umd.edu`

The Problem Session

9 September, 1993

Proposed by M. Gromov

(1) Look at coarse Lipschitz maps from uniformly contractible spaces or contractible coverings of compact manifolds to euclidean space. One wants them to have nonzero degree, but we don't want them to collapse things too much. Translate to the language of C^*-algebras. When do you have maps to euclidean space? Points which are far away to begin with should stay far away.

(2) Can one compute (or express) something like L^p-cohomology by C^*-techniques or C^*-invariants?

(3) Does an amenable discrete group admit a proper isometric action on a Hilbert space (in the *metric* sense of proper)?[1] A suitable generalization of the notion of almost flat bundles might yield a proof of the Novikov Conjecture for amenable groups.[2]

(4) Does every finitely generated or finitely presented group admit a uniformly metrically proper Lipschitz embedding into a Hilbert space? Even such an embedding into a reflexive uniformly convex Banach space would be interesting. This seems hard.

(5) Can one give a new proof using the above philosophy (of mapping to Euclidean space or Hilbert space) of the Strong Novikov conjecture (injectivity of the assembly map for the K-theory of the group C^*-algebra) or the Baum-Connes Conjecture for discrete subgroups of $SO(n,1)$, $SU(n,1)$.

(6) Is there a discrete group Γ (other than $\mathbb{Z}$) such that $B\Gamma$ is finite, and such that some compactification of $E\Gamma$, satisfying the "compact sets become small at infinity" condition,[3] maps to a compactification of

[1] Soon after the conference, this problem was solved. See the paper by Bekka, Cherix, and Valette in these proceedings for the solution.

[2] See the paper by Gromov in these proceedings for more details on what was intended by this question.

[3] This is the condition assumed in [CP], and also later used in the paper of Hurder in these proceedings.

$\mathbb{R}$? A homological version of this would be to ask if there is a finite $B\Gamma$ so that the coarse cohomology of Γ is zero in all dimensions.

(7) One can put a "norm" on K_0 of a ring R (assuming that the "rank" of a finitely generated projective R-module is well-defined) by letting

$$\|x\| = \inf\{\operatorname{rank} P + \operatorname{rank} Q : x = [P] - [Q], \quad P, Q \text{ f. gen. projective}\},$$

and similarly for $L_0(R)$ or $L^0(R)$ if R has an involution. These can be made into actual norms on the vector spaces $K_0(R)\otimes_{\mathbb{Z}}\mathbb{R}$, $K_0(R)\otimes_{\mathbb{Z}}\mathbb{R}$. Compute the norms in $L_0(\mathbb{R}\pi)\otimes_{\mathbb{Z}}\mathbb{R}$ or in $K_0(C^*(\pi))\otimes_{\mathbb{Z}}\mathbb{R}$ of the images of rational homology classes on $B\pi$ under the assembly maps. It seems that one gets something non-zero for (at least some) manifolds with strictly negative curvature.

REFERENCES:

[CP] G. Carlsson and E. K. Pedersen, *Controlled algebra and the Novikov conjectures for K- and L-theory*, Topology (to appear).

Proposed by C. Stark

(1) Let X be uniformly contractible open manifold and let $C^*(X)$ be its Roe C^*-algebra (as defined in [Ro]). How does analysis of $C^*(X)$ relate to flexibility of X [DFW]? What does uniform contractibility of X correspond to in $C^*(X)$? What properties of metrics guarantee that uniformly contractible $\mathbb{R}^n$'s are boundedly rigid?

(2) Is there a splitting theory for manifolds whose fundamental group is given by a complex of groups as described by Haefliger [H]? One hopes that the categorical descriptions of Nil and UNil of Connolly-Koźniewski [CK] and of Ranicki [R] will generalize.

REFERENCES:

[CK] F. X. Connolly and T. Koźniewski, *Nil groups in K-theory and surgery theory*, Forum Math. **7** (1995), 45–76.

[DFW] A. Dranishnikov, S. Ferry and S. Weinberger, *A large Riemannian manifold which is flexible*, preprint.

[H] A. Haefliger, *Extension of complexes of groups*, Ann. Inst. Fourier, Grenoble **42** (1992), 275–311.

[R] A. Ranicki, *Exact Sequences in the Algebraic Theory of Surgery*, Mathematical Notes 26, Princeton Univ. Press, Princeton, NJ (1981).

[Ro] J. Roe *Coarse cohomology and index theory on complete Riemannian manifolds*, Mem. Amer. Math. Soc. **497** (1993).

Proposed by A. Ranicki

(1) Translate any solution of the Novikov conjecture (for some group π) into an effective algorithm for deforming a nonsingular quadratic form over $\mathbb{Z}\pi$ to a sheaf over $B\pi$ of nonsingular quadratic forms over $\mathbb{Z}$, i.e., explain "disassembly" algebraically. In other words, construct an inverse of the assembly map in L-theory on the level of representatives, starting from the geometry. This would already be of interest for the case $\pi = \mathbb{Z}^k$ free abelian, if it were possible to do this all at once, without induction on k.

(2) Similarly for the algebraic K-theory Novikov conjecture and K_0 of the group ring. How do you see that projectives are stably free?

(3) Use the automatic structure on hyperbolic groups [CDP, Ch. 12, Théorème 7.1] to extend solutions of word problems to modules and quadratic forms over the group rings.

(4) Novikov originally wanted to construct a map (algebraically!) from L-theory to rational homology. Can one obtain a combinatorial formula for Pontrjagin classes by a constructing a map of spectra from the algebraic L-spectrum of $\mathbb{Q}$ to an Eilenberg-MacLane spectrum realizing the Hirzebruch L-class? Note that there are papers of Connes-Sullivan-Teleman [CST] and Moscovici and Wu [MW], giving semi-local formulæ for (dual) Pontrjagin classes as explicit Alexander-Spanier cycles, using Lipschitz analysis and cyclic cohomology machinery.

REFERENCES:

[CDP] M. Coornaert, T. Delzant and A. Papadopoulos, *Géométrie et théorie des groupes: Les groupes hyperboliques de Gromov*, Lecture Notes in Math. 1441, Springer-Verlag, Berlin, Heidelberg, New York (1990).

[CST] A. Connes, D. Sullivan and N. Teleman, *Formules locales pour les classes de Pontrjagin topologiques*, C. R. Acad. Sci. Paris (Sér. I) **317** (1993), 521–526.

[MW] H. Moscovici and F. Wu, *Localization of Pontrjagin classes via finite propagation speed*, C. R. Acad. Sci. Paris (Sér. I) **317** (1993), 661–665.

Proposed by S. Ferry and S. Weinberger

(1) Let M be a proper metric space, and let $C^*(M)$ be its Roe C^*-algebra (as defined in [Ro]). Let $L^{bdd}_{M*}(\mathbb{C})$ be the L-theory of the M-bounded $\mathbb{C}$-module category of [FP]. Construct an explicit natural transformation $L^{bdd}_{M*} \to K_*(C^*(M))$.[4] Use this to prove an analogue of the Kaminker-Miller Theorem [KM] in the bounded case.

(2) Let π be a group such that $B\pi$ has the homotopy type of a simple Poincaré complex. The "existence Borel" conjecture for π asserts the existence of an aspherical compact ANR homology manifold with fundamental group π. Is the conjecture true for word hyperbolic groups? A compact ANR manifold, however, need *not* have a topological manifold resolution. A resolution exists if and only if the local index invariant of Quinn [Q1, Q2] is 1, but there are cases where it is $\neq 1$ [BFMW]. (See the paper by Ferry and Pedersen in these proceedings, as well as [W2], for more information.) But it is hard to construct aspherical examples with Quinn invariant $\neq 1$. Is there a word hyperbolic Poincaré duality group realizing a nontrivial Quinn invariant?

(3) An open n-manifold is called *hyperspherical* if it admits a proper Lipschitz map to $\mathbb{R}^n$ (of degree one). Is every uniformly contractible manifold M hyperspherical? (Compare [DFW].)

(4) Is there a Riemannian metric on $\mathbb{R}^n$ so that $K^{\ell f}_*(\mathbb{R}^n) = \widetilde{K}_*(S^n) \to K_*(C^*(\mathbb{R}^n))$ fails to be injective? (See the paper by Higson and Roe in these proceedings for the construction of the map.)

(5) O. Attie has developed a surgery theory for open manifolds with bounded geometry [A1, A2]. A bounded geometry Novikov conjecture would assert the injectivity of the assembly map in this theory. Is this true under reasonable hypotheses?

(6) "Uniformly finite" homology groups of a metric space are defined in [BW]. Is the "uniformly finite" (homology) L-genus $L(M) \in H^{uf}_0(M; \mathbb{Z})$ a bounded geometry homotopy invariant?

(7) To what extent are analogues of the Borel and Novikov conjectures valid in the context of stratified spaces? See the appendix to Chapter 9 of Weinberger's book [W1] for preliminary results.

REFERENCES:

[A1] O. Attie, *Quasi-isometry classification of some manifolds of bounded*

[4] See the paper by Pedersen, Roe and Weinberger in these proceedings for a partial solution.

geometry, Math. Z. **216** (1994), 501–527.

[A2] O. Attie, *A surgery theory for manifolds of bounded geometry*, preprint.

[BW] J. Block and S. Weinberger, *Aperiodic tilings, positive scalar curvature and amenability of spaces*, J. Amer. Math. Soc. **5** (1992), 907–918.

[BFMW] J. Bryant, S. Ferry, W. Mio and S. Weinberger, *The topology of homology manifolds*, Bull. Amer. Math. Soc. **28** (1993), 324–328.

[DFW] A. Dranishnikov, S. Ferry and S. Weinberger, *A large Riemannian manifold which is flexible*, preprint.

[FP] S. Ferry and E. K. Pedersen, *Epsilon surgery theory*, in these proceedings.

[KM] J. Kaminker and J. Miller, *Homotopy invariance of the analytic index of signature operators over C^*-algebras*, J. of Operator Theory **14** (1985), 113–127.

[PRW] E. K. Pedersen, J. Roe and S. Weinberger, *On the homotopy invariance of the boundedly controlled analytic signature of a manifold over an open cone*, in these proceedings

[Q1] F. Quinn, *Resolutions of homology manifolds, and the topological characterization of manifolds*, Inventiones Math. **72** (1983), 267–284; Corrigendum: *ibid.* **85** (1986), 653.

[Q2] F. Quinn, *An obstruction to the resolution of homology manifolds*, Michigan Math. J. **34** (1987), 284–291.

[Ro] J. Roe *Coarse cohomology and index theory on complete Riemannian manifolds*, Mem. Amer. Math. Soc. **497** (1993).

[W1] S. Weinberger, *The Topological Classification of Stratified Spaces*, Univ. of Chicago Press, Chicago (1994).

[W2] S. Weinberger, *Non locally-linear manifolds and orbifolds*, Proc. Inter. Cong. Math., Zürich, 1994 (to appear).

Proposed by M. Rothenberg

(1) Can one do smoothing theory on C^*-algebras? For example, can one classify "distinct differentiable structures" on C^*-algebras by operator-theoretic methods?[5]

[5] If X is a compact topological manifold, then a differentiable structure on X amounts

(2) If the answer to (1) is yes, can one attack a relative version (corresponding to the non-commutative analogue of compact manifolds with boundary)?

REFERENCES:

[C] A. Connes, *Noncommutative Geometry*, Academic Press, San Diego (1994).

Proposed by J. Rosenberg

(1) There is a known C^*-analogue of the L-theory assembly map, namely the assembly map in topological K-theory (for the group C^*-algebra). However, there is no known C^*-analogue of the surgery exact sequence, where each homotopy group of the fiber of the assembly map has a concrete interpretation as a "structure set." What is the C^*-analogue of $\mathcal{S}(M)$ or of $\mathcal{S}^{bdd}(X)$? Consider, for example, the case where M is a simply connected manifold. Then the assembly map $K_*(M) \to K_*(\mathbb{C}) = KU_*$ or $KO_*(M) \to KO_*$ is the map induced by the "collapse" map $M \to$ point. The kernel of the assembly map consists of classes of "generalized elliptic operators" on M with vanishing index, just as (modulo 2-torsion) the class in the structure set of M corresponding to a homotopy equivalence $h : M' \to M$ of manifolds can be detected by $h_*([D_{M'}]) - [D_M]$, where D_M and $D_{M'}$ are the classes of the signature operators of the manifolds. The analogue of the "structure set" also seems to appear in the problem of classifying manifolds of positive scalar curvature, where it measures those classes in $KO_*(B\pi)$ coming from spin manifolds with fundamental group π admitting positive scalar curvature. Try to give a purely C^*-algebraic analogue of the structure set[6] and an analogue of the surgery sequence in more general settings.

(2) Can one give a direct operator-theoretic proof of the homotopy invariance, in the bounded category, of the image of the signature class under the C^*-assembly map? This would help in giving a direct C^*-proof of "bounded Novikov." (See also the first question of Ferry of Weinberger, and the second question of Roe).

(3) What are "good" ways to compute "homology" and "cohomology" of

to a choice of a distinguished dense subalgebra $C^\infty(X)$ of the commutative C^*-algebra $C(X)$, satisfying certain natural axioms. Connes has proposed studying analogous dense subalgebras of non-commutative C^*-algebras [C].

[6] The paper of Higson and Roe in these proceedings constructs a C^*-analogue of the bounded structure set.

C^*-algebras?

Proposed by J. Davis

(1) Let Γ be a discrete group. Is there a sense in which all of $\tilde{K}_0(\mathbb{Z}\Gamma)$ or of $\tilde{K}_0(\mathbb{Q}\Gamma)$ comes from the *torsion* subgroups $H \subset \Gamma$ of Γ? (The answer to this might involve a rather complicated "induction" procedure, since for example, if Γ is a product of a finite group H with $\mathbb{Z}$, then $K_{-1}(\mathbb{Z}H)$ shows up as a direct summand in $\tilde{K}_0(\mathbb{Z}\Gamma)$, even if $\tilde{K}_0(\mathbb{Z}H)$ vanishes.)

(2) Is $H_*(\mathcal{E}\Gamma/\Gamma; K(\mathbb{Q}\Gamma_x)) \to K_*(\mathbb{Q}\Gamma)$ an isomorphism for a discrete group Γ? Here $\mathcal{E}\Gamma$ is the proper Γ-space with $\mathcal{E}\Gamma^H \simeq *$ if H torsion, $\mathcal{E}\Gamma^H = \emptyset$ otherwise. This is a version of Baum-Connes. Translate the known geometry of this problem into algebra.

Proposed by J. Roe

(1) Develop an appropriate index theory for manifolds parametrized by metric spaces, $\begin{array}{c} M \\ \downarrow \\ X \end{array}$. (See the paper by Pedersen, Roe, and Weinberger in these proccedings.) Develop a good theory to get an analytic signature $sign_X(M) \in K_*(C^*(X))$, where $C^*(X)$ is the C^*-completion of the locally traceable operators over X with bounded propagation.

(2) Given a unital C^*-algebra A, and a proper metric space X, let $L_*^{bdd}(X; A)$ be the L-theory of the X-bounded A-module category of [FP]. Find a natural transformation $L_*^{bdd}(X;A) \to K_*(C^*(X;A))$ so that the diagram

$$\begin{array}{ccc} H_*^{\ell f}(X;\mathbb{L}(A)) & \xrightarrow{A} & L_*^{bdd}(X;A) \\ \downarrow & & \downarrow \\ H_*^{\ell f}(X;\mathbb{K}(A)) & \xrightarrow{A} & K_*(C^*(X;A)) \end{array}$$

commutes. (Maybe one needs to invert 2.) Here the maps marked A are the assembly maps. What are the kernels? (On the spectrum level, what is the fiber of the assembly map? See also the first question posed by Rosenberg above.) Here, A is an arbitrary C^*-algebra.

(3) Is there a connection between being able to find a "good" compactification X of $E\Gamma$ and being able to find a good dense subalgebra of $C^*(\Gamma)$

consisting of "rapidly decaying" functions? What is the analogue of the assumption of contractibility of X?

REFERENCES:

[FP] S. Ferry and E. K. Pedersen, *Epsilon surgery theory*, in these proceedings.

Proposed by J. Cuntz and M. Puschnigg

For C^*-algebras A given by generators and relations, characterize classes of dense subalgebras $\mathcal{A}$ such that $K_*(\mathcal{A}) \to K_*(A)$ is an isomorphism and cyclic homology of $\mathcal{A}$ is "big enough" to relate to cyclic homology of the $\mathbb{C}$-algebra generated by the generators, independent of the algebra in the class.

EXAMPLE: Let $A = C(S^1)$, $\mathcal{A} = C^\alpha(S^1)$ $0 < \alpha \leq 1$. The cyclic homology fundamental class of $C^\infty(S^1)$ lives in dim 1, but for $\mathcal{A}_\alpha$ lives in degree the smallest odd integer $\geq \frac{1}{2\alpha}$.

Proposed by A. Valette

(1) Let Γ be a finitely generated group. Connes and Moscovici [CM] have shown that if Γ satisfies the two following conditions (PC) and (RD) (with respect to some word length function L on Γ), then the Strong Novikov conjecture (injectivity of the assembly map for the group C^*-algebra) holds for Γ:

(PC) (Polynomial cohomology) Any element in $H^*(\Gamma, \mathbb{C})$ can be represented by a cocycle with polynomial growth.

(RD) (Rapid decay) There exist constants $C > 0$, $r \geq 0$ such that for any $f \in \mathbb{C}\Gamma$: $\|\lambda(f)\| \leq C\|f(1+L)^r\|_2$, where $\|\lambda(f)\|$ is the operator norm of the operator of left convolution by f on $\ell^2(\Gamma)$.

We shall focus here on (RD), which is relevant for some problems in harmonic analysis (see [JoV]). The following is known:

- Any hyperbolic group has (RD) (see [Har], [JoV] for proofs).
- For $n \geq 3$, $Sl_n(\mathbb{Z})$ does not have (RD) (see [Jol]).

CONJECTURE: If Γ acts properly, cocompactly either on a Riemannian symmetric space or on an affine building, then Γ has (RD).

(2) Is RD a quasi-isometry invariant?

(3) (Connes) Suppose there is on Γ a conditionally negative-type function

ϕ (i.e.,

$$\phi(g) = \|\alpha(g)\xi - \xi\|^2$$

for some affine isometric action α of Γ on a Hilbert space) such that $e^{-t\phi}$ is integrable for t big. Is Γ K-amenable?

(4) What is the relationship between uniformly bounded representations and uniform embedding in a Hilbert space?

REFERENCES:

[CM] A. Connes and H. Moscovici, *Cyclic cohomology, the Novikov conjecture, and hyperbolic groups*, Topology **29** (1990), 345–388.

[Har] P. de la Harpe, *Groupes hyperboliques, algèbres d'opérateurs et un théorème de Jolissaint*, C. R. Acad. Sci. Paris (Sér. I) **307** (1988), 771–774.

[Jol] P. Jolissaint, *Rapidly decreasing functions in reduced C^*-algebras of groups*, Trans. Amer. Math. Soc. **317** (1990), 167–196.

[JoV] P. Jolissaint & A. Valette, *Normes de Sobolev et convoluteurs bornés sur $L^2(G)$*, Ann. Inst. Fourier, Grenoble **41** (1991), 797–822.

Proposed by S. Hurder

(1) Is there a discrete group Γ with a classifying space $B\Gamma$ whose universal cover $E\Gamma$ admits a good compactification, but not a good Γ-invariant compactification?

(2) Is there a discrete group Γ with a classifying space $B\Gamma$ whose universal cover $E\Gamma$ has a good compactification, but not a Z-compactification in the sense of Bestvina-Mess [BM]? (In other words, is it always possible to arrange for the compactification $\overline{E\Gamma}$ of $E\Gamma$ to have a one-parameter family h_t of self-maps with $h_0 = id$, $h_t(\overline{E\Gamma}) \subset E\Gamma$ for $t > 0$?) The outer automorphism group of a free group is a potential candidate.

REFERENCES:

[BM] M. Bestvina and G. Mess, *The boundary of negatively curved groups*, J. Amer. Math. Soc. **4** (1991), 469–481.

Proposed by F. Connolly

(1) Let Γ be a uniform lattice in a Lie group. Assume $\hat{H}^*(\Gamma;\mathbb{Z})$ is periodic, but Γ contains no finite dihedral subgroup. Is there a compact M such that $\Gamma = \pi_1(M)$ and $\tilde{M} = \mathbb{R}^m \times S^{n-1}$?

(2) If G is finitely presented and VFL (having a subgroup H of finite index for which the trivial $\mathbb{Z}H$-module $\mathbb{Z}$ has a finite free $\mathbb{Z}H$-resolution) without elements of order 2, let $\varepsilon : \mathbb{Z}G \to \mathbb{F}_2$ be the augmentation, mod 2. Is $\varepsilon_* : UNil_n^h(\mathbb{Z}G; \mathbb{Z}G, \mathbb{Z}G) \to UNil_n(\mathbb{Z}; \mathbb{Z}, \mathbb{Z})$ an isomorphism $\forall n$, i.e., are the Connolly-Koźniewski Arf invariants the only splitting obstructions?

(3) Is $2 \cdot UNil_{2n}^h(H; G_1, G_{-1}) = 0$ for any three finitely presented groups G_1, G_{-1}, and H, where $G_{-1} \supset H \subset G_1$?

(4) Let Γ be an arithmetic subgroup of a linear semi-simple Lie group G with maximal compact subgroup K. Let M be a locally flat Γ manifold. Let $f : M \to G/K$ be a stratified proper Γ-homotopy equivalence. Assume f is a topologically simple homotopy equivalence. Does the stratified space $\Gamma\backslash M$ compactify and $\Gamma\backslash f$ extend to a *simple* homotopy equivalence $\overline{\Gamma\backslash f} : \left(\overline{\Gamma\backslash M}, \partial\right) \to (\overline{\Gamma\backslash G/K}, \partial)$, where $\overline{\Gamma\backslash G/K}$ denotes the compactification of $(\Gamma\backslash G/K)$ due to Borel and Serre [BS]?

REFERENCES:

[BS] A. Borel and J.-P. Serre, *Corners and arithmetic groups*, Comment. Math. Helv. **48** (1973), 436–491.

[CK] F. X. Connolly and T. Koźniewski, *Nil groups in K-theory and surgery theory*, Forum Math. **7** (1995), 45–76.

Proposed by T. Koźniewski

Waldhausen [W] defined his Nil groups as K-groups of a category with endomorphisms. Which endomorphisms represent 0 in the Nil group? In the usual Nil groups, these are stably triangular. What happens in fancier Nil groups?

REFERENCES:

[W] F. Waldhausen, *Algebraic K-theory of generalized free products*, Ann. of Math. **108** (1978), 135–256.

Proposed by M. Yan

How does one get the rational symmetric signature of a manifold out of the de Rham complex of the covering space? The hope would be to construct an algebraic recipe. Is the real symmetric signature easier? This is another algorithmic question.

Proposed by E. Pedersen

How is the C^*-algebra of a foliation related to foliated control? What are the C^*-analogues of Farrell-Jones type arguments (as in [FJ1, FJ2])? One may have to extend the foliated theory to an asymptotic theory. Then one should use this to prove Borel-type isomorphism results.

REFERENCES:

[FJ1] F. T. Farrell and L. Jones, *K-theory and dynamics, I*, Ann. of Math. **124** (1986), 531–569; *ibid., II*, Ann. of Math. **126** (1987), 451–493.

[FJ2] F. T. Farrell and L. E. Jones, *Foliated control theory, I*, K-theory **2** (1988), 357–399; *ibid., II*, K-theory **2** (1988), 401–430.

[JR] J. Roe, *From foliations to coarse geometry and back*, Proc. Conf. on foliations, Santiago, 1994 (World Scientific).

Proposed by B. Williams

(1) Given a fiber bundle $\begin{array}{ccc} F & \to & E \\ & & \downarrow \\ & & B \end{array}$ with compact finitely dominated fiber, are there transfer maps $K_n(\mathbb{Z}\pi_1(B)) \to K_n(\mathbb{Z}\pi_1(E))$ for all n analogous to the ones for $n = 0, 1$ and in A-theory? Such a transfer should be compatible with the A-theory transfer.

(2) If we replace $\mathbb{Z}\pi_1(B)$ by a suitable C^*-algebra, is there an analogous transfer?

Proposed by W. Lück

(1) There is a paper of Peter Linnell [L] where he proves for a certain class of groups Γ that there is a division ring D satisfying

$$\mathbb{C}\Gamma \subset D \subset \mathcal{U}(\Gamma)$$

where $\mathcal{U}(\Gamma)$ is the algebra of operators affiliated to the von Neumann algebra. This implies in particular the zero-divisor conjecture. Can this be generalized to other torsion-free groups?

(2) Does the isomorphism conjecture for algebraic K-theory imply the rationality of L^2-Betti numbers of compact manifolds and the zero-divisors conjecture? (The paper of Gromov in these proceedings is relevant here.)

(3) (proposed jointly with R. Jung) The elliptic homology $\mathcal{E}\ell\ell_*(X)$ is obtained [KS] from $\Omega_*(X)$ by dividing out $\mathbb{CP}^2$-bundles over zero-bordant bases and then inverting the class of $\mathbb{CP}^2$. By composing the map from oriented bordism to symmetric L-theory given by the symmetric signature and the map from symmetric L-theory to quadratic L-theory given by crossing with a simply connected surgery problem realizing signature 8, one obtains a natural transformation

$$s : \mathcal{E}\ell\ell_*(M)[\tfrac{1}{2}] \to L_*(\mathbb{Z}\pi_1(M))[\tfrac{1}{2}].$$

(This is well-defined using, say, the results of [LR].) The map s factors through the assembly map A as

$$\mathcal{E}\ell\ell_*(M)[\tfrac{1}{2}] \to H_*(M, \mathbb{L}(\mathbb{Z})[\tfrac{1}{2}]) \xrightarrow{A} L_*(\mathbb{Z}\pi_1(M))[\tfrac{1}{2}],$$

and Rainer Jung has shown that the first map is an isomorphism. Thus the injectivity of A (after inverting 2) is equivalent to asking how much of $\mathcal{E}\ell\ell_*(B\pi)$ can be detected by symmetric signatures. Does this shed any new light on the Novikov conjecture? Does the class of an oriented manifold in $\mathcal{E}\ell\ell_*(B\pi)$ only depend on its oriented homotopy type?

REFERENCES:

[KS] M. Kreck and S. Stolz, $\mathbb{HP}^2$-*bundles and elliptic homology*, Acta Math. **171** (1993), 231–261.

[L] P. Linnell, *Division rings and group von Neumann algebras*, Forum Math. **5** (1993), 561–576.

[LR] W. Lück and A. Ranicki, *Surgery obstructions of fibre bundles*, J. Pure and Appl. Algebra **81** (1992), 139–189.

On the Steenrod homology theory

John Milnor

In 1940 Steenrod defined homology groups for compact metric spaces based on "regular cycles". (See Steenrod [13], [14]). The object of this note is to give an axiomatic characterization of Steenrod's homology theory.* It is assumed that the reader is familiar with Eilenberg and Steenrod [4], Foundations of Algebraic Topology.

Let $\mathcal{A}_{CM}$ denote the category of compact metric pairs and continuous maps. Let G be an abelian group, and let p be a point.

Main Theorem. *There exists one and only one homology theory** on $\mathcal{A}_{CM}$ (up to natural equivalence) which satisfies the following two Axioms as well as the seven Eilenberg-Steenrod Axioms and which satisfies $H_0(p) = G$.*

The new axioms are:

Axiom 8. *Invariance under relative homeomorphism.*

Axiom 9. (Cluster Axiom.) *If X is the union of countable many compact subsets $X_1, X_2, \cdots$ which intersect pairwise at a single point b, and which have diameters tending to zero; then $H_q(X, b)$ is naturally isomorphic to the direct product of the groups $H_q(X_i, b)$.*

Precise statements will be given in §1.

The Čech homology theory satisfies all of these axioms except number 4, the Exactness Axiom. In fact, Axioms 1,2,3 and 5 through 8 are proved by Eilenberg and Steenrod, while Axiom 9 follows easily from the continuity property of the Čech theory. (See [4, Chapters IX, X]).

Actually, for a wide variety of coefficient groups G, the Čech homology theory is exact; and therefore coincides with the Steenrod homology theory. This is the case if G is infinitely divisible, or has finite exponent, or can be topologized as a compact group. (For G compact this is proved in [4, pg. 248]. The other cases follow since every infinitely divisible group can be imbedded as a direct summand of a torus $S^1 \times S^1 \times \cdots$; while every group with finite exponent can be imbedded as a direct summand of a direct product of finite groups.)

This paper was originally issued in the form of mimeographed notes at Berkeley, 1961.

* We will work with non-reduced homology theory, although Steenrod defined only reduced homology groups. Furthermore we will shift homology dimension indices by 1.

** Such a theory is constructed on the category $\mathcal{A}_C$ of all compact pairs. However, the uniqueness proof only works in the subcategory $\mathcal{A}_{CM}$. A similar homology theory has been defined by Borel and Moore [2].

The singular homology theory does not satisfy either Axiom 8 or Axiom 9. As an example, it is shown in [1] that the singular homology group $H_3(S^2 \vee S^2 \vee \cdots; \mathbb{Z})$ of an infinite cluster of 2-spheres with diameter tending to zero is non-trivial.

Analogous axioms for cohomology theory are formulated in §3. It turns out that these axioms uniquely characterize the Čech cohomology theory.

An important tool in working with Steenrod homology groups is a modified form of the Continuity Axiom, which involves not only the inverse limit operation (as in the usual Continuity Axiom) but also the first derived functor of inverse limit. (See Theorem 4.) This first derived functor has been studied by Yeh [15]. Higher derived functors of the inverse limit functor are defined in an Appendix, but the more general construction is not needed for this paper.

It is natural to ask whether Steenrod homology groups can be defined for a wider variety of spaces. Sitnikov [10], [11] has proposed a definition for metric spaces which amounts to the following.

Define the **Steenrod-Sitnikov homology groups** of a metric pair (X, X') to be the direct limit of the Steenrod homology groups of all compact pairs (C, C') contained in (X, X'). That is, define

$$H_q(X, X'; G) = \varinjlim H_q(C, C'; G) .$$

More generally, this definition makes sense whenever X is a Hausdorff space.

It is easily verified that this homology theory satisfies the Eilenberg-Steenrod axioms, including a strong version of the Excision Axiom. (Compare [4, pg. 255].) Another useful property is the following:

Assertion. *Every exact coefficient sequence* $0 \longrightarrow G' \longrightarrow G \longrightarrow G'' \longrightarrow 0$ *gives rise to an exact sequence of Steenrod-Sitnikov homology groups*

$$\cdots \longrightarrow H_q(X; G') \longrightarrow H_q(X; G) \longrightarrow H_q(X; G'') \longrightarrow H_{q-1}(X; G') \longrightarrow \cdots .$$

This can be proved using the techniques of §1 or §4. The analogous statement for Čech homology would be false. As a final argument in favor of the Steenrod-Sitnikov theory we mention the following:

Sitnikov duality theorem. *If A is an arbitrary subset of the sphere S^{n+1} then for $0 < q < n$ the Steenrod-Sitnikov homology group $H_q(A; G)$ is isomorphic to the Čech cohomology group $H^{n-q}(S^{n+1} - A; G)$.*

(For A compact this theorem was proved by Steenrod [13]. A proof is outlined in §1. The more general case follows using the definition

$$H_q(A) = \varinjlim H_q(C) ;$$

together with the following lemma, due to Sitnikov: *The Čech cohomology group of an arbitrary subset of Euclidean space is equal to the direct limit of the Čech cohomology groups of its open neighborhoods.*)

§1. The Uniqueness Theorem

Let $\mathcal{A}_{CM}$ denote the category of all compact metric pairs, and all continuous maps between such pairs. Assume that a homology theory is given on $\mathcal{A}_{CM}$ so as to satisfy the seven Eilenberg-Steenrod axioms, as well as the following two.

Axiom 8. *If $f : (X, X') \longrightarrow (Y, Y')$ is a map in $\mathcal{A}_{CM}$ which carries $X - X'$ homeomorphically onto $Y - Y'$, then*

$$f_* \; : \; H_q(X, X') \longrightarrow H_q(Y, Y')$$

is an isomorphism.

(This is a strengthened form of the Excision Axiom, number 6.)

Axiom 9. *Suppose that the compact metric space X is the union of compact subsets $X_1, X_2, \cdots$ with diameters tending to zero and suppose that $X_i \cap X_j = \{b\}$ for $i \neq j$. Let*

$$r_i \; : \; (X, b) \longrightarrow (X_i, b)$$

denote the unique retraction which carries X_j into the base point b for $j \neq i$. Then the correspondence

$$u \longrightarrow ((r_1)_*(u), (r_2)_*(u), \cdots)$$

defines an isomorphism of $H_q(X, b)$ onto the direct product of the groups $H_q(X_i, b)$.

These two axioms can be formulated in a different way, using the locally compact homology theory associated with H_*. (See [4, pg.271]). Given any locally compact space Y let $Y^\bullet = Y \cup \omega$ denote the one-point compactification of Y. If $Y^\bullet$ is metrizable, **define** the group $H_q^{\mathrm{LF}}(Y)$ to be equal to $H_q(Y^\bullet, \omega)$. Here the 'LF' is supposed to suggest a theory based on locally finite chains. (More generally, if Y_1 is a closed subset, define $H_q^{\mathrm{LF}}(Y, Y_1) = H_q(Y^\bullet, Y_1^\bullet)$.) With this notation we can reformulate as follows.

Axiom 8^{LF}. *For any compact metric pair (X, X') the natural homomorphism*

$$H_q(X, X') \longrightarrow H_q^{\mathrm{LF}}(X - X')$$

is an isomorphism.

Axiom 9^{LF}. *If Y is the disjoint union of open subsets $Y_1, Y_2, \cdots$, then $H_q^{\mathrm{LF}}(Y)$ is naturally isomorphic to the direct products of the groups $H_q^{\mathrm{LF}}(Y_i)$.*

These are clearly equivalent to Axioms 8 and 9 respectively.

Given any star-finite cell complex K let $C_q^{inf}(K; G)$ denote the q-th chain group of K based on infinite chains, with coefficients in $G = H_0(p)$. Let $|K|$ denote the underlying topological space.

Theorem 1. *If K is a countable, locally finite CW complex then $H_q^{\mathrm{LF}}(|K|)$ is naturally isomorphic to the q-th homology group of the chain complex $C_*^{inf}(K;G)$. A corresponding assertion holds for pairs (K,L).*

For the moment we will only be able to give the proof for the special case in which K is finite dimensional.

Let K^n denote the n-skeleton of K. The difference $|K^n| - |K^{n-1}|$ is a disjoint union of n-cells. hence, using Axiom 9^{LF}, it follows that

$$H_q^{\mathrm{LF}}(|K^n|,|K^{n-1}|) \;=\; H_q^{\mathrm{LF}}(|K^n|-|K^{n-1}|)$$

is zero for $q \neq n$, and is isomorphic to $C_n^{inf}(K;G)$ for $q=n$. (That is, this group is a direct product of copies of G, one copy for each oriented n-cell.)

Assertion. *The boundary operator*

$$\partial \;:\; H_n^{\mathrm{LF}}(|K^n|,|K^{n-1}|) \longrightarrow H_{n-1}^{\mathrm{LF}}(|K^{n-1}|,|K^{n-2}|)$$

corresponds under this isomorphism to the usual boundary operator in $C_*^{inf}(K;G)$, based on incidence numbers between oriented cells.

We will assume that this assertion is known for the case of a finite CW complex. The more general assertion then follows easily, using the fact that every cell is contained in a finite subcomplex.

Lemma 1. *The group $H_q^{\mathrm{LF}}(|K^n|)$ is isomorphic to $H_q(C_*^{inf}(K;G))$ for $n > q$.*

This is proved by induction on n. The appropriate induction hypothesis is that:

$$H_q^{\mathrm{LF}}(|K^n|) \;=\; \begin{cases} 0 & \text{for } q > n \\ Z_q(C_*^{inf}(K;G)) & \text{for } q = n, \\ H_q(C_*^{inf}(K;G)) & \text{for } q < n. \end{cases}$$

This is certainly satisfied for $n=0$. Assuming that it is true for $n-1$, and considering the exact sequence of the pair $(|K^n|,|K^{n-1}|)$ we obtain a proof for the given integer n.

If K is finite dimensional, then this completes the proof of Theorem 1. The proof in the general case will be given in §2.

(As a corollary to Theorem 1 we have the **Steenrod duality theorem.** Let X be a compact subset of S^{n+1}. Then $S^{n+1}-X$ is the underlying space of a simplicial complex K. Thus

$$H_q(X) \;\cong\; H_{q+1}(S^{n+1},X) \;\cong\; H_{q+1}(C_*^{inf}(K;G))$$

for $0 < q < n$. But a standard argument shows that the $(q+1)$-st homology group of K based on infinite chains is isomorphic to $H^{n-q}(C^*(K;G))$, the usual cohomology group based on infinite cochains. This is isomorphic to the $(n-q)$-th singular or Čech cohomology group of $S^{n+1}-X$.)

Theorem 2. *Any compact metric pair (X, X') can be imbedded in a compact metric pair (T, T') so that*

(1) *T and T' are contractible, and*

(2) *$T - X$ is the underlying space of a CW complex with subcomplex $T' - X'$.*

Furthermore, any map $f : (X, X') \longrightarrow (X_1, X_1')$ can be extended to a map $\overline{f} : (T, T') \longrightarrow (T_1, T_1')$, which carries the subset $T - X$ into $T_1 - X_1$.

The proof will be based on the idea of the "fundamental complex" of a compact metric space. (Lefschetz [6]). To simplify the notation, consider a single space X instead of a pair. For any finite open covering α of X the nerve will be denoted by X_α and the open sets in the covering by α_v. Let β be a refinement of α. A map

$$p \ : \ |X_\beta| \longrightarrow |X_\alpha|$$

will be called a projection if p maps simplexes linearly into simplexes; and if, for each vertex w of $|X_\beta|$, the image $p(w)$ is a linear combination of vertices v for which $\alpha_v \supset \beta_w$. The basic properties of this concept are:

(1) Projections from $|X_\beta|$ to $|X_\alpha|$ exist and any two such projections are homotopic.

(2) The composition of two projections $|X_\gamma| \longrightarrow |X_\beta| \longrightarrow |X_\alpha|$ is again a projection.

Let $\alpha(1), \alpha(2), \cdots$ be a sequence of coverings of X with mesh tending to zero such that each $\alpha(i+1)$ is a refinement of $\alpha(i)$. Let $\alpha(0)$ denote the trivial covering consisting of a single set. The abbreviation X_i will be used for the underlying space $X_{\alpha(i)}$. Choose projections $p_i : X_{i+1} \longrightarrow X_i$ and let M_i denote the mapping cylinder of p_i. We will assume that these mapping cylinders are disjoint, except that $M_i \cap M_{i-1}$ is equal to X_i. Hence the infinite union

$$F \ = \ M_0 \cup M_1 \cup M_2 \cup \cdots$$

is a well defined locally compact space.

Definition. *F is called a* **fundamental complex** *for X.*

The finite union $M_0 \cup M_1 \cup \cdots \cup M_{i-1}$ will be denoted by F_i; and the base point X_0 by F_0. Note that F is contractible. This follows since each F_i is a deformation retract of F_{i+1}.

Assertion. *F is the underlying space of a CW complex.*

Proof. The linearity property of p_i guarantees that each p_i is a cellular map from X_{i+1} to some subdivision of X_i. It follows that M_i is the underlying space of a CW complex. Now by subdividing the subcomplex X_{i+1} of M_i we can make the CW structure of M_i compatible with that of M_{i+1}. Hence the union F has a compatible CW structure.

The fundamental complex F can be "compactified" as follows. Let $r_i : F_{i+1} \longrightarrow F_i$ denote the canonical retraction, and define T as the inverse limit of the sequence

$$F_0 \stackrel{r_0}{\longleftarrow} F_1 \stackrel{r_1}{\longleftarrow} F_2 \stackrel{r_2}{\longleftarrow} \cdots .$$

It is clear that F is imbedded as a dense open subset of T, and that $T - F$ can be identified with the inverse limit of the sequence

$$X_0 \stackrel{p_0}{\longleftarrow} X_1 \stackrel{p_1}{\longleftarrow} X_2 \stackrel{p_2}{\longleftarrow} \cdots .$$

Note that T is contractible. A contraction $c : T \times [0,1] \longrightarrow T$ is defined as follows. Let $c(t,0) = t$ and let $c(t, 1/n)$ denote the image of t under the projection map $T \longrightarrow F_{n-1} \subset T$. The deformation retraction of F_n onto F_{n-1} is used to define $c(t,u)$ for $\frac{1}{n+1} \leq u \leq \frac{1}{n}$. Continuity as $u \to 0$ is easily verified.

Definition. *The sequence of coverings $\alpha(i)$ and projections p_i will be called* **convergent** *if:*

Condition 1. *The mesh of $\alpha(i)$ tends to zero as $i \to \infty$, and:*

Condition 2. *For each i and each $\epsilon > 0$ there exists $j = j(i, \epsilon)$ so that the composition*

$$p_i p_{i+1} \cdots p_{j-1} \ : \ X_j \longrightarrow X_i$$

carries each simplex into a set of diameter less than ϵ.

Lemma 2. *There exists a convergent sequence of coverings $\alpha(i)$ and projections p_i. If the sequence is convergent then X can be identified with the inverse limit of the sequence $X_0 \stackrel{p_0}{\longleftarrow} X_1 \stackrel{p_1}{\longleftarrow} \cdots$.*

The proof is not difficult. This Lemma, together with the preceding discussion proves the first part of Theorem 2. The extension to a pair of spaces is straightforward.

Now consider a map $f : X \longrightarrow Y$. Let $\{\alpha(i), p_i\}$ be a convergent sequence for X with fundamental complex $F(X)$, and let $\{\beta(i), q_i\}$ be a convergent sequence for Y with fundamental complex $F(Y)$. First suppose that each $\alpha(i)$ happens to be a refinement of the induced covering $f^{-1}\beta(i)$. Let

$$s_i \ : \ X_i \longrightarrow |X_{f^{-1}\beta(i)}| \subset Y_i$$

be a projection map. Since both $s_i p_i$ and $q_i s_{i+1}$ are projection maps, it follows that they are canonically homotopic. In fact for each $x \in X_{i+1}$ the two images $s_i p_i(x)$ and $q_i s_{i+1}(x)$ lie in a common simplex of Y_i.

Extend s_i and s_{i+1} to a map $s'_i : M(p_i) \longrightarrow M(q_i)$ as follows. For each $x \in X_{i+1}$ the line segment joining x to $p_i(x)$ in $M(p_i)$ should map into the broken line segment joining $s_{i+1}(x)$ to $q_i s_{i+1}(x)$ to $s_i p_i(x)$. Piecing these maps s'_i together we obtain a map $s : F(X) \longrightarrow F(Y)$.

Next we will prove that s extends to a map $T(X) \longrightarrow T(Y)$. For each $i \leq j \leq k$ consider the diagram

$$\begin{array}{ccccc} T(X) & \xrightarrow{\ r\ } & F_k(X) & \xrightarrow{\ r'\ } & F_j(X) & & \\ & & \downarrow s & & \downarrow s & & \\ & & F_k(Y) & \xrightarrow{\ r''\ } & F_j(Y) & \xrightarrow{\ r'''\ } & F_i(Y) \end{array}$$

where r, r', r'', r''' denote canonical retractions. This diagram is "almost" commutative in the following sense. For each $x \in F_k(X)$ the two images $sr'(x)$ and $r''s(x)$ are either equal, or lie in a common simplex of Y_j.

Now if $j \geq j(i, \epsilon)$ so that r''' carries each simplex of Y_j into a set of diameter $< \epsilon$ in Y_i, it follows that the two composite maps $T(X) \longrightarrow F_i(Y)$ are ϵ-homotopic. Letting j and k tend to infinity, it follows that these composite maps tend uniformly to a limit

$$f_i \ : \ T(X) \longrightarrow F_i(Y) \ .$$

Since each diagram

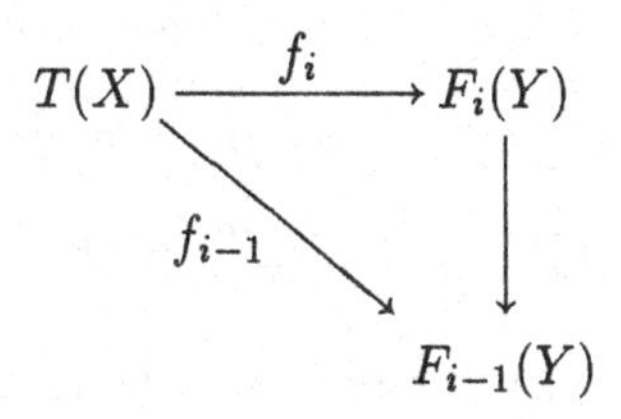

is commutative, it follows that the f_i define a map $\overline{f} : T(X) \longrightarrow T(Y)$. It is easily verified that $\overline{f}|X = f$, $\overline{f}|F(X) = s$.

Now consider the more general case where $\alpha(i)$ is not assumed to be a refinement of $f^{-1}\beta(i)$. For each i it is still possible to choose $j = j_i$ so that $\alpha(j_i)$ is a refinement of $f^{-1}\beta(i)$. Let $T'(X)$ denote the compactified fundamental complex corresponding to the sequence of coverings $\alpha(0), \alpha(j_1)$, $\alpha(j_2), \cdots$ and the composite projections $p_0 \cdots p_{j_1 - 1} \cdots$. Then it is not difficult to extend the identity map of X to a map $T(X) \longrightarrow T'(X)$. The above argument then constructs a map $T'(X) \longrightarrow T(Y)$.

Again the relative case presents no new difficulty. This completes the proof of Theorem 2.

Still assuming Theorem 1, we can now prove the Uniqueness Theorem.

Theorem 3. *Given two homology theories H and $\overline{H}$ on the category $\mathcal{A}_{CM}$, both satisfying the nine axioms, any coefficient isomorphism $H_0(b) \cong \overline{H}_0(b) \cong G$ extends uniquely to an equivalence e between the two homology theories.*

Proof. First consider the case of a single space X. Let $t_0 = F_0$ denote the base point of T. The symbol $+$ will denote topological sum. From the exact sequence of the triple $(T, X + t_0, t_0)$ we see that $H_q(X) = H_q(X + t_0, t_0)$ is isomorphic to $H_{q+1}(T, X + t_0) = H^{\mathrm{LF}}_{q+1}(T - X, t_0)$. Now $T - X$ is the underlying space of a CW complex K, hence this last group is isomorphic to $H_{q+1}C^{inf}_*(K, t_0; G)$. Similar isomorphisms can be constructed for the homology theory $\overline{H}$. Define e as the composite isomorphism

$$H_q(X) \longrightarrow H_{q+1}C^{inf}_*(K, t_0; G) \longrightarrow \overline{H}_q(X) .$$

Using the last part of Theorem 2 it can be verified that e is natural with respect to mappings. That is, any map $f : X \longrightarrow Y$ induces a commutative diagram

$$\begin{array}{ccc} H_q(X) & \xrightarrow{\ e\ } & \overline{H}_q(X) \\ \downarrow & & \downarrow \\ H_q(Y) & \xrightarrow{\ e\ } & \overline{H}_q(Y) . \end{array}$$

It follows that e does not depend on the choice of T.

The proof for a pair (X, X') is similar, making use of isomorphisms

$$\begin{aligned} H_q(X, X') \cong\ & H_q(X \cup T', T') \cong H_{q+1}(T, X \cup T') \\ \cong\ & H^{\mathrm{LF}}_{q+1}(T - X, T' - X') \cong H_{q+1}C^{inf}_*(K, L; G) . \end{aligned}$$

Since e clearly commutes with the boundary homomorphisms

$$\partial\ :\ H_q(X, X') \longrightarrow H_{q-1}(X')$$

this completes the argument.

§2. A continuity property of the Steenrod homology

Again let H be a homology theory satisfying the nine axioms. The object of this section is to show that H satisfies a modified form of the Continuity Axiom. (See [4, pg. 260].) This fact will be used to complete the proof of Theorem 1, and hence of Theorem 3.

It is first necessary to describe the first derived functor of the inverse limit functor.

Let $A_1 \longleftarrow A_2 \longleftarrow A_3 \longleftarrow \cdots$ be an inverse system of abelian groups; briefly denoted by $\{A_i\}$. A homomorphism

$$h\ :\ \{A_i\} \longrightarrow \{B_i\}$$

will mean a sequence of homomorphisms $h_i : A_i \longrightarrow B_i$ such that each square

$$\begin{array}{ccc} A_i & \longleftarrow & A_{i+1} \\ {\scriptstyle h_i}\downarrow & & \downarrow{\scriptstyle h_{i+1}} \\ B_i & \longleftarrow & B_{i+1} \end{array}$$

is commutative. The inverse limit functor L assigns to each inverse sequence $\{A_i\}$ a group $L(A_i)$, and to each homomorphism $h : \{A_i\} \longrightarrow \{B_i\}$ a homomorphism $Lh : L\{A_i\} \longrightarrow L\{B_i\}$.

Both L and its first derived functor L' can be obtained from the following construction, which was communicated to the author by Steenrod.

Let Π denote the direct product of the groups A_i and define $d : \Pi \longrightarrow \Pi$ by

$$d(a_1, a_2, a_3, \cdots) = (a_1 - p_1 a_2, a_2 - p_2 a_3, a_3 - p_3 a_4, \cdots) ,$$

where $p_i : A_{i+1} \longrightarrow A_i$ denotes the projection homomorphism. Define $L\{A_i\}$ as the kernel of d (this clearly coincides with the usual definition); and **define** $L'\{A_i\}$ as the cokernel $\Pi/d\Pi$.

The two important properties of the derived functor L' are:

(1) If each projection $p_i : A_{i+1} \longrightarrow A_i$ is an epimorphism, then $L'\{A_i\} = 0$.

(2) Every exact sequence $0 \longrightarrow \{A_i\} \longrightarrow \{B_i\} \longrightarrow \{C_i\} \longrightarrow 0$ gives rise to an exact sequence

$$\begin{aligned} 0 \longrightarrow L\{A_i\} \longrightarrow L\{B_i\} \longrightarrow L\{C_i\} & \\ \longrightarrow L'\{A_i\} \longrightarrow L'\{B_i\} \longrightarrow L'\{C_i\} \longrightarrow 0 \, . & \end{aligned}$$

Proofs are easily supplied.

Theorem 4. *Let H be a homology theory satisfying the nine axioms, and let $X_1 \longleftarrow X_2 \longleftarrow X_3 \longleftarrow \cdots$ be an inverse system of compact metric spaces with inverse limit X. Then there is an exact sequence*

$$0 \longrightarrow L'\{H_{q+1}(X_i)\} \longrightarrow H_q(X) \longrightarrow L\{H_q(X_i)\} \longrightarrow 0$$

for each integer q. A corresponding assertion holds if each space is replaced by a pair.

The proof will be based on a construction similar to that used in the proof of Theorem 2. Let M_i denote the mapping cylinder of the map $X_{i+1} \longrightarrow X_i$, and let M_0 be the cone over X_1 with vertex t_0. Then the union

$$F = M_0 \cup M_1 \cup \cdots$$

can be imbedded in a contractible compact metric space T with $T - F = X$.

Using the homology sequence of the triple $(T, X + t_0, t_0)$ it is seen that

$$H_q(X) \cong H_q(X + t_0, t_0) \cong H_{q+1}(T, X + t_0) \cong H_{q+1}^{\mathrm{LF}}(F - t_0) .$$

(The symbol + denotes topological sum.)

Let F_1 denote the union $M_1 + M_3 + M_5 + \cdots$ and let F_0 denote the union $(M_0 - t_0) + M_2 + M_4 + \cdots$. Thus:

$$F_0 \cup F_1 = F - t_0 , \text{ and } F_0 \cap F_1 = X_1 + X_2 + \cdots .$$

Note that:

$$H_q^{\mathrm{LF}}(M_i) = H_q(M_i) = H_q(X_i) , \text{ and } H_q^{\mathrm{LF}}(M_0 - t_0) = 0 .$$

Now consider the Mayer-Vietoris sequence of the triad $(F - t_0; F_0, F_1)$, using the homology theory H^{LF}. The group $H_q^{\mathrm{LF}}(F_0 \cap F_1)$ is isomorphic to the direct product

$$\Pi_q = H_q(X_1) \times H_q(X_2) \times \cdots ;$$

and the direct sum $H_q^{\mathrm{LF}}(F_0) \oplus H_q^{\mathrm{LF}}(F_1)$ is isomorphic to this same direct product. Thus the Mayer-Vietoris sequence reduces to:

$$\cdots \longrightarrow \Pi_{q+1} \xrightarrow{\psi} \Pi_{q+1} \longrightarrow H_q(X) \longrightarrow \Pi_q \xrightarrow{\psi} \Pi_q \longrightarrow \cdots .$$

The homomorphism ψ can easily be computed. It turns out that

$$\psi(a_1, a_2, \cdots) = (-a_1 - p_{1*}a_2, a_2 + p_{2*}a_3, -a_3 - p_{3*}a_4, \cdots) .$$

Perform an automorphism on the domain of ψ as follows: replace each a_i by $(-1)^i a_i$. This has the effect of replacing ψ by d, where

$$d(a_1, a_2, \cdots) = (a_1 - p_{1*}a_2, a_2 - p_{2*}a_3, \cdots) .$$

Thus the cokernel of d_{q+1} is equal to $L'\{H_{q+1}(X_i)\}$ and the kernel of d_q is $L\{H_q X_i\}$. Hence the Mayer-Vietoris sequence reduces to

$$0 \longrightarrow L'\{H_{q+1}(X_i)\} \longrightarrow H_q(X) \longrightarrow L\{H_q(X_i)\} \longrightarrow 0$$

as required.

This can be generalized to pairs (X_i, X_i') as follows. Using the Relative Homeomorphism Axiom, it is sufficient to consider the case where X_i' is a single point x_i. But there is a canonical direct sum decomposition

$$H_q(X) = H_q(X, x) \oplus H_q(x) .$$

Since Theorem 4 is known for single spaces, it follows easily for pairs; which completes the proof.

Proof of Theorem 1, concluded. Let X be the underlying space of a countable, locally finite CW complex, with n-skeleton Y. For $n > q+1$ we will prove that the inclusion homomorphism

$$H_q^{\mathrm{LF}}(Y) \longrightarrow H_q^{\mathrm{LF}}(X)$$

is an isomorphism. The corresponding assertion for a pair of CW complexes then follows from the Five Lemma. This will clearly complete the proof of Theorem 1.

Let $N_1 \supset N_2 \supset \cdots$ be a sequence of "neighborhoods of infinity" in X. Assume that

(1) each N_i is the underlying space of a subcomplex,
(2) each $X - N_i$ contains only finitely many cells, and
(3) $\cap N_i$ is vacuous.

Then the one-point compactification $(X \cup \omega, Y \cup \omega)$ is the inverse limit of the sequence $(X \cup \omega, Y \cup N_i \cup \omega)$. Hence Theorem 4 can be applied.

But the groups

$$H_q(X \cup \omega, Y \cup N_i \cup \omega) \;=\; H_q^{\mathrm{LF}}(X, Y \cup N_i)$$

are zero for $q \leq n$. This follows directly from the portion of Theorem 1 which has already been proved, since $X - (Y \cup N_i)$ is contained in a finite subcomplex of X, and contains no cells of dimension $\leq n$.

Applying Theorem 4 it follows that $H_q^{\mathrm{LF}}(X, Y) = 0$ for $q+1 \leq n$; and hence that

$$H_q^{\mathrm{LF}}(Y) \;\cong\; H_q^{\mathrm{LF}}(X)$$

for $q+1 < n$. This completes the proof of Theorem 1, and hence of Theorem 3.

§3. Axioms for cohomology

Consider analogous axioms for a cohomology theory on the category $\mathcal{A}_{CM}$ of compact metric pairs. That is, consider the seven Eilenberg-Steenrod axioms together with:

Axiom 8c. *Invariance under relative homeomorphism.*

Axiom 9c. *The cohomology group $H^q(X_1 \vee X_2 \vee \cdots, b)$ of a cluster is naturally isomorphic to the direct sum of the groups $H^q(X_i, b)$.*

Assertion. *The Čech cohomology theory satisfies these nine cohomology axioms. Any other cohomology theory on the category $\mathcal{A}_{CM}$ which satisfies these axioms is naturally isomorphic to the Čech theory.*

Proof. The Čech cohomology theory satisfies all nine axioms, since 8c and 9c are easy consequences of the continuity property of the Čech groups.

In order to prove uniqueness it is sufficient to prove the following continuity theorem:

Theorem 4c. *If X is the inverse limit of an inverse sequence of compact spaces X_i then $H^q X$ is the direct limit of the groups $H^q X_i$.*

This is proved by an argument similar to that in §2. No derived functors of direct limits come in, since the direct limit functor is exact.

Theorem 1 can also be dualized. In this case the appropriate statement is:

Theorem 1c. *If K is a locally finite CW complex then $H^*(|K| \cup \omega, \omega)$ is isomorphic to the cohomology ring of the formal complex K based on finite cochains.*

§4. Construction of a homology theory

This section will construct a homology theory on the category $\mathcal{A}_C$ of all compact pairs and continuous maps. It will first be shown that the Čech cohomology groups of a compact space X can be obtained from a free cochain complex $\overline{C}^* X$. The desired homology groups $\overline{H}_q(X; G)$ are then obtained as the homology groups of the dual chain complex $\mathrm{Hom}(\overline{C}^* X, G)$.

Let S^m denote the m-sphere with base point b. For any compact pair (X, A) let $F^m(X, A)$ denote the function space consisting of all maps

$$(X, A) \longrightarrow (S^m, b) .$$

The symbol b will also be used for the constant map, considered as base point in $F^m(X, A)$. Integer coefficients are to be understood.

Theorem of Moore [9]. *There is a canonical homomorphism*

$$\phi \ : \ H_q(F^m(X, A), b) \longrightarrow H^{m-q}(X, A)$$

using singular homology and Čech cohomology groups. If X has finite covering dimension k, then ϕ is an isomorphism for $q < 2(m - k)$.

Proof. As Moore remarks, it is sufficient to consider the case when (X, A) is a triangulable pair. (See Mardešić [8, pg. 221] together with Mardešić [7, Theorem 6].) First suppose that X is connected, and that A is non-vacuous. If $q \geq 0$ and $k < m$ the assertion follows from [9, Theorem 4]. (Moore excluded the case $q = 0$ since he worked with the group $H_q F^m(X, A)$, without base point.) If $q < 0$ or $k \geq m$ the assertion is trivial, since both groups are zero. The case A vacuous follows, using [9, pg. 187]. Now if X is disconnected then (X, A) splits into a topological sum and $F^m(X, A)$ splits into a cartesian product. Using the Künneth theorem, the proof can be completed.

This result can be extended to arbitrary compact pairs as follows. Let S stand for the reduced suspension operation, and identify the sphere S^{m+1} with SS^m. Then there is a canonical imbedding

$$SF^m(X,A) \subset F^{m+1}(X,A) ,$$

and hence a canonical homomorphism

$$H_{m-q}(F^m(X,A),b) \longrightarrow H_{m+1-q}(F^{m+1}(X,A),b) .$$

Lemma 3. *The direct limit of the singular groups $H_{m-q}(F^m(X,A),b)$ as m tends to infinity is canonically isomorphic to the Čech cohomology groups $H^q(X,A)$, for any compact pair (X,A).*

Proof. For X finite dimensional this follows from Moore's result, since the homomorphism ϕ as defined by Moore commutes up to sign with the suspension homomorphism. In particular, this is the case if (X,A) is triangulable.

Now consider an inverse system of triangulable pairs (X_u, A_u) with inverse limit (X,A). According to Mardešić [7] we have

$$H_*F^m(X,A) \cong \varinjlim_u H_*F^m(X_u,A_u) .$$

Therefore

$$\begin{aligned}
\varinjlim_m H_{m-q}(F^m(X,A),b) &\cong \varinjlim_m \varinjlim_u H_{m-q}(F^m(X_u,A_u),b) \\
&\cong \varinjlim_u \varinjlim_m H_{m-q}(F^m(X_u,A_u),b) \\
&\cong \varinjlim_u H^q(X_u,A_u) \cong H^q(X,A) .
\end{aligned}$$

This completes the proof of Lemma 3.

For any compact space X we will construct a free cochain complex $\overline{C}^*X$ such that the homology of this cochain complex is just the Čech cohomology of X.

First observe that there is a canonical imbedding of the singular chain group $C_k(Y,b)$ into the singular chain group $C_{k+1}(SY,b)$ of the suspension. [Given a map f from the standard k-simplex Δ^k into Y, extend to the map $f*(\text{identity})$ of the join $\Delta^{k+1} = \Delta^k * v$ into the join $Y*v$, and then project into

$$SY = Y*v/(Y \cup (b*v)) .]$$

Hence there is a canonical imbedding

$$C_{m-q}(F^mX,B) \subset C_{m+1-q}(F^{m+1}X,b) .$$

Define

$$\overline{C}^q X = \varinjlim_m C_{m-q}(F^m X, b)$$

to be the union of these groups. (Caution: q can be positive or negative). It is clear that:

(1) $\overline{C}^* X$ is a contravariant functor of X,
(2) each $\overline{C}^q X$ is a free abelian group, and
(3) $H^q(\overline{C}^* X)$ is isomorphic to the Čech group $H^q X$.

In particular, this cohomology group is zero for $q < 0$.

This construction extends to a pair (X, A) as follows:

Lemma 4. *For a compact pair (X, A), the natural homomorphism $\overline{C}^q X \longrightarrow \overline{C}^q A$ is onto. Defining $\overline{C}^q(X, A)$ as the kernel of this homomorphism, the groups $H^q\overline{C}^*(X, A)$ are canonically isomorphic to the Čech cohomology groups of (X, A).*

Proof. Any generator of $\overline{C}^q A$ is given by a map $f : \Delta^{m-q} \times A \longrightarrow S^m$. This is identified with the generator corresponding to a certain map $f' : \Delta^{m+1-q} \times A \longrightarrow S^{m+1}$. It is easily seen that f' is null-homotopic, and hence can be extended to a map $\Delta^{m+1-q} \times X \longrightarrow S^{m+1}$. Thus every generator of $\overline{C}^q A$ comes from a generator of $\overline{C}^q X$.

To prove the second assertion, note that $\varinjlim_m C_{m-q}(F^m(X, A), b)$ is imbedded as a subgroup of $\overline{C}^q(X, A)$. It follows from Lemma 3 that this first chain complex has homology isomorphic to the Čech cohomology of (X, A). We must show that the inclusion map induces an isomorphism of homology. But each homology group fits into an exact sequence:

$$\begin{array}{ccccccccc} H^{q-1}X & \longrightarrow & H^{q-1}A & \longrightarrow & H^q(X,A) & \longrightarrow & H^qX & \longrightarrow & H^qA \\ \downarrow\cong & & \downarrow\cong & & \downarrow & & \downarrow\cong & & \downarrow\cong \\ H^{q-1}X & \longrightarrow & H^{q-1}A & \longrightarrow & H^q\overline{C}^*(X,A) & \longrightarrow & H^qX & \longrightarrow & H^qA \end{array}$$

and it can be verified that the resulting diagram is commutative. Using the Five Lemma, this completes the proof.

Definition. *For any compact pair (X, A) define*

$$\overline{C}_q(X, A; G) = \operatorname{Hom}(\overline{C}^q(X, A), G) .$$

The homology groups of this chain complex will be denoted by $\overline{H}_q(X, A; G)$, and called the **Steenrod homology groups** *of (X, A).*

Since $\overline{C}^*$ is a free cochain complex we clearly have the following. (See Eilenberg and MacLane [3, pg. 824].)

Lemma 5. *The homology theory $\overline{H}$ is related to Čech cohomology theory by a split exact sequence*

$$0 \longrightarrow \operatorname{Ext}(H^{q+1}(X, A); G) \longrightarrow \overline{H}_q(X, A; G) \longrightarrow \operatorname{Hom}(H^q(X, A); G) \longrightarrow 0 .$$

Theorem 5. *The homology theory $\overline{H}$, defined on the category $\mathcal{A}_C$ of compact pairs, satisfies the nine axioms of §1.*

Proof. For any compact pair (X, A) we have a short exact sequence

$$0 \longrightarrow \overline{C}_q(A; G) \longrightarrow \overline{C}_q(X; G) \longrightarrow \overline{C}_q(X, A; G) \longrightarrow 0$$

and therefore an infinite exact sequence of homology groups. Thus Axioms 1,2,3,4 can be verified.

Any map $f : (X, A) \longrightarrow (Y, B)$ induces a homomorphism from the exact sequence of Lemma 5 for the pair (X, A) to the corresponding exact sequence for (Y, B). Hence if f induces an isomorphism of Čech cohomology groups it follows that

$$f_* : \overline{H}_q(X, A; G) \longrightarrow \overline{H}_q(Y, B; G)$$

is an isomorphism.

In particular this argument applies to the inclusion maps

$$i_0 \text{ and } i_1 : (X, A) \longrightarrow (X \times [0,1], A \times [0,1]) ,$$

and to the projection map in the opposite direction. The Homotopy Axiom (number 5) follows immediately.

This argument also applies if f is a relative homeomorphism. This proves the Excision Axiom (number 6) and the Relative Homeomorphism Axiom (number 8).

Taking x to be a single point, Lemma 5 implies that

$$\overline{H}_0(x; G) \cong G \ , \ \overline{H}_q(x; G) = 0 \text{ for } q \neq 0$$

which proves the Dimension Axiom (number 7).

We will prove a version of the Cluster Axiom which involves an uncountable cluster. Let X be the union of compact subsets X_u which intersect pairwise at the point b. Assume that every neighborhood of b contains all but a finite number of the X_u. Let $r_u : X \longrightarrow X_u$ denote the canonical retraction.

Using the continuity property of Čech cohomology, it is seen that the homomorphisms

$$r_u^* : H^q(X_u; b) \longrightarrow H^q(X, b)$$

provide an injection representation of $H^q(X, b)$ as a direct sum. But the functors Hom($, G$) and Ext($, G$) carry direct sums into direct products. Using Lemma 5 it is seen that the homomorphisms

$$r_{u*} : \overline{H}_q(X, b) \longrightarrow \overline{H}_q(X_u, b)$$

provide a projection representation of $\overline{H}_q(X, b)$ as a direct product. This proves the Cluster Axiom (number 9), and completes the proof of Theorem 5.

One embarrassing feature of this homology theory is the possible presence of negative dimensional homology groups. It follows from Lemma 5 that $\overline{H}_q(C, A; G)$ is zero for $q \leq -2$. If the Čech cohomology group $H^0(X, A)$ is free abelian, it follows also that $\overline{H}_{-1}(X, A; G)$ is zero.

Problem 1. *Is the Čech cohomology group $H^0(X, A)$ of a compact pair always free abelian?*

Spanier has pointed out to me that this question is equivalent to the following:

Problem 2.* *Given a set S, is the group of all bounded functions from S to the integers always free abelian?*

In fact this group is just the cohomology group $H^0(\widetilde{S})$ of the Tychonoff compactification of S, considered as a discrete space. Conversely $H^0(X, A)$ is a subgroup of the group of all bounded functions $X \longrightarrow \mathbb{Z}$.

If the set S is countable, then an affirmative answer to Problem 2 has been given by Specker [12], assuming the continuum hypothesis. It follows that $H^0(X, A)$ is free abelian whenever $X - A$ has a countable dense subset, This proves:

Assertion. *If $X - A$ has a countable dense subset then $\overline{H}_q(X, A; G)$ is zero for $q < 0$.*

Appendix. Derived functors of inverse limits.

For any directed set D let $\mathcal{G}(D)$ denote the category of all inverse systems of abelian groups indexed by D, the concept of homomorphism being defined as in §2. The inverse limit operation L is a functor from $\mathcal{G}(D)$ to the category $\mathcal{G}$ of abelian groups and homomorphisms. This functor is left exact. To measure the deviation of L from right exactness one introduces derived functors L^r.

The following trick can be used to define L^r in terms of more familiar constructions. A subset U of D will be called an **initial segment** if, whenever $d_1 < d_2$ with $d_2 \in U$ it follows that $d_1 \in U$. Topologize D by taking the initial segments as the open sets. (Caution: D will never be a Hausdorff space.) Given an initial segment U and an inverse system $\{A_d\}$ define $L_U\{A_d\}$ as the subset of the direct product $\prod_{d \in U} A_d$ consisting of elements $\{a_d\}$ such that

$$\pi_{d_1}^{d_2}(a_{d_2}) = a_{d_1}$$

for all $d_1 < d_2 \in U$, where $\pi_{d_1}^{d_2}$ is the projection map in the inverse system.

* See the addendum at the end of this paper for references to the solution of this problem.

Assertion. *The correspondence $U \longrightarrow L_U\{A_d\}$ defines a sheaf $\mathcal{A}$ of abelian groups over D. (See Godement [5].) Conversely every sheaf over D comes in this way from a uniquely defined inverse system. The inverse limit $L\{A_d\}$ is just the group of globally defined sections of $\mathcal{A}$.*

The proof is straightforward.

Yeh calls an inverse system $\{A_d\}$ **star epimorphic** if, for each initial segment U, the natural homomorphism

$$L\{A_d\} \longrightarrow L_U\{A_d\}$$

is onto. Evidently $\{A_d\}$ is star epimorphic if and only if the corresponding sheaf $\mathcal{A}$ is flabby.

Now define $L^r\{A_d\}$ to be the r-th cohomology group of D with coefficients in $\mathcal{A}$. Evidently:

(1) $L^0 = L$.

(2) Each short exact sequence $0 \longrightarrow \{A_d\} \longrightarrow \{B_d\} \longrightarrow \{C_d\} \longrightarrow 0$ gives rise to an infinite exact sequence

$$0 \longrightarrow L^0\{A_d\} \longrightarrow L^0\{B_d\} \longrightarrow L^0\{C_d\} \longrightarrow L^1\{A_d\} \longrightarrow \cdots .$$

(3) If $\{A_d\}$ is star epimorphic then $L^r\{A_d\} = 0$ for $r > 0$.

It can be shown by rather complicated examples that the functors L^r are all non-trivial in general. However:

Assertion. *If D is the directed set of positive integers then the functor L^r is zero for $r > 1$, and L^1 coincides with the functor L' defined in §2.*

The proof is not difficult.

References

[1] M. Barratt and J. Milnor, *An example of anomalous singular homology.* Proc. Amer. Math. Soc. 13 (1962), 293–297.

[2] A. Borel and J. C. Moore, *Homology theory for locally compact spaces.* Michigan Math. J. 7 (1960), 137–159.

[3] S. Eilenberg and S. MacLane, *Group extensions and homology.* Annals of Math. 43 (1942), 757–831. (Also in *Eilenberg-MacLane: Collected Works*, Academic Press, Orlando, 1986, pp. 15–90.)

[4] S. Eilenberg and N. Steenrod, *Foundations of algebraic topology.* Princeton 1952.

[5] R. Godement, *Topologie algébrique et théorie des faisceaux.* Paris, Hermann 1958.

[6] S. Lefschetz, *Topology.* New York, A. M. S. Colloq. Publ., 1930.

[7] S. Mardešić, *On inverse limits of compact spaces.* Glasnik Mat. 13 (1958), 249–266.

[8] S. Mardešić, *Chainable continua and inverse limits.* Glasnik Mat. 14 (1959), 219–232.

[9] J. C. Moore, *On a theorem of Borsuk.* Fundamenta Math. 43 (1956) 195–201.

[10] K. Sitnikov, *The duality law for non-closed sets.* Doklady Akad. Nauk SSSR 81 (1951), 359–372 (Russian).

[11] K. Sitnikov, *Combinatorial homology of non-closed sets I.* Mat. Sbornik 34 (1954), 3–34 (Russian).

[12] E. Specker, *Additive Gruppen von Folgen ganzer Zahlen.* Portugaliae Math. 9 (1950), 131–140.

[13] N. Steenrod, *Regular cycles of compact metric spaces.* Annals of Math, 41 (1940), 833–851.

[14] N. Steenrod, *Regular cycles of compact metric spaces.* Lectures in topology pp. 43–55, University of Michigan Press, 1941.

[15] Z.-Z. Yeh, *Higher inverse limits and homology theories.* Thesis, Princeton University, 1959.

Addendum (May 1995)

I am indebted to K. Kunen for the information that Problem 2 has long since been answered affirmatively by G. Nöbeling. In fact, more generally: *Any commutative ring which is generated by idempotents, and is torsion free as an additive group, is necessarily free as an additive group.*
Here are some references.

[1] G. Nöbeling, *Verallgemeinerung eines Satzes von Herrn E. Specker*, Invent. Math. 6 (1968), 41–55.

[2] G. Bergman, *Boolean rings of projection maps*, J. London Math. Soc. 4 (1972), 593–598.

[3] P. Hill, *The additive group of commutative rings generated by idempotents*, Proc. Amer. Math. Soc. 38 (1973), 499–502.

[4] P. Hill, *Criteria for freeness in groups and valuated vector spaces in Abelian Group Theory*, Springer Lecture Notes #616 (1977), 140–157.

SUNY at Stony Brook, Institute for Mathematical Sciences, Math Bldg, Stony Brook NY 11794-3660, USA

email: jack@math.sunysb.edu

Homotopy type of differentiable manifolds

William Browder

It is our aim to give some homotopical conditions on a space which are necessary and sufficient under certain circumstances for it to be the homotopy type of a differentiable manifold.

We first mention some necessary conditions:

Let M^n be a closed differentiable manifold.

(1) M satisfies the Poincaré Duality Theorem, i.e. $H_n(M) = \mathbb{Z}$, with generator g, and $\cap g : H^i(M) \to H_{n-i}(M)$ is an isomorphism.

By a theorem of Whitney, M may be differentiably embedded in a sphere, say S^{n+k}. Let ν be the normal bundle which is oriented if M is oriented. A tubular neighborhood of M in S^{n+k} is diffeomorphic to a neighborhood of the zero cross section in $E(\nu)$, the total space of ν. If we collapse the exterior of this neighborhood to a point, we get the Thom complex of ν, $T(\nu)$. The collapsing map $c : S^{n+k} \to T(\nu)$ has the property that $c_*(\iota) = \Phi(g)$, where ι generates $H_{n+k}(S^{n+k})$ and Φ is the Thom isomorphism $\Phi : H_j(M) \to H_{j+k}(T(\nu))$. Hence we get:

(2) There exists an oriented vector bundle ν over M, such that $\Phi(g)$ is spherical in $H_{n+k}(T(\nu))$.

It turns out that in some circumstances the conditions (1) and (2) may be sufficient.

Theorem 1. *Let X be a connected finite polyhedron, with $\pi_1(X) = 0$. Suppose the following two conditions are satisfied:*

(1) *X satisfies the Poincaré Duality Theorem, i.e. for some n $H_n(X) = \mathbb{Z}$, and if g is a generator, $\cap g : H^i(X) \to H_{n-i}(X)$ is an isomorphism for all i.*

(2) *There exists an oriented vector bundle ξ^k over X, such that $\Phi(g) \in H_{n+k}(T(\xi))$ is spherical. Then, if n is odd, X is the homotopy type of an n-dimensional closed differentiable manifold, whose stable normal bundle is induced from ξ by the homotopy equivalence.*

In case M^n is a closed differentiable manifold of dimension $n = 4k$, the Hirzebruch Index Theorem is another property. In case $n = 4k$, this

This paper first appeared in the mimeographed proceedings of the 1962 Århus Conference.

condition together with (1) and (2) is sufficient.

Theorem 2. *Let X be as in Theorem 1, except that $n = 4q \neq 4$. In addition, suppose:*

(3) $I(X) = \langle L_q(\overline{p}_1, \ldots, \overline{p}_q), g\rangle$, where $I(X)$= index (signature) of the cup product bilinear form $H^{2q}(X;\mathbb{R}) \times H^{2q}(X;\mathbb{R}) \to H^{4q}(X;\mathbb{R}) = \mathbb{R}$, L_q is the Hirzebruch polynomial, and $\overline{p}_1, \ldots, \overline{p}_q$ are the dual classes to the Pontrjagin classes of ξ.

Then X is the homotopy type of an n-dimensional closed differentiable manifold with normal bundle induced from ξ.

In case ξ is the trivial bundle, condition (2) is equivalent to:

(2′) $\Sigma^k(g) \in H_{n+k}(\Sigma^k X)$ is spherical, where Σ denotes suspension.

Then the conclusions to Theorems 1 and 2 in this case, give that X is the homotopy type of a π-manifold.

M. Kervaire [1] has given an example of a 10-dimensional combinatorial manifold K which is not the homotopy type of a differentiable manifold. K has properties (1) and (2′). One may deduce from Theorems 1 and 2 then, that $K \times S^q$ *is* the homotopy type of a π-manifold if $4 \not| q$, $q > 1$. It follows from a theorem of S. Cairns and M. Hirsch that $K \times S^q$ is not itself a differentiable manifold. Similarly $K \times K$ is the homotopy type of a π-manifold.

(It also follows from Theorems 1 and 2 that if X is a connected polyhedral H-space, $\pi_1(X) = 0$, and $H_n(X) \neq 0$, $H_i(X) = 0$, $i > n$, $n \not\equiv 2 (\mathrm{mod}\, 4)$, then X is the homotopy type of a π-manifold).

We will sketch the proof of Theorem 1.

By embedding X in an Euclidean space and taking a small neighborhood of it, we may replace X by a homotopically equivalent space which is an open manifold. We assume then that X is an open manifold. Let $f : S^{n+k} \to T(\xi)$ be a map, such that $f_*(\iota) = \Phi(g)$. By a theorem of Thom [4], f is homotopic to a map transverse regular on $X \subset T(\xi)$, so we may assume that f is transverse regular on X. Then $N = f^{-1}(X)$ is a closed n-submanifold of S^{n+k} with normal bundle induced from ξ.

Lemma 1. *$f : N \to X$ is of degree 1, i.e. $f_*(\iota) = g$, where $\iota \in H_n(N)$ is the canonical generator given by the orientation induced from the orientation of the normal bundle.*

This follows easily from the naturality of the Thom isomorphism with respect to bundle maps.

Lemma 2. *If $f : H_*(A) \to H_*(B)$ is a map of degree 1, where A, B are spaces satisfying* (1), *then* Kernel(f_*) *is a direct summand of $H_*(A)$.*

Proof. $\alpha = P_A f^*(P_B)^{-1}$ is a map $\alpha : H_*(B) \to H_*(A)$ such that $f_*\alpha = 1$, where P_A, P_B are the isomorphisms given by Poincaré Duality.

Note also that Kernel(f_*) is orthogonal to $\alpha(H_*(B))$ under the intersection pairing.

We wish that N were connected and simply connected and that Kernel(f_*) $= 0$, so that we might apply J. H. C. Whitehead's Theorem to conclude that f is a homotopy equivalence. These things are not true, but we may make them true by changing N by surgery in certain cases. This is where the extra assumption that n is odd (or the assumption on the index) will come in.

Surgery is the following process: We have a differentiable embedding $\phi : S^p \times D^{q+1} \to M^n$, $n = p + q + 1$. We remove interior $\phi(S^p \times D^{q+1})$ and replace it by $D^{p+1} \times S^q$, which has the same boundary $S^p \times S^q$. This is again a differentiable manifold.

We should like to do surgery to make N connected, then simply connected, and then kill the Kernel(f_*). At each stage we produce a new manifold, and we must verify that we still have a map of degree 1 of the new manifold $N' = \chi(N, \phi)$ into X which induces the normal bundle of N' from ξ.

Lemma 3. *Let $\phi : S^p \times D^{q+1} \to M^n$ be an embedding, and let $\phi | S^p \times 0 = \phi'$. Suppose that $f \circ \phi'$ is homotopic to a constant map. Then there is a map $f'' : N' \to X$, $N' = \chi(N, \phi)$ of degree 1.*

Proof. $S^p \times 0$ is a deformation retract of $S^p \times D^{q+1}$, so that since $f \circ \phi' \simeq *$, then $f \circ \phi \simeq *$, $f'(\phi(S^p \times D^{q+1})) = *$. Then we set $f'' = f'$ outside $D^{p+1} \times S^q$ and $f''(D^{p+1} \times S^q) = *$, and $f'' : N' \to X$ is clearly of degree 1 if f was.

It can be shown that by choosing the product structure carefully on $S^p \times D^{q+1}$, the new manifold N' still has its normal bundle induced from ξ, cf. [2,3].

Then by surgery we may make N into a connected manifold, ($p = 0$ in the surgery) and simply connected, ($p = 1$). It follows from the relative Hurewicz Theorem that Kernel(f_*) in the lowest non-zero dimension consists of spherical classes.

To make sure we can do surgery to kill a spherical class we must first find an embedding of a sphere representing the class. This we can do by Whitney's Theorem in dimensions $\leq \frac{1}{2}n$, if $n \neq 4$. We must also have the normal bundle to the embedded sphere trivial.

Lemma 4. *Let $\phi : S^p \to N^n$ with $f \circ \phi \simeq *$, $p < \frac{1}{2}n$. Then the normal bundle γ to S^p in N^n is trivial.*

Proof. $\phi^*(\tau(N)) = \tau(S^p) \oplus \gamma$, and $f^*(\xi) = \nu$, the normal bundle to N in S^{n+k}. Then $\tau(N) \oplus \nu$ is trivial. Hence $\phi^*(\tau(N) \oplus \nu) = \tau(S^p) \oplus \gamma \oplus \phi^*(\nu)$ is trivial. Since $\nu = f^*(\xi)$, $\phi^*(\nu) = \phi^* f^* \xi = (f \circ \phi)^*(\xi)$ is trivial. But $\tau(S^p) \oplus \epsilon$ is trivial if ϵ is trivial. Hence $\gamma \oplus \epsilon'$ is trivial, where ϵ' is trivial, $\epsilon' = \tau(S^p) \oplus \phi^*(\nu)$. Since γ is an $(n-p)$-dimensional bundle over S^p, and

$n - p > p$, this implies that γ is trivial.

If n is odd, we only have to kill Kernel(f_*) in dimensions $< \frac{1}{2}n$, and by Poincaré Duality we have killed the whole kernel. (If n is even, we have also dimension $\frac{1}{2}n$ to consider, and here the extra hypothesis on the index is necessary to take care of questions about triviality of γ).

It remains to show that the surgery can be used to kill all of Kernel(f_*), particularly around the middle dimension. With the aid of Lemma 2, however, the techniques of [1], [2], and [3] show that this can be done in the various cases.

References

[1] M. Kervaire *A manifold which does not admit any differentiable structure*, Comment. Math. Helv. 34 (1960), 257–270.

[2] J. Milnor *A procedure for killing the homotopy groups of differentiable manifolds*, Proc. Symp. Pure Math. 3 (1960, Differentiable Geometry), 39–55, Amer. Math. Soc., Providence, R.I., 1961.

[3] M. Kervaire and J. Milnor, *Groups of homotopy spheres, I.*, Ann. of Math. 77 (1963), 504–537.

[4] R. Thom, *Quelques proprietés globales des variétés différentiables*, Comment. Math. Helv. 28 (1954), 17–86.

Mathematics Department, Princeton University, Princeton NJ 08544-0001, USA

email: `browder@math.princeton.edu`

K-theory, group C^*-algebras, and higher signatures (Conspectus)

G. G. Kasparov

Preface

The text which follows is a preprint version of my paper [0]. I am grateful to the editors of these Proceedings for their suggestion to publish it. I must explain here that the published version of the paper [0] differs significantly from the text of the preprint version. Actually at the time when the preprint was written (1981) many technical points of this work were still quite cumbersome. For this reason I assumed that the length of the proof would not allow me to publish it in a journal. I was thinking about writing a book. But to begin with I decided to make a short version (a conspectus), with only sketches of proofs, and distributed it as a preprint.

Fortunately, at the time when the final version was written (1987), a number of mathematicians had already contributed much to the subject. Some essential technical points, especially related to the construction of the product in KK-theory, were significantly simplified. So I benefited greatly from the existing publications which allowed me to publish the final text as a journal paper.

However, I find it reasonable to publish now also the preprint version. Actually there are two reasons. First of all, the exposition of the published version contains, in addition to technical improvements, also some generalizations of the main constructions of the preprint. So some people find it more difficult to read the published version. Maybe it would be better if a person interested in the subject could first look at the preprint. Secondly, some of the results contained in the preprint were not included into the published version. This concerns Theorems 3 and 4 of section 6 of the preprint describing the stable structure of some extensions of group C^*-algebras related with nilpotent Lie groups, and also the proof of Theorem 2 of section 9 (see Theorem 6.3 of [0] which was given without proof in [0]). I believe it will be good now to have these results published.

Originally distributed 1981. Preface written in 1994.

I would like to express my particular gratitude to one of the editors, Jonathan Rosenberg, who prepared the text in $\mathcal{A}\mathcal{M}\mathcal{S}$-TEX, made some small corrections (including some corrections of the language) and provided it with Editor's footnotes.

From my part, I also contributed some footnotes containing comments and additional explanations. In some places I made some minor corrections in the text in order to improve the language or to clarify the exposition. Unfortunately, as it was mentioned in [0], in the preprint version there were some small errors in the proof of the Novikov conjecture. For this reason I had to include in the text of sections 8 and 9 a couple of *Notes added* designed to correct the argument. For the same reason the statement and the proof of Theorem 1 in section 9 were also a little bit changed. The list of References of the preprint remains in its original form except that Russian titles are translated into English and reference [0] is added. Some additional references are given in footnotes.

Introduction

In the theory of operator algebras, a problem of considerable interest is the study of group C^*-algebras and covariance algebras. Some new methods for this are provided now by the theory of extensions of C^*-algebras and by the operator K-theory. So quite naturally there arises a question of computing the operator K-functor for a sufficiently large class of group C^*-algebras. In this paper we compute the K-functor of group C^*-algebras and covariance algebras of all connected amenable Lie groups and their discrete subgroups. For general connected Lie groups we compute some distinguished part of the corresponding K-functor.

As an example of applications of these results we study in detail the stable structure of some extensions of C^*-algebras associated with extensions of nilpotent Lie groups. Another application is the problem of the homotopy invariance of higher signatures of smooth manifolds. We prove the homotopy invariance of higher signatures for all manifolds having a fundamental group which can be embedded as a closed discrete subgroup into some connected Lie group.

The main instrument in obtaining these results is the operator K-bifunctor defined in [15]. In this paper all definitions and results concerning the K-bifunctor are extended to the category of C^*-algebras with an action of a separable locally compact group (in [15] only compact groups were allowed).

This paper may be regarded as a continuation of [15], as well as the second part of the earlier paper [12]. The results concerning C^*-algebras of simply connected solvable groups were announced in the introduction to [15]

(p. 574). The results concerning higher signatures for discrete subgroups of $GL(n, \mathbb{C})$ were announced in [12], §8.

This text is a *conspectus* containing the results and only sketches of proofs.

§1. Notation

We will use the notation of [15] (see [15], §§1 and 2). Here we recall some basic points and make some assumptions.

1. All results are valid for complex and real C^*-algebras. An algebra with a (norm) continuous action[1] of a group G will be called a *G-algebra*. All homomorphisms in the category of algebras with the action of G are assumed to be G-equivariant. In §§1–4 the group G will be fixed. Beginning with §5 we shall often consider simultaneously the categories of algebras with actions of different groups (the trivial group also is not excluded). It will be clear from the context which category is being considered. All groups acting on C^*-algebras are supposed to be *locally compact* and *separable* (i.e. second countable) and homomorphisms between groups continuous, all subgroups are assumed to be closed. "Real" C^*-algebras[2] will not be considered in this paper.

2. All algebras, subalgebras, homomorphisms and Hilbert modules that we consider are *$\mathbb{Z}_2$-graded* ([15], §2). A linear subspace $\mathcal{F}$ of an algebra A is called *graded* if $\mathcal{F} = \mathcal{F} \cap A^{(0)} + \mathcal{F} \cap A^{(1)}$. We always consider tensor products of algebras or modules with the minimal (spacial) C^*-norm and the group action on a tensor product is supposed to be *diagonal*, i.e. $\forall g \in G$, $g(x_1 \hat{\otimes} x_2) = g(x_1) \hat{\otimes} g(x_2)$. The group action on Hilbert modules is always assumed to be continuous in norm. Under a *homomorphism of Hilbert B-modules* $\chi : E_1 \to E_2$ we understand an element of $\mathcal{L}(E_1, E_2)$ (see [14], where are also given the definitions of the G-action on $\mathcal{L}(E_1, E_2)$ and of the subspace of *compact* homomorphisms $\mathcal{K}(E_1, E_2) \subset \mathcal{L}(E_1, E_2)$). An isomorphism of B-modules is called *isometric* if it preserves the scalar product.

3. An element x of an algebra with a G-action will be called *G-continuous* if the function $x \mapsto g(x)$ is continuous in norm on G. Quite similarly, one defines a *G-continuous homomorphism* $\chi \in \mathcal{L}(E_1, E_2)$.[3]

---

[1] Editor's note: Groups are assumed to act on a C^*-algebra A by $*$-automorphisms. The map $G \times A \to A$ defined by the action is not always assumed to be continuous but it will be assumed continuous for all algebras appearing as arguments of the K-bifunctor.

[2] Editor's note: in the sense of Atiyah, as generalized to the C^*-algebra context in the author's earlier work.

[3] Editor's note: Caution: this is a much weaker condition than G-equivariance.

4. The scalar field (i.e. the algebra $\mathbb{R}$ or $\mathbb{C}$) will be denoted by C. Put $\mathcal{H}^{(0)} \simeq \mathcal{H}^{(1)} \simeq \bigoplus_1^\infty L^2(G)$, and $\forall x \in \mathcal{H}^{(i)}$, put $\deg x = i$. The graded Hilbert space $\mathcal{H}^{(0)} \oplus \mathcal{H}^{(1)}$ will be denoted by $\mathcal{H}$. The tensor product $\mathcal{H} \otimes B$ considered as a (graded) Hilbert B-module will be denoted by $\mathcal{H}_B$. A Hilbert B-module E will be called *stable* if $E \oplus \mathcal{H}_B \simeq E$.

5. The Clifford algebra of a linear $*$-space V (cf. [15], §2, 11) is denoted by C_V. For a locally compact space X, we denote by $C(X)$ the algebra of continuous functions on X *tending to* 0 *at* ∞. If τ is a vector $*$-bundle over X (cf. [15], §2, 12), then $C_\tau(X)$ is the algebra of continuous sections, tending to 0 at ∞, of the Clifford bundle associated with τ. In this paper, until otherwise specified, we shall consider $C_\tau(X)$ only in the case when X is a smooth Riemannian manifold and τ is the cotangent bundle to X in the real case and the complexification of the cotangent bundle to X in the complex case. The involution $*$ on τ is trivial in the real case and is fiberwise complex conjugation in the complex case.

6. The left Haar measure on G is denoted by $dg = d_G g$ and the modular function of G by $\mu = \mu_G$.

7. For a connected group G we denote by G_c its maximal compact subgroup.[4]

8. The group C^*-algebra of G is denoted by $C^*(G)$. The covariance C^*-algebra of a G-algebra A (i.e. the crossed product of G and A) is denoted by $C^*(G, A)$.

§2. The stabilization of Hilbert modules

The stabilization theorem of [14] (Theorem 2) was used in [15] in several basic constructions including the product-intersection. We need the following generalization of this theorem[5] to the case of the actions of non-compact groups.

Theorem 1. *Let E be a countably generated Hilbert B-module. (The G-action on B and on E is supposed to be continuous.) Then there exists a G-continuous isometric isomorphism $\chi : E \oplus \mathcal{H}_B \simeq \mathcal{H}_B$ of degree 0.*

[4] Editor's note: This is unique up to conjugacy.

[5] The stabilization theorem of this section eventually appeared to be unnecessary in the final version [0] of the paper. (See [0], section 2, where the product was constructed without ever mentioning this theorem.) However, the main result of this section has some independent interest. It was later proved in a much simpler way. (See J. A. Mingo and W. J. Phillips, *Equivariant triviality theorems for Hilbert C^*-modules*, Proc. Amer. Math. Soc. **91** (1984), 225–230. The same simple proof was also communicated to me by G. Skandalis in 1982.)

(Recall that in the case of a compact group G there existed a G-equivariant isomorphism χ, i.e. $g(\chi) = \chi$, $\forall g \in G$.)

Sketch of proof. The main difficulty in generalizing the proof of Theorem 2 of [14] is the absence of Mostow's theorem on periodic vectors. We will indicate a way to get around this difficulty. Let us call an element of a Hilbert B-module E *integrable* if

$$c(x) = \int_G \|\langle x, g(x)\rangle\| \cdot \mu(g)^{-\frac{1}{2}}\, dg < \infty.$$

If this condition is satisfied then $\forall f \in L^2(G)$, there exists an element $y = \int_G f(g)g(x)\, dg \in E$ with $\|y\|_E \leq \sqrt{c(x)}\|f\|_{L^2}$.

Choosing f in an appropriate way we can make y close enough to x. The action of continuous compactly supported functions on y be the formula: $f_1 \mapsto \int_G f_1(g)g(y)\, dg$ can be extended to the action of the reduced W^*-algebra $W^*_{\text{red}}(G)$ on y (because $W^*_{\text{red}}(G)$ acts on $L^2(G)$). In fact we need only the action of $W^*_{\text{red}}(G_1)$, where G_1 is some *open, almost connected* subgroup of G. (A group is called almost connected if its factor group by the connected component of the identity is compact.) We can choose G_1 in such a way that it is generated by a neighbourhood U of the identity in G with a compact closure $\bar{U} = K$. From the results of [5] and [7] it follows that $W^*_{\text{red}}(G_1)$ is injective and semifinite. Using the structure theory of injective semifinite algebras [5] we can prove the following main lemma (which is the replacement of Mostow's theorem in this context):

Lemma 1. *For any homogeneous integrable element x of a Hilbert B-module E, there exists a real number $c > 0$ such that $\forall \varepsilon > 0$ there is a homomorphism $\varphi \in \mathcal{K}(\mathcal{H}_B, E)$ of degree 0 satisfying the conditions:*

(1) $\exists z \in \mathcal{H}_B$ *such that* $\|z\| \leq c$, $\|x - \varphi(z)\| \leq \varepsilon$.
(2) $\|\varphi\| \leq c$.
(3) $\forall g \in K$, $\|g(\varphi) - \varphi\| \leq \varepsilon$. □

After that a slightly modified scheme of the proof of Theorem 2 of [14] can be applied to obtain Theorem 1 from Lemma 1 in the case when the set of integrable homogeneous elements is dense in $E^{(0)}$ and $E^{(1)}$. The general case follows from this by an application of the embedding $\pi : E \hookrightarrow L^2(G) \otimes E = E_1$, $\pi(x) = f \otimes x$, where $f \in L^2(G)$ is an arbitrary element of norm 1. It is easily verified that integrable elements of E_1 are dense in $E_1^{(0)}$ and $E_1^{(1)}$. □

§3. The technique of product-intersection

This section is in full analogy with the §3 of [15]. Here is for comparison the analogue of Theorem 4, §3 of [15].

Theorem 1. *Let $\mathcal{E}$, $\mathcal{E}_1$ be subalgebras with strictly positive elements and $\mathcal{E}_2$, $\mathcal{F}$ graded separable linear subspaces in the algebra $\mathcal{M}(\mathcal{D})$. Assume that all elements of $\mathcal{E}$, $\mathcal{E}_1$, $\mathcal{E}_2$, and $\mathcal{F}$ are G-continuous, $\mathcal{E}$ is an ideal in $\mathcal{E}_1$, $\mathcal{E}_1 \cdot \mathcal{E}_2 \subset \mathcal{E}$, $\mathcal{D} \subset \mathcal{E}_1 + \mathcal{E}_2$, $[\mathcal{F}, \mathcal{E}_1] \subset \mathcal{E}_1$, $[\mathcal{F}, \mathcal{E}] \subset \mathcal{E}$. Then there exists a pair of G-continuous elements M, N of degree 0 in $\mathcal{M}(\mathcal{D})$ satisfying the conditions:*

(1) $M + N = 1$, $M \geq 0$, $N \geq 0$.
(2) $M \cdot \mathcal{E}_1 \subset \mathcal{E}$, $N \cdot \mathcal{E}_2 \subset \mathcal{E}$, $N \cdot \mathcal{E}_1 \subset \mathcal{E}_1$.
(3) $[\mathcal{F}, M] \subset \mathcal{E}$, $[\mathcal{F}, N] \subset \mathcal{E}$.
(4) $\forall g \in G$, $g(M) - M \in \mathcal{E}$, $g(N) - N \in \mathcal{E}$. □

Similar changes, mainly related with the replacement of G-invariant elements by G-continuous ones, must be made also in Theorem 5, §3 of [15]. In the analogues of Theorems 1, 2, 3 of [15], §3, in addition to the family of linear maps: $\varphi_x : A \to A$, $\varphi_x(a) = [F_x, a]$, $x \in X$, it is necessary to consider also the family of linear maps: $\varphi_g : A \to A$, $\varphi_g(a) = \beta(g) \cdot (g(a) - a)$, $g \in G$, where $\beta \in C(G)$. The inequality (1) of §3 of [15] must be replaced by

$$\varphi_y(h) \cdot \varphi_y(h)^* + \varphi_y(h)^* \cdot \varphi_y(h) \leq ch^m, \tag{1}$$

where $y \in X \cup G$. In the item 3) of Theorem 3 of [15], §3, the commutators $[M, F_x]$, $[N, F_x]$ must be replaced by $\varphi_y(M)$ and $\varphi_y(N)$ with $y \in X \cup G$. Proofs of all these theorems need only minor changes.

§4. K-bifunctor

This section is analogous to the §4 in [15], although the main definitions are slightly different. To avoid unnecessary repetitions we point out here that only *algebras with a continuous G-action and countable approximate units will be the arguments of the K-bifunctor $KK(A, B) = KK_G(A, B)$. All Hilbert C^*-modules that we consider will be countably generated.*

Definition 1. Let A and B be G-algebras. Denote by $\mathcal{E}(A, B) = \mathcal{E}_G(A, B)$ the collection of triples (E, φ, T), where E is a Hilbert B-module, $\varphi : A \to \mathcal{L}(E)$ a homomorphism, and $T \in \mathcal{L}(E)$ a G-continuous operator[6] of degree 1, such that $\forall a \in A$, $\forall g \in G$, the elements

$$[\varphi(a), T], \quad (T^2 - 1)\,\varphi(a), \quad (T - T^*)\,\varphi(a), \quad (g(T) - T)\,\varphi(a) \tag{1}$$

[6] In the Definition 2.2 of [0], we did not impose the condition of G-continuity on the operator T but only a weaker condition that all elements $\varphi(a)T$ and $T\varphi(a)$, $\forall a \in A$, should be G-continuous. In fact, the present Definition and the Definition 2.2 of [0] give isomorphic KK-groups if A has a countable approximate unit. In this case one can easily construct a G-continuous T in a weaker assumption of [0], 2.2, by applying Theorem 1.4 of [0].

belong to $\mathcal{K}(E)$. By $\mathcal{D}(A, B)$ we will denote the collection of *degenerate* triples, i.e. those ones for which all elements (1) are equal to 0.

Definition 2.

1°. The triples (E_1, φ_1, T_1) and $(E_2, \varphi_2, T_2) \in \mathcal{E}(A, B)$ are called *unitarily equivalent* if there is a G-equivariant isometric isomorphism $u : E_1 \to E_2$ of degree 0 such that $\forall a \in A$, $\varphi_2(a) = u\varphi_1(a)u^{-1}$ and $T_2 = uT_1u^{-1}$.

2°. A triple $x = (E, \varphi, T) \in \mathcal{E}(A, B[0,1])$ is called a *homotopy* connecting the triples $x_0 = (E_0, \varphi_0, T_0)$ and $x_1 = (E_1, \varphi_1, T_1) \in \mathcal{E}(A, B)$ if the restrictions of x to the endpoints of $[0, 1]$, i.e. the triples

$$\left(E_t = E \otimes_{B[0,1]} B([t]),\, (r_t)_* \circ \varphi,\, (r_t)_*(T)\right),$$

where $(r_t)_* : \mathcal{L}(E) \to \mathcal{L}(E_t)$ is the restriction homomorphism, for $t = 0$ and $t = 1$ coincide with x_0 and x_1, respectively.

Definition 3. Let $\bar{\mathcal{E}}(A, B)$ be the set of equivalence classes of $\mathcal{E}(A, B)$ modulo unitary equivalence and homotopy, $\bar{\mathcal{D}}(A, B)$ the image of $\mathcal{D}(A, B)$ in $\bar{\mathcal{E}}(A, B)$. The operation of addition is introduced into $\bar{\mathcal{E}}(A, B)$ by the direct sum:

$$(E_1, \varphi_1, T_1) \oplus (E_2, \varphi_2, T_2) = (E_1 \oplus E_2, \varphi_1 \oplus \varphi_2, T_1 \oplus T_2).$$

The factor semigroup $\bar{\mathcal{E}}(A, B)/\bar{\mathcal{D}}(A, B)$ will be denoted by $KK(A, B)$.

Definition 4. A homomorphism $f : A_2 \to A_1$ induces a homomorphism of groups $f^* : KK(A_1, B) \to KK(A_2, B)$ by $f^*(E, \varphi, T) = (E, \varphi \circ f, T)$. A homomorphism $g : B_1 \to B_2$ gives rise to $g_* : \mathcal{L}(E) \to \mathcal{L}(E \otimes_{B_1} B_2)$ and induces a homomorphism of groups $g_* : KK(A, B_1) \to KK(A, B_2)$ by $g_*(E, \varphi, T) = (E \otimes_{B_1} B_2, g_* \circ \varphi, g_*(T))$. For any algebra D, the homomorphism $\sigma_D : KK(A, B) \to KK(A\hat{\otimes}D, B\hat{\otimes}D)$ is defined by $\sigma_D(E, \varphi, T) = (E\hat{\otimes}D, \varphi\hat{\otimes}id, T\hat{\otimes}1)$. (*Note that in* [15] *this last homomorphism was denoted by* τ_D.)

Theorem 1. *$KK(A, B)$ is a group, homotopy-invariant in A and B.* □

If the group G is compact then applying the stabilization theorem of [14] and averaging T over G one can easily verify that the definitions of $KK(A, B)$ given here and in [15] coincide.

Theorem 2. *Let A, B, A_1, B_1 be algebras with a continuous G-action and countable approximate units, such that A is an ideal in A_1, B an ideal in B_1.*

1°. The group $KK(A, B)$ will not be changed if in Definitions 1–3 Hilbert B-modules are replaced by Hilbert B_1-modules and the condition that the elements (1) belong to $\mathcal{K}(E)$ is replaced by the condition that they belong to the ideal $\mathcal{K}(\overline{E \cdot B}) \subset \mathcal{K}(E)$.

2°. The group $KK(A, B)$ will not be changed if in Definitions 1–3 it is additionally required that the homomorphism φ can be extended to A_1. More precisely, we can consider triples $(E, \varphi : A_1 \to \mathcal{L}(E), T)$ as elements of $\mathcal{E}(A, B)$ if we preserve the list (1) as it is with $a \in A$, but if in the definition of $\mathcal{D}(A, B)$ and in the Definition 2, $a \in A_1$.

Both these two changes can be made simultaneously without changing $KK(A, B)$.

Sketch of proof. The proof of the first part is straightforward. The proof of the second part goes in the same way as the proof of the second part of Theorem 2 in §4 of [15]. The main changes are the following ones: $\mathcal{H}_{\tilde{A}}$ must be replaced by $F = (\bigoplus_1^\infty A) \oplus \left(\bigoplus_1^\infty \tilde{A}\right) \oplus \mathcal{H}_{\tilde{A}}$ and $\mathcal{H} \otimes \mathcal{H}_B$ must be replaced by $E_1 = F \otimes_{\tilde{A}} E$. To construct the operator $1 \otimes T$ in $\mathcal{L}(E_1)$, one applies Theorem 1 of §2 to the $\tilde{A}$-module F. □

Theorem 3.

$$KK(A, B_1 \oplus B_2) \simeq KK(A, B_1) \oplus KK(A, B_2),$$

$$KK(A_1 \oplus A_2, B) \simeq KK(A_1, B) \oplus KK(A_2, B). \quad \square$$

Theorem 4 (existence of the product). *The statement of this theorem coincides with the statement of Theorem 4 in §4 of [15].*

Sketch of proof. Fix elements $x_1 = (\tilde{E}_1, \varphi_1, T_1) \in \mathcal{E}(A_1, B_1 \hat{\otimes} D)$, $x_2 = (\tilde{E}_2, \varphi_2, T_2) \in \mathcal{E}(D \hat{\otimes} A_2, B_2)$. In view of Theorem 2 we may assume that $\tilde{E}_1$ is a stable $\tilde{B}_1 \hat{\otimes} \tilde{D}$-module, $\tilde{E}_2$ is a stable $\tilde{B}_2$-module, and the homomorphisms $\varphi_1 : \tilde{A}_1 \to \mathcal{L}(\tilde{E}_1)$, $\varphi_2 : \tilde{D} \hat{\otimes} \tilde{A}_2 \to \mathcal{L}(\tilde{E}_2)$ are unital. By E_1 and E_2 we will denote the closures of $\tilde{E}_1 \cdot (B_1 \hat{\otimes} D)$ and $\tilde{E}_2 \cdot B_2$ in $\tilde{E}_1$ and $\tilde{E}_2$ respectively. Put

$$\tilde{E}_{12} = \left(\tilde{E}_1 \hat{\otimes} \tilde{A}_2\right) \hat{\otimes}_{(\tilde{B}_1 \hat{\otimes} \tilde{D} \hat{\otimes} \tilde{A}_2)} \left(\tilde{B}_1 \hat{\otimes} \tilde{E}_2\right),$$

$E_{12} = \tilde{E}_{12} \hat{\otimes}_{(\tilde{B}_1 \hat{\otimes} \tilde{B}_2)} (B_1 \hat{\otimes} B_2)$, and let

$$\Phi_2 : \mathcal{L}(\tilde{E}_1) \hat{\otimes} \tilde{A}_2 \to \mathcal{L}(\tilde{E}_{12}) \to \mathcal{L}(E_{12})$$

be the natural homomorphism. Define

$$\varphi_1 \otimes_D \varphi_2 : A_1 \hat{\otimes} A_2 \to \mathcal{L}(E_{12})$$

as the composition $\Phi_2 \circ (\varphi_1 \hat{\otimes} id)$.

According to Theorem 1 of §2, there exists a G-continuous isometric isomorphism $\chi : \mathcal{H}_{\tilde{B}_1 \hat{\otimes} \tilde{D}} \simeq \tilde{E}_1$. Therefore $\tilde{E}_{12}$ is G-continuously isomorphic to

$$\tilde{E}'_{12} = \left(\mathcal{H}_{\tilde{B}_1 \hat{\otimes} \tilde{D}} \hat{\otimes} \tilde{A}_2\right) \hat{\otimes}_{\left(\tilde{B}_1 \hat{\otimes} \tilde{D} \hat{\otimes} \tilde{A}_2\right)} \left(\tilde{B}_1 \hat{\otimes} \tilde{E}_2\right)$$

via $\left(\chi^{-1} \hat{\otimes} id\right) \hat{\otimes} id$. Noticing that $\tilde{E}'_{12}$ can be identified with $\mathcal{H}_{\tilde{B}_1} \hat{\otimes} \tilde{E}_2$, we get a G-continuous isomorphism of $\tilde{B}_1 \hat{\otimes} \tilde{B}_2$-modules $\zeta : \mathcal{H}_{\tilde{B}_1} \hat{\otimes} \tilde{E}_2 \xrightarrow{\simeq} \tilde{E}_{12}$

The restriction of the G-continuous operator

$$\zeta \left(1 \hat{\otimes} T_2\right) \zeta^{-1} \in \mathcal{L}(\tilde{E}_{12})$$

to $\mathcal{L}(E_{12})$ will be denoted by $\tilde{T}_2$. The operator $\Phi_2 \left(T_1 \hat{\otimes} 1\right)$ will be denoted by $\tilde{T}_1$.

We will not repeat Definition 5 of §4 of [15]. Its generalization is quite obvious: the algebra $\mathfrak{A}_1$ contains all "irregularities" related with x_1 (i.e. with the elements of the list (1) for x_1) and the algebra $\mathfrak{A}_2$ all "irregularities" related with x_2 as well as with the commutator $\left[\tilde{T}_1, \tilde{T}_2\right]$. (Note here that the definition of $\mathfrak{A}_2$ in [15] needs a little correction. All elements of the list given there must be multiplied by $(\varphi_1 \otimes_D \varphi_2)\left(A_1 \hat{\otimes} A_2\right)$. Otherwise, in the case of a degenerate x_2 one will not get $\mathfrak{A}_2 = 0$.) The pair of G-continuous operators $M_1, M_2 \in \mathcal{L}(E_{12})$ of degree 0, in addition to the conditions 1) – 3) of Definition 5, §4, [15], must satisfy also the condition

$$4) \qquad \forall g \in G, \qquad g(M_i) - M_i \in \mathcal{K}(E_{12}), \quad i = 1, 2.$$

The set of such pairs (M_1, M_2) will be denoted by $S(x_1, x_2; \chi)$.

For $(M_1, M_2) \in S(x_1, x_2; \chi)$, put

$$T_1 \divideontimes_D T_2 = \sqrt{M_1} \cdot T_1 + \sqrt{M_2} \cdot T_2.$$

Define $x_1 \otimes_D x_2$ as the triple $(E_{12}, \varphi_1 \otimes_D \varphi_2, T_1 \divideontimes_D T_2)$. The independence of $x_1 \otimes_D x_2$ of the choice of χ can be established as follows. Put

$$\mathcal{E}_1 = \Phi_2 \left(\mathcal{K}\left(E_1\right) \hat{\otimes} A_2\right) + \mathcal{K}(E_{12}).$$

If χ' is another isomorphism then the corresponding operator $\tilde{T}_2'$ satisfies the conditions:

$$\mathcal{E}_1 \cdot \left(\tilde{T}_2 - \tilde{T}_2'\right) \subset \mathcal{K}(E_{12}), \qquad \left(\tilde{T}_2 - \tilde{T}_2'\right) \cdot \mathcal{E}_1 \subset \mathcal{K}(E_{12}).$$

In the construction of the pair (M_1, M_2) we can use Theorem 1 of §3 and include $\left(\tilde{T}_2 - \tilde{T}_2'\right)$ into $\mathcal{E}_2$. Then

$$(M_1, M_2) \in S(x_1, x_2; \chi) \cap S(x_1, x_2; \chi')$$

and

$$(T_1 \circledast_D T_2) - (T_1 \circledast_D T_2)' \in \mathcal{K}(E_{12}).$$

The remaining part of the proof of Theorem 4, §4, [15], goes through without essential changes. □

Theorem 5. *Let $H = H^{(0)} \oplus H^{(1)}$ be a graded G-Hilbert space and $F : H^{(0)} \to H^{(1)}$ a G-invariant operator with $1 - FF^* \in \mathcal{K}(H^{(1)})$, $1 - F^*F \in \mathcal{K}(H^{(0)})$. Assume that $\operatorname{ind} F = 1$ and all elements of $\ker F$ and $\ker F^*$ are G-invariant. Put $T = \begin{pmatrix} 0 & F^* \\ F & 0 \end{pmatrix} \in \mathcal{L}(H)$ and denote the element $(H, T) \in KK_G(C, C)$ by 1. This element does not depend on a particular choice of H and F and is the identity of the product-intersection, i.e. $\forall x \in KK_G(A, B)$, $x \otimes_C 1 = x$, $1 \otimes_C x = x$.*

Proof is the same as that of Theorem 5, §4, [15]. □

Corollary 1. *For any separable G-algebra D, the group $\Lambda_D = KK_G(D, D)$ is an associative ring, with multiplication given by the product-intersection. The element $1_D = \sigma_D(1)$ is the unit of the ring Λ_D. For any separable G-algebra A and any G-algebra$B$, the group $KK_G(D\hat{\otimes}A, B)$ is a left Λ_D-module and $KK_G(A, B\hat{\otimes}D)$ is a right Λ_D-module.* □

Next we will need (and use) the definition of the groups $K_i(A)$, $K^i(A)$ and $KK^i(A, B)$ given in §5, [15].[7]

[7] However, we have changed the convention about upper and lower indices from the one used in [15] and in the original version of this preprint to the one in more common use, so that $K_i(A)$ is covariant in A and $K^i(A)$ is contravariant in A. Note that the change of the upper position of an index to the lower one or in the opposite direction results in the change of the sign of the index. See more about notational conventions for the indices in [0], 2.22.

Definition 5. Put $R^i(G) = KK_G^i(C, C)$. In the complex case, put $R^*(G) = \bigoplus_{i=0}^{1} R^i(G)$, and in the real case, put $R^*(G) = \bigoplus_{i=0}^{7} R^i(G)$. The ring $R^*(G)$ may be called the *representation ring* of G.

The ring $R^*(G)$ is associative and skew-commutative (cf. Theorem 6, §5, [15]). The left and right actions of the *commutative* ring $R^0(G)$ on the groups $KK_G^i(A, B)$ coincide. If G is compact then $R^0(G)$ is the usual representation ring of G.[8] For a discrete group G, the ring $R^*(G)$ was defined in [12] by $R^*(G) = K^*(C^*(G))$. The two definitions obviously coincide. In fact, it is straightforward from the definitions that for a discrete group G and any G-algebra A, $K_G^i(A) \simeq K^i(C^*(G, A))$ where $C^*(G, A)$ is the crossed product of G and A.[9] The computation of the rings $R^*(G)$ is a problem of great interest for connected groups, as well as for discrete groups.

The next theorem is a generalization of Theorem 6, §4, [15]. (Note that a correction is necessary in the statement of Theorem 6, §4, [15]: the algebras D and E must be separable from the beginning.) For brevity we will restrict ourselves to the generalization of the statement of only the first part of Theorem 6, §4, [15].

Theorem 6. *Let algebras A, D, E be separable. Assume that there are elements $\alpha \in KK(D, E)$, $\beta \in KK(E, D)$ such that $\alpha \otimes_E \beta = 1_D$. Then the element $\gamma = \beta \otimes_D \alpha$ is an idempotent in the ring $KK(E, E)$. The homomorphism*

$$\_\_ \otimes_D \alpha : KK(A, B\hat{\otimes}D) \to KK(A, B\hat{\otimes}E)$$

is a monomorphism and

$$\mathrm{Im}(\_\_ \otimes_D \alpha) = \mathrm{Im}\,\gamma \stackrel{\mathrm{def}}{=} \{x \in KK(A, B\hat{\otimes}E) \mid x \cdot \gamma = x\}.$$

The homomorphism

$$\_\_ \otimes_E \beta : KK(A, B\hat{\otimes}E) \to KK(A, B\hat{\otimes}D)$$

is an epimorphism and

$$\ker(\_\_ \otimes_E \beta) = \ker\gamma \stackrel{\mathrm{def}}{=} \{x \in KK(A, B\hat{\otimes}E) \mid x \cdot \gamma = 0\}.$$

The group $KK(A, B\hat{\otimes}E)$ is a direct sum $\mathrm{Im}\,\gamma \oplus \ker\gamma$. If $\gamma = 1_E$, then $\_\_ \otimes_D \alpha$ and $\_\_ \otimes_E \beta$ are isomorphisms. A similar statement holds for the groups $KK(D\hat{\otimes}A, B)$ and $KK(E\hat{\otimes}A, B)$. □

---

[8] Editor's note: And in the real case $R^*(G)$ is the graded representation ring computed in the last section of [1].

[9] Editor's note: Dually, when G is compact, $K_i^G(A) \simeq K_i(C^*(G, A))$; see P. Julg, *K-théorie équivariante et produits croisés*, C. R. Acad. Sci. Paris Sér. I Math. **292** (1981), 629–632.

Theorem 7. *If separable algebras D and E are strongly Morita equivalent*[10] *([25]) then for any separable algebra A there are isomorphisms:*

$$KK(A\hat{\otimes}D,\, B) \simeq KK(A\hat{\otimes}E,\, B), \qquad KK(A,\, B\hat{\otimes}D) \simeq KK(A,\, B\hat{\otimes}E).$$

Sketch of proof. If X is a D-E-imprimitivity bimodule, define $\alpha \in KK(D,\, E)$ by the triple $(H\hat{\otimes}X,\, 1\hat{\otimes}id : D \to \mathcal{L}(H\hat{\otimes}X),\, T\hat{\otimes}1)$, where $T \in \mathcal{L}(H)$ is the operator from Theorem 5. An element $\beta \in KK(E,\, D)$ is defined similarly by means of the dual bimodule X^*. Then $\alpha \otimes_E \beta = 1_D$, $\beta \otimes_D \alpha = 1_E$. □

§5. The reduction of G to the maximal compact subgroup

In this section we will consider the functorial properties of the K-bifunctor with respect to change of the group G.

Definition 1. A homomorphism of groups $f : \Gamma \to G$ induces in an obvious way *the restriction homomorphism* $r_{G,\Gamma} : KK_G(A,\, B) \to KK_\Gamma(A,\, B)$.

Let Γ be a subgroup of G. If Γ acts on a topological space X, then $G\times_\Gamma X$ is the factor space of $G\times X$ by the action of Γ: $h(g,\, x) = (gh^{-1},\, h(x))$ $(h \in \Gamma,$ $g \in G,\, x \in X)$. The group G acts on $G \times_\Gamma X$ by $g(g_1,\, x) = (gg_1,\, x)$. If the space X is locally compact then the G-algebra $C(G \times_\Gamma X)$ can easily be constructed out of the Γ-algebra $C(X)$. Here is the construction.

Definition 2. Let B be a Γ-algebra. Denote by $C(G \times_\Gamma B)$ the set of continuous functions $f : G \to B$ satisfying the conditions:

(1) $\forall h \in \Gamma$, $f(gh) = h^{-1}(f(g))$.
(2) If $g\Gamma \to \infty$ in G/Γ, then $\|f(g)\| \to 0$.

This set is a subalgebra of the algebra of all bounded continuous functions $C_b(G,\, B)$, hence a C^*-algebra. The group G acts on $C(G \times_\Gamma B)$ by $(gf)(g_1) = f(g^{-1}g_1)$. If $B = C(X)$ then $C(G \times_\Gamma B) \simeq C(G \times_\Gamma X)$. Quite similarly, for a Hilbert B-module E, one can define a Hilbert $C(G \times_\Gamma B)$-module $C(G \times_\Gamma E)$.

Lemma 1. *Let G be a subgroup in $\mathcal{G}$, Γ a subgroup in G, and B and D Γ-algebras.*

(1) *If E is a Hilbert B-module then*

$$\mathcal{K}(C(G \times_\Gamma E)) \simeq C(G \times_\Gamma \mathcal{K}(E)).$$

[10] Editor's note: in a G-equivariant way.

(2) *If B_1 is a G-algebra then there is an isomorphism:*

$$C(G\times_\Gamma B)\hat{\otimes}B_1 \simeq C(G\times_\Gamma(B\hat{\otimes}B_1)) : \sum_i f_i(g)\hat{\otimes}b_i \mapsto \sum_i f_i(g)\hat{\otimes}g^{-1}(b_i).$$

(3)

$$C(\mathcal{G}\times_G C(G\times_\Gamma B)) \simeq C(\mathcal{G}\times_\Gamma B).$$

(4) *If E is a D-module, F a B-module, then*

$$C(G\times_\Gamma (E\otimes_D F)) \simeq C(G\times_\Gamma E)\otimes_{C(G\times_\Gamma D)} C(G\times_\Gamma F). \quad \square$$

Lemma 2. *Let Γ be a subgroup in G. There exists a non-negative continuous (cut-off) function λ on G satisfying the conditions:*

(1) $\int_\Gamma \lambda(gh)\, d_\Gamma h = 1,\ \forall g\in G.$

(2) $\forall\varepsilon>0$ *there exists a neighbourhood U of the identity in G such that $\forall g_1\in U$, $\forall g\in G$,*

$$\int_\Gamma |\lambda(g_1gh)-\lambda(gh)|\, d_\Gamma h < \varepsilon. \quad \square$$

Theorem 1. *Let Γ be a subgroup in G, A and B Γ-algebras. There exists an induction homomorphism, functorial in A and B:*

$$i_{\Gamma,G}: KK_\Gamma(A,B)\to KK_G(C(G\times_\Gamma A), C(G\times_\Gamma B))$$

having the following properties:

(1) *If A is separable, $x\in KK_\Gamma(A,D)$, $y\in KK_\Gamma(D,B)$, then*

$$i_{\Gamma,G}(x\otimes_D y) = i_{\Gamma,G}(x)\otimes_{C(G\times_\Gamma D)} i_{\Gamma,G}(y);$$

(2) *If A and B are G-algebras, D a Γ-algebra and $x\in KK_G(A,B)$, then*

$$i_{\Gamma,G}\circ\sigma_D\circ r_{G,\Gamma}(x) = \sigma_{C(G\times_\Gamma D)}(x);$$

(3) *If G is a subgroup in $\mathcal{G}$ then*

$$i_{G,\mathcal{G}}\circ i_{\Gamma,G} = i_{\Gamma,\mathcal{G}};$$

(4) *If $A=B$ then*

$$i_{\Gamma,G}(1_A) = 1_{C(G\times_\Gamma A)}.$$

Sketch of proof. Let $z = (E, \varphi, T) \in \mathcal{E}_\Gamma(A, B)$. The homomorphism φ obviously induces

$$\psi : C(G \times_\Gamma A) \to \mathcal{L}(C(G \times_\Gamma E)).$$

Put $\forall g \in G$,

$$S(g) = \int_\Gamma \lambda(gh) \cdot h(T) \, d_\Gamma h,$$

where λ is the function from Lemma 2. It is easily verified that $S : G \to \mathcal{L}(E)$ defines an element of $\mathcal{L}(C(G \times_\Gamma E))$. The triple $(C(G \times_\Gamma E), \psi, S)$ gives the required element $i_{\Gamma, G}(z) \in KK_G(C(G \times_\Gamma A), C(G \times_\Gamma B))$. This element does not depend on the choice of λ. The properties (1) – (4) are checked with the use of Lemma 1. □

Definition 3. Let X be a complete Riemannian manifold, G a group acting on X by isometries. We will construct *the canonical element* $\alpha \in K^0_G(C_\tau(X))$. Consider the Hilbert space of L^2-forms $H = L^2(\bigwedge^*(X))$ graded by the usual decomposition $\bigwedge^* = \bigwedge^{\mathrm{ev}} \oplus \bigwedge^{\mathrm{od}}$. The homomorphism $\varphi : C_\tau(X) \to \mathcal{L}(H)$ is first defined on 1-forms by $\varphi(\omega) = \lambda_\omega + \lambda^*_{\omega^*}$, where λ_ω is exterior multiplication by ω, λ^*_ω the operator adjoint to λ_ω (internal multiplication), and ω^* the complex conjugate 1-form. After that φ is extended by multiplicativity on $C_\tau(X)$. Let d be the operator of exterior derivation on smooth compactly supported forms on X, δ the operator formally adjoint to d, and $\Delta = d\delta + \delta d = (d + \delta)^2$ the Laplace operator. For complete Riemannian manifolds, it is known that the operator $d + \delta$ is essentially self-adjoint on smooth forms with compact support.[11] This allows one to define the operator

$$\frac{d + \delta}{\sqrt{1 + \Delta}} \in \mathcal{L}(H).$$

The element α is defined as the triple $\left(H, \varphi, \frac{d+\delta}{\sqrt{1+\Delta}}\right)$.

Definition 4. In the assumptions of the Definition 3, let additionally X be connected, simply connected and have non-positive sectional curvature. Under these assumptions we will construct *the canonical element*

$$\beta \in K_0^G(C_\tau(X)).$$

[11] See J. A. Wolf, *Essential self-adjointness for the Dirac operator and its square*, Indiana Univ. Math. J. **22** (1973), 611–640.

Let $\rho(x)$ be the geodesic distance from some fixed point $o \in X$ to the point $x \in X$. The covector field $\theta(x)$ on X is defined by

$$\theta = \frac{\rho d\rho}{\sqrt{1+\rho^2}}.$$

The operator of left (Clifford) multiplication by θ on the Hilbert $C_\tau(X)$-module $E = C_\tau(X)$ will be denoted again by θ. Using the cosine theorem (1.13.2) of [9] it is not difficult to verify that the pair $\beta = (E, \theta)$ defines an element of the group $K_0^G(C_\tau(X))$.

The element $\beta \otimes_{C_\tau(X)} \alpha \in R^0(G)$ will be denoted by γ.

Lemma 3. $\alpha \otimes_C \beta = \gamma \cdot 1_{C_\tau(X)}$.[12]

Sketch of proof. In analogy with the proof of Theorem 7, §5 of [15], we will use the method of "rotation." In this situation however it will be necessary to construct separately a family of homomorphisms

$$\{\psi_t : C_\tau(X) \to \mathcal{M}\left(C_\tau(X)\hat{\otimes} C_\tau(X)\right)\}$$

and a family of covector fields $\{\theta_t(x, y)\}$ on $X \times X$ $(0 \leq t \leq 1)$. To do this, consider the map $p_t : X \times X \to X$ sending a point $(x, y) \in X \times X$ into the point in X which divides the geodesic segment joining x and $y \in X$ in the proportion $t : (1-t)$. After the appropriate normalization we can extend the corresponding map of covectors $p_t^* : \Omega^1(X) \to \Omega^1(X \times X)$ to the required homomorphism ψ_t.

To construct the covector field $\theta_t(x, y)$ note that the fiber of p_t over each point $z \in X$ is homeomorphic to X. For $t \leq \frac{1}{2}$, this homeomorphism is given by the projection $q_2 : X \times X \to X$ to the second direct factor, and for $t \geq \frac{1}{2}$, by the projection q_1 to the first factor. Let $t \leq \frac{1}{2}$, $z \in X$. Denote by z_t the point dividing the geodesic segment in X joining the fixed point $o \in X$ with the point z in the proportion $2t : (1-2t)$. Let $\theta_{z_t}(x)$ be the covector field on X defined in the same way as $\theta(x)$ but with the center at z_t. Next, define the covector field $\theta_t'(x, y)$ on $X \times X$ by putting $\theta_t'(x, y)(\xi) = \theta_{z_t}(y)(q_2(\xi))$ for vectors ξ tangent to the fiber of p_t over the point z and $\theta_t'(x, y)(\xi) = 0$ for vectors ξ orthogonal to the fibers of p_t. After the appropriate normalization of θ_t', in order to assure that its length tends to 1 at ∞, we get the required field θ on $X \times X$ for $t \leq \frac{1}{2}$. For $t \geq \frac{1}{2}$ the construction is similar. The construction of the pair $\{M_t\}$, $\{N_t\}$ is the same as in Theorem 7, §5 of [15]. (Note here that in the proof of Theorem 7, §5 of [15], conjugation by the element U is introduced by mistake. It is unnecessary.) □

[12] Under the same assumptions one can prove also that $\gamma \cdot 1_{C_\tau(X)} = 1_{C_\tau(X)}$ — see [0], definition 5.1 and theorem 5.3.

Lemma 4. *Let G be the proper motion group of the Euclidean space $\mathbb{R}^n$, $G_c = SO(n)$ its maximal compact subgroup, $X = G/G_c = \mathbb{R}^n$. Then one has the following relations:*

$$\gamma = \beta \otimes_{C_\tau(X)} \alpha = 1, \qquad \alpha \otimes_C \beta = 1_{C_\tau(X)}.$$

Sketch of proof. [13] The second relation follows from the first one and Lemma 3. To prove the first relation consider for $0 < t \leq 1$ the differential operator

$$\mathcal{D}_t = d + \delta + t(\lambda_{\rho\, d\rho} + \lambda^*_{\rho\, d\rho})$$

acting on the space $\Omega^*_c(X)$ of smooth forms with compact support (λ_ω and λ^*_ω here again are the exterior and internal products respectively, ρ the distance from 0 in $\mathbb{R}^n$). Let $H = L^2(\bigwedge^*(X))$. The operator $\mathcal{D}_t$ is essentially self-adjoint on $\Omega^*_c(X)$, so $T_t = \frac{\mathcal{D}_t}{\sqrt{t^2+\mathcal{D}_t^2}}$ is a bounded self-adjoint operator on H. Moreover, $T_t^2 = 1 - \frac{t^2}{t^2+\mathcal{D}_t{}^2} \leq 1$ and $1 - T_t^2 \in \mathcal{K}(H)$ for any $t > 0$. The strong limit $T_0 = \lim_{t\to 0} T_t$ is equal to $\frac{d+\delta}{\sqrt{\Delta}}$.

We can identify the product of operators

$$\theta \divideontimes_{C_\tau(X)} \frac{d+\delta}{\sqrt{1+\Delta}}$$

with T_1. If the action of G on H is replaced by the *trivial* one, the element $(H, T_1) \in R^0(G)$ becomes equal to 1.[14] Denote the space H with the *trivial G-action and the opposite grading* by H'. Consider in the space $H \oplus H'$ the homotopy of operators

$$S_t = \begin{pmatrix} T_t & \sqrt{1-T_t^2} \\ \sqrt{1-T_t^2} & -T_t \end{pmatrix}, \qquad 0 \leq t \leq 1.$$

In the group $R^0(G)$ one has:

$$(H \oplus H', S_1) = (H, T_1) - 1, \qquad (H \oplus H', S_0) = 0,$$

because the last pair is obviously degenerate. It appears that

$$(H \oplus H', S_t)_{0 \leq t \leq 1}$$

[13] A much simpler proof of this lemma comes from the existence of the retraction of G onto G_c — see [0], 5.9.

[14] Editor's note: $\ker T_1$ is one-dimensional and contained in the even-graded part. This is is proved below.

is a homotopy in the sense of the Definition 2, §4, and the proof will be finished after checking this assertion.

The check is quite standard. The only place which is worth mentioning is the relation:

$$\forall g \in G, \qquad \lim_{t \to 0} \|(g-1)(1-T_t^2)\| = 0,$$

where g denotes both the element of G and the corresponding operator in $\mathcal{L}(H)$. To prove this relation note that according to [10],

$$\forall f \in \Omega_c^q(X), \qquad \langle \mathcal{D}_t{}^2 f, f \rangle_H \geq 2tq \cdot \|f\|_H^2.$$

The kernel of $\mathcal{D}_t{}^2$ on $L^2(X)$ is spanned by the function $e^{-tx^2/2}$. It follows from the harmonic oscillator theory that all non-zero eigenvalues of $\mathcal{D}_t{}^2$ on $L^2(X)$ are not less than $2t$. Therefore, on the orthogonal complement to $\ker \mathcal{D}_t{}^2$ in H we have:

$$1 - T_t^2 \leq \frac{t^2}{t^2 + 2t} \leq \frac{t}{2}.$$

On the other hand,

$$\forall g \in G, \qquad \lim_{t \to 0} \left\| (g-1) \left(\frac{e^{-tx^2/2}}{\|e^{-tx^2/2}\|} \right) \right\| = 0. \quad \square$$

Lemma 5. *Let G be a connected semisimple Lie group, $\mathfrak{g} = \mathfrak{k} + \mathfrak{p}$ a Cartan decomposition of its Lie algebra, Γ the connected subgroup in G with the Lie algebra $\mathfrak{k}$,*[15] *$X = G/\Gamma$. Then one has the relation:*

$$\alpha \otimes_C \beta = 1_{C_\tau(X)}.$$

Proof. According to [9], X possesses the properties required by Definitions 3 and 4 and the group $\Gamma_1 = \mathrm{Ad}_G(\Gamma)$ is compact. The action of Γ on X factors through Γ_1. So identifying X with $\mathfrak{p}$ by $\exp^{-1}$ and applying Theorem 7, §5, [15], we obtain that $\beta \otimes_{C_\tau(X)} \alpha = 1$ in $R^0(\Gamma)$, i.e. $r_{G,\Gamma}(\gamma) = 1$ in $R^0(\Gamma)$. Now let V be the cotangent space to $X = G/\Gamma$ at the point (Γ). Then $C_\tau(X) = C(G \times_\Gamma C_V)$. Applying the relation (2) of Theorem 1 with $D = C_V$ we get

$$\gamma \cdot 1_{C_\tau(X)} = \sigma_{C_\tau(X)}(\gamma) = i_{\Gamma,G} \circ \sigma_{C_V} \circ r_{G,\Gamma}(\gamma) = 1_{C_\tau(X)}.$$

The application of Lemma 3 finishes the proof. $\square$

[15] Editor's note: Thus $\mathrm{Ad}\,\Gamma$ is maximal compact in the adjoint group $\mathrm{Ad}\,G$. The group Γ will be maximal compact in G if G is linear.

Theorem 2. *Let G be a connected Lie group, G_c its maximal compact subgroup, A and B G-algebras, and assume that A is separable.*

(1) *There exist elements $\alpha = \alpha_G \in K^0_G(C_\tau(G/G_c))$ and $\beta = \beta_G \in K^G_0(C_\tau(G/G_c))$ with the properties:*

$$\alpha \otimes_C \beta = 1_{C_\tau(G/G_c)}, \qquad r_{G,G_c}(\beta \otimes_{C_\tau(G/G_c)} \alpha) = 1;$$

$\gamma = \gamma_G = \beta \otimes_{C_\tau(G/G_c)} \alpha$ is an idempotent in $R^0(G)$, and for any connected subgroup $G_1 \subset G$, $r_{G,G_1}(\gamma_G) = \gamma_{G_1}$. If G is amenable then $\gamma = 1$.

(2) *If G is amenable then the restriction homomorphism*

$$r_{G,G_c} : KK^i_G(A, B) \to KK^i_{G_c}(A, B)$$

is an isomorphism. In the general case r_{G,G_c} is an epimorphism with

$$\ker r_{G,G_c} = \ker \gamma \stackrel{\text{def}}{=} \{x \in KK^i_G(A, B) \mid \gamma \cdot x = 0\}.$$

The group $KK^i_G(A, B)$ is the direct sum of $\ker \gamma$ and the subgroup

$$\operatorname{Im} \gamma \stackrel{\text{def}}{=} \{x \in KK^i_G(A, B) \mid \gamma \cdot x = x\};$$

the latter maps isomorphically onto $KK^i_{G_c}(A, B)$ via r_{G,G_c}.

(3) *For any subgroup $G_1 \subset G$, the homomorphisms*

$$\_\otimes_C \alpha : KK^i_{G_1}(A, B) \to KK^i_{G_1}(A \hat{\otimes} C_\tau(G/G_c), B)$$

$$\_\otimes_C \beta : KK^i_{G_1}(A, B) \to KK^i_{G_1}(A, B \hat{\otimes} C_\tau(G/G_c))$$

are epimorphisms and their kernels coincide with the kernel of the multiplication by $r_{G,G_1}(\gamma)$. The inverse homomorphisms

$$(\beta \otimes_{C_\tau(G/G_c)} \_) \quad \text{and} \quad (\_ \otimes_{C_\tau(G/G_c)} \alpha)$$

are monomorphisms and their images coincide with the image of the multiplication by $r_{G,G_1}(\gamma)$. If G is amenable, all these homomorphisms are isomorphisms.

(4) *Let V be the cotangent space to G/G_c at the point (G_c). If the action of G_c on V is spin*[16] *then there exist elements*

$$\alpha' \in K^{\dim(G/G_c)}_G(C(G/G_c)) \text{ and } \beta' \in K^G_{\dim(G/G_c)}(C(G/G_c))$$

[16] Editor's note: i.e., the map $\mathrm{Ad}^* : G_c \to SO(V)$ factors through the double cover $\mathrm{Spin}(V)$ of $SO(V)$.

with the properties analogous to those listed in (1). Moreover,

$$\beta' \otimes_{C(G/G_c)} \alpha' = \gamma.$$

The homomorphisms of product and intersection with α' and β' possess the properties listed in (3).

Sketch of proof. First we shall construct α and β. We begin with finding a compact normal subgroup Γ in G such that G/Γ is a Lie group. Then $\Gamma \subset G_c$ because Γ is contained in some maximal compact subgroup of G and all maximal compact subgroups are conjugate to each other. If α and β are already constructed for G/Γ then for G they can be obtained by restriction $r_{G/\Gamma,G}$. Therefore we may assume that G is a Lie group. Using the construction of the maximal compact subgroup given in [30], Ch. 17, Theorem 6, we will carry out an induction on $\dim G$. If G is not semisimple then, as shown in [30], G has a normal subgroup Γ isomorphic to a torus T^p or to a Euclidean space $\mathbb{R}^n$.

In the case $\Gamma = T^p$, again clearly $\Gamma \subset G_c$. The induction step (from G/Γ to G) here is performed by restriction $r_{G/\Gamma,G}$. In the case $\Gamma = \mathbb{R}^n$, applying the homomorphism $(i_{G_c\Gamma,G}) \circ \sigma_{C_W}$ (where W is the cotangent space to $G/G_c\Gamma$ at the point $(G_c\Gamma)$) to the elements

$$\alpha_{G_c\Gamma} \in K^0_{G_c\Gamma}(C_\tau(G_c\Gamma/G_c)), \qquad \beta_{G_c\Gamma} \in K_0^{G_c\Gamma}(C_\tau(G_c\Gamma/G_c))$$

we get elements

$$\alpha'_{G_c\Gamma} \in KK_G(C_\tau(G/G_c), C_\tau(G/G_c\Gamma)),$$

$$\beta'_{G_c\Gamma} \in KK_G(C_\tau(G/G_c\Gamma), C_\tau(G/G_c)).$$

The desired α and β are defined by

$$\alpha = \alpha'_{G_c\Gamma} \otimes_{C_\tau(G/G_c\Gamma)} r_{G/\Gamma,G}(\alpha_{G/\Gamma}),$$

$$\beta = r_{G/\Gamma,G}(\beta_{G/\Gamma}) \otimes_{C_\tau(G/G_c\Gamma)} \beta'_{G_c\Gamma},$$

where $\alpha_{G/\Gamma}$ and $\alpha_{G/\Gamma}$ exist by the induction hypothesis. If at some step G is semisimple and non-compact (the case that does not occur if the initial G was amenable), the induction step can be carried out using Lemma 5. This finishes the construction of α and β.

Now note that item (3) follows from Theorem 6, §4, and item (4) follows from (2), (3), and Lemma 1, §5, [15]. It remains to verify (2). Let V be the cotangent space to G/G_c at the point (G_c). Denote the algebra $C_\tau(G/G_c) = C(G \times_{G_c} C_V)$ by D. Consider the composition of homomorphisms:

$$\eta = (\beta \otimes_D \_\_) \circ (\_\_ \otimes_D \alpha) \circ (i_{G_c, G}) \circ \sigma_{C_V} :$$
$$KK_{G_c}(A, B) \to KK_{G_c}(A \hat{\otimes} C_V, B \hat{\otimes} C_V)$$
$$\to KK_G(A \hat{\otimes} D, B \hat{\otimes} D) \to KK_G(A \hat{\otimes} D, B) \to KK_G(A, B).$$

It follows from item (2) of Theorem 1 that $\eta \circ r_{G, G_c}$ coincides with the multiplication by γ. On the other hand, $r_{G, G_c} \circ \eta = 1$. This can be verified with the use of a G_c-equivariant contraction of G/G_c, which is constructed by the same induction on $\dim G$ as in the first part of the proof. □

Theorem 3. *Let G be a connected Lie group and let π be a discrete subgroup of G. Then the subgroup*

$$\operatorname{Im} \gamma = \{x \in R^i(\pi) \mid r_{G, \pi}(\gamma) \cdot x = x\}$$

of the group $R^i(\pi)$ is isomorphic to $K^i_{G_c}(C_\tau(G/\pi))$.

Sketch of proof. All stationary subgroups of the action of G on $G/\pi \times G/G_c$ are finite. Hence there exists a G-invariant Riemannian metric on $G/\pi \times G/G_c$ and $C_\tau(G/\pi \times G/G_c)$ is a G-algebra. The restriction of the cotangent bundle of $G/\pi \times G/G_c$ to the submanifold $(\pi) \times G/G_c$ is a π-bundle over G/G_c. It will be denoted by ξ, and we will denote by Y a small π-invariant tubular neighbourhood of the submanifold $(\pi) \times G/G_c$ in $G/\pi \times G/G_c$.

Definition 3 gives us an element $\alpha_1 \in K^0_G(C_\tau(G/\pi \times G/G_c))$. Define the element $\beta_1'' \in KK_\pi(C_\tau(G/G_c), C_\tau(Y))$ as the triple $(C_\tau(Y), \varphi_1, \theta_1)$, where φ_1 is left multiplication by elements of $C_\tau(G/G_c)$ (lifted to Y via the projection to G/G_c) and θ_1 is left multiplication by the fiberwise radial covector field[17] on Y (orthogonal to G/G_c). Put

$$\beta_1' = r_{G, \pi}(\beta_G) \otimes_{C_\tau(G/G_c)} \beta_1'' \in K_0^\pi(C_\tau(Y)),$$

and let $\beta_1 \in K_0^\pi(C_\tau(G/\pi \times G/G_c))$ be the image of β_1' under the homomorphism induced by the inclusion $Y \hookrightarrow G/\pi \times G/G_c$.

Consider the composition of homomorphisms:

$$\left(\_\_ \otimes_{C_\tau(G/\pi \times G/G_c)} \alpha_1\right) \circ (i_{\pi, G}) \circ \left(\sigma_{C_\xi(G/G_c)}\right) :$$
$$R^i(\pi) \to KK^i_\pi(C_\xi(G/G_c), C_\xi(G/G_c))$$
$$\to KK^i_G(C_\tau(G/\pi \times G/G_c), C_\tau(G/\pi \times G/G_c))$$
$$\to K^i_G(C_\tau(G/\pi \times G/G_c)).$$

[17]See the precise formula for a similar covector field θ_x in the proof of Theorem 2 of section 8 below.

Define the inverse homomorphism as the composition

$$(\beta_1 \otimes_{C_\tau(G/\pi\times G/G_c)} \_\_) \circ r_{G,\pi}.$$

These homomorphisms establish an isomorphism of the subgroup $\operatorname{Im}\gamma$ of $R^i(\pi)$ and the group $K^i_G(C_\tau(G/\pi \times G/G_c))$, which itself is isomorphic to $K^i_{G_c}(C_\tau(G/\pi))$ by Theorem 2. □

§6. C^*-algebras of connected groups

In this section, when we consider a connected group[18] G with the maximal compact subgroup G_c, we will denote by $V = V_G$ the cotangent space to G/G_c at the point (G_c). Notation $C_c(G, D)$ will be used for the space of compactly supported continuous functions on G with values in any Banach space D.

Definition 1. Let B be a G-algebra, E a Hilbert B-module. We introduce the right action of $C_c(G, B)$ on $C_c(G, E)$ by the formula:

$$(eb)(t) = \int_G e(s) \cdot s(b(s^{-1}t))\, ds,$$

where $e \in C_c(G, E)$, $b \in C_c(G, B)$, and we define the scalar product on $C_c(G, E)$ with values in $C_c(G, B)$ by

$$(e,\, f)(t) = \int_G s^{-1}(e(s),\, f(st))_E\, ds,$$

where $e, f \in C_c(G, E)$. The completion of $C_c(G, E)$ in the norm

$$\|e\| = \|(e, e)\|^{\frac{1}{2}}_{C^*(G,B)}$$

is a Hilbert $C^*(G, B)$-module. We will denote it by $C^*(G, E)$.

Theorem 1. *Let A and B be G-algebras. There exists a homomorphism, functorial in A and B,*

$$j_G : KK_G(A, B) \to KK(C^*(G, A), C^*(G, B))$$

with the properties:

[18] Editor's note: However, for purposes of Definition 1 and Theorem 1, G can be any separable locally compact group.

(1) *if A is separable, $x \in KK_G(A, D)$, $y \in KK_G(D, B)$, then*

$$j_G(x \otimes_D y) = j_G(x) \otimes_{C^*(G, D)} j_G(y);$$

(2) *if $A = B$ then $j_G(1_A) = 1_{C^*(G, A)}$.*

Sketch of proof. Let $z = (E, \varphi, T) \in \mathcal{E}_G(A, B)$. The homomorphism φ induces $\psi : C^*(G, A) \to \mathcal{L}(C^*(G, E))$ by the formula:

$$(\psi(a)e)(t) = \int_G \varphi(a(s)) \cdot s(e(s^{-1}t))\, ds,$$

where $a \in C_c(G, A)$, $e \in C_c(G, E)$. Define the operator $R \in \mathcal{L}(C^*(G, E))$ by $(Re)(s) = T(e(s))$, where $e \in C_c(G, E)$. From the easily verified isomorphism $\mathcal{K}(C^*(G, E)) \simeq C^*(G, \mathcal{K}(E))$, it follows that the triple $w = (C^*(G, E), \psi, R)$ belongs to

$$\mathcal{E}(C^*(G, A), C^*(G, B)).$$

Put $j_G(z) = w$. □

Theorem 2. *Let G be a connected group, A a separable G-algebra, B a separable algebra, $\gamma \in R^0(G)$ the element constructed in Theorem 2, §5. Put*

$$\tilde{\gamma} = j_G \sigma_A(\gamma) \in KK(C^*(G, A), C^*(G, A)).$$

(1) *There exist elements*

$$\tilde{\alpha} \in KK(C^*(G_c, A \hat{\otimes} C_V), C^*(G, A)),$$

$$\tilde{\beta} \in KK(C^*(G, A), C^*(G_c, A \hat{\otimes} C_V))$$

with the properties:

$$\tilde{\alpha} \otimes_{C^*(G, A)} \tilde{\beta} = 1_{C^*(G_c, A \hat{\otimes} C_V)}; \qquad \tilde{\beta} \otimes_{C^*(G_c, A \hat{\otimes} C_V)} \tilde{\alpha} = \tilde{\gamma}.$$

If the action of G_c on V is spin then the algebra $C^*(G_c, A \hat{\otimes} C_V)$ can everywhere be replaced by $C^*(G_c, A) \hat{\otimes} C_V$.

(2) *For an amenable group G the homomorphisms*

$$\tilde{\alpha} \otimes_{C^*(G, A)} \_\_ : KK^i(C^*(G, A), B) \to KK^i(C^*(G_c, A \hat{\otimes} C_V), B),$$

$$\_\_ \otimes_{C^*(G, A)} \tilde{\beta} : KK^i(B, C^*(G, A)) \to KK^i(B, C^*(G_c, A \hat{\otimes} C_V))$$

are isomorphisms. In general they are epimorphisms and their kernels coincide with the kernels of the left and right intersection with $\tilde{\gamma}$, respectively. The inverse homomorphisms

$$\tilde{\beta}\otimes_{C^*(G_c,\,A\hat{\otimes}C_V)}\text{—} : KK^i(C^*(G_c,\, A\hat{\otimes}C_V),\, B) \to KK^i(C^*(G,\, A),\, B),$$

$$\text{—}\otimes_{C^*(G_c,\,A\hat{\otimes}C_V)}\tilde{\alpha} : KK^i(B,\, C^*(G_c,\, A\hat{\otimes}C_V)) \to KK^i(B,\, C^*(G,\, A))$$

are in general monomorphisms with images equal to the subgroups on which $\tilde{\gamma}$ acts as the identity. Moreover,

$$KK^i(C^*(G,\, A),\, B) \simeq \ker\left(\tilde{\alpha}\otimes_{C^*(G,A)}\text{—}\right)\oplus\mathrm{Im}\left(\tilde{\beta}\otimes_{C^*(G_c,\,A\hat{\otimes}C_V)}\text{—}\right),$$

$$KK^i(B,\, C^*(G,\, A)) \simeq \ker\left(\text{—}\otimes_{C^*(G,A)}\tilde{\beta}\right)\oplus\mathrm{Im}\left(\text{—}\otimes_{C^*(G_c,\,A\hat{\otimes}C_V)}\tilde{\alpha}\right).$$

(3) *If the action of G_c on V is spin then in item (2) the groups*

$$KK^i(C^*(G_c,\, A\hat{\otimes}C_V),\, B) \text{ and } KK^i(B,\, C^*(G_c,\, A\hat{\otimes}C_V))$$

can everywhere be replaced by

$$KK^{i-\dim G/G_c}(C^*(G_c,\, A),\, B) \text{ and } KK^{i+\dim G/G_c}(B,\, C^*(G_c,\, A)),$$

respectively.[19]

Sketch of proof. Applying the composition of homomorphisms $j_G\circ\sigma_A$ to the elements α and β constructed in Theorem 2, §5, we get elements

$$\alpha' \in KK(C^*(G,\, C(G\times_{G_c}(A\hat{\otimes}C_V))),\, C^*(G,\, A)),$$

$$\beta' \in KK(C^*(G,\, A),\, C^*(G,\, C(G\times_{G_c}(A\hat{\otimes}C_V)))).$$

According to Theorem (2.13) of [8], the algebras

$$C^*(G,\, C(G\times_{G_c}(A\hat{\otimes}C_V))) \text{ and } C^*(G_c,\, A\hat{\otimes}C_V)$$

are strongly Morita equivalent. This implies the existence of $\tilde{\alpha}$ and $\tilde{\beta}$. The remaining statements follow from Theorem 6, §4 and Lemma 1, §5, [15]. □

[19] The isomorphism $K_i(C^*(\mathbb{R},\, A)) \simeq K_{i-1}(A)$ constructed by A. Connes in his paper *An analogue of the Thom isomorphism for crossed products of a C^*-algebra by an action of $\mathbb{R}$*, Adv. Math. **39** (1981), 31–55, is a special case of the above isomorphisms.

Remark 1. For the category of complex algebras, in the statement of Theorem 2, the algebra C_V can be everywhere replaced by the algebra of complex functions $C(V)$. Indeed, from the Theorem 7, §5 of [15] it follows that there are canonical elements[20]

$$\alpha_0 \in KK_{G_c}(C(V), C_V), \qquad \beta_0 \in KK_{G_c}(C_V, C(V)).$$

Applying the composition of homomorphisms $j_{G_c} \circ \sigma_A$, we get elements

$$\tilde{\alpha}_0 \in KK(C^*(G_c, A\hat{\otimes}C(V)), C^*(G_c, A\hat{\otimes}C_V)),$$

$$\tilde{\beta}_0 \in KK(C^*(G_c, A\hat{\otimes}C_V), C^*(G_c, A\hat{\otimes}C(V)))$$

which are inverses of each other and this according to Theorem 6, §4, allows one to replace C_V by $C(V)$.

Remark 2. If $A = C$ and $\beta = (E, T) \in K_0^G(C_\tau(G/G_c))$ is the element constructed in Theorem 2, §5, then it can be easily verified that the natural homomorphism $C^*(G) \to \mathcal{L}(C^*(G, E))$ factors through the reduced C^*-algebra $C^*_{\text{red}}(G)$. Therefore, $\tilde{\beta} = f^*(\tilde{\beta}_1)$, where

$$\tilde{\beta}_1 \in KK(C^*_{\text{red}}(G), C^*(G_c, C_V))$$

and $f : C^*(G) \to C^*_{\text{red}}(G)$ is the quotient map. Hence, with

$$\tilde{\gamma}_1 = \tilde{\beta}_1 \otimes_{C^*(G_c, C_V)} \tilde{\alpha} \in KK(C^*_{\text{red}}(G), C^*(G))$$

we have: $\tilde{\gamma} = f^*(\tilde{\gamma}_1)$.

Now we can formulate the following *strengthening* of the conjecture (4.1) of [27].

Conjecture. *The element $f_*(\tilde{\gamma}_1) \in KK(C^*_{\text{red}}(G), C^*_{\text{red}}(G))$ coincides with $1_{C^*_{\text{red}}(G)}$.*

We must note however that *if* this conjecture is true then the element $\tilde{\gamma}_1$ defines a cross-section in the exact sequence of the groups K_* for the extension

$$0 \to \ker f \to C^*(G) \to C^*_{\text{red}}(G) \to 0,$$

which gives the *canonical* splitting

$$K_*(C^*(G)) \simeq K_*(C^*_{\text{red}}(G)) \oplus K_*(\ker f).$$

[20] Editor's note: which together with the algebraic periodicity of the Clifford algebras give rise to Bott periodicity.

I do not know to what extent this last assertion may be true. In fact, it is true in the examples where $C^*(G)$ was calculated.[21]

Theorem 2 can be applied to the study of the discrete series representations of connected Lie groups, as was suggested in [27]. Besides that, it can be used in the study of the structure of group C^*-algebras. We will consider here mainly the case of C^*-algebras of nilpotent Lie groups (announced in [13] and [15]). *Up to the end of this section all C^*-algebras will be complex.*

Theorem 3. *Let G be a connected amenable group, Γ a normal subgroup in G. Assume that the action of G_c on V_G is spin, and $\Gamma \simeq \mathbb{R}$. Then the natural extension of C^*-algebras*

$$0 \to B \to C^*(G) \to C^*(G/\Gamma) \to 0 \tag{1}$$

does not split.[22]

Proof. Applying Theorem 2 we get:

$$KK(C^*(G/\Gamma), C^*(G)) = KK^1(C^*(G/\Gamma), C^*(G/\Gamma)) = 0.$$

Therefore there is a short exact sequence (in the second variable of the K-bifunctor):

$$\begin{aligned}(2)\quad 0 \to KK^0(C^*(G/\Gamma), C^*(G/\Gamma)) \xrightarrow{\delta} KK^1(C^*(G/\Gamma), B)\\ \to KK^1(C^*(G/\Gamma), C^*(G)) \to 0.\end{aligned}$$

Both left- and right-hand KK-groups are isomorphic to

$$KK^0(C^*(G_c), C^*(G_c)) \neq 0.$$

[21] For the complex K-theory, the conjecture is proved now for all connected real linear reductive Lie groups (see A. Wassermann, *Une démonstration de la conjecture de Connes–Kasparov pour les groupes de Lie linéaires connexes réductifs*, C. R. Acad. Sci. Paris, Sér. I Math. **304** (1987), 559–562) and also for the class of K-amenable Lie groups which is characterized by the property $\gamma = 1$ and includes, aside from amenable Lie groups, also Lorentz groups $SO(n,1)$ and complex Lorentz groups $SU(n,1)$. (For the result that $\gamma = 1$ for $SU(n,1)$ see P. Julg, G. Kasparov, *L'anneau $KK_G(\mathbb{C}, \mathbb{C})$ pour $G = SU(n,1)$*, C. R. Acad. Sci. Paris, Sér. I Math. **313** (1991), 259–264.)

[22] The result on the non-splittability of the extension (1) remains true if G is any almost connected separable locally compact group and Γ solvable and simply connected (without any assumptions on the spin action); see N. V. Gorbachev and G. G. Kasparov, *On extensions related with group C^*-algebras*, Uspekhi Mat. Nauk **40** (1985), no. 2, 173–174; English translation: Russian Math. Surveys **40** (1985), no. 2, 213–214.

According to Lemma 6, §7 of [15], the element of $KK^1(C^*(G/\Gamma), B)$ corresponding to the extension (1) is equal to $\delta(1_{C^*(G/\Gamma)})$, which is obviously non-zero. $\square$

There exists a class of groups G satisfying the conditions of Theorem 3 for which the algebra B admits a simple description. This is the class of all connected, simply connected nilpotent Lie groups such that the orbits of the maximal dimension in the coadjoint representation of the group fill the whole complement to the hyperplane orthogonal to the Lie algebra of the normal subgroup $\Gamma \simeq \mathbb{R}$. In this case by the Kirillov trace formula ([16], (7.3) and [17], §15), B is a continuous trace algebra. From [6], (10.9.6), and [4], it follows that $B \simeq \mathcal{K} \otimes C(X)$, where X is the space of orbits of the maximal dimension in the coadjoint representation.

Let $\mathfrak{g}$ be the Lie algebra of G, $\mathfrak{p}$ the hyperplane in $\mathfrak{g}^*$ orthogonal to the Lie algebra of Γ. Then $(\mathfrak{g}^* \setminus \mathfrak{p})$ is a fiber bundle over X. Since each of its fibers is homeomorphic to a Euclidean space, this fiber bundle has a global cross-section $s : X \to (\mathfrak{g}^* \setminus \mathfrak{p})$. Now notice that there is a vector bundle ξ over $(\mathfrak{g}^* \setminus \mathfrak{p})$ whose fiber over an arbitrary point $f \in (\mathfrak{g}^* \setminus \mathfrak{p})$ is the factor algebra $\mathfrak{g}/\mathfrak{g}_f$, where $\mathfrak{g}_f$ is the Lie algebra of the stability subgroup $G_f = \{g \in G \mid g(f) = f\}$. The total space of the vector bundle $s^*(\xi)$ over X is homeomorphic to $(\mathfrak{g}^* \setminus \mathfrak{p})$.

The vector bundle ξ has a symplectic structure, with the canonical 2-form

$$Q_f(x, y) = \langle f, [x, y] \rangle, \qquad \text{where } f \in \mathfrak{g}^*, \ , x, y \in \mathfrak{g}.$$

This symplectic structure defines a complex structure on ξ and, as a consequence, on $s^*(\xi)$. By the Thom isomorphism ([15], §5, Theorem 8), the K-functors of X and $(\mathfrak{g}^* \setminus \mathfrak{p})$ are isomorphic:

$$KK^1(C^*(G/\Gamma), B) \simeq KK^1(C^*(G/\Gamma), C(\mathfrak{g}^* \setminus \mathfrak{p})) \simeq \mathbb{Z} \oplus \mathbb{Z}.$$

The isomorphism of $KK^1(C^*(G/\Gamma), B)$ with $\mathbb{Z} \oplus \mathbb{Z}$ follows also from (2) but now it is chosen in such a way that it does not depend of the choice of splitting in (2).

Next we will calculate the element of $KK^1(C^*(G/\Gamma), B)$ corresponding to the extension (1). We need to make one more additional assumption on G (which is possibly always true, but I could not find any information about it in the literature). Our assumption is related with polarizations. It is known that for any $f \in \mathfrak{g}^*$, there exists a Lie subalgebra $\mathfrak{m}_f \subset \mathfrak{g}$, called polarization, such that its dimension is $\frac{1}{2} \cdot (\dim \mathfrak{g} + \dim \mathfrak{g}_f)$ and $f([\mathfrak{m}_f, \mathfrak{m}_f]) = 0$. We will assume that for $f \in (\mathfrak{g}^* \setminus \mathfrak{p})$, there is a *continuous* function $f \mapsto \mathfrak{m}_f$ sending a functional f to a polarization $\mathfrak{m}_f$.

Theorem 4. *Assume that G is a connected, simply connected nilpotent Lie group satisfying the two conditions imposed above: that the orbits of the maximal dimension fill the complement to the hyperplane $\mathfrak{p}$ and that there exists a continuous function $f \mapsto \mathfrak{m}_f$ defined on this complement. Denote by n the maximal dimension of orbits in $\mathfrak{g}^*$. Then under the isomorphism $KK^1(C^*(G/\Gamma), B) \simeq \mathbb{Z} \oplus \mathbb{Z}$ which was fixed above, the element corresponding to the extension (1) equals $\pm\left(1, (-1)^{\frac{n}{2}+1}\right)$. (The sign $\pm$ depends on the choice of orientations on G and Γ.)*

Proof. Let M_f be the connected subgroup in G with Lie algebra $\mathfrak{m}_f$. Put $H_0 = \mathbb{C}$. The group M_f acts on H_0 via the representation $\exp(y) \mapsto e^{i\langle f, y\rangle}$. Consider the linear fiber bundle $\eta_f = G \times_{M_f} H_0$ over G/M_f and the natural representation of G in the Hilbert space $H_f = L^2(\eta_f)$. This is just the irreducible representation corresponding to the orbit containing f. The family of representations of the group G in the spaces $\left\{H_{s(x)}\right\}_{x\in X}$, where s is the cross-section considered above, defines a homomorphism $\psi : C^*(G) \to \mathcal{L}(\mathcal{H}_{C(X)})$. And again by the Kirillov trace formula applied to the function $(a^*) * a$ (where a is any smooth function with compact support on G), we have: $\psi(B) \subset \mathcal{K}_{C(X)}$. The homomorphism $\varphi : C^*(G/\Gamma) \to \mathcal{L}(\mathcal{H}_{C(X)})/\mathcal{K}_{C(X)}$ induced by ψ is precisely the Busby homomorphism for the extension (1).

The cross-section s can be chosen from the beginning in such a way that $s(-x) = -s(x)$. Moreover, we can choose the function $f \mapsto \mathfrak{m}_f$ so that $\mathfrak{m}_{-f} = \mathfrak{m}_f$. Obviously, $\eta_{-f} = \bar{\eta}_f$, the fiber bundle complex conjugate to η_f. Given the complex conjugation $H_0 \to H_0$, we get a fiberwise antilinear isomorphism $\eta_f \to \bar{\eta}_f$ and an antilinear isometric isomorphism $\omega_f : L^2(\eta_f) \to L^2(\bar{\eta}_f)$. Clearly,

$$\varphi_{-x}(\bar{a}) = \omega_{s(x)}\varphi_x(a)\omega^{-1}_{s(x)}, \qquad \text{where } x \in X, \quad a \in C_c(G),$$

and $\bar{a}$ is the complex conjugate function.

Now consider the following general situation. Assume that the algebras A and B are complexifications of real algebras (i.e. "real" algebras in the sense of [14] and [15]). Then there is also a "real" structure on $\mathcal{H}_B$ and $\mathcal{L}(\mathcal{H}_B)$ (cf. [14] and [15]). Sending a triple $(\mathcal{H}_B, \varphi, T) \in \mathcal{E}(A, B)$ into $(\mathcal{H}_B, \bar{\varphi}, \bar{T})$, where $\bar{\varphi}(a) = \overline{\varphi(\bar{a})}$ (and the upper line means the "real" involution), we get an automorphism of $KK(A, B)$ which will be called the *complex conjugation*. Since all Clifford algebras $C_{p,q}$ are "real," we get a complex conjugation on all $KK^i(A, B)$. Returning to our concrete situation, we have established that the automorphism $X \to X : x \mapsto -x$, followed by the complex conjugation on $KK^1(C^*(G/\Gamma), B)$, keeps the element corresponding to the extension (1) unchanged.

The automorphism $x \mapsto -x$ corresponds on $KK^1(C^*(G/\Gamma), B) \simeq \mathbb{Z} \oplus \mathbb{Z}$ to the composition of multiplication by $(-1)^{\dim X} = (-1)^{\dim G}$ and the transposition of the two copies of $\mathbb{Z}$. As to the complex conjugation, we first point out that it commutes with the product-intersection. The elements $\tilde{\alpha}$ and $\tilde{\beta}$ from Theorem 2 are "real," i.e., remain unchanged under complex conjugation. The Clifford periodicity $KK^{p,q+1} \to KK^{p+1,q}$ after the complex conjugation changes its sign (i.e. anticommutes with the complex conjugation). Therefore in the composition of isomorphisms

$$KK^1(C^*(G/\Gamma), B) \simeq KK^1(C^*(G/\Gamma), C(\mathfrak{g}^* \setminus \mathfrak{p})) \simeq \mathbb{Z} \oplus \mathbb{Z},$$

the complex conjugation results in the change of sign by $(-1)^{(\dim G - \dim X)/2} = (-1)^{\frac{n}{2}}$ at the first stage and by $(-1)^{\dim G - 1}$ at the second stage. Consequently, the element of the group $\mathbb{Z} \oplus \mathbb{Z}$ corresponding to the extension (1) must be of the type $(d, (-1)^{\frac{n}{2}+1} d)$, where $d \in \mathbb{Z}$. But in view of the exact sequence (2), the subgroup of $\mathbb{Z} \oplus \mathbb{Z}$ generated by this element is a direct summand in $\mathbb{Z} \oplus \mathbb{Z}$. Hence $d = \pm 1$. □

Example. For the $(2k+1)$-dimensional Heisenberg group $G = \Gamma_{2k+1}$ and its central subgroup $\Gamma = \Gamma_1$, the element corresponding to the extension (1) is equal to $(-1, (-1)^k)$ for the natural choice of orientations. This calculation for Γ_{2k+1} was first carried out at the request of Professor A. A. Kirillov in 1976 (even before the product-intersection was constructed and the Bott periodicity and the isomorphism $KK^1(C^*(\Gamma_{2k+1}/\Gamma), B) \simeq \mathbb{Z} \oplus \mathbb{Z}$ were rigorously proved). The result for Γ_3 was later announced in [13]. Briefly, the method we used in that calculation was the following straightforward one. The canonical generator of the group

$$K_0(C^*(\Gamma_{2k+1}/\Gamma)) \simeq K_0(C(\mathbb{R}^{2k})) \simeq \tilde{K}_0(C(S^{2k}))$$

can be precisely written as a projection in $M_N(C(S^{2k}))$ up to stable equivalence (for some N). This projection can be lifted to some element $p \in M_N(C^*(\Gamma_{2k+1}))$. The element $\psi(p) \in M_N(\mathcal{L}(\mathcal{H}_{C(X)}))$, where $X = \mathbb{R} \setminus 0$, defines an element of $K_{-1}(C(X)) \simeq \mathbb{Z} \oplus \mathbb{Z}$. By a precise homotopy of (pseudodifferential) operators this last element can be identified with $(-1, (-1)^k) \in \mathbb{Z} \oplus \mathbb{Z}$.

§7. C^*-algebras of discrete groups

Definition 1. Let A be a Γ-algebra, X a locally compact right Γ-space with X/Γ locally compact. Define $C(X \times_\Gamma A)$ to be the algebra of continuous functions $f : X \to A$ satisfying the conditions (1) and (2) of Definition 2, §5 (with G replaced by X). The subspace of functions $f \in C(X \times_\Gamma A)$ with $f(x) = 0$ when $(x\Gamma)$ is out of some compact set in X/Γ (depending on f) will be denoted by $C_c(X \times_\Gamma A)$.

Lemma 1. *Assume that Γ_1 and Γ_2 are subgroups in G such that the left action of Γ_1 on G/Γ_2 is free and proper. If A is a Γ_2-algebra, then the algebras $C^*(\Gamma_1, C(G \times_{\Gamma_2} A))$ and $C((\Gamma_1\backslash G) \times_{\Gamma_2} A)$ are strongly Morita equivalent.*

Sketch of proof. (Cf. [26], example 2.) Define on $E = C_c(G \times_{\Gamma_2} A)$ the right action of

$$C((\Gamma_1\backslash G) \times_{\Gamma_2} A)$$

by pointwise multiplication and the left action of $C_c(\Gamma_1, C(G \times_{\Gamma_2} A))$ by the formula:

$$(a \cdot e)(t) = \int_{\Gamma_1} a(s, t)\, e(s^{-1}t)\, \mu_{\Gamma_1}(s)^{\frac{1}{2}}\, d_{\Gamma_1}s,$$

where $a \in C_c(\Gamma_1, C(G \times_{\Gamma_2} A))$, $e \in E$, $s \in \Gamma_1$, $t \in G$. The left scalar product on E with values in $C_c(\Gamma_1, C(G \times_{\Gamma_2} A))$ is defined by

$$\langle e_1, e_2 \rangle (s, t) = \mu_{\Gamma_1}(s)^{-\frac{1}{2}} e_1(t) e_2^*(s^{-1}t)$$

and the right scalar product with values in $C((\Gamma_1\backslash G) \times_{\Gamma_2} A)$ by

$$\langle e_1, e_2 \rangle (\Gamma_1 t) = \int_{\Gamma_1} e_1^*(s^{-1}t) e_2(s^{-1}t)\, d_{\Gamma_1}s,$$

where $e_1, e_2 \in E$, $s \in \Gamma_1$, $t \in G$. Completing E we get the required imprimitivity bimodule. □

Theorem 1. *Assume that G is a connected group, G_c its maximal compact subgroup, V the cotangent space to G/G_c at the point (G_c), π a discrete subgroup in G without torsion, A a separable π-algebra, D a separable algebra, $\gamma \in R^0(G)$ the element constructed in Theorem 2, §5. Put*

$$\tilde{\gamma} = j_\pi \circ \sigma_A \circ r_{G,\pi}(\gamma) \in KK(C^*(\pi, A), C^*(\pi, A)).$$

Then the subgroup

$$\operatorname{Im} \tilde{\gamma}_\ell = \{x \in KK^i(C^*(\pi, A), D) \mid \tilde{\gamma} \otimes_{C^*(\pi, A)} x = x\}$$

of the group $KK^i(C^*(\pi, A), D)$ *is isomorphic to*

$$KK^i(C(G \times_{(\pi \times G_c)} (A \hat{\otimes} C_V)), D)$$

and the subgroup

$$\operatorname{Im} \tilde{\gamma}_\rho = \{x \in KK^i(D, C^*(\pi, A)) \mid x \otimes_{C^*(\pi, A)} \tilde{\gamma} = x\}$$

of the group $KK^i(D, C^*(\pi, A))$ *is isomorphic to*

$$KK^i(D, C(G \times_{(\pi \times G_c)} (A \hat{\otimes} C_V))).$$

Here the action of $(\pi \times G_c)$ *on* $(A \hat{\otimes} C_V)$ *is defined by* $(s, t)(a \hat{\otimes} c) = s(a) \hat{\otimes} t(c)$ *and the right action of* $(\pi \times G_c)$ *on* G *by* $(s, t)(g) = s^{-1} g t$. *If the action of* G_c *on* V *is spin then*

$$\operatorname{Im} \tilde{\gamma}_\ell \simeq KK^{i - \dim G/G_c}(C(G \times_\pi A), D),$$

$$\operatorname{Im} \tilde{\gamma}_\rho \simeq KK^{i + \dim G/G_c}(D, C(G \times_\pi A)).$$

In the case of an amenable group G, *the subgroups* $\operatorname{Im} \tilde{\gamma}_\ell$ *and* $\operatorname{Im} \tilde{\gamma}_\rho$ *coincide with the whole groups* $KK^i(C^*(\pi, A), D)$ *and* $KK^i(D, C^*(\pi, A))$, *respectively.*

Sketch of proof. In view of [8], (2.13), the algebra $C^*(\pi, A)$ is strongly Morita equivalent to $C^*(G, C(G \times_\pi A))$. By Theorem 2, §6, the "$\tilde{\gamma}$-part" of the K-bifunctor of D and $C^*(G, C(G \times_\pi A))$ is isomorphic to the K-bifunctor of D and $C^*(G_c, C(G \times_\pi A) \hat{\otimes} C_V)$. The latter algebra is strongly Morita equivalent to

$$C^*(G, C(G \times_\pi A) \hat{\otimes} C(G \times_{G_c} C_V)) \simeq C^*\Big(G, C\big((G \times G)_{(\pi \times G_c)} (A \hat{\otimes} C_V)\big)\Big).$$

It remains to apply Lemma 1 and to notice that the right $(\pi \times G_c)$-space $G \backslash (G \times G)$ is homeomorphic to G via $(g_1, g_2) \mapsto g_1^{-1} g_2$. □

For the next theorem we will fix a separable algebra D and denote by $k_i(A)$ the group $KK^{-i}(D, A)$ and by $k^i(A)$ the group $KK^i(A, D)$. Let $H_*(\pi, M)$ and $H^*(\pi, M)$ be as usual the homology and cohomology of a discrete group π with values in a π-module M (cf. [18]). For a π-algebra A, the groups $k_*(A)$ and $k^*(A)$ are clearly π-modules.

Theorem 2. *Let G be a connected group, π its discrete subgroup without torsion, A a separable π-algebra. If the algebra D is nuclear*[23] *then there exists a (homological) spectral sequence (E^r, d^r) with differentials $d^r : E^r_{p,q} \to E^r_{p-r,q+r-1}$ and the term $E^2_{p,q} = H_p(\pi, k_q(A))$, converging to the subgroup $\operatorname{Im}\tilde{\gamma}_\rho$ of $k_*(C^*(\pi, A))$. If the algebra A is nuclear then there exists a (cohomological) spectral sequence (E_r, d_r) with differentials $d_r : E_r^{p,q} \to E_r^{p+r,q-r+1}$ and the term $E_2^{p,q} = H^p(\pi, k^q(A))$, converging to the subgroup $\operatorname{Im}\tilde{\gamma}_\ell$ of $k^*(C^*(\pi, A))$.*

Sketch of proof. In the notation of Theorem 1, the smooth manifold $\pi\backslash G/G_c$ may be considered as a classifying space $B\pi$ for the group π (cf. §9 below). Let

$$X^0 \subset X^1 \subset \cdots \subset X^n = B\pi$$

be its filtration by the skeleta of some smooth triangulation ($\dim X^i = i$). Denote by Y^i a small open neighbourhood of X^i such that $\forall i$, $(Y^i \setminus Y^{i-1})$ is a disjoint sum $\bigcup_m \left(S^i_m \times \operatorname{Int}(I^{n-i})\right)$, where each S^i_m is a closed subset of X^i homeomorphic to the i-dimensional simplex and $\operatorname{Int}(I^{n-i})$ is the interior of the $(n-i)$-dimensional cube. Define a filtration

$$A_0 \subset A_1 \subset \cdots \subset A_n = C\left(G \times_{(\pi\times G_c)} (A\hat{\otimes}C_V)\right)$$

by

$$A_i = \left\{f \in C\left(G \times_{(\pi\times G_c)} (A\hat{\otimes}C_V)\right) \mid f(g) = 0 \text{ for } \pi g G_c \notin Y^i\right\}.$$

Now one applies the standard construction of a spectral sequence associated with a filtration (cf. [18] or [28]). To identify the term E^2 notice

[23] The condition of nuclearity of D and A was included into the statement of the Theorem in order to have the exact sequences used in the proof. From this point of view, the condition of nuclearity of D and A is unnecessary because the exact sequences used in the proof are available without this assumption – see [0], 6.10. On the other hand, for the part of the Theorem related with $k_*(A)$ one really needs some additional assumption concerning D. What is really necessary is countable additivity for k^* and k_*. Whereas for k^* this is automatic, for k_* this is true, for example, if D is type I with $K_*(D)$ finitely generated — see J. Rosenberg and C. Schochet, *The Künneth theorem and the universal coefficient theorem for Kasparov's generalized K-functor,* Duke Math. J. **55** (1987), 431–474.

that

$$\begin{aligned} E^1_{p,q} &= k_{p+q}(A_p/A_{p-1}) \\ &= KK^{-p-q}\left(D, \bigoplus_m C\left(S^p_m \times \operatorname{Int}(I^{n-p})\right)\hat{\otimes} A \hat{\otimes} C_{n,0}\right) \\ &= \bigoplus_m KK^{-q}(D, A) \\ &= \bigoplus_m k_q(A). \end{aligned}$$

It is easily verified that the differential d^1 coincides with the boundary homomorphism from the p-chains with values in $k_q(A)$ to the $(p-1)$-chains. The same argument applies to E_2 as well. □

Remark 1. The exact sequences of [24] for $\pi = \mathbb{Z}$ are immediate consequences of the spectral sequences of Theorem 2. (All differentials in this case are 0 beginning with the second one.)

§8. Representable K-functor and Poincaré duality

In this section the group Γ will be compact. The representable cohomological K-functor of a Γ-space X, as defined by Fredholm complexes in [29], will be denoted here by $RK^*_\Gamma(X)$. If X is compact then $RK^*_\Gamma(X) \simeq K^\Gamma_{-*}(C(X))$. Define the representable homological K-functor $RK^\Gamma_*(X)$ as the direct limit

$$\varinjlim_{Y \subset X} K^{-*}_\Gamma(C(Y))$$

over the inductive system of all compact Γ-subsets $Y \subset X$.

For compact manifolds with boundary, the Poincaré duality can be formulated as follows.

Theorem 1. *Assume that $\overline{M}$ is a smooth compact manifold with the boundary $\partial\overline{M}$, and a compact group Γ acts on $\overline{M}$ smoothly. Put $M = \overline{M} \setminus \partial\overline{M}$. Then there exist elements*

$$\alpha_{M\times\overline{M}} \in K^0_\Gamma\left(C_\tau(M) \otimes C(\overline{M})\right), \qquad \beta_{M\times\overline{M}} \in K^\Gamma_0\left(C_\tau(M) \otimes C(\overline{M})\right)$$

with the properties:

$$\beta_{M\times\overline{M}} \otimes_{C_\tau(M)} \alpha_{M\times\overline{M}} = 1_{C(\overline{M})}, \qquad \beta_{M\times\overline{M}} \otimes_{C(\overline{M})} \alpha_{M\times\overline{M}} = 1_{C_\tau(M)}.$$

The homomorphisms

$$\begin{aligned}
(_\otimes_{C_\tau(M)} \alpha_{M\times\overline{M}}) &: K_i^\Gamma\left(C_\tau(M)\right) \to K_\Gamma^{-i}\left(C(\overline{M})\right), \\
(\beta_{M\times\overline{M}} \otimes_{C_\tau(M)} _) &: K_\Gamma^{-i}\left(C_\tau(M)\right) \to K_i^\Gamma\left(C(\overline{M})\right), \\
\left(_\otimes_{C(\overline{M})} \alpha_{M\times\overline{M}}\right) &: K_i^\Gamma\left(C(\overline{M})\right) \to K_\Gamma^{-i}\left(C_\tau(M)\right), \\
\left(\beta_{M\times\overline{M}} \otimes_{C(\overline{M})} _\right) &: K_\Gamma^{-i}\left(C(\overline{M})\right) \to K_i^\Gamma\left(C_\tau(M)\right)
\end{aligned}$$

are isomorphisms.

Sketch of proof. Choose a Γ-invariant Riemannian metric on $\overline{M}$. The element $\alpha_{M\times\overline{M}}$ is the image of the element $\alpha \in K_\Gamma^0\left(C_\tau(M)\right)$ constructed in Definition 3, §5, under the homomorphism induced by the multiplication

$$C_\tau(M) \otimes C(\overline{M}) \to C_\tau(M).$$

To construct $\beta_{M\times\overline{M}}$, identify M with

$$\overline{M} \cup_{\partial\overline{M}} (\partial\overline{M} \times [0, 1))$$

and embed $\overline{M}$ diagonally into $M \times \overline{M}$. The conormal bundle to $\overline{M}$ in $M \times \overline{M}$ will be denoted by η. It is isomorphic to the cotangent bundle τ to $\overline{M}$. Denote the total space of η by Y and the lifting of η to Y by ξ. Let $1_{\overline{M}} \in K_0^\Gamma\left(C(\overline{M})\right)$ be the image of $1 \in K_0^\Gamma(C)$ under the inclusion $C \hookrightarrow C(\overline{M})$.[24] The Thom isomorphism ([15], §5, Theorem 8)

$$K_0^\Gamma\left(C(\overline{M})\right) \xrightarrow{\cong} K_0^\Gamma\left(C_\xi(Y)\right)$$

transfers $1_{\overline{M}}$ into some element $\beta' \in K_0^\Gamma\left(C_\xi(Y)\right)$. The required $\beta_{M\times\overline{M}}$ is the image of β' under the inclusion

$$C_\xi(Y) \hookrightarrow C_\tau(M) \otimes C(\overline{M}). \quad \square$$

For the application to higher signatures we need the Poincaré duality for non-compact manifolds.

[24] Editor's note: In other words, $1_{\overline{M}} \in K_\Gamma^0\left(\overline{M}\right)$ is simply the class of the trivial line bundle over $\overline{M}$.

Theorem 2. *Assume that X is a complete Riemannian manifold (without boundary) and a compact group Γ acts on X isometrically. Then one has the following natural isomorphisms:*

$$\mathcal{PD}_* : K_i^\Gamma(C_\tau(X)) \xrightarrow{\cong} RK_i^\Gamma(X),$$

$$\mathcal{PD}^* : K_\Gamma^i(C_\tau(X)) \xrightarrow{\cong} RK_\Gamma^i(X).$$

Sketch of proof. Take in X an exhaustive increasing system of compact Γ-manifolds with boundaries:

$$\overline{M}_1 \subset \overline{M}_2 \subset \cdots \subset X.$$

Theorem 1 gives us a concordant system of isomorphisms:

$$K_i^\Gamma(C_\tau(M_p)) \to K_\Gamma^{-i}(C(\overline{M}_p)) \quad \text{and} \quad K_\Gamma^{-i}(C_\tau(M_p)) \to K_i^\Gamma(C(\overline{M}_p)).$$

The isomorphism $\mathcal{PD}_*$ is obtained as a direct limit isomorphism. To obtain $\mathcal{PD}^*$, choose a smooth Γ-invariant function $\varepsilon(x)$ on X such that $\forall x \in X$, there is a geodesic coordinate system in the $\varepsilon(x)$-neighbourhood U_x of x. Define $\forall x \in X$ a covector field θ_x in U_x by

$$\theta_x(y) = \frac{\rho_x(y) \cdot d\rho_x(y)}{\varepsilon(x)},$$

where $\rho_x(y)$ is the geodesic distance from x to $y \in U_x$ and d is the exterior derivative in the y variable.

Given an arbitrary $z = (E, \varphi, T) \in K_\Gamma^{-i}(C_\tau(X))$, where $E = \mathcal{H}\hat{\otimes}C_{0,i}$, it is not difficult to construct a continuous family of pairs $M_x, N_x \in \mathcal{L}(E)$ such that $\forall g \in \Gamma$, $g(M_x) = M_{gx}$, $g(N_x) = N_{gx}$, and the family of "intersection operators"

$$F_x = \theta_x \divideontimes_{C_\tau(U_x)} T = \sqrt{M_x} \cdot \varphi_*(\theta_x) + \sqrt{N_x} \cdot T$$

forms a Fredholm complex

$$\{F_x\} : (\mathcal{H}\hat{\otimes}C_{0,i}) \times X \to (\mathcal{H}\hat{\otimes}C_{0,i}) \times X$$

over X. Put $\mathcal{PD}^*(z) = \{F_x\}$. Now the statement about $\mathcal{PD}^*$ follows from Milnor's $\varprojlim^1$ exact sequence [19]. □

As a corollary of Theorem 2 and Theorem 3, §5, we get:

Theorem 3. *Let G be a connected Lie group, π a discrete subgroup. Then the subgroup*

$$\operatorname{Im}\gamma = \{x \in R^i(\pi) \mid r_{G,\pi}(\gamma)\cdot x = x\}$$

of the group $R^i(\pi)$ is isomorphic to $RK^i_{G_c}(G/\pi)$. □

Note added

As it was pointed out in the Preface, in the preprint distributed in 1981 there were some small errors in the proof of the Novikov conjecture given in section 9. We indicate here some modification to section 8 that will be necessary in order to give a correct argument in section 9.

The modification that is needed is the replacement of the integer coefficients by rational ones. For this we fix some C^*-algebra D such that $K_0(D) \simeq \mathbb{Q}$, $K_1(D) = 0$.[25] Let us consider this algebra with the trivial action of any group (G, or π, or Γ). We can define the rational version of the KK-groups simply by putting

$$KK_G(A, B;\, \mathbb{Q}) = KK_G(A, B \otimes D).$$

The representation ring of G should be replaced by the group

$$R^i(G;\, \mathbb{Q}) = KK^i_G(\mathbb{C}, D).$$

Similarly, one can define the rational representable cohomological K-functor using complexes of Fredholm operators in Hilbert modules over D parametrized by a paracompact space X. Alternatively, one can use the RKK-groups defined in [0]. Then one can put

$$RK^*_\Gamma(X;\, \mathbb{Q}) = RKK^*_\Gamma(X; \mathbb{C}, D).$$

The rational representable homological K-functor $RK^\Gamma_*(X;\, \mathbb{Q})$ is defined as the direct limit

$$\varinjlim_{Y \subset X} K_\Gamma^{-*}(C(Y), D)$$

over the inductive system of all compact Γ-subsets $Y \subset X$.

Both Theorems 1 and 2 remain valid, with their proofs, for the rational K-theory groups (cf. section 4 of [0]). Theorem 3 also remains valid, together

[25] Editor's note: For example, one can use the "universal UHF algebra," the infinite tensor product of all finite-dimensional matrix algebras.

with Theorem 3 of section 5, but of course the element γ in the statement of these theorems should be understood as the element of the usual (integer) $R^0(\pi)$.

§9. Higher signatures

Let π be an arbitrary topological group, $E\pi \to B\pi$ a principal locally trivial π-bundle with the space $E\pi$ contractible and $B\pi$ paracompact (π acts on $E\pi$ on the left). Then the space $B\pi$ is called *the classifying space*[26] of the group π.

Let us fix the discrete group π. Consider a connected oriented closed smooth manifold M^n. If we fix a homomorphism $\iota : \pi_1(M^n) \to \pi$ then there is a map $f : M^n \to B\pi$ inducing the homomorphism ι on fundamental groups. (f is the classifying map for the π-bundle on M^n associated with the universal covering via ι.) Denote by $[M^n]$ the fundamental cycle of M^n in $H_n(M^n)$ and by $L_*(M^n)$ the Pontryagin-Hirzebruch characteristic class of M^n.

The problem of homotopy invariance of higher signatures [23] may be formulated as follows. If $h : N^n \to M^n$ is an orientation-preserving homotopy equivalence of oriented smooth manifolds, then is it true that $\forall x \in H^*(B\pi; \mathbb{Q})$ the intersection indices

$$\langle L_*(M^n) \cdot f^*(x), [M^n]\rangle \quad \text{and} \quad \langle L_*(N^n) \cdot h^*f^*(x), [N^n]\rangle$$

are equal? If we consider the homology class

$$\mathcal{D}(L_*(M^n)) = L_*(M^n) \cap [M^n] \in H_*(M^n; \mathbb{Q})$$

Poincaré dual to $L_*(M^n)$, then the equivalent question is whether

$$f_*(\mathcal{D}(L_*(M^n))) \in H_*(B\pi; \mathbb{Q})$$

is a homotopy invariant of M^n.

In [11], [21], [22], [12] the author and A. S. Mishchenko independently have suggested a general method of reducing this problem to another one dealing with the K-functor of the space $B\pi$. Following [12], §8, we will prove here the homotopy invariance of higher signatures for any group π which is a discrete subgroup of some connected Lie group.

[26] Editor's note: One can justify the definite article by the fact that $B\pi$ is unique up to homotopy equivalence.

Multiplying M^n by the torus T^n, we can reduce to the case of an even-dimensional manifold M^{2n}. First we will construct some distinguished element in the complex homological K-functor $K_0(M^{2n})$. Fix a Riemannian metric on M^{2n}. Put $H = L^2\left(\bigwedge^*\left(M^{2n}\right)\right)$ with the grading defined by the grading operator ε which on p-forms is equal to $i^{p(p-1)+n} \cdot *$, where $* : \bigwedge^p\left(M^{2n}\right) \to \bigwedge^{2n-p}\left(M^{2n}\right)$ is the Hodge $*$-operator. Define the homomorphism $\varphi : C(M^{2n}) \to \mathcal{L}(H)$ to be given by multiplication by functions, and let the operator $\frac{d+\delta}{\sqrt{1+\Delta}}$ be defined as in §5. Denote the element

$$\left(H, \varphi, \frac{d+\delta}{\sqrt{1+\Delta}}\right) \in K_0(M^{2n})$$

by $[d+\delta]$. (In all constructions and statements dealing with the element $[d+\delta]$ we shall confine ourselves to the case of the complex K-functor.)

Now we will apply the Atiyah-Singer index theorem to the signature operator $(d+\delta)$ (cf. [2], §6) lifted to the vector bundle $\bigwedge^*\left(M^{2n}\right) \otimes \xi$, where ξ is an arbitrary vector bundle on M^{2n}. As was proved in [12], the analytical index will be equal to

$$[\xi] \otimes_{C(M^{2n})} [d+\delta] \in K_0(\mathbb{C}) = \mathbb{Z}.$$

The topological index is equal to

$$2^n \cdot \langle \operatorname{ch} \xi \cdot \mathcal{L}_*(M^{2n}),\, [M^{2n}] \rangle,$$

where $\mathcal{L}_*(M^{2n})$ is the modified Pontryagin-Hirzebruch class (see [2]. §6). Since the vector bundle ξ is arbitrary and $(\operatorname{ch} \otimes \mathbb{Q})$ is an isomorphism, the homotopy invariance of $f_*\left(\mathcal{D}(L_*(M^n))\right)$ is equivalent to the homotopy invariance of $f_*([d+\delta]) \in RK_0(B\pi) \otimes \mathbb{Q}$.

Definition 1. Let G be an arbitrary locally compact group. We will construct a homomorphism $\alpha : R^i(G) \to RK^i(BG)$. Given $z = (\mathcal{H}\hat{\otimes}C_{i,0}, T) \in R^i(G)$, consider the family of Fredholm operators

$$T_x = \int_G \lambda(g^{-1}x) \cdot g(T)\, dg \in \mathcal{L}\left(\mathcal{H}\hat{\otimes}C_{i,0}\right)$$

parameterized by the points $x \in EG$, where $\lambda(x)$ is a cut-off function on EG (see [3], Ch. 7, §2, Proposition 8).[27] Since $\forall g \in G$, $T_{gx} = g(T_x)$, the family $\{T_x\}$ defines a Fredholm complex

$$\{\widetilde{T}_x\} : \left(\left(\mathcal{H}\hat{\otimes}C_{i,0}\right) \times EG\right)/G \to \left(\left(\mathcal{H}\hat{\otimes}C_{i,0}\right) \times EG\right)/G$$

on BG. (The action of G on $\left(\mathcal{H}\hat{\otimes}C_{i,0}\right) \times EG$ is diagonal.) Put $\alpha(z) = \{\widetilde{T}_x\}$.

---

[27] Editor's note: This means $\lambda \geq 0$ is continuous, $\operatorname{supp}\lambda$ has compact intersection with $G \cdot K$, for any compact $K \subset EG$, and the integral of λ over G-orbits is identically 1.

Definition 2. Let π be a discrete group. We will construct a homomorphism $\beta : RK_i(B\pi) \to K_i(C^*(\pi))$. Consider a compact set $X \subset B\pi$. The inclusion $X \hookrightarrow B\pi$ induces a regular covering $\widetilde{X} \to X$ with the group π. Consider the vector space F_X of continuous functions $\mu : \widetilde{X} \to C^*(\pi)$ satisfying the condition: $\forall x \in \widetilde{X}$, $\forall g \in \pi$, $\mu(gx) = g \cdot \mu(x)$. (On the right side we use the *product* of the element g and $\mu(x) \in C^*(\pi)$.) We introduce on F_X the right action of $C(X) \otimes C^*(\pi)$ as the right multiplication by continuous functions $\nu : X \to C^*(\pi)$ and we define on F_X the scalar product with values in $C(X) \otimes C^*(\pi)$ by the formula:

$$\langle \mu_1, \mu_2 \rangle(x) = \mu_1^*(x) \cdot \mu_2(x).$$

(Clearly, the last product only depends on the class $\pi \cdot x \in \widetilde{X}/\pi = X$ and therefore defines a function $X \to C^*(\pi)$.) Endowed with these structures F_X becomes a Hilbert $C(X) \otimes C^*(\pi)$-module. It is easily verified that F_X is a finitely generated projective $C(X) \otimes C^*(\pi)$-module. Hence, it defines some element $[F_X] \in K_0(C(X) \otimes C^*(\pi))$. Considering the intersection product

$$[F_X] \otimes_{C(X)} \_\_ : K^{-i}(C(X)) \to K_i(C^*(\pi))$$

and passing to the direct limit over the inductive system of all compact subsets $X \subset B\pi$, we obtain the homomorphism β.

Note that the alternative intersection product

$$[F_X] \otimes_{C(X)} \_\_ : K^i(C^*(\pi)) \to K_{-i}(C(X))$$

after passing to the inverse limit gives the homomorphism

$$\widetilde{\alpha} : R^i(\pi) = K^i(C^*(\pi)) \to LK^i(B\pi),$$

where $LK^i(B\pi)$ is $\varprojlim_{X \subset B\pi} K^i(C(X))$ over the same system of all compact subsets $X \subset B\pi$.[28] It is easily verified that $\widetilde{\alpha}$ is the composition of α and the natural restriction $RK^i(B\pi) \to LK^i(B\pi)$. The homomorphisms $\widetilde{\alpha}$ and β are obviously dual to each other in the sense that

$$\widetilde{\alpha}(x) \otimes_{B\pi} y = x \otimes_{C^*(\pi)} \beta(y) \in K_0(C) = \mathbb{Z}.$$

[28] Editor's note: However, in general the theory LK^i is not well-behaved since the Mittag-Leffler condition may not be satisfied; see [1].

Note added

The homomorphisms α, β and $\tilde{\alpha}$ can also be defined for the rational K-theory groups (see *Note added* at the end of section 8). Actually for the application to the Novikov conjecture we will need just these rational homomorphisms. Note that for locally finite countable CW-complexes X one has $RK_*(X;\ \mathbb{Q}) \simeq RK_*(X) \otimes \mathbb{Q}$ and $RK^*(X;\ \mathbb{Q}) \simeq \mathrm{Hom}(RK_*(X), \mathbb{Q})$.[29] Therefore, for the rational K-groups, if the image of $\tilde{\alpha} \otimes \mathbb{Q}$ is dense in $LK^*(B\pi;\ \mathbb{Q})$ in the projective limit topology, then $\beta \otimes \mathbb{Q}$ is a monomorphism. For the integer K-groups this may be wrong. In fact, the pairing between $RK^*(B\pi)$ and $RK_*(B\pi)$ may be highly degenerate: take, for example, $\pi = \mathbb{Z}[\frac{1}{2}]$.

In the next Theorem 1, we changed all K-theory groups to the rational ones for two reasons: because for the Novikov conjecture we need just the rational version of this theorem and also because the proof of the integer version, which was stated in the preprint distributed in 1981, contained a gap, whereas for the rational version of this result the argument goes well.

Theorem 1. *If π is a discrete subgroup of a connected Lie group G, then $\forall i$ the image of*

$$\tilde{\alpha} : R^i(\pi;\ \mathbb{Q}) \to LK^i(B\pi;\ \mathbb{Q})$$

is dense in the projective limit topology on $LK^i(B\pi;\ \mathbb{Q})$. Moreover, if π has no torsion, then $\forall i$

$$\alpha : R^i(\pi;\ \mathbb{Q}) \to RK^i(B\pi;\ \mathbb{Q})$$

is an epimorphism.

Sketch of proof. Denote the epimorphism $R^i(\pi;\ \mathbb{Q}) \to RK^i_{G_c}(G/\pi;\ \mathbb{Q})$ of Theorem 3, §8, by δ. If π has no torsion then $G_c \backslash G/\pi \simeq B\pi$ and $RK^i_{G_c}(G/\pi;\ \mathbb{Q}) \simeq RK^i(B\pi;\ \mathbb{Q})$. It can be proved that $\alpha = \delta$. We omit this.

In general it is necessary to prove that for any compact subset $X \subset B\pi$, the images of the homomorphisms

$$RK^i_{G_c}(G/\pi;\ \mathbb{Q}) \to RK^i_{G_c}(G \times_\pi \widetilde{X};\ \mathbb{Q})$$

and

$$RK^i_{G_c}(G \times_\pi E\pi;\ \mathbb{Q}) \to RK^i_{G_c}(G \times_\pi \widetilde{X};\ \mathbb{Q})$$

[29] This is essentially the uniqueness theorem for homology and cohomology. See some more details about this in G. Kasparov and G. Skandalis, *Groups acting on buildings, Operator K-theory, and the Novikov conjecture*, K-theory **4** (1991), 303–337, Lemma 3.4 and Remark on pp. 314–315.

coincide. (Here $\widetilde{X}$ is the regular covering of X with the group π induced by the inclusion $X \hookrightarrow B\pi$.)

Note that the projection

$$(G_c\backslash G) \times_\pi E\pi \to B\pi$$

is a fiber bundle with contractible fibers (homeomorphic to $G_c\backslash G$). Hence there is a cross-section $s : B\pi \to (G_c\backslash G) \times_\pi E\pi$. Let $Y \to X$ be the G_c-bundle over X induced from the G_c-bundle

$$G \times_\pi E\pi \to (G_c\backslash G) \times_\pi E\pi$$

via the map s combined with the inclusion $X \hookrightarrow B\pi$. Clearly the G_c-spaces $G \times_\pi \widetilde{X}$ and Y are homotopy equivalent, and Y is compact. Moreover, there is a homotopy equivalence of G_c-spaces:

$$G \times_\pi E\pi \simeq G \times_\pi EG \simeq (G/\pi) \times EG \simeq (G/\pi) \times EG_c.$$

Hence it is sufficient to verify that for any compact G_c-subspace $Y \subset (G/\pi) \times EG_c$, the images of the homomorphisms

$$RK^*_{G_c}(G/\pi;\ \mathbb{Q}) \to RK^*_{G_c}(Y;\ \mathbb{Q})$$

and

$$RK^*_{G_c}(G/\pi \times EG_c;\ \mathbb{Q}) \to RK^*_{G_c}(Y;\ \mathbb{Q})$$

coincide.

Note that all stability subgroups of the action of G_c on G/π are finite. Since the rational cohomology of finite groups is trivial, using the G. Segal spectral sequence,[30] it is not difficult to see that the image of

$$RK^*_{G_c}(G/\pi \times EG_c;\ \mathbb{Q}) \to RK^*_{G_c}(Y;\ \mathbb{Q})$$

coincides with the image of

$$RK^*(G_c\backslash G/\pi;\ \mathbb{Q}) \to RK^*(G_c\backslash Y;\ \mathbb{Q}).$$

Since the map

$$RK^*_{G_c}(G/\pi;\ \mathbb{Q}) \to RK^*(G_c\backslash G/\pi;\ \mathbb{Q})$$

is surjective we get the required assertion. $\square$

[30]See G. Segal, *Classifying spaces and spectral sequences* and *Equivariant K-theory*, Publ. Math. Inst. des Hautes Études Scient. **34** (1968), 105–112, 129–151.

Corollary 1. *If π is a discrete subgroup of a connected Lie group G, then*

$$\beta \otimes \mathbb{Q} : RK_i(B\pi) \otimes \mathbb{Q} \to K_i(C^*(\pi)) \otimes \mathbb{Q}$$

is a monomorphism. □

Remark 1. In the first proof of Theorem 1 (as it was announced in [12]), the assertion that δ is epimorphic was proved by a straightforward construction of elliptic (Fredholm) complexes and their homotopies. Because of that the whole proof was rather cumbersome.

Theorem 2. *Assume that M^{2n} is a closed oriented smooth manifold and $f : M^{2n} \to B\pi$ a continuous map. Then the element $\beta(f_*([d+\delta])) \in K_0(C^*(\pi))$ is a homotopy invariant of the manifold M^{2n}.*

Sketch of proof. Put $B = C^*(\pi)$ and let $\widetilde{M}$ be the regular covering of M^{2n} with the group π induced by f. Fix a Riemannian metric on M and lift it also to $\widetilde{M}$. Denote by $\Omega^p(M;\, B)$ the space of smooth forms ω on $\widetilde{M}$ with values in B satisfying $\forall g \in \pi$, $\forall x \in \widetilde{M}$, for any tangent vectors $v_1, \ldots, v_p$ at the point x the condition

$$\omega(gx)(gv_1, \ldots, gv_p) = g \cdot \omega(x)(v_1, \ldots, v_p).$$

The space $\Omega^p_c(\widetilde{M})$ of $\mathbb{C}$-valued compactly supported smooth forms on $\widetilde{M}$ is a subspace in $\Omega^p(M;\, B)$. The inclusion $\sigma : \Omega^p_c(\widetilde{M}) \hookrightarrow \Omega^p(M;\, B)$ is given by

$$\sigma(\omega_c)(x)(v_1, \ldots, v_p) = \sum_{g \in \pi} \omega_c(gx)(gv_1, \ldots, gv_p) \cdot g^{-1}, \tag{1}$$

where g^{-1} is considered as an element of $C^*(\pi)$. Define the operator

$$\eta : \Omega^p(M;\, B) \to \Omega^p(M;\, B) \qquad \text{by} \qquad \eta(\omega) = i^{p(p-1)+n} \cdot \omega$$

and let $* : \Omega^p(M;\, B) \to \Omega^{2n-p}(M;\, B)$ be the Hodge $*$-operator (defined pointwise by the Riemannian metric). The space

$$\Omega^*(M;\, B) = \bigoplus_{p=0}^{2n} \Omega^p(M;\, B),$$

endowed with the right multiplication by elements of B and the B-scalar product defined by

$$(\omega_1, \omega_2) = \int_{M^{2n}} *(\omega_1^*) \wedge \omega_2,$$

where

$$\omega_1^*(x)(v_1, \ldots, v_p) = (\omega_1(x)(v_1, \ldots, v_p))^*,$$

becomes a pre-Hilbert B-module. The grading on $\Omega^*(M; B)$ is introduced by the grading operator $\varepsilon = * \cdot \eta$. Completing $\Omega^*(M; B)$ we get a Hilbert B-module E.

Operators d, $\delta = -\varepsilon d \varepsilon$ and $\Delta = d\delta + \delta d$ on $\Omega^*(M; B)$ are defined as usual. Moreover, there is a *norm bounded* operator $(1+\Delta)^{-1}$ on $\Omega^*(M; B)$. To define this operator we notice that the equation $(1+\Delta)\omega_1 = \omega_2$ can be solved in the space of L^2-forms on $\widetilde{M}$. By the regularity theorem for elliptic equations, ω_1 is smooth whenever ω_2 is smooth. If $\omega_2 \in \Omega_c^*(\widetilde{M})$ then ω_1 defines an element of $\Omega^*(M; B)$ by the same formula (1). (We omit the proof.[31]) Now the operator $\frac{d+\delta}{\sqrt{1+\Delta}}$ can be defined on E, and one has:

$$\beta(f_*([d+\delta])) = \left(E, \frac{d+\delta}{\sqrt{1+\Delta}}\right) \in K_0(B).$$

We will apply an analogue of the algebraic surgery technique (cf. [20]) to the Poincaré complex $(\Omega^*(M; B), d)$ in order to reduce the element $\left(E, \frac{d+\delta}{\sqrt{1+\Delta}}\right)$. To make the first surgery we want to find elements $x_1, \ldots, x_m \in \Omega^0(M; B)$ such that

$$\Delta + \sum_{i=1}^{m} \theta_{x_i, x_i} \geq \text{const} > 0 \quad \text{on } \Omega^0(M; B).$$

(Recall from [15] that $\theta_{x,y}(z) \stackrel{\text{def}}{=} x \cdot (y, z)$.) Put $k = (1+\Delta)^{-1}$. Since $\Delta \geq 1-k$, it is enough to have the above inequality with $1-k$ instead of Δ. Denote the completion of $\Omega^0(M; B)$ by E^0. Since $k \in \mathcal{K}(E^0)$ is positive and $\Omega^0(M; B)$ is dense in E^0, one can find elements $x_1, \ldots, x_m \in \Omega^0(M; B)$ such that $k - \sum_{i=1}^m \theta_{x_i, x_i} \leq \frac{1}{2}$. Then $(1-k) + \sum_{i=1}^m \theta_{x_i, x_i} \geq \frac{1}{2}$; therefore

$$\Delta + \sum_{i=1}^{m} \theta_{x_i, x_i} \geq \frac{1}{2}.$$

[31] See Theorem 2 in my article: *An index for invariant elliptic operators, K-theory, and representations of Lie groups,* Dokl. Akad. Nauk SSSR **268** (1983), 533–537; English translation: Soviet Math. Dokl. **27** (1983), 105–109. In this article also the calculus is described for group invariant pseudodifferential operators, which allows to prove that the operator $(1+\Delta)^{-1}$ belongs to $\mathcal{K}(E)$.

Now consider the new complex $\widetilde{\Omega}^*(M;\, B)$ with all $\widetilde{\Omega}^p$ the same as before except for $\widetilde{\Omega}^1 = \Omega^1 \oplus B^m$ and $\widetilde{\Omega}^{2n-1} = \Omega^{2n-1} \oplus B^m$. (In the case $n = 1$ we put $\widetilde{\Omega}^1 = \Omega^1 \oplus B^m \oplus B^m$.) The grading operator ε extends to $\widetilde{\varepsilon} : \widetilde{\Omega}^* \to \widetilde{\Omega}^*$, which identifies the first direct summand B^m with the second one. The operators $\widetilde{d} : \widetilde{\Omega}^0 \to \widetilde{\Omega}^1$ and $\widetilde{\delta} : \widetilde{\Omega}^1 \to \widetilde{\Omega}^0$ are defined by the formulas:

$$\widetilde{d}(y) = (d(y),\, (x_1,\, y),\, \ldots,\, (x_m,\, y))\,,$$

$$\widetilde{\delta}(y,\, b_1,\, \ldots,\, b_m) = \delta(y) + \sum_{i=1}^{m} x_i b_i.$$

Put $\widetilde{d} = d \oplus 0 : \widetilde{\Omega}^1 \to \widetilde{\Omega}^2$, $\widetilde{\delta} = \delta \oplus 0 : \widetilde{\Omega}^2 \to \widetilde{\Omega}^1$. Define $\widetilde{d}$ and $\widetilde{\delta}$ on $\widetilde{\Omega}^{2n-1}$ by applying $\widetilde{\varepsilon}$. Clearly, the equality $\widetilde{d}^2 = 0$ remains valid. Moreover, the element $\left(E,\, \frac{d+\delta}{\sqrt{1+\Delta}}\right) \in K_0(B)$ does not change. Indeed, there is a homotopy of $\widetilde{d} \to d \oplus 0$ and $\widetilde{\delta} \to \delta \oplus 0$ which comes from the homotopy of the elements x_i to 0.

In the complex $\widetilde{\Omega}^*$ the operator $\widetilde{\Delta}$ on $\widetilde{\Omega}^0$ is equal to $\Delta + \sum_{i=1}^{m} \theta_{x_i,\, x_i} \geq \frac{1}{2}$. Now we can change our complex once more into $\widetilde{\widetilde{\Omega}}^*(M;\, B)$ having $\widetilde{\widetilde{\Omega}}^0 = 0$ and $\widetilde{\widetilde{\Omega}}^{2n} = 0$ and

$$\widetilde{\widetilde{\Omega}}^1 = \ker\left(\widetilde{\delta}|_{\widetilde{\Omega}^1}\right), \qquad \widetilde{\widetilde{\Omega}}^{2n-1} = \ker\left(\widetilde{d}|_{\widetilde{\Omega}^{2n-1}}\right).$$

The remaining $\widetilde{\widetilde{\Omega}}^p$ are equal to $\widetilde{\Omega}^p$. (In the case $n = 1$ we put

$$\widetilde{\widetilde{\Omega}}^1 = \ker\left(\widetilde{\delta}|_{\widetilde{\Omega}^1}\right) \cap \ker\left(\widetilde{d}|_{\widetilde{\Omega}^1}\right).)$$

From $\widetilde{\Delta} \geq \frac{1}{2}$ on $\widetilde{\Omega}^0$ it follows that the new element

$$\left(\widetilde{\widetilde{E}},\, \frac{\widetilde{\widetilde{d}} + \widetilde{\widetilde{\delta}}}{\sqrt{1 + \widetilde{\widetilde{\Delta}}}}\right) \in K_0(B)$$

coincides with the former one.

One application of the described surgery reduces the length of $\Omega^*(M;\, B)$ by 2. As a result of n surgeries we arrive at a complex having only the middle B-module $\bar{\Omega}^n$ non-zero. Clearly, in this case $d = \delta = 0$. Completing $\bar{\Omega}^n$, we get a Hilbert B-module F for which the zero operator $\frac{\Delta}{1+\Delta}$ belongs to

$1-\mathcal{K}(F)$, i.e., $1 \in \mathcal{K}(F)$. Considering the stabilization $F \oplus \mathcal{H}_B$ and applying Lemma 5, §6, [15], we conclude that F is finitely generated and projective. In particular, it follows that $F = \bar{\Omega}^n$. With the grading $F = F^{(0)} \oplus F^{(1)}$ defined by ε we have:

$$\beta(f_*([d+\delta])) = \left[F^{(0)}\right] - \left[F^{(1)}\right] \in K_0(C^*(\pi)).$$

Now we can give another realization of the same element. Consider a triangulation of M^{2n}. Integration of differential forms over simplices gives a homomorphism of the complex $(\Omega^*(M;\, B),\, d)$ into the complex of simplicial cochains

$$C^0(M;\, B) \xrightarrow{\partial} \dots \xrightarrow{\partial} C^{2n}(M;\, B)$$

of the manifold M^{2n} with values in the local system of coefficients B. This homomorphism is an isomorphism on homology. Moreover, the hermitian from on $\Omega^*(M;\, B)$ given by

$$\varphi(\omega_1,\, \omega_2) = \int_{M^{2n}} \tilde{\eta}(\omega_1^*) \wedge \omega_2 = (\varepsilon(\omega_1),\, \omega_2),$$

with $\tilde{\eta}(\omega) = i^{\deg\omega(\deg\omega+1)-n}\omega$ defines on the homology of $\Omega^*(M;\, B)$ an intersection index corresponding to the usual intersection index $\tilde{\varphi}$ on $H^*(M;\, B)$ (which is defined by the product of simplicial cochains with the same correction factor $\tilde{\eta} = i^{p(p+1)-n}$ in the dimension p).

Parallel to the sequence of surgeries that we have applied to $\Omega^*(M;\, B)$, a sequence of purely algebraic surgeries of $C^*(M;\, B)$ (cf. [20]) can be constructed. On every step of this sequence there will be a homomorphism $\Omega^*(M;\, B) \to C^*(M;\, B)$ inducing an isomorphism on homology. After n such surgeries only the middle homology group of $C^*(M;\, B)$ will be non-zero. It will be a finitely generated projective B-module $\widetilde{F}$ with a non-degenerate Hermitian form $\tilde{\varphi}$ given by the intersection index. Let $\widetilde{F}_+$ and $\widetilde{F}_-$ be the positive and negative (spectral) submodules of $\widetilde{F}$ defined by $\tilde{\varphi}$. Clearly

$$\beta(f_*([d+\delta])) = \left[\widetilde{F}_+\right] - \left[\widetilde{F}_-\right] \in K_0(C^*(\pi)).$$

It follows from [20] that this last element is a homotopy invariant of M^{2n}. □

Combining Corollary 1 and Theorem 2 we get:

Theorem 3. *Assume that π is a discrete subgroup of a connected Lie group. Then all higher signatures of smooth manifolds with the group π are homotopy invariant.* □

REFERENCES

0. G. G. Kasparov, *Equivariant KK-theory and the Novikov conjecture*, Invent. Math. **91** (1988), 147–201.
1. M. F. Atiyah and G. B. Segal, *Equivariant K-theory and completion*, J. Diff. Geometry **3** (1969), 1–18.
2. M. F. Atiyah and I. M. Singer, *The index of elliptic operators, III*, Ann. of Math. **87** (1968), 546–604.
3. N. Bourbaki, *Intégration*, Chap. VII–VIII, Hermann, Paris, 1963.
4. I. D. Brown, *Dual topology of a nilpotent Lie group*, Ann. Sci. École Norm. Sup. **6** (1973), 407–411.
5. A. Connes, *Classification of injective factors*, Ann. of Math. **104** (1976), 73–115.
6. J. Dixmier, *Les C^*-algèbres et leurs représentations*, Gauthier-Villars, Paris, 1969.
7. J. Dixmier, *Sur la représentation régulière d'un groupe localement compact connexe*, Ann. Sci. École Norm. Sup. **2** (1969), 423–436.
8. P. Green, *The structure of imprimitivity algebras*, J. Funct. Anal. **36** (1980), 88–104.
9. S. Helgason, *Differential Geometry and Symmetric Spaces*, Academic Press, New York and London, 1962.
10. L. Hörmander, *On the index of pseudo-differential operators*, Elliptische Differentialgleichungen, Band II, Akademie-Verlag, Berlin, 1971, pp. 127–146.
11. G. G. Kasparov, *A generalized index for elliptic operators*, Funk. Anal. i Prilozhen. **7** no. 3 (1973), 82–83; English translation, Funct. Anal. Appl. **7** no. 3 (1973), 238–240.
12. G. G. Kasparov, *Topological invariants of elliptic operators, I: K-homology*, Izv. Akad. Nauk SSSR, Ser. Mat. **39** (1975), 796–838; English translation, Math. USSR–Izv. **9** (1975), 751–792.
13. G. G. Kasparov, *The K-functor in the theory of extensions of C^*-algebras*, Funk. Anal. i Prilozhen. **13** no. 4 (1979), 73–74; English translation, Funct. Anal. and Appl. **13** (1979), 296–297.
14. G. G. Kasparov, *Hilbert C^*-modules: theorems of Stinespring and Voiculescu*, J. Operator Theory **4** (1980), 133–150.
15. G. G. Kasparov, *The operator K-functor and extensions of C^*-algebras*, Izv. Akad. Nauk SSSR, Ser. Mat. **44** (1980), 571–636; English translation, Math. USSR–Izv. **16** (1981), 513–572.
16. A. A. Kirillov, *Unitary representations of nilpotent Lie groups*, Uspekhi Mat. Nauk **17** no. 4 (1962), 57–110; English translation, Russian Math. Surv. **17** no. 4 (1962), 53–104.
17. A. A. Kirillov, *Elements of the Theory of Representations*, Nauka, Moscow, 1972; English translation, Springer-Verlag, Berlin, Heidelberg, and New York.
18. S. MacLane, *Homology*, Springer-Verlag, Berlin, Heidelberg, and New York, 1963.
19. J. Milnor, *On axiomatic homology theory*, Pacific J. Math. **12** (1962), 337–341.
20. A. S. Mishchenko, *Homotopy invariants of nonsimply connected manifolds, I: Rational invariants*, Izv. Akad. Nauk SSSR Ser. Mat. **34** (1970), 501–514; English translation, Math. U.S.S.R.-Izv. **4** (1970), 506–519.

21. A. S. Mishchenko, *Infinite dimensional representations of discrete groups, and homotopy invariants of nonsimply connected manifolds*, Uspekhi Mat. Nauk **28** no. 2 (1973), 239–240.

22. A. S. Mishchenko, *Infinite dimensional representations of discrete groups, and higher signatures*, Izv. Akad. Nauk SSSR Ser. Mat. **38** (1974), 81–106; English translation, Math. U.S.S.R.–Izv. **8** (1974), 85–111.

23. S. P. Novikov, *Algebraic construction and properties of Hermitian analogs of K-theory over rings with involution from the viewpoint of the Hamiltonian formalism. Applications to differential topology and the theory of characteristic classes*, Izv. Akad. Nauk SSSR, Ser. Mat. **34** (1970), 253–288, 475–500; English translation, Math. USSR–Izv. **4** (1970), 257–292, 479–505.

24. M. Pimsner and D. Voiculescu, *Exact sequences for K-groups and* Ext*-groups of certain cross-product C^*-algebras*, J. Operator Theory **4** (1980), 93–118.

25. M. A. Rieffel, *Strong Morita equivalence of certain transformation group C^*-algebras*, Math. Ann. **222** (1976), 7–22.

26. M. A. Rieffel, *Applications of strong Morita equivalence to transformation group C^*-algebras*, Operator algebras and applications (Kingston, Ont., 1980), Part I (R. V. Kadison, ed.), Proc. Symp. Pure Math., vol. 38, Amer. Math. Soc., Providence, R.I., 1984, pp. 299–310.

27. J. Rosenberg, *Group C^*-algebras and topological invariants*, Operator algebras and group representations, Vol. II (Neptun, 1980), Monographs Stud. Math., vol. 18, Pitman, Boston and London, 1984, pp. 95–115.

28. C. Schochet, *Topological methods for C^*-algebras, I: spectral sequences*, Pacific J. Math. **96** (1981), 193–211.

29. G. Segal, *Fredholm complexes*, Quart. J. Math. **21** (1970), 385–402.

30. Théorie des algèbres de Lie, Topologie des groupes de Lie, *Séminaire "Sophus Lie"*, Paris, 1955.

Département Mathématique–Informatique, Université de Marseille–Luminy, 13288 Marseille Cedex 9, France

email: `kasparov@lumimath.univ-mrs.fr`

A coarse approach to the Novikov Conjecture

Steven C. Ferry and Shmuel Weinberger

ABSTRACT. This is an expository paper explaining coarse analogues of the Novikov Conjecture and describing how information on the original Novikov Conjecture can be derived from these. For instance, we will explain how Novikov's theorem on the topological invariance of rational Pontrjagin classes is a consequence of a coarse theorem (whose proof we sketch in an appendix) that in turn also implies the Novikov Conjecture for nonpositively curved manifolds. We also formalize the technique so that it can be applied in a wide variety of other contexts. Thus, besides a few purely geometric results, we also discuss equivariant, A-theoretic, stratified, and foliated versions of the higher signature problem. Closely related papers are [GL], [CGM], [CP], [KaS], [HR], [Hu]. See also the surveys [Wel], [FRW], for wider perspectives.

1. Background

Recall that the *signature* of an oriented $4k$-dimensional manifold is the difference of the dimensions of the maximal positive and negative definite subspaces of the inner product space given by cup product on the middle-dimensional cohomology. Hirzebruch's signature formula describes the signature of M as a universal polynomial L in the Pontrjagin classes of M:

$$sign(M) = \langle L(M), [M] \rangle \in \mathbb{Q}.$$

The L-polynomial is a graded polynomial with pieces in every fourth dimension, but only the top-dimensional piece is given a homotopy-theoretic interpretation by Hirzebruch's formula. For simply-connected manifolds, this top component is the only homotopy invariant piece of $L(M)$. Novikov's conjecture says that for nonsimply connected manifolds the largest conceivably homotopy invariant piece of $L(M)$ is in fact homotopy invariant.

Ferry was partially supported by NSF Grants DMS 9003746 and DMS 9305758, and Weinberger was partially supported by an NSF Grant and a Presidential Young Investigator Award.

More precisely, let $B\pi$ be the universal space for principal π-bundles, where $\pi = \pi_1(M)$, that is, a space with fundamental group π and contractible universal cover. The universal cover of M is a principal π-bundle, so classification gives a well-defined homotopy class of maps $\tau : M \to B\pi$.

For each $\alpha \in H^*(B\pi; \mathbb{Q})$, one can form the quantity:

$$sign_\alpha(M) = \langle \tau^*\alpha \cup L(M), [M] \rangle.$$

Novikov's conjecture is the statement that this element is a homotopy invariant of M. In the simply connected case, this is a consequence of Hirzebruch's formula.

It is sometimes more convenient to deal with an equivalent dual form of the conjecture, which is that the *higher signature*

$$\tau_*(L(M) \cap [M]) \in H_*(B\pi; \mathbb{Q})$$

is a homotopy invariant. This version of the conjecture allows us to work with all of the cohomology of $B\pi$ simultaneously, rather than proceeding one element at a time.

We remark that the rational version of the conjecture is equivalent in its smooth, PL, and topological versions. The reader can work exclusively with smooth manifolds except for one (of two) arguments given for the integral version in the final section. By that point the reader will have seen a proof of the topological invariance of rational Pontrjagin classes and should be willing to believe that there is a knowable theory of topological manifolds!

For topologists, the significance of the conjecture is rooted in surgery theory, which classifies manifolds within a fixed homotopy type [B], [W]. The first three chapters of [We1] contain a survey of surgery theory which is more than adequate for the purposes of this paper.

We sketch two arguments showing that the higher signatures are the only possible homotopy invariant characteristic classes. This result is due to Peter Kahn [Ka] in the simply connected case and to Mishchenko in general. Our first argument is bare-handed and conceptual, while the second is quite simple but assumes a certain amount of machinery.

The first ingredient is Wall's π-π theorem, which shows that surgery obstructions are invariant under bordism of the domain and target spaces over the fundamental group. Here is a statement:

Theorem (Wall [W]). *Let $f : (M, \partial_1 M, \partial_2 M) \to (X, \partial_1 X, \partial_2 X)$ be a surgery problem with $f| : \partial_2 M \to \partial_2 X$ a homotopy equivalence. If*

$\pi_1(\partial_1 X) \to \pi_1(X)$ is an isomorphism, then f can be surgered rel ∂_2 to a homotopy equivalence of triples.

Thus, if $f : M \to X$ is a surgery problem and $F : (W, M, M') \to (Z, X, X')$ is a surgery problem with $F|M = f$ and $F| : M' \to X'$ a homotopy equivalence, then f can be surgered to a homotopy equivalence. In this formulation, the π-π theorem can be viewed as a statement of cobordism invariance of intersection numbers.

The product theorem says that if one crosses a surgery problem with a simply connected manifold of dimension $4k$, the surgery obstruction is multiplied by the signature. This implies that {signature 0 things} $\times$ {arbitrary bordism classes} arise as differences of bordism classes of homotopy equivalent manifolds. According to [CoF], bordism of manifolds with maps to X is determined rationally by associating to the pair (M, f) the element $f_*(L(M) \cap [M])$ of the graded rational homology of X. Furthermore, all elements of rational cohomology arise. On modding out by the elements of the form $V \times (N, g)$ with $sign(V) = 0$, all that one has left is the image of the top piece of the L-class. As a consequence of this and some elementary bordism arguments, any element of the kernel of $H_*(X) \to H_*(B\pi)$ occurs as the difference of pushed forward homology L-classes for manifolds with maps to X, so the only conceivable invariant of cobordism and homotopy invariance is the homology higher signature.

Here is the second argument: Consider the surgery exact sequence

$$\cdots \to \mathcal{S}(M) \to [M; G/TOP] \to L_n(\mathbb{Z}\pi),$$

where $\pi = \pi_1 M$ and $\mathcal{S}(M)$ is the *structure set* consisting of homotopy equivalences $f : M' \to M$ modulo the equivalence relation in which $f_1 : M_1 \to M$ is equivalent to $f_2 : M_2 \to M$ if there is a homeomorphism $h : M_1 \to M_2$ such that $f_2 \circ h$ is homotopic to f_1. The topological case is somewhat easier than the smooth case, since in that case $\mathcal{S}(M)$ and $[M; G/TOP]$ are groups and the maps are homomorphisms. Otherwise, we have only an exact sequence of sets. Now, $[M; G/CAT] \otimes \mathbb{Q} \cong \oplus H^{4i}(M; \mathbb{Q})$. The map $\mathcal{S}(M) \to H^{4i}(M; \mathbb{Q})$ sends a structure to the difference of the L-classes of domain and range. The homomorphism $A : H^{4i}(M; \mathbb{Q}) \to L_n(\mathbb{Z}\pi) \otimes \mathbb{Q}$ factors through a universal homomorphism $\oplus H_{n-4i}(B\pi, \mathbb{Q}) \to L_n(\mathbb{Z}\pi) \otimes \mathbb{Q}$ by Poincaré dualizing and then pushing forward. Again the upshot is the same – the variation of characteristic classes is subject only to restrictions determined by the image in group homology. Novikov's conjecture says that A is a monomorphism and that the higher signature is homotopy invariant. The PL version of this argument works similarly, but the smooth case requires some more effort. See [We3] for details in the smooth case.

How does one prove the conjecture? Most recent proofs use a reformulation of the problem due to Mishchenko (or an analytic variant of this). The basic homotopy invariant of an n-manifold is the equivariant cellular chain complex of the universal cover of the manifold, together with its Poincaré duality structure. This is an element of a suitable Grothendieck group, $L^n(\mathbb{Z}\pi)$, and is called the *symmetric signature* of M [Ra1]. Away from the prime 2, this $L^n(\mathbb{Z}\pi)$ is the same as the group $L_n(\mathbb{Z}\pi)$ that occurs in the surgery exact sequence. The assembly map $H_*(B\pi;\mathbb{Q}) \to L^n(\mathbb{Z}\pi)\otimes\mathbb{Q}$ sends the higher signature $\tau_*(L(M) \cap [M]) \in H_*(B\pi;\mathbb{Q})$ to the symmetric signature of M. The image in $L^n(\mathbb{Z}\pi)$ of the surgery obstruction of a normal map is the difference of the symmetric signatures of the domain and range. As a result, surgery theory [Ra2] asserts that the kernel of the composition of this map with the natural map $H_*(M;\mathbb{Q}) \to H_*(B\pi;\mathbb{Q})$ represents the differences of L-classes of manifolds homotopy equivalent to M. The cokernel of the map, with a dimension shift, is related to the manifolds homotopy equivalent to M which have the same characteristic classes as M.

This strategy of proof can be integralized; that is, we might try to show that in special cases the map A is integrally injective. The main interesting point is that the characteristic classes live most naturally in a generalized homology theory (the domain of an improved A) which is ordinary homology at 2 and KO theory away from 2. The spectrum representing this theory is written $\mathbb{L}(e)$. This is the surgery spectrum of the trivial group. It is an Ω spectrum with 0th space $\mathbb{Z} \times G/TOP$. The map to KO theory is given by considering the KO-homology class given by the signature operator. The integral Novikov Conjecture inverting 2 then is that this signature operator class is homotopy invariant in $KO(B\pi)[\frac{1}{2}]$. At 2, it is a homotopy invariance statement involving the Morgan-Sullivan class [MoS] (see [RW2]). We will see in the last section that for certain torsion-free π this is a homotopy invariant.

2. The bounded category

The category of manifolds bounded over a metric space X is defined as follows: Objects are manifolds M with not-necessarily continuous maps $p : M \to X$ such that closures of inverse images of balls are compact. A morphism $(M_1, p_1) \to (M_2, p_2)$ is a continuous map $f : M_1 \to M_2$ which is bounded over X in the sense that there is a $k > 0$ for which $d(p_1(m), p_2 \circ f_1(m)) < k$ for all $m \in M_1$. In particular, we take as understood that one can form the (bounded) homotopy category over X where all maps and homotopies are bounded over X.

For $X = pt$, this is the usual category of compact manifolds and continuous maps. We shall see that there are other interesting examples. An example to keep in mind is that a homotopy equivalence between finite polyhedra lifts to a bounded homotopy equivalence of universal covers (over either universal cover) where the covers are given path metrics so that each 1-simplex has length 1. It is possible to redo topology (or differential geometry and analysis) in this setting using appropriate metrics. This is not entirely straightforward, but most of what one needs has been established for suitably nice control spaces X.

Definition 2.1. We will say that X is *uniformly contractible* if there is a function $f : (0, \infty) \to (0, \infty)$ so that for each $x \in X$ and $t > 0$, the ball of radius t centered at x contracts to a point in the concentric ball of radius $f(t)$.

In this situation, Ferry and Pedersen [FP] have shown that the Product Theorem and the π-π Theorem – the two basic principles of surgery theory discussed in §1 – still hold. Note that some care is needed in formulating the bounded π-π condition. Nonetheless, not much will be lost in this section by assuming that all manifolds are simply connected over X. In light of this, we formulate a bounded Novikov conjecture.

BOUNDED NOVIKOV CONJECTURE. If X is uniformly contractible and M is a manifold over X, then $f_*(L(M) \cap [M]) \in H_*^{\ell f}(X; \mathbb{Q})$ is a bounded homotopy invariant, where $H_*^{\ell f}(X; \mathbb{Q})$ is locally finite homology.

The same reasoning as above shows that this is the most general bounded-homotopy invariant piece of the L polynomial in the simply-connected case. In general, one would need to take into account the fundamental group system over X.

This conjecture is true in many cases. If $X = cP$, the open cone on a finite polyhedron, this is verified in [FP]. If X has a nice compactification, this is verified in [FW2]. These proofs are extensions of the original splitting proof of the Novikov Conjecture for $\mathbb{Z}^n$ given by Farrell-Hsiang in [FH1] or via the formula of Shaneson in [Sh].[1] See also [Ca], where Cappell gave an extension of the splitting method to prove the Novikov Conjecture for a wide class of groups. Of course, many of the other proofs of Novikov for particular groups apply to particular metric spaces. Much of the book [We2] is devoted to applications of these ideas.

Another approach which is often quite useful (see [Car], [CP], [FW2],

[1] which is also proved by splitting.

[Hu]) is to compactify the metric space X (i.e. embed it into a compact space Z) in such a way that sequences that are a bounded distance apart in X have the same limit points in $Z - X$. Then one can embed the bounded category of X into the continuously controlled theory of Z: one simply adds on to all objects the ideal set and insists that the map so given be continuous! This theory is simpler in many ways. See [ACFP] for the case of K-theory. If Z is contractible, then the assembly map maps isomorphically to the continuously controlled theory and one has split the assembly map.

Remark 2.2. The correct context for defining assembly maps in the bounded setting is exotic homology with coefficients in the relevant spectra. Often, but not always, this coincides with the locally finite homology of X with coefficients in the spectrum and is conceivably always isomorphic to the bounded K- or L- theory. For related material, see [BW], [DFW], and the Hurder and Higson-Roe papers in these proceedings. For more information on the Novikov and Borel conjectures, see the discussion in [We1].

3. Topological invariance of rational Pontrjagin classes

We show that the bounded Novikov Conjecture over $X = \mathbb{R}^n$ implies Novikov's celebrated theorem on topological invariance of rational Pontrjagin classes. In [We4] this approach is applied in an analytic context. (See also [PRW]).

We will use the definition of the L-class and, equivalently, the rational Pontrjagin classes, given by Thom and Milnor. To evaluate the L-class on a homology class, dualize the homology class and represent it, if possible, by an embedded submanifold with trivial normal bundle. The value of the L-class on the given homology class is the signature of the submanifold. Replacing M by $M \times T$ for some torus T guarantees that there will be enough submanifolds with trivial normal bundle to generate the cohomology, so the collection of signatures determines the L-class and therefore the rational Pontrjagin classes. This is explained in §20 of [MS].

The topological invariance of rational Pontrjagin classes says that if $f : M' \to M$ is a topological homeomorphism between smooth manifolds, then $f^*(L(M)) = L(M')$. If $\alpha \in H_*(M'; \mathbb{Q})$ is a homology class, $\langle f^*(L(M)), \alpha\rangle = \langle L(M), f_*(\alpha)\rangle$. If $V \times \mathbb{R}^k \subset M$ is a framed submanifold dual to $f_*(\alpha)$, $\tilde{f}^{-1}(V \times \{0\})$ is a framed submanifold of M' dual to α, where $\tilde{f}$ is a smooth approximation to f transverse to $V \times \{0\}$. Interpreting the L-classes as signatures, the topological invariance of rational Pontrjagin classes follows from the claim below.

CLAIM. If $W' \to W = V \times \mathbb{R}^k$ is a homeomorphism between smooth manifolds, V compact, then the signature of the transverse inverse image of V is that of V. By the "transverse inverse image of V," we mean the manifold obtained by approximating $W' \to W$ by a smooth map which is transverse to $V \times \{0\}$.

Change the metric to make the Euclidean direction genuinely flat and complete, rather than just a small normal ball to V in W. Choosing the smooth approximation $\tilde{f}$ to be a bounded homotopy equivalence in the new metric, the claim follows from the bounded homotopy invariance property in the bounded Novikov conjecture for the case $X = \mathbb{R}^k$, as the homeomorphism is certainly a bounded homotopy equivalence. The bounded homotopy invariance of the codimension n signature is explained in the appendix at the end of this paper.

4. Application of the principle of descent to the Novikov Conjecture

In this section, we show how to use the bounded Novikov Conjecture on the universal cover of a manifold M to deduce the usual Novikov Conjecture on the manifold itself. We first describe the method in a purely geometric form, following [FW1], where we use a reformulation of bounded simply connected surgery to prove that certain homotopy equivalences are tangential. Next, we rewrite the argument more algebraically so that it applies to more fundamental groups, functors besides L-theory, etc.

Let E' be a complete simply connected manifold of nonpositive curvature, and let Γ be a group acting on E' with compact[2] manifold quotient. In the geometric version of our method, we assume that $f : E'/\Gamma \to E/\Gamma$ is a homotopy equivalence of closed aspherical manifolds and try to show that f is tangential. Knowing this, surgical machinery shows that the assembly map is injective and that the Novikov Conjecture is true (integrally!!). In a somewhat disguised form, the strategy goes back to [FH2].

Consider the diagram given via covering space theory:

$$\begin{array}{ccc}
E' \times_\Gamma E' & \xrightarrow{\hat{f} = f \times_\Gamma f} & E \times_\Gamma E \\
\downarrow{\scriptstyle proj_2} & & \downarrow{\scriptstyle proj_2} \\
E'/\Gamma & \xrightarrow{f} & E/\Gamma
\end{array}$$

[2] Compactness is not essential in this argument, as was shown in [FW1]. However, the algebraic reformulation given below becomes more complicated without this assumption.

Note that the vertical arrows are equivalent to the tangent bundle. There is a zero section given by $[e] \to [e, e]$ and $proj_2$ maps a neighborhood of $[e, e]$ homeomorphically to a neighborhood of $[e]$. The restriction of $\hat{f}$ to each fiber is a lift of f to the universal cover, so we have a family of (uniformly) bounded homotopy equivalences of the fibers parameterized over E'/Γ. If we could continuously change these bounded homotopy equivalences into homeomorphisms, we would be done, since the resulting fiber-preserving homeomorphism would exhibit the desired equivalence of tangent bundles.

In light of bounded surgery theory, the statement that parameterized[3] families of uniformly bounded homotopy equivalences can be turned into families of homeomorphisms is equivalent to the assertion that the "bounded assembly map" is an isomorphism. This is essentially proved for compact nonpositively curved manifolds in [FH2], [FW2], [HTW]. In [FW1], we deduce a slightly weaker fact directly from the α-approximation theorem of [CF]. A little extra care, see [FW1], enabled us to use this method to handle noncompact complete manifolds of nonpositive curvature.

We extend the utility of this method with an algebraic formulation. We begin by constructing a diagram:

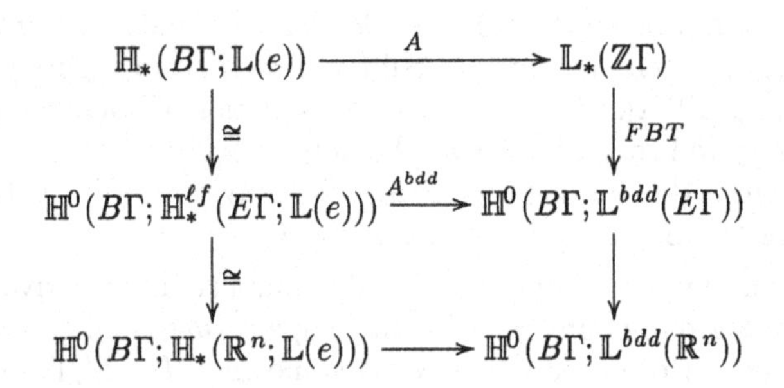

$$\begin{array}{ccc} \mathbb{H}_*(B\Gamma;\mathbb{L}(e)) & \xrightarrow{A} & \mathbb{L}_*(\mathbb{Z}\Gamma) \\ \downarrow \cong & & \downarrow FBT \\ \mathbb{H}^0(B\Gamma;\mathbb{H}_*^{\ell f}(E\Gamma;\mathbb{L}(e))) & \xrightarrow{A^{bdd}} & \mathbb{H}^0(B\Gamma;\mathbb{L}^{bdd}(E\Gamma)) \\ \downarrow \cong & & \downarrow \\ \mathbb{H}^0(B\Gamma;\mathbb{H}_*(\mathbb{R}^n;\mathbb{L}(e))) & \longrightarrow & \mathbb{H}^0(B\Gamma;\mathbb{L}^{bdd}(\mathbb{R}^n)) \end{array}$$

The first line is the assembly map in the form relevant to surgery theory for pure classification problems mentioned at the end of §1. The second is the map on the level of the space of sections of assembly maps associated to the fibration $E \times_\Gamma E \to B\Gamma$. The right side is therefore "twisted generalized cohomology." The bottom line is the same sequence after identifying E with $\mathbb{R}^n$ via the logarithm map. The first set of vertical arrows are "family bounded transfers." Geometrically, a chain or surgery problem with target $B\Gamma$ gives rise, for each point of $B\Gamma$, to the transfer of that chain or problem to the copy of $E = E\Gamma$ based at a $B\Gamma$-lift at that point.

[3]Unlike [FW1], here "parameterized families" will mean "blocked families" which suffice for these problems. The reader should probably just ignore this point. For some of the discussion in §5, though, parameterization must be taken literally.

The fact that the left hand vertical arrows are equivalences when $B\Gamma$ is a finite complex is due to a form of Spanier-Whitehead duality and the fact that $E\Gamma \to \mathbb{R}^n$ is a proper homotopy equivalence.

Putting together the maps

$$\mathbb{L}_*(\mathbb{Z}\Gamma) \to \mathbb{H}^0(B\Gamma; L^{bdd}(E\Gamma)) \to \mathbb{H}^0(B\Gamma; \mathbb{L}^{bdd}(\mathbb{R}^n)) \\ \to \mathbb{H}^0(B\Gamma; \mathbb{H}_*^{\ell f}(\mathbb{R}^n; \mathbb{L}(e))) \to \mathbb{H}_*(B\Gamma; \mathbb{L}(e))$$

we have split the assembly map.

There are several possible extensions of the argument. First, we would not need to compare to euclidean space if we knew that $\mathbb{H}_*(E\Gamma; \mathbb{L}(e)) \to \mathbb{L}^{bdd}(E\Gamma)$ were a homotopy equivalence. Alternatively, it would suffice to know that this map was split injective in a Γ-equivariant fashion. If we had an equivariant compactification, for instance, the continuously controlled theory would provide this. Without the equivariance, this is the Bounded Novikov Conjecture of §2.

Remark 4.1. It is worth noting that the paper [GL] already implicitly contains a version of the descent argument. The index theorem for Dirac operators with coefficients in almost flat bundles is used to prove the analog of the bounded Novikov conjecture for Euclidean space for the positive scalar curvature problem, and then deduces the "Gromov-Lawson conjecture" for fundamental groups of nonpositively curved manifolds by a families (relative) index theorem argument.

In any case, this approach can be carried out for groups which admit suitable compactifications. This includes fundamental groups of complete manifolds of nonpositive curvature [K], groups that act on Tits buildings [Sk], CAT 0 groups, torsion-free word hyperbolic groups [CM], etc. The method applies to many other functors, such as the K- and A- functors. In other words, the assembly map with coefficients in $K(R)$ or $A(X)$ is split injective for an arbitrary ring R or space X for all of the groups just mentioned. The method applies, in addition, to extensions of one such group by another and the results are integral. Considering groups acting properly discontinuously on such metric spaces gives integral results of a more complicated nature for groups which are not necessarily torsion-free (relating the value of the functor for a group to its homology and the placement of the torsion elements within it). This last is closely connected to results announced in [FRW] and comes out of considerations of the next section.

5. Remarks on extensions

The abstract reformulation of the tangentiality argument as descent, while powerful and of wide applicability, does lose some geometric flavor which is useful for other applications. We will follow the train of thought underlying [FRW]. Some, but not all, of the following considerations can be rephrased algebraically as before, but for us they *arise* most naturally geometrically.

TERMINOLOGY. We will say that M is *good* if the tangentiality argument applies to it. In particular, the classes of manifolds mentioned in the last paragraph of §4 are all "good."

OBSERVATION. If M is good and smooth and $M' \to M$ is a homotopy equivalence, then M' is smoothable.

This follows from everything we've said so far. If M is good, the the surgery exact sequence for M splits and therefore M' is normally cobordant to M', i.e., the corresponding structure lies in the image of L_{n+1}. Consequently, if M is smooth, then M' will be as well – compare with the smooth surgery exact sequence and use the fact that the surgery obstruction groups are the same in all categories.[4]

However, there is a more perspicacious and logically simpler (smoothing theory is almost a prerequisite for topological surgery) way to go about proving this observation, and that is by smoothing theory. The main result of smoothing theory is that isotopy classes of smoothings of a manifold correspond in a 1-1 fashion to smooth vector bundle reductions of the tangent microbundle. In particular, there are no such reductions unless the manifold is smoothable. Since we know that the tangent bundle of M' is the pullback of the tangent bundle of the smooth manifold M, we see that M' is smoothable.

This argument is very adaptable. By carefully arguing using families of smooth manifolds, we obtain:

Theorem 5.1. *If M is good enough, then there is a natural splitting of*

$$G(M)/\mathit{Diff}(M) \to G(M)/\mathit{Homeo}(M)$$

which can be taken as close to the identity as you like. Here $G(M)$ denotes the space of homotopy automorphisms of M.

[4] With some interpretations, this statement is false in the equivariant setting.

Corollary 5.2. *(Smooth approximation of homeomorphisms) If M is an irreducible compact locally symmetric space of non-compact type, then there are continuous sections of Diff$(M) \to$ Homeo(M) arbitrarily close to the identity.*

Note the irony of this result. While there are point set theoretically more homeomorphisms, the space *Diff*(M) breaks up as *Homeo*$(M)\times$?. One can show that ? has a very rich rational homotopy theory and typically infinitely generated homotopy groups. As far as we know, *Homeo*(M) is rationally contractible (though this seems unlikely). It does have rich homotopy theory, but this is taking us far afield.

Remark 5.3. The tangentiality method is the only one we know for proving the above approximation theorem. The main difficulty with other approaches is that we have no machinery available for the analysis of *unstable* homeomorphism and diffeomorphism groups.

However, *Diff*(M) and *Homeo*(M) are difficult to visualize. A simpler way to probe the geometry here is to consider M with a G-action. Here there are phenomena which are plainly observed at the level of objects (components) rather than involving the higher homotopy of function spaces.

In the equivariant situation, the h-cobordism theorem takes a different form in the smooth and topological categories. There are many smoothly nontrivial h-cobordisms which are topological products and there are nonsmoothable h-cobordisms. However, there is no essential difficulty (see [SW]) in repeating the tangentiality argument we gave in §3 equivariantly.

Proposition 5.4. *(announced in* [FRW]*) If $G \times M \to M$ is an action of a compact group by isometries on (say) a nonpositively curved closed*[5] *manifold and $\phi : M \to M'$ is an equivariant map which is a homotopy equivalence, then ϕ is topologically tangential.*

Not so relevant for the remaining part of our discussion, but interesting nonetheless, is that the tangential representations at corresponding points of fixed points of M and M' are topologically conjugate. It is not true that they are necessarily differentiably conjugate, even if the actions are smooth. More relevant is the following:

Corollary 5.5. *M' is equivariantly smoothable.*

[5] Of course, noncompact manifolds are also allowed with appropriate conditions at infinity.

For equivariant smoothing theory, see [LR]. Now, using a relative version of the theorem, which is in any case necessary for the infinite volume applications, we obtain:

Corollary 5.6. *Any equivariant h-cobordism on a nonpositively curved G-manifold is smoothable.*

We remark that they are not all necessarily products because of Nil problems. See [SW], [Q]. In other words, the natural map

$$Wh^{diff}(M) \to Wh^{top}(M)$$

is split surjective. The G-action is subsumed in the notation here – M is a G-manifold. Of course, this can be rephrased purely algebraically and this is done in in [FRW]. It is tantalizingly similar to the terms arising in the Baum-Connes Conjecture for the orbifold fundamental group of M.[6] Indeed, it was the K-theoretic reformulation of this result that led us to the proof of the splitting of the A-theory assembly map (via smooth approximation of homeomorphisms) given above. Of course, the modern approach to the A-theory result would be by directly applying descent to A-theory. See [CP]. The nonpositively curved cases here are subsumed in [FJ], but the method here has advantages of generality (substantial functor independence), simplicity, and unstability.

Moreover, the same train of thought also takes one quite close to the L-theory version of the same phenomenon. Topologically equivariantly tangential maps preserve the symbol of the equivariant signature operator, aside from 2-local phenomena. See [MR], [RW1], [RW2], [CSW]. In light of the equivariant surgery theory of [We2], this says that the L-group of the orbifold π_1 must contain $H_*(M/G; \mathbb{L}(\pi_0(G_m)))$, or away from 2 for G finite, $K^G(M)$.

Aside. For G positive dimensional, the topological normal invariant term

$$H_*(M/G; \mathbb{L}(\pi_0(G_m)))$$

is quite different from what one obtains by thinking along more conventional terms $K^G(M)$. For some more discussion, see [We2]. It can be viewed as the simplest explanation of the failures of equivariant Novikov discussed in the introduction to [RW1].

[6] Needless to say, everything we've said is equally transparently correct for proper actions of a group on an appropriate manifold. This point has been emphasized, in a slightly different context, by [BDO].

The geometric view can be pushed further in these directions. One recognizes that orbifolds are just special stratified spaces and one can repeat the whole game for nonpositively curved strata of a stratified space – this works even better when the whole space has such properties. One can apply stratified surgery and/or the generalized triangulation theory of [AH] to analyze what the implications of these results are. The type of result subsumed by this is that, say, the K- or L-theory of an amalgamated free product contains as a direct summand the thing predicted by the Mayer-Vietoris sequence. See [W], [Ca] (and [PP] for the C^*-algebraic version). The results of that announcement which come from examining just the bottom stratum follow from descent, so we will not belabor the point here.

The analysts have taken the lead in applying these ideas to foliations. See [BC], [Hu]. The difficulty for the algebraic versions of descent is again that there is not yet a convenient foliated version of all of these theories, though analytically this has long ago been dealt with. Indeed, the parameterized algebraic versions are still under vigorous investigation. The following is a mild generalization of the pretty interpolation between Novikov's theorem and conjecture in [BC].

Theorem 5.7. *If $\phi : M' \to M$ is a foliated homotopy equivalence where M is foliated by nonpositively curved leaves, then ϕ is topologically foliated tangential.*

This shows the homotopy invariance in this setting of any characteristic classes that are topological characteristic classes of foliations. Unfortunately, at present not so much seems to be known about which classes are of this form.

6. Appendix

As some kind of service, we outline one approach to the bounded Novikov conjecture over $\mathbb{R}^k$ which will complete a sketch of Novikov's theorem. We use the time-honored technique of splitting.

Recall Browder's $M \times \mathbb{R}$ theorem which says that if a manifold N is proper homotopy equivalent to $M \times \mathbb{R}$, M simply connected and high-dimensional, then N is diffeomorphic to $M' \times \mathbb{R}$ for some M'. In case V is simply connected, we use a bounded version of the $M \times \mathbb{R}$ theorem to show that W' is diffeomorphic to $W'' \times \mathbb{R}$ where W'' is bounded homotopy equivalent over

$\mathbb{R}^{k-1}$ to $V \times \mathbb{R}^{k-1}$ and where the diagram

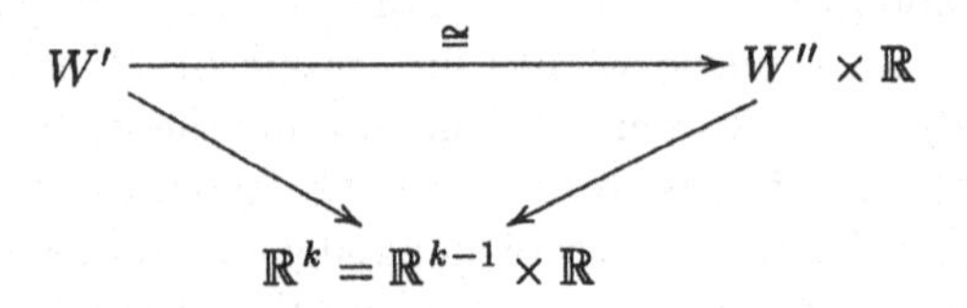

boundedly commutes.

Continuing this process shows that W' is boundedly diffeomorphic to $V' \times \mathbb{R}^k$ where V' is a closed smooth manifold homotopy equivalent to V. Since V' is a transverse inverse image of $V \times \{0\}$ under a bounded approximation to the original map, the result follows from homotopy invariance of the signature.

In case V is not simply connected, we can do smooth surgery on circles in V to make it simply connected. Doing surgery on $S^1 \times \mathbb{R}^k$'s in the domain reduces us to the simply connected case without altering signatures. The reader worried about approximating topologically embedded $S^1 \times \mathbb{R}^k$'s by smooth ones in the domain can cross with $\mathbb{CP}^\ell$ for some large ℓ in both the domain and range to get back into the trivial range.

Remark.

(1) The splitting argument given above is taken from the calculation of the the bounded L-group of $\mathbb{R}^n$ given in [FP], where it is shown that the bounded simply-connected surgery obstruction over $\mathbb{R}^k$ is $\mathbb{Z}$, where the $\mathbb{Z}$ is the codimension k signature. In the presence of bounded surgery theory, this special case implies the bounded Novikov Conjecture over $\mathbb{R}^k$. The bounded splitting theorem needed is a rather special case of Quinn's End Theorem and is used in [FP]. A short proof of the bounded h-cobordism theorem parameterized over $\mathbb{R}^n$ is given in [P]. The knowledgable reader should have little difficult modifying the argument in [P] to prove the required bounded $M \times \mathbb{R}$ theorem.

(2) Chapman [Ch] uses a similar controlled splitting theorem to deduce the topological invariance of Whitehead torsion. The idea is to show that a sufficiently controlled homotopy equivalence between PL manifolds can be split into controlled homotopy equivalences between PL manifolds with fewer handles. The result then follows by induction from the sum theorem for Whitehead torsion.

(3) These bounded/controlled arguments are based on ideas similar to Novikov's and Chapman's original arguments. The bounded/controlled versions are simpler technically than the originals in part because a smooth or PL approximation to a bounded/controlled homotopy

equivalence is still bounded or controlled, while an approximation to a homeomorphism is no longer a homeomorphism. In fact, the α-approximation theorem of [CF] says that every sufficiently controlled homotopy equivalence between high-dimensional manifolds is close to a homeomorphism, so in the end a controlled approximation to a homeomorphism loses no essential information.

REFERENCES

[ACFP] D. R. Anderson, F. X. Connolly, S. C. Ferry, and E. K. Pedersen, *Algebraic K-theory with continuous control at infinity*, J. of Pure and Appl. Alg. (1994), 25–47.

[AH] D. R. Anderson and W.-C. Hsiang, *Extending combinatorial piecewise linear structures on stratified spaces II*, Trans. Amer. Math. Soc. **260** (1980), 223–253.

[BC] P. Baum and A. Connes, *Leafwise homotopy equivalence and rational Pontrjagin classes*, Foliations (Tokyo 1983), Adv. Stud. Pure Math., vol. 5, 1985, pp. 1–14.

[BDO] P. Baum, M. Davis, and C. Ogle, *Novikov conjecture for proper actions of discrete groups (1990 preprint).*

[B] W. Browder, *Surgery on Simply Connected Manifolds*, Springer-Verlag, Berlin-New York, 1970.

[BW] J. Block and S. Weinberger, *Large scale homology theories and geometry*, (in preparation).

[Ca] S. Cappell, *Mayer-Vietoris sequences in Hermitian K-theory*, Algebraic K-Theory, III, Lecture Notes in Math., vol. 343, Springer-Verlag, Berlin-New York, 1972, pp. 478–512.

[CSW] S. Cappell, J. Shaneson, and S. Weinberger, *Classes topologiques caractéristiques pour les actions de groupes sur les espaces singuliers*, C. R. Acad. Sci. Paris Sér. I Math. **313** (1991), 293–295.

[Car] G. Carlsson, *Applications of bounded K-theory*, 1990 Princeton preprint.

[CP] G. Carlsson and E. K. Pedersen, *Controlled algebra and the Novikov conjectures for K- and L- theory*, Topology (to appear).

[Ch] T. A. Chapman, *Controlled Simple Homotopy Theory and Applications*, Lecture Notes in Mathematics, vol. 1009, Springer-Verlag, Berlin-New York.

[CF] T. A. Chapman and S. C. Ferry, *Approximating homotopy equivalences by homeomorphisms*, Amer. J. Math. **101** (1979), 583–607.

[CoF] P. E. Conner and E. E. Floyd, *Differentiable Periodic Maps*, Springer-Verlag, Berlin-New York, 1964.

[CM] A. Connes and H. Moscovici, *Cyclic cohomology, the Novikov conjecture, and hyperbolic groups*, Topology **29** (1990), 345–388.

[CGM] A. Connes, M. Gromov and H. Moscovici, *Conjecture de Novikov et fibrés presque plats*, C. R. Acad. Sci. Paris Sér. I Math. (1990), 273–282.

[DFW] A. N. Dranishnikov, S. Ferry, and S. Weinberger, *Large Riemannian manifolds which are flexible*, (preprint).

[FH1] F. T. Farrell and W.-C. Hsiang, *Manifolds with* $\pi_1 = G \times_\alpha T$, Amer. J. Math. **95**

(1973), 813–848.

[FH2] ______, *On Novikov's Conjecture for nonpositively curved manifolds, I*, Ann. of Math. **113** (1981), 197–209.

[FJ] F. T. Farrell and L. E. Jones, *Rigidity and other topological aspects of compact non-positively curved manifolds*, Bull. Amer. Math. Soc. **22** (1990), 59–64.

[FP] S. C. Ferry and E. K. Pedersen, *Epsilon surgery theory*, these proceedings.

[FRW] S. C. Ferry, J. Rosenberg and S. Weinberger, *Equivariant topological rigidity phenomena*, C. R. Acad. Sci. Paris Sér. I Math. **306** (1988), 777–782.

[FW1] S. C. Ferry and S. Weinberger, *Curvature, tangentiality, and controlled topology*, Invent. Math. **105** (1991), 401 – 414.

[FW2] ______, *in preparation.*

[GL] M. Gromov and H. B. Lawson, *Positive scalar curvature in the presence of a fundamental group*, Ann. of Math. **111** (1980), 209–230.

[HR] N. Higson and J. Roe, *On the coarse Baum-Connes conjecture*, these proceedings.

[HTW] B. Hughes, L. Taylor, and E. B. Williams, *Controlled topology over manifolds of nonpositive curvature*, preprint.

[Hu] S. Hurder, *Exotic index theory and the Novikov conjecture*, these proceedings.

[Ka] P. Kahn, *A note on topological Pontrjagin classes and the Hirzebruch index formula*, Illinois J. Math. **16** (1972), 243–256.

[K] G. Kasparov, *Equivariant KK-theory and the Novikov conjecture*, Invent. Math. **91**, 147–201.

[KaS] G. Kasparov and G. Skandalis, *Groupes "boliques" et conjecture de Novikov*, C. R. Acad. Sci. Paris Sér. I Math. **319** (1994), 815–820.

[LR] R. Lashof and M. Rothenberg, *G-smoothing theory*, Algebraic and Geometric Topology (Stanford, 1976), Proc. Symp. Pure Math., vol. 32, Part 1, Amer. Math. Soc., Providence, RI, 1978, pp. 211–266.

[MR1] I. Madsen and M. Rothenberg, *On the classification of G-spheres I*, Acta. Math. **160** (1988), 65–104.

[MR2] ______, *On the classification of G-spheres II*, Math. Scand. **64** (1983).

[MR3] ______, *On the classification of G-spheres III*, Proc. Northwestern Homotopy Theory Conf., Contemp. Math., vol. 19, Amer. Math. Soc., Providence, RI, 1983, pp. 193–226.

[MS] J. Milnor and J. Stasheff, *Characteristic Classes*, Princeton University Press, Princeton, NJ, 1974.

[MoS] J. Morgan and D. Sullivan, *The transversality characteristic class and linking cycles in surgery theory*, Ann. of Math. **99** (1974), 384–463.

[PP] M. Pimsner, *K-groups of crossed products by groups acting on trees*, Invent. Math. **86** (1986), 603–634.

[P] E. K. Pedersen, *On the bounded and thin h-cobordism theorem parameterized by* $\mathbb{R}^k$, Transformation Groups, Poznan 1985, Lecture Notes in Math., vol. 1217, Springer-Verlag, Berlin-New York, 1986, pp. 306–320.

[PRW] E. K. Pedersen, J. Roe, and S. Weinberger, *On the homotopy invariance of the boundedly controlled signature of a manifold over an open cone*, these proceedings.

[Q] F. Quinn, *Homotopically stratified spaces*, J. Amer. Math. Soc. **1** (1988), 441–499.

[Ra1] A. A. Ranicki, *The total surgery obstruction*, Algebraic Topology, Aarhus 1978, Lecture Notes in Math., vol. 1217, Springer-Verlag, Berlin-New York, 1979, pp. 275–316.

[Ra2] ———, *The algebraic theory of surgery, I, II*, Proc. Lond. Math. Soc. **40** (1980), 87-192, 193–287.

[RW1] J. Rosenberg and S. Weinberger, *The signature operator at 2, (in preparation)*.

[RW2] ———, *Higher G-indices for Lipschitz manifolds*, K-theory **7** (1993), 100–132.

[RW3] ———, *An equivariant Novikov conjecture*, K-theory **4** (1990), 29–53.

[RtW] M. Rothenberg and S. Weinberger, *Group actions and equivariant Lipschitz analysis*, Bull. Amer. Math. Soc. **17** (1987), 109–112.

[Sh] J. Shaneson, *Wall's surgery obstruction groups for $\mathbb{Z} \times G$*, Ann. of Math. **90** (1969), 296–334.

[Sk] G. Skandalis, *Approche de la conjecture de Novikov par la cohomologie cyclique (d'après A. Connes, M. Gromov et H. Moscovici)*, Séminaire Bourbaki, 1990–91, exposé no. 739, Astérisque **201–203** (1992), 299–320.

[St] M. Steinberger, *The equivariant topological s-cobordism theorem,*, Invent. Math. **91** (1988), 61–104.

[SW] M. Steinberger and J. West, *Approximation by equivariant homeomorphisms*, Trans. Amer. Math. Soc. **302** (1987), 297–317.

[Wa] F. Waldhausen, *Algebraic K-theory of amalgamated free products*, Ann of Math **108** (1978), 135-256.

[W] C. T. C. Wall, *Surgery on Compact Manifolds*, Academic Press, New York, 1971.

[We1] S. Weinberger, *Aspects of the Novikov conjecture*, Geometric invariants of elliptic operators, Contemp. Math., vol. 105, Amer. Math. Soc., Providence, RI, 1990, pp. 281–297.

[We2] ———, *The Topological Classification of Stratified Spaces*, U. of Chicago Press, Chicago, 1994.

[We3] ———, *On smooth surgery*, Comm. Pure and Appl. Math. **43** (1990), 695–696.

[We4] ———, *An analytic proof of the topological invariance of rational Pontrjagin classes*, preprint.

Department of Mathematical Sciences, SUNY at Binghamton, Binghamton, NY 13901, USA

email: `steve@math.binghamton.edu`

Department of Mathematics, University of Pennsylvania, Philadelphia, PA 19104, USA

email: `shmuel@archimedes.math.upenn.edu`

Geometric Reflections on the Novikov conjecture

Mikhael Gromov

ABSTRACT. The simplest manifestation of rough geometry in the Novikov conjecture appears when one looks at a fiberwise rough (coarse) equivalence between two vector bundles over a given base, where the fibers are equipped with metrics of negative curvature. Such an equivalence induces a fiberwise homeomorphism of the associated sphere bundles and hence, by Novikov's theorem, an isomorphism of the rational Pontrjagin classes of the bundles.

What is relevant of the negative curvature, as we see it nowadays, is a certain largeness of such spaces. More generally, if X^n is a contractible manifold admitting a cocompact group of isometries (or more general uniformly contractible space) one expects it to be rather large in many cases, e.g. admitting a proper Lipschitz map into $\mathbb{R}^n$ of non-zero degree. An example of an X without such maps to $\mathbb{R}^n$ may eventually lead to a counterexample to the Novikov conjecture. On the other hand, there is a growing list of spaces where such a map is available.

A purely analytico-geometric counterpart of the Novikov conjecture for X is the claim that the non-reduced L^2-cohomology $L^2H^*(X)$ does not vanish. A similar conjecture can be stated for the Dirac operator (instead of the de Rham complex): the square of the Dirac operator on X contains zero in its spectrum. Both properties express the idea of the "spectral largeness" of X, and the latter is closely related to the non-existence of a metric with positive scalar curvature quasi-isometric to X. The non-existence of a positive-scalar-curvature metric on X is yet another version of the Novikov conjecture which is often somewhat easier than the original Novikov conjecture, as one can combine here operator-theoretic techniques with the minimal surface approach of Schoen and Yau.

Most of the work on the Novikov conjecture (see [FRR]) is an outgrowth of the original ideas of Lusztig [Lu] and Mishchenko [M] when they first started working on the problem. In its usual formulation, the Novikov conjecture is a problem about the topology of compact manifolds with large fundamental group, but one is inevitably led to the study of certain aspects

---

Notes by Jonathan Rosenberg of a lecture given at the Oberwolfach meeting on "Novikov conjectures, index theorems and rigidity," 8 September, 1993. An expanded version of these remarks will be found in [Gr3].

of the geometry of the (non-compact) universal cover. There are many different code-words for these ideas: asymptotic geometry, coarse geometry, quasi-isometry, pseudo-isometry, etc. We begin by indicating (through a representative example, similar to that treated in [FH], and not as complicated as the most general ones for which one can verify the conjecture) how these ideas come into play. Suppose $M' \xrightarrow{h} M$ is a homotopy equivalence of compact (necessarily aspherical) hyperbolic manifolds (without boundary). Let Γ be the common fundamental group. Then associated to h is the commutative diagram

$$\begin{array}{ccc} \tilde{M}' \times_\Gamma \tilde{M}' & \xrightarrow{\tilde{h}} & \tilde{M} \times_\Gamma \tilde{M} \\ {\scriptstyle p'}\downarrow & & \downarrow{\scriptstyle p} \\ M' & \xrightarrow{h} & M, \end{array}$$

which can be viewed as a map of bundles where the fibers are copies of hyperbolic space H^n. These bundles are equivalent to the (topological) tangent bundles of M' and of M, respectively. Thus a strong version of the Novikov conjecture is equivalent to the statement that p' and h^*p have the same rational Pontrjagin classes. One way of proving this is to first use the fact that the exponential map $\mathbb{R}^n \to H^n$ is a diffeomorphism, whose inverse $\log : H^n \to \mathbb{R}^n$ is a Lipschitz map to Euclidean space. One can use this to compactify the fibers of the bundles by adding spheres at infinity, and $\tilde{h}$ extends to a fiberwise homeomorphism of the sphere bundles. Now apply to the sphere bundles Novikov's theorem on the topological invariance of rational Pontrjagin classes. We conclude that the sphere bundles of p' and h^*p have the same rational Pontrjagin classes, and thus p' and h^*p have the same rational Pontrjagin classes. (For further discussion of the relationship between Novikov's theorem on the topological invariance of rational Pontrjagin classs and the Novikov conjecture, see [FW].) A similar idea can be used in a combinatorial or discrete setting: if X_1 and X_2 are hyperbolic spaces (in the sense of [GdlH, Ch. 1]) on which a group Γ acts freely and properly with compact quotients, and if $h : X_1/\Gamma \to X_2/\Gamma$ is a homotopy equivalence, then h lifts to a quasi-isometry $\tilde{h} : X_1 \to X_2$ of the universal covers, sending some ε_1-dense net in X_1 to an ε_2-dense net in X_2, and by extending to the boundary (constructed in [GdlH, Ch. 7]), one sees $\tilde{h}$ gives a homeomorphism "at infinity."

In discussions with Blaine Lawson a number of years ago (see [GrL]), we realized there is a close parallel between the Novikov conjecture and the positive scalar curvature problem. We pause now to discuss the latter and then we will return to the subject of the Novikov conjecture. A conjecture parallel to the Novikov conjecture (see [GrL]) is that a closed aspherical manifold $V = X/\Gamma$ should not admit a Riemannian metric of positive scalar curvature. In the Novikov conjecture we (ostensibly) study properties of the group Γ, but here it's really the coarse geometry of the simply connected open manifold X that is immediately relevant. If we can show that the Dirac operator on X is non-invertible, this will imply (for reasons to be explained in a moment) that X cannot have a Riemannian metric of uniformly positive scalar curvature, and thus that V cannot have a Riemannian metric of positive scalar curvature. A similar conjecture is that the non-reduced L^2 cohomology of X must be non-zero. (The *non-reduced L^2 cohomology* $L^2H^*(X)$ is the quotient of the L^2 closed forms by the L^2 exact forms which are exterior derivatives of L^2 forms. The *reduced L^2 cohomology* is defined similarly, but dividing out by the *closure* of the L^2 exact forms which are exterior derivatives of L^2 forms. These are in general different; for example, $X = \mathbb{R}$ has vanishing reduced L^2 cohomology but non-vanishing non-reduced L^2 cohomology in degree 1, since not every function in $L^2(\mathbb{R})$ has an antiderivative in $L^2(\mathbb{R})$.) As pointed out by Lott [Lo, §8 in the preprint version], this conjecture about non-reduced L^2 cohomology can be deduced from a suitable strengthening of the Novikov conjecture.[1] But these conjectures about spectral geometry of X (say for X an open, uniformly contractible Riemannian manifold) are still valid even when there is no group around, or if the group has no non-trivial homology (so that there are no higher signatures to discuss).

The connection between the Dirac operator and positive scalar curvature

[1] Here is a quick version of the argument. Indeed, the non-reduced L^2 cohomology on any complete Riemannian manifold X is non-zero if and only if the signature operator $d + d^*$ on X has 0 in its spectrum (for its action on L^2 forms with respect to the Riemannian metric). (The "only if" direction is clear, and the "if" direction follows from the fact that if 0 is in the spectrum, then either 0 is an eigenvalue, in which case even the reduced L^2 cohomology is non-zero, or else $d+d^*$ does not have a gap in its spectrum near zero, and the non-reduced L^2 cohomology is non-Hausdorff and thus infinite dimensional.)

Now suppose that X is the universal cover of a compact aspherical manifold $V = X/\Gamma$ and that the "Strong Novikov Conjecture" (SNC) holds for Γ in the sense of [R]. For simplicity assume V is oriented. Arguing as in [R], but with the signature operator in place of the Dirac operator, we see that if 0 is not in the spectrum of the signature operator on X, then its generalized index in $K_*(C^*_r(\Gamma))$ must vanish. But then the SNC gives as in [R, §2]:

$$\langle \mathbf{L}(V) \cup id^*(x),\, [V]\rangle = 0$$

for any cohomology class x on $B\Gamma = V$. Taking $x \neq 0$ in the top-degree cohomology, we get a contradiction. The orientability condition can be removed as in [R, §3].

comes from the simple formula

$$\text{(Lichn)} \qquad \mathcal{D}^2 = \Delta + \frac{1}{4}S,$$

first discovered by Lichnerowicz [Li] (see also [LaM, Theorem 8.8]), valid on any Riemannian spin manifold. Here $\mathcal{D}$ is the Dirac operator acting on the spinor bundle, S is the scalar curvature, and Δ is a non-negative elliptic operator that in local coordinates looks like the Laplacian. Thus if S is bounded below by a positive constant $s > 0$, $\mathcal{D}^2 \geq \frac{s}{4} > 0$, and so $\mathcal{D}$ has a bounded inverse. If S is strictly positive but not uniformly so, then at least we immediately see from (Lichn) that $\mathcal{D}$ cannot have a non-zero kernel in the L^2 spinors.

There are basically two known approaches to proving non-existence of positive scalar curvature metrics: the *Dirac operator method* and the *minimal hypersurface method*. (For a complete survey of these and of some of the results one can obtain from them, see [RS].) The minimal hypersurface method of [SY] is in some sense parallel to the codimension-one splitting methods for attacking the Novikov conjecture (see notably [C]), whereas the Dirac operator method parallels the work on the Novikov conjecture using higher index theory of the signature operator. The idea of the minimal hypersurface method is based on the fact that one can show, using the stability condition, that a stable minimal hypersurface in a manifold of positive scalar curvature can be given a new metric (conformally equivalent to the induced metric) in which it has positive scalar curvature. Thus sometimes a manifold with suitable codimension-one submanifolds which do not admit metrics of positive scalar curvature cannot admit such a metric, either. This method has the advantage that it applies even in the absence of a spin structure, but big technical problems arise from the fact that minimal currents in high-dimensional manifolds usually have singularities, so that when one tries to represent a homology class in codimension one by a stable minimal current, one often does not get a smooth submanifold. It is for this reason that many results obtained from the minimal hypersurface method are stated in the literature only for manifolds of dimension ≤ 7. While Schoen and Yau ([S], [Y]) have announced that they can get around the dimensional restriction in the method, no details have appeared yet.

Let us return again to the operator-theoretic approach. To show that a spin manifold cannot have a Riemannian metric of positive scalar curvature, we need to show that the Dirac operator is non-invertible, and to do this one usually needs to perturb the operator a bit by adding a suitable "vector potential." How to make this idea precise is suggested by a theorem of

Vafa and Witten [VW]. We briefly review their result and then discuss how similar techniques can be applied to *non-compact* manifolds.

Theorem (Vafa and Witten [VW]**).** *Let V be a compact even-dimensional Riemannian spin manifold (without boundary), let S be the spinor bundle on V (with its usual connection coming from the Riemannian metric), and let $\mathcal{D}$ denote the Dirac operator of V (acting on sections of S). Then there is a constant $\varepsilon(V)$, depending only on V and its Riemannian metric, with the property that if E is an* **arbitrary** *vector bundle on V with a connection and a metric, and if $\mathcal{D}_E$ denotes the twisted Dirac operator acting on sections of $S \otimes E$, then $\mathcal{D}_E$ has an eigenvalue with absolute value $\leq \varepsilon(V)$.*

Proof (Quick sketch). Recall that since we're assuming that V is even-dimensional, the spinor bundle has a canonical splitting $S = S^+ \oplus S^-$ into "half-spinor bundles," and that $\mathcal{D}$ sends sections of S^+ to sections of S^- and vice versa. Also, the Dirac operator is self-adjoint and elliptic, and so has a good spectral decomposition. (For details of all these facts, see [LaM, Chapter II].) The Atiyah-Singer Index Theorem gives a formula for the index of $\mathcal{D}_E$ (as an operator from sections of $S^+ \otimes E$ to sections of $S^- \otimes E$):

$$\text{(A-S)} \qquad \operatorname{ind} \mathcal{D}_E = \langle \hat{A}(V) \cup \operatorname{Ch}[E], [V] \rangle.$$

If this quantity is non-zero, then $\mathcal{D}_E$ must have zero as an eigenvalue and we're done. However, there will be some bundles E for which the formula (A-S) gives 0, in which case there is no reason why $\mathcal{D}_E$ should have a non-zero kernel. It's for these bundles that we need an estimate on the spectrum.

The idea for getting this estimate is that if a differential operator has spectrum far away from zero, then a small perturbation in the operator cannot suddenly create a non-zero kernel. Therefore, assuming that $\operatorname{ind} \mathcal{D}_E = 0$, we first find another bundle E' such that

$$\langle \hat{A}(V) \cup \operatorname{Ch}[E] \cup \operatorname{Ch}[E'], [V] \rangle \neq 0.$$

Because of (A-S), this means that $\operatorname{ind} \mathcal{D}_{E \otimes E'} \neq 0$, so that $\mathcal{D}_{E \otimes E'}$ has zero as an eigenvalue. Now if the bundle E' were topologically trivial, $E \otimes E'$ would just be a direct sum of $\dim E'$ copies of E, but with a different connection (since we're not assuming E' is flat). So $\mathcal{D}_{E \otimes E'}$ would be a perturbation of a direct sum of $\dim E'$ copies of $\mathcal{D}_E$ by lower-order terms. Thus, since $\mathcal{D}_{E \otimes E'}$ has zero as an eigenvalue, $\mathcal{D}_E$ cannot have too large a gap in its spectrum around zero.

Of course, we've oversimplified too much, since if ind $\mathcal{D}_E = 0$ and ind $\mathcal{D}_{E\otimes E'} \neq 0$, then E' could not be topologically trivial. However, we can always find another bundle E'' such that $E \oplus E''$ is topologically trivial, and then

$$\mathcal{D}_{E\otimes(E'\oplus E'')} = \mathcal{D}_{E\otimes E'} \oplus \mathcal{D}_{E\otimes E''}.$$

So giving $E \oplus E''$ the direct sum of the connection on E' with a connection on E'', we see that $\mathcal{D}_{E\otimes(E'\oplus E'')}$ has zero as an eigenvalue, and we can argue as before. □

Now let's discuss how we could apply the same ideas to open manifolds. (For an example of an application of these ideas in the different context of Kähler geometry, see [Gr2].) The same philosophy ought to apply, but since an elliptic operator on a non-compact manifold need not have discrete spectrum, we need to be much more careful about what constitutes a "small" perturbation of an operator. Before we get to this, we'll pursue some more consequences of these ideas with respect to compact manifolds. Suppose that $V = X/\Gamma$ is compact and aspherical, with fundamental group Γ. Intuitively, if we had a bundle E on V which were "almost flat," and for which we could compute by (A-S) that ind $\mathcal{D}_E \neq 0$, then $\mathcal{D}_E$ would have to have non-zero kernel. On the other hand, since E is "almost flat," the formula (Lichn) (with $\mathcal{D}$ replaced by $\mathcal{D}_E$) would not be off by very much, and also $\mathcal{D}_E$ would be only a small perturbation of $\dim E$ copies of $\mathcal{D}$. So if V had a Riemannian metric of positive scalar curvature, we could conclude that $\mathcal{D}$ would have to have an eigenvalue very close to 0, contradicting the estimate coming from (Lichn). So V could not have a positive scalar curvature metric after all.

This argument can indeed often be made to work, but the correct notion of "almost flat bundle" [CGM] is not really a single vector bundle but rather a sequence of $\mathbb{Z}/2$-graded vector bundles $E_i = E_i^0 \oplus E_i^1$, such that the formal differences $[E_i^0] - [E_i^1]$ all represent the same element of $K^0(V)$, and such that all the bundles are equipped with connections whose curvatures tend to 0 in the right sense as $i \to \infty$. One will also usually have $\dim E_i \to \infty$ if this is the case, since there is a limit on how flat we can make a representative of bounded dimension of a fixed K-theory class. (This limitation comes from Chern-Weil theory, which relates the curvature of the bundle to the Chern classes. Thus if any of the Chern classes is non-zero, the curvature must be non-zero.)

As an example of how to make this precise, suppose the fundamental group Γ of a compact non-positively curved manifold V^{2n} is residually finite (i.e., Γ has a faithful family of homomorphisms to finite groups). This condition is satisfied, for example, by arithmetic groups. Then we can use the finite quotients of Γ to construct a tower of finite coverings V_i of V which "converge" in some sense to the universal covering X. Pull the non-positively curved Riemannian metric on V back to each V_i, so that $\operatorname{diam} V_i \to \infty$. Choose smooth degree-one maps $V_i \to S^{2n}$ which have Lipschitz constants tending to 0 on bigger and bigger balls. (One can do this because of the non-positive curvature condition, since the inverse of the exponential map at a basepoint is well-defined and has Lipschitz constant 1 on bigger and bigger balls as $i \to \infty$. See for example [GL, §5].) Then we can pull back a fixed non-trivial bundle on S^{2n} by means of these maps, and we get a sequence of bundles (of course each defined on a different manifold V_i) whose curvatures tend to 0 in the operator norm (on L^2 sections). Pushing the bundles back down by means of the covering maps $V_i \to V$, we get an almost flat sequence of bundles (now with ranks tending to infinity) on V, which show that V cannot also admit a metric of positive scalar curvature. In fact, if one does this argument carefully, one can get an estimate on the Novikov-Shubin invariants of V, in other words, of the spectral density near 0 of the Dirac operator or Laplacian on the universal cover.[2]

Heuristically, we would expect a similar argument to prove the Novikov conjecture in similar cases, by using the analogue of (A-S) for the signature operator with coefficients in E:

$$\operatorname{ind} D_E^{\text{sign}} = \langle \mathcal{L}(V) \cup \operatorname{Ch}[E], [V] \rangle$$

and controlling the error terms in the expansion of $(D_E^{\text{sign}})^2$ coming from the curvature of E. Some arguments of this sort are carried out in [HiS]. However, there are complications in carrying out the details—one needs the connection, and not just the curvature, to be small, and one needs some "spectral purity" condition in order to define "almost Betti numbers."

In fact, if one generalizes the notion of an almost flat K-theory class,[3] one can show that the class of any complex line bundle over $B\Gamma$ is almost flat. This implies the Novikov conjecture for any higher signature coming from a 2-dimensional cohomology class. (Of course, Novikov himself [N] proved the homotopy invariance of any higher signature coming from a 1-dimensional cohomology class.) To prove this, consider a compact manifold V with a map to $B\Gamma$. The line bundle pulls back on the non-compact universal cover X of

[2] The possible width of a spectral gap is estimated by the norm of an index-changing perturbation. Similar ideas appear in the study of the "non-commutative isoperimetric function" in [Hu], and in §3 of [Lü].

[3] See the comment in [CGM]: "il est d'ailleurs nécessaire dans les applications ... de reformuler cette notion [de fibré presque plat] en l'adaptant aux fibrés de dimension infinie munis d'une superconnexion en un sens convenable."

V to a topologically trivial line bundle L, and thus we can take arbitrary "roots" $L^{\frac{1}{n}}$ of L which provide the necessary approximating sequence $(L^{\frac{1}{n}})^n$. See [Gr].

In the final analysis, then, the Novikov conjecture seems to be a statement about the "largeness" of the universal covers X of compact aspherical manifolds V. It seems to be important to quantify this; in particular to determine in what sense X "dominates" Euclidean space. The best situation would be if we could always show that X is strongly hyper-Euclidean, i.e., that X admits a Lipschitz map to a Euclidean space of the same dimension.[4] One very weak partial result in this direction is that if X is both complete and uniformly contractible in some Riemannian metric, then it has to have infinite volume. But it's hard to quantify its "growth" without additional assumptions. See [Gr1] for a discussion of different notions of "largeness."

What would be the way to construct a counterexample to the Novikov conjecture? Evidently we would need a rather pathological group. Most of the standard group-theoretical constructions rely on amalgamated free products and HNN extensions, which would lead to strange geometry in dimensions 1 and 2. But we know that the Novikov conjecture is by its nature a high-dimensional problem (the conjecture is true for cohomology classes on $B\Gamma$ in dimensions 1 and 2), so quite different techniques are needed.

We could try to measure the rate of contractibility of $E\Gamma$ in various dimensions, that is, to see how big a chain is needed to bound a given cycle. In dimension 1, this question is related to the solvability of the word problem in Γ; however, we're interested in this problem in higher dimensions. In most cases for which one can compute anything, the contractibility rate seems to grow exponentially. Perhaps to get a counterexample to the Novikov conjecture one should look for a case of very non-uniform growth in different directions.

References

[C] S. E. Cappell, *On homotopy invariance of higher signatures*, Invent. Math. **33** (1976), 171–179.

[CGM] A. Connes, M. Gromov and H. Moscovici, *Conjecture de Novikov et fibrés presques plats*, C. R. Acad. Sci. Paris Sér. I Math. **310** (1990), 273–277.

[FH] F. T. Farrell and W.-C. Hsiang, *On Novikov's conjecture for nonpositively curved*

[4] Note added since the original lecture: Dranishnikov, Ferry, and Weinberger have now constructed a uniformly contractible manifold with no such map, but it is not clear if one can arrange for such a uniformly contractible manifold to be the universal cover of a compact aspherical manifold.

manifolds, I, Ann. Math. **113** (1981), 197–209.

[FRR] S. Ferry, A. Ranicki, and J. Rosenberg, *A history and survey of the Novikov conjecture*, these Proceedings.

[FW] S. Ferry and S. Weinberger, *A coarse approach to the Novikov conjecture*, these Proceedings.

[GdlH] E. Ghys, P. de la Harpe, et al., *Sur les groupes hyperboliques d'après Mikhael Gromov*, Progress in Math., vol. 83, Birkhäuser, Boston, 1990.

[Gr1] M. Gromov, *Large Riemannian manifolds*, Curvature and Topology of Riemannian Manifolds, Proc., Katata, 1985 (K. Shiohama, T. Sakai and T. Sunada, eds.), Lecture Notes in Math., vol. 1201, 1986, pp. 108–121.

[Gr2] M. Gromov, *Kähler hyperbolicity and L_2-Hodge theory*, J. Diff. Geometry **33** (1991), 263–292.

[Gr3] M. Gromov, *Positive curvature, macroscopic dimension, spectral gaps and higher signatures*, Functional Analysis on the Eve of the 21st Century (Proc. conf. in honor of I. M. Gelfand's 80th birthday) (S. Gindikin, J. Lepowsky and R. Wilson, eds.), Progress in Math., Birkhäuser, Boston, 1995 (to appear).

[GrL] M. Gromov and H. B. Lawson, Jr., *Positive scalar curvature and the Dirac operator on complete Riemannian manifolds*, Publ. Math. Inst. Hautes Études Sci. **58** (1983), 83–196.

[HiS] M. Hilsum and G. Skandalis, *Invariance par homotopie de la signature dans un fibré presque plat*, J. reine angew. Math. **423** (1992), 73–99.

[Hu] S. Hurder, *Topology of covers and spectral theory of geometric operators*, Index Theory and Operator Algebras (J. Fox and P. Haskell, eds.), Contemp. Math., vol. 148, Amer. Math. Soc., Providence, 1993, pp. 87–119.

[LaM] H. B. Lawson, Jr. and M.-L. Michelson, *Spin Geometry*, Princeton Math. Series, vol. 38, Princeton Univ. Press, Princeton, N. J., 1989.

[Li] A. Lichnerowicz, *Spineurs harmoniques*, C. R. Acad. Sci. Paris Sér. A–B **257** (1963), 7–9.

[Lo] J. Lott, *private communication,* contained also in the preprint with W. Lück: *L^2-topological invariants of 3-manifolds*, MPI, Bonn.

[Lu] G. Lusztig, *Novikov's higher signatures and families of elliptic operators*, J. Diff. Geom. **7** (1971), 229–256.

[Lü] W. Lück, *Approximating L^2-invariants by their finite-dimensional analogues*, preprint, U. Mainz.

[M] A. S. Mishchenko, *Infinite dimensional representations of discrete groups and higher signatures*, Math. USSR–Izv. **8** (1974), 85–111.

[N] S. P. Novikov, *On manifolds with free abelian fundamental group and applications*, Izv. Akad. Nauk SSSR, Ser. Mat. **30** (1966), 208–246.

[R] J. Rosenberg, *C^*-algebras, positive scalar curvature, and the Novikov conjecture*, Publ. Math. Inst. Hautes Études Sci. **58** (1983), 197–212.

[RS] J. Rosenberg and S. Stolz, *Manifolds of positive scalar curvature*, Algebraic Topology and its Applications (G. Carlsson, R. Cohen, W.-C. Hsiang, and J. D. S. Jones, eds.), M. S. R. I. Publications, vol. 27, Springer, New York, 1994, pp. 241–267.

[S] R. Schoen, *Minimal manifolds and positive scalar curvature*, Proc. Internat. Congress of Mathematicians, Warsaw, 1983, Polish Scientific Publishers and North Holland, Warsaw, 1984, pp. 575–578.

[SY] R. Schoen and S.-T. Yau, *On the structure of manifolds with positive scalar curvature*, Manuscripta Math. **28** (1979), 159–183.

[VW] C. Vafa and E. Witten, *Eigenvalue inequalities for fermions in gauge theories*, Comm. Math. Physics **95** (1984), 257–276.

[Y] S.-T. Yau, *Minimal surfaces and their role in differential geometry*, Global Riemannian Geometry, T. J. Willmore and N. J. Hitchin, eds., Ellis Horwood and Halsted Press, Chichester, England, and New York, 1984, pp. 99–103.

INSTITUT DES HAUTES ÉTUDES SCIENTIFIQUES, 35, ROUTE DE CHARTRES, 91440 BURES-SUR-YVETTE, FRANCE (May–February)
DEPARTMENT OF MATHEMATICS, UNIVERSITY OF MARYLAND, COLLEGE PARK, MD 20742, U. S. A. (February–May)

email: `gromov@ihes.fr; gromov@math.umd.edu`

Controlled Fredholm representations

Alexander S. Mishchenko [1]

Abstract

A notion of family of Fredholm representations controlled at infinity and a new topology of the space of Fredholm representations which differs from the Kasparov one are introduced. These are used to obtain a new proof of Novikov's conjecture on the homotopy invariance of higher signatures for complete non-positively curved Riemannian manifolds. The conjecture is proved for a class of manifolds larger than the class of special manifolds considered by H. D. Rees.

1 Introduction

One of the most significant concepts that describes the smooth structure of manifolds is that of the characteristic classes of a manifold. From the moment it was introduced the following problem was posed and taken up: to what extent do particular characteristic classes depend on the smooth structure of the manifold by means of which they are determined?

In particular one of the questions is: which characteristic classes are homotopy invariant? For which characteristic classes α do we have the equality

$$\alpha(X) = f^*(\alpha(Y))$$

for every homotopy equivalence of manifolds $f : X \to Y$?

It is well known that the Stiefel-Whitney classes are homotopy invariant for any smooth compact closed manifold. The proof is based on the expression of the characteristic Stiefel-Whitney classes in terms of the fundamental cycle of the manifolds, by means of cohomology operations (the Steenrod squares).

The problem of the homotopy invariance of the Pontrjagin classes turns out to be more difficult and, accordingly, more significant from the standpoint of application to various problems of algebraic and differential topology. Numerous attempts to solve the problem have led to new interesting relations among topology, representation theory, discrete groups, C^*-algebra

[1] Partially supported by SFB No. 343, Bielefeld, Germany, and a Visiting Professorship at Brown University, Providence, R.I., USA.

and others. The importance of this problem stems, in particular, from the fact that in the problem of classifying smooth structures on a manifold it is necessary to have a description of all the homotopy invariant Pontrjagin classes.

I shall recall that the Pontrjagin class $p_i(X)$ of a manifold X is a cohomology class

$$p_i(X) \in H^{4i}(X; \mathbf{Z}).$$

Here we shall consider only rational Pontrjagin classes, that is, we shall assume that Pontrjagin classes are cohomology classes with rational coefficients or even with real coefficients:

$$p_i(X) \in H^{4i}(X; \mathbf{R}).$$

Each rational characteristic class can be represented as a polynomial in the classes $p_i(X)$.

Let $L(X)$ denote the Hirzebruch L-genus of the manifold X, which is described in terms of the Wu generators t_k as the symmetric polynomial

$$L(X) = \prod_k \frac{t_k/2}{\tanh(t_k/2)}.$$

The first result obtained along these lines was the formula for computing the signature of a 4-dimensional manifold (Rokhlin, [28], 1952) in terms of the characteristic classes. Subsequently, F. Hirzebruch ([8], 1953) established a similar formula for the signature of a $4k$-dimensional manifold:

$$\operatorname{sign} M^{4k} = 2^{2k}\langle L(M^{4k}), [M^{4k}]\rangle. \tag{1.1}$$

Here $\operatorname{sign} M^{4k}$ means the signature of the quadratic form which is determined on the cohomology group of the middle dimension $H^{2k}(M^{4k}; \mathbf{Q})$. Therefore the signature is a homotopy invariant, and hence the Hirzebruch number $\langle L(M^{4k}), [M^{4k}]\rangle$ is also homotopy invariant.

The relationship of homotopy-invariant Pontrjagin classes to the problem of classification of smooth structures on a manifold involves the following. Assume that

$$f : X \to Y$$

is a smooth mapping of degree 1 of oriented manifolds. The map f will be bordant to a homotopy equivalence if and only if a number of conditions are met. For simply connected manifolds these conditions include, in particular, the requirement that the signatures of manifolds X and Y coincide. Due to results on the classification of smooth structures on simply connected manifolds (W. Browder [2], 1962, S. Novikov [24], 1964), it follows that for simply connected manifolds the Hirzebruch number is (up to a constant) the unique homotopy-invariant rational characteristic number.

For nonsimply connected manifolds the problem of surgering f to a homotopy equivalence turns out to be less trivial; it can be reduced to a surgery obstruction which is an element of an analog of algebraic K-theory (so-called Hermitian K-theory). The groups of Hermitian K-theory were constructed in stable algebra using nondegenerate quadratic forms on free modules. In 1970, the author ([16]) expressed the surgery obstruction for f modulo 2-primary torsion as the difference of generalized symmetric signatures of the manifolds X and Y, and showed that symmetric signatures are analogs of the signature of manifolds and can be represented by nondegenerate Hermitian forms on free modules over the rational group ring of the fundamental group.

As a particular consequence, if some rational characteristic Pontrjagin number is a homotopy invariant then it necessarily has the form

$$\sigma_x(X) = \langle L(X)g^*(x), [X]\rangle, \tag{1.2}$$

where $g : X \to B\pi_1(X)$ is the characteristic mapping of the manifold X into the Eilenberg-MacLane complex, while $x \in H^*(B\pi_1(X); \mathbf{Q})$ is some cohomology class. Numbers of this form were called higher signatures.

Thus, the problem was reduced to the determination of homotopy invariance of higher signatures. The problem has yet to be fully solved, although it has attracted the attention of topologists for the last 25 years. In 1965, S. Novikov ([25], [26]) established the homotopy invariance of higher signatures in dimension 1, and then V. Rokhlin ([29], 1966) established it in dimension 2 for decomposable cohomology classes. Subsequently, G. G. Kasparov ([10], 1970) established the homotopy invariance of higher signatures for cohomology classes that are decomposable into a product of one-dimensional classes. This result is essentially equivalent to the homotopy invariance of all higher signatures for free abelian groups. S. E. Cappell ([3], 1974) employed the same method of surgery on codimension 1 submanifolds to expand the class of fundamental groups for which all higher signatures are homotopy-invariant.

Another approach to the solution of the problem of homotopy invariance of higher signature involves the use of representation theory, by means of which higher signatures can be represented in manifestly homotopy-invariant terms. For example, if

$$\rho : \pi_1(X) \to U(n)$$

is a unitary representation, then cohomology with respect to the local system of coefficients generated by the representation ρ admits a nondegenerate quadratic form, and the signature $\operatorname{sign}_\rho(X)$ is defined. One can verify that

$$\operatorname{sign}_\rho M^{4k} = 2^{2k}\langle \operatorname{ch}\rho\, L(M^{4k}), [M^{4k}]\rangle = \sigma_{\operatorname{ch}\rho}(M^{4k}), \tag{1.3}$$

where $\mathrm{ch}\rho$ is the Chern character of representation ρ. Since $\mathrm{sign}_\rho M^{4k}$ is a homotopy invariant, the right side of the equality, $\sigma_{\mathrm{ch}\rho}(M^{4k})$, is also homotopy-invariant. Unfortunately the Chern character $\mathrm{ch}\,\rho$ is trivial for any unitary representation ρ. In 1972 G. Lusztig ([15]) generalized this formula for representations

$$\rho : \pi_1(X) \to U(p, q),$$

where $U(p, q)$ is the unitary group for an indefinite metric of type (p, q). But in the case considered by G. Lusztig the set

$$\{\mathrm{ch}\,\rho\} \subset H^*(B\pi_1(X); \mathbf{Q})$$

does not cover all higher signatures. The next approach was to expand the class of representations of fundamental group $\pi_1(X)$ to a class of infinite-dimensional representations, the so-called Fredholm representations.

In [17], [18], Novikov's conjecture on the homotopy invariance of higher signatures

$$\mathrm{sign}_x(M) = \langle L_*(M) \cdot f^*(x), [X] \rangle$$

of non-simply connected manifolds M was proven in the case when the fundamental group $\pi_1(M)$ of the compact smooth (oriented) manifold M has a classifying space $B\pi$ (that is the Eilenberg-MacLane complex $K(\pi, 1)$) that is a compact Riemannian manifold with non-positive sectional curvature. In [18] it was also claimed that the proof is valid for all fundamental groups π whose classifying space $B\pi$ is a (noncompact) complete non-positively curved manifold. In [7] the proof of Novikov's conjecture was extended to the case where the $B\pi$ is a complete Riemannian manifold of finite volume and with sectional curvature strongly negative. In [11] G. G. Kasparov presented a proof of Novikov's conjecture for all complete non-positive curved manifolds. This last proof was provided by the method of representation of special elements in Kasparov KK-theory and based on the hard technique of intersection theory in KK-theory.

The class of fundamental groups for which the Novikov conjecture holds was enlarged in the papers of A. Connes and H. Moscovici ([6]) and A. Connes, M. Gromov and H. Moscovici ([4], [5]). In the first paper [6] the authors presented new methods of description of higher signatures via the index of an elliptic operator on the universal covering of non-simply connected manifolds, and reduced the problem to the description of an arbitrary cohomology class of the fundamental group by means of cyclic cohomology of suitable subalgebras of the C^*-algebra $C^*[\pi]$ obtained by completing the group algebra $\mathbf{C}[\pi]$. This gives a proof of the Novikov conjecture for hyperbolic groups π.

The second paper [4] was devoted to the development of the technique of so-called almost flat bundles and the related notion of semi-representations of groups. In some cases when a semi-representation has small norm it is possible to construct in a natural way both a vector bundle over the classifying space $B\pi$ and a signature of a quadratic form over the group algebra $\mathbf{C}[\pi]$. The authors stated that their methods apply in all the cases for which the Novikov conjectures are known to hold.

In the third paper ([5]) the authors introduced a notion of Lipschitz cohomology classes of the group π, which are detected by a family of Lipschitz maps from the group π to a Euclidean space $\mathbf{R}^N$. It seems that this construction is a generalization of a vector field from the paper of H. Rees ([27]). In any case in [5] the theorem about the homotopy invariance of higher signatures which are represented by Lipschitz cohomology classes is proved. This means in particular that yet a third proof of the Novikov conjecture for hyperbolic groups was presented.

Therefore it is interesting to clarify how much one can obtain in this direction using Fredholm representations. In 1982 in [9] it was shown that the method of Fredholm representations is valid at least for groups π such that the classifying space $B\pi$ is a complete non-positive curved manifold, that can be compactified to a manifold with boundary. It was proved there that, in this case, Novikov's conjecture is true for all cohomology classes $x \in$ Image (i^*) where

$$i^* : H^*(M^+; \mathbf{Q}) \longrightarrow H^*(M; \mathbf{Q})$$

is the natural homomorphism generated by the inclusion

$$i : M \longrightarrow M^+$$

of M into its one point compactification M^+. Moreover in [9] it was claimed that the method of Fredholm representation yields only the homotopy invariance of the higher signatures $\mathrm{sign}_x(M)$ with $x \in \mathrm{Image}(i^*)$.

We shall show that a good notion of trivial elements for Fredholm representations yields the proof of Novikov's conjecture for all cohomology classes of complete non-positively curved manifolds without requiring the existence of a compactification to a manifold with boundary. It seems useful to analyze the Fredholm representation method to see if it is possible to avoid the difficult techniques of Kasparov KK-theory.

2 Scheme of application of Fredholm representations controlled at infinity

Let M be a closed oriented non simply connected manifold with fundamental group π. Let $B\pi$ be the classifying space for the group π and let

$$f_M : M \longrightarrow B\pi,$$

be a map inducing the isomorphism of fundamental groups.

Let $\Omega_*(B\pi)$ denote the bordism group of pairs (M, f_M). Recall that $\Omega_*(B\pi)$ is a module over the ring $\Omega_* = \Omega_*(\text{ pt })$.

In [16] a homomorphism was constructed

$$\sigma : \Omega_*(B\pi) \longrightarrow L_*(\mathbf{C}\pi) \tag{2.1}$$

which for every manifold (M, f_M) assigns the element $\sigma(M) \in L_*(\mathbf{C}\pi)$, where $L_*(\mathbf{C}\pi)$ is the Wall group for the group ring $\mathbf{C}\pi$.

The homomorphism σ satisfies the following conditions:

(a) σ is homotopy invariant,

(b) if N is a simply connected manifold and $\tau(N)$ is its signature then

$$\sigma(M \times N) = \sigma(M)\tau(N) \in L_*(\mathbf{C}\pi).$$

We shall be interested only in the groups after tensor multiplication with the field $\mathbf{Q}$, in other words in the homomorphism

$$\sigma : \Omega_*(B\pi) \otimes \mathbf{Q} \longrightarrow L_*(\mathbf{C}\pi) \otimes \mathbf{Q}.$$

However

$$\Omega_*(B\pi) \otimes \mathbf{Q} \approx H_*(B\pi; \mathbf{Q}) \otimes \Omega_*.$$

Hence one has

$$\sigma : H_*(B\pi; \mathbf{Q}) \longrightarrow L_*(\mathbf{C}\pi) \otimes \mathbf{Q}.$$

Therefore the homomorphism σ represents the cohomology class

$$\bar{\sigma} \in H^*(B\pi; L_*(\mathbf{C}\pi) \otimes \mathbf{Q}).$$

Then for any manifold (M, f_M) one has

$$\sigma(M, f_M) = \langle L(M) f_M^*(\bar{\sigma}),\, [M] \rangle \in L_*(\mathbf{C}\pi) \otimes \mathbf{Q}. \tag{2.2}$$

Hence if $\alpha : L_*(\mathbf{C}\pi) \otimes \mathbf{Q} \longrightarrow \mathbf{Q}$ is an additive functional and $\alpha(\bar{\sigma}) = x \in H^*(B\pi; \mathbf{Q})$ then

$$\operatorname{sign}_x(M, f_M) = \langle L(M) f_M^*(x), [M] \rangle \in \mathbf{Q}$$

should be a homotopy-invariant higher signature. This gives a description of the family of all homotopy-invariant higher signatures. Therefore one should study the cohomology class

$$\bar{\sigma} \in H^*(B\pi; L_*(\mathbf{C}\pi) \otimes \mathbf{Q}) = H^*(B\pi; \mathbf{Q}) \otimes L_*(\mathbf{C}\pi) \otimes \mathbf{Q}.$$

2.1 The first step

The first step consists in changing from the group algebra $\mathbf{C}\pi$ to the (reduced) C^*-algebra $C^*[\pi]$. The natural inclusion generates the homomorphism

$$\hat{}: L_*(\mathbf{C}\pi) \otimes \mathbf{Q} \longrightarrow L_*(C^*[\pi]) \otimes \mathbf{Q} = K_*(C^*[\pi]) \otimes \mathbf{Q}.$$

Then using $\hat{}$ one can obtain the new formula

$$\hat{\sigma}(M, f_M) = \langle L(M) f_M^*(\hat{\bar{\sigma}}), [M] \rangle \in K_*(C^*[\pi]).$$

The left side turns out to be homotopy-invariant. The right side loses a part of the information about the higher signatures. But at the crucial moment, it is possible to write a new formula

$$\hat{\sigma}(M, f_M) = \langle L(M) f_M^*(\mathrm{ch}_A \xi_A), [M] \rangle \in K_*(C^*[\pi]) \tag{2.3}$$

where ξ_A is the canonical A-bundle over $B\pi$ which is determined by the natural representation of the fundamental group

$$\pi \longrightarrow \mathbf{C}\pi \longrightarrow C^*[\pi] = A,$$

$$K_A^* = K_A^*(\mathrm{pt}) = K_*(A),$$

$$\mathrm{ch}_A : K_A^*(X) \longrightarrow H^*(X; K_A^* \otimes \mathbf{Q}).$$

The formula (2.3) is called the Hirzebruch formula for non-simply connected manifolds. It was proved in [23].

I do not know if it is possible to write a formula similar to (2.3) in any sense for the algebra $\mathbf{C}\pi$. If so then one could use the technique due to A. Connes and H. Moscovici ([6]).

2.2 The second step

The second step consists of natural representations of the algebra A. The first example of the representation of commutative algebras as functions on the character group was considered by G. Lusztig ([15]). The most effective way proved to be considering the Fredholm representations.

A Fredholm representation means a pair

$$\rho = (F, T),$$

where

$$T : \pi \longrightarrow B(H)$$

is a (unitary) representation into the algebra of bounded operators of the Hilbert space H, and

$$F : H \longrightarrow H$$

is a Fredholm operator such that for any $g \in \pi$

$$T(g)F - FT(g) \in C(H),$$

where $C(H)$ denotes the family of compact operators. It is useful to consider all objects in a graded category. The representation T can be extended to a symmetric representation

$$T : C^*[\pi] \longrightarrow B(H)$$

with the same condition on F. Any Fredholm representation gives a well-defined homomorphism

$$\rho^* : K_A^*(X) \longrightarrow K^*(X)$$

such that the following diagram commutes:

$$\begin{array}{ccc} K_A^*(X) & \xrightarrow{\rho^*} & K^*(X) \\ \downarrow \mathrm{ch}_A & & \downarrow \mathrm{ch} \\ H^*(X; K_A^* \otimes \mathbf{Q}) & \xrightarrow{H(\rho^*)} & H^*(X; \mathbf{Q}) \end{array}$$

Hence if one has a family of Fredholm representations

$$\rho = (F_y, T_y), \quad y \in Y,$$

which is continuous in the uniform norm, then it induces a homomorphism

$$\rho^* : K_A^*(X) \longrightarrow K^*(X \times Y) \tag{2.4}$$

giving the commutative diagram

$$\begin{array}{ccc} K_A^*(X) & \xrightarrow{\rho^*} & K^*(X \times Y) \\ \downarrow \mathrm{ch}_A & & \downarrow \mathrm{ch} \\ H_A^*(X; K_A^* \otimes \mathbf{Q}) & & H^*(X \times Y; \mathbf{Q}) \\ \downarrow H(\rho^*) & & \uparrow \cup \\ H^*(X; K^*(Y) \otimes \mathbf{Q}) & \xrightarrow{\mathrm{ch}} & H^*(X; H^*(Y) \otimes \mathbf{Q}). \end{array}$$

Applying the homomorphism (2.4) to (2.3), one has

$$\mathrm{ch}\, \rho^*(\hat{\sigma}(M, f_M)) \in H^*(Y; \mathbf{Q}),$$

$$\begin{array}{rcl} \mathrm{ch}\, \rho^* \langle L(M) f_M^*(\mathrm{ch}_A \xi_A), [M] \rangle & = & \\ = \langle L(M) f_M^*(\mathrm{ch} \rho^*(\xi_A), [M] \rangle & \in & H^*(Y; \mathbf{Q}). \end{array}$$

Let $\{y_\alpha\}_{\alpha \in \Lambda}$ be a basis in $H^*(Y; \mathbf{Q})$. Then

$$f_M^*(\mathrm{ch}\, \rho^*(\xi_A)) \in H^*(M \times Y; \mathbf{Q}) = H^*(M; \mathbf{Q}) \otimes H^*(Y; \mathbf{Q})$$

and we can decompose

$$f_M^*(\operatorname{ch}\rho^*(\xi_A)) = \sum_{i\in\Lambda_0} u^i \otimes y_i$$

where $\Lambda_0 \subset \Lambda$ is a finite subset, $u^i \in H^*(M;\mathbf{Q})$, $y_i \in H^*(Y;\mathbf{Q})$. Therefore

$$\operatorname{ch}\rho^*(\hat{\sigma}(M, f_M)) = \sum_{i\in\Lambda_0} \langle L(M)u^i, [M]\rangle y_i \in H^*(Y;\mathbf{Q}).$$

Hence $\langle L(M)u^i, [M]\rangle$ is a homotopy invariant for any i. It is not difficult to show that $u^i = f_M^*(u_\infty^i)$ for some $u_\infty^i \in H^*(B\pi;\mathbf{Q})$.

2.3 The third step

The third step consists of defining a *family of Fredholm representations with compact supports.*

Definition 1 *Let Y be a locally compact space, and let π be a group. A family $\rho = (F_y, T_y)$, $y \in Y$ is called a family of Fredholm representations with compact supports if*

(a) for each $y \in Y$, (F_y, T_y) is a Fredholm representation of π, and the maps $y \mapsto F_y$ and $y \mapsto T_y(g)$ (for any $g \in \pi$) are continuous in the operator norm;

(b) there exists a compact $K \subset Y$ and a constant C such that for any $y \in Y\backslash K$ the operator F_y is invertible and

$$\|F_y\| < C, \quad \|F_y^{-1}\| < C;$$

(c) for any $g \in \pi$ and $\varepsilon > 0$ there exists a compact set $K = K(\varepsilon, g) \subset Y$ such that for any $y \in Y\backslash K$

$$\|F_y T_y(g) - T_y(g)F_y\| < \varepsilon.$$

One can construct a well-defined homomorphism

$$\rho^* : K_A^*(X) \longrightarrow K_{comp}^*(X \times Y)$$

for any finite CW-complex X. Let $\xi = \{E \longrightarrow X\}$ be a locally trivial vector bundle, the fiber of which is a finitely generated Hilbert A-module V and the structure group of which is $\operatorname{Aut}_A(V)$. Let $\{U_\alpha\}$ be a finite covering and

$$\varphi_{\alpha\beta} : U_{\alpha\beta} \longrightarrow \operatorname{Aut}_A(V), \quad U_{\alpha\beta} = U_\alpha \cap U_\beta,$$

be the transition functions,

$$\varphi_{\alpha\beta}(x)\varphi_{\beta\gamma}(x)\varphi_{\gamma\alpha}(x) = 1, \quad x \in U_{\alpha\beta\gamma} = U_\alpha \cap U_\beta \cap U_\gamma.$$

The representation T_y determines the homomorphism

$$T_y^* : \mathrm{Aut}_A(V) \longrightarrow B(\bar{H}), \quad \bar{H} = V \otimes_A H$$

and therefore determines the Hilbert bundle $\underline{H}$ associated with ξ over the base space $X \times Y$, $\underline{H} = E \otimes_A H$. Then one should consider a homomorphism

$$\Phi : \underline{H} \longrightarrow \underline{H}$$

as a family of maps

$$\Phi_\alpha(x, y) : \bar{H} \longrightarrow \bar{H}, \quad x \in U_\alpha, \quad y \in Y$$

such that

$$T_y^*(\varphi_{\alpha\beta}(x))\Phi_\alpha(x, y) = \Phi_\beta(x, y)T_y^*(\varphi_{\alpha\beta}(x)), \quad x \in U_{\alpha\beta}, \quad y \in Y. \quad (2.5)$$

We say that the homomorphism

$$\Phi : \underline{H} \longrightarrow \underline{H}$$

is a realization of the representation ρ^* on the element ξ, $[\xi] \in K_A^*(X)$ if

$$\text{(a) } \Phi_\alpha(x, y) - 1_V \otimes F_y \in C(\bar{H}), \quad x \in U_\alpha, \quad y \in Y; \quad (2.6)$$

(b) For any $\varepsilon > 0$ there exists a compact $K \subset Y$, such that for any $x \in U_\alpha$, $y \in Y \backslash K$ one has

$$\|\Phi_\alpha(x, y) - 1_V \otimes F_y\| < \varepsilon. \quad (2.7)$$

Theorem 1 *Let $X_0 \subset X$ and $\Phi^0(x, y)$, $x \in X_0$, be a realization of the representation ρ^* on the bundle ξ over the subspace $X_0 \subset X$. Then there exists a realization $\Phi(x, y)$, $x \in X$, such that*

$$\Phi(x, y)|_{X_0} = \Phi^0(x, y).$$

Corollary 1 *If Φ^0 and Φ^1 are two realizations of the representation ρ^* on the bundle ξ then there exists a homotopy Φ^t, $0 \le t \le 1$, in the class of realizations of the representation ρ^* which extends Φ^0 and Φ^1.*

Corollary 2 *The correspondence*

$$\xi \mapsto [\Phi] \in K^0_{comp}(X \times Y)$$

is a well-defined homomorphism

$$\rho^* : K_A(X) \longrightarrow K^0_{comp}(X \times Y)$$

for any finite CW-complex X.

Corollary 3 *If X is a non-compact CW-complex then* ρ^* *can be extended to a homomorphism*

$$\rho^* : K_A(X) \longrightarrow K^0(X \times Y^+, X \times (+)) \tag{2.8}$$

where Y^+ *is one point compactification of Y. Moreover, the diagram*

$$\begin{array}{ccc} K^*_A(X) & \xrightarrow{\rho^*} & K^*(X \times Y^+, X \times (+)) \\ \big\downarrow ch_A & & \big\downarrow ch \\ H^*_A(X; K^*_A \otimes \mathbf{Q}) & & H^*(X \times Y^+, X \times (+); \mathbf{Q}) \\ \big\downarrow H(\rho^*) & & \big\uparrow \cup \\ H^*(X; K^*(Y^+, (+)) \otimes \mathbf{Q}) & \xrightarrow{H(ch)} & H^*(X; H^*(Y^+, (+); \mathbf{Q})) \end{array}$$

is commutative.

Therefore one can repeat the technique in the absolute case.

2.4 The fourth step

The fourth step is standard (see [18]). Let $B\pi$ be a finite-dimensional complete Riemannian manifold with nonpositive sectional curvature. Let

$$Y = T^* B\pi \xrightarrow{p} B\pi$$

be the cotangent bundle and let

$$\tilde{Y} \xrightarrow{\tilde{p}} \tilde{B\pi}$$

be its universal covering. Consider the bundle $\Lambda^*(\tilde{Y})$, the bundle of differential forms over $\tilde{Y}$, which is a complex vector bundle. The complex structure corresponds to the natural almost complex structure in the tangent bundle of the manifold Y. Fix a basepoint $\tilde{y}_0$ in $\tilde{Y}$. Define the homomorphism

$$\tilde{F} : \Lambda^*(\tilde{Y}) \longrightarrow \Lambda^*(\tilde{Y}), \quad \tilde{y} \in \tilde{Y},$$

$$\tilde{F}_{\tilde{y}}(\omega) = \omega \wedge d_c \varphi(\tilde{y}),$$

where d_c is the complex differential of the function

$$\varphi(\tilde{y}) = \sqrt{d^2(\tilde{p}(\tilde{y}_0), \tilde{p}(\tilde{y})) + \|\tilde{y}\|^2 + 1}.$$

The differential d_c is defined by

$$d_c = d - idI$$

where I determines the complex structure in the bundle $T(T^*\tilde{B\pi})$. Let $y \in Y$ and

$$H_y = q_!(\Lambda^*(\tilde{Y})) = \oplus_{\tilde{y} \in q^{-1}(y)} \Lambda^*_{\tilde{y}},$$

$$F_y = q_!(\tilde{F})_y, \quad y \in Y,$$

where $q : \tilde{Y} \longrightarrow Y$ is the covering map. It is clear that $F_y + F_y^*$ is a Fredholm operator. Let $g : \tilde{Y} \longrightarrow \tilde{Y}$ be the diffeomorphism representing the action of the element g of the monodromy group π, and let $\Lambda^* dg$ be its differential. Put

$$T_y(g) = (q_!(\Lambda^* dg))_y.$$

Then the family $\rho^* = (F_y, T_y)$, $y \in \tilde{Y}$ satisfies the conditions of a Fredholm representation with compact supports.

The realization of ρ^* on the bundle ξ_A can be determined by the following. In this case $V = A$, $\mathrm{Aut}_A(A) = A$, and the transition functions are given by right multiplication by the elements of group π.

Put

$$\tilde{\Phi}(\tilde{x}, \tilde{y})(\omega) = \omega \wedge d_c \tilde{\varphi}(\tilde{y}, \tilde{x}),$$

where

$$\tilde{\varphi}(\tilde{y}, \tilde{x}) = \sqrt{d^2(\tilde{p}(\tilde{x}), \tilde{p}(\tilde{y})) + \|\tilde{y}\|^2 + 1}.$$

Then

$$\bar{\Phi}(\tilde{x}, y) = q_!(\tilde{\Phi})_{\tilde{x}, y}, \quad \tilde{x} \in \tilde{Y}, \quad y \in Y.$$

is an equivariant family of operators such that for any $\varepsilon > 0$ and compact set $K_1 \subset Y$ there exist a compact $K_2 \subset Y$ such that

$$\|\bar{\Phi}(\tilde{x}, y) - F_y\| < \varepsilon$$

for any $\tilde{x} \in q^{-1}(K_1)$, $\quad y \in Y \backslash K_2$. This means that the element

$$\rho^*(\xi_A) \in K(Y \times Y^+, Y \times (+))$$

is represented by the family of operators

$$\Phi(x, y) = \bar{\Phi}(\tilde{x}, y)/\pi.$$

In fact the family $\Phi(x, y)$ is an isomorphism for $x \neq y \in Y$. Hence the element $\rho^*(\xi_A)$ lies in the image of the homomorphism

$$K(Y \times Y, Y \times Y \backslash \Delta) = K(Y \times Y^+, Y \times Y^+ \backslash \Delta) \longrightarrow K(Y \times Y^+, Y \times (+)).$$

This last condition means that for any $\zeta \in K(Y)$, one has

$$(1 \otimes \zeta)\rho^*(\xi_A) = (\zeta \otimes 1)\rho^*(\xi_A).$$

Thus we should consider the class

$$f_M^*(\mathrm{ch}\rho^*(\xi_A)) = \sum u^i \otimes y^i \in H^*(M; \mathbf{Q}) \otimes H^*(Y^+, (+); \mathbf{Q})$$

such that for any $x \in H^*(Y; \mathbf{Q})$ one has

$$\sum_{i \in \Lambda_0} (f_M^*(x) u^i) \otimes y_i = \sum_{i \in \Lambda_0} u^i \otimes (y^i x). \tag{2.9}$$

Let

$$V = \mathrm{Im}(H^*(Y; \mathbf{Q}) \to H^*(M; \mathbf{Q})).$$

Then there exists a compact $Y^\alpha \subset Y$ such that

$$V = \mathrm{Im}(H^*(Y^\alpha; \mathbf{Q}) \to H^*(M; \mathbf{Q})),$$

and therefore $u^i \in V$ for all $i \in \Lambda_0$. One can easily show that the y^i are independent. This concludes the proof of the Novikov conjecture for complete non-positively curved Riemannian manifolds, assuming Theorem 1 and its Corollaries (which will be proved below).

3 Subsidiary considerations

3.1

All spaces below are considered to be finite or infinite CW-complexes with smooth simplicial divisions on smooth manifolds. If a CW-complex X is infinite, then the topology on X is considered to be the direct limit of the topologies of the family of all its finite subcomplexes. The subspace $Y \subset X$ is considered as a simplicial subcomplex of X or one of its simplicial subdivisions. A compact subspace $Y \subset X$ means that Y is a finite subcomplex.

The cohomology groups $H^*(X; G)$ are considered as the homology groups of the simplicial cochain complex, whose cochain groups consist of all functions from the family of all simplexes (of a given dimension) to the coefficient group G. For our purposes, it will be sufficient to consider the case where

G is a vector space over $\mathbf{Q}$. Therefore if X is an infinite CW-complex, then the cohomology group $H^*(X;G)$ is the inverse limit

$$H^*(X;G) = \lim_{\longleftarrow}(H^*(X^\alpha;G))$$

where X^α runs over the family of all finite subcomplexes of X. (This would not necessarily be true if we didn't use rational coefficients.) Moreover, the family $\{H^*(X^\alpha;G)\}$ satisfies the Mittag-Leffler condition for stabilization of images, that is, for any α, there exists $\beta_0 > \alpha$, such that for any $\beta > \beta_0$, one has

$$\text{Image}\{H^*(X^\beta;G)\longrightarrow H^*(X^\alpha;G)\} = \text{Image}\{H^*(X^{\beta_0};G)\longrightarrow H^*(X^\alpha;G)\}.$$

Respectively $H^*_{comp}(X;G)$ denotes the homology groups of the simplicial cochain complex consisting of all cochains which vanish everywhere with the exception of a finite number of simplexes. Recall that there exists a graded pairing

$$H^*(X;G_1) \otimes H^*_{comp}(X;G_2) \to H^*_{comp}(X;G_1 \otimes G_2)$$

which is determined by the cup-product of the cochains.

In the case where X is a finite complex the topological K-theory that we prefer to use is complex $\mathbf{Z}_2$-graded K-cohomology theory. If the space X is an infinite CW-complex, then it is convenient to deal with the inverse limit

$$K^*(X) = \lim_{\leftarrow}(K^*(Y))$$

where Y runs over the directed set of all finite subcomplexes $Y \subset X$. Certainly any locally trivial complex vector bundle ξ determines an element $[\xi] \in K^0(X)$. Also the elements of the group $K^*(X)$ are represented by continuous families of Fredholm operators F_x, $x \in X$ acting on a Hilbert space. Moreover if we have two locally trivial Hilbert bundles $\tilde{H}_1$ and $\tilde{H}_2$ whose fibers are the Hilbert spaces H_1 and H_2 and, if F is a continuous homomorphism of the bundles

$$F : \tilde{H}_1 \to \tilde{H}_2$$

such that over an arbitrary point $x \in X$ the bounded operator

$$F_x : (\tilde{H}_1)_x \to (\tilde{H}_2)_x$$

is a Fredholm operator then the triple $\xi = (\tilde{H}_1, F, \tilde{H}_2)$ determines an element $[\xi] \in K^0(X)$. Recall that the Chern character is a mapping

$$\text{ch} : K^*(X) \to H^*(X;\mathbf{Q})$$

where $\mathbf{Q}$ is the rational number field.

Let X be an (infinite) CW-complex and let ξ and η be two locally trivial vector bundles over X. Let

$$\varphi : \xi \to \eta$$

be a homomorphism such that there exists a compact subset $K \subset X$, such that for any $x \in X \backslash K$, the fiber homomorphism

$$\varphi_x : \xi_x \to \eta_x$$

is an isomorphism. It is known (see for example [1]) that the family of triples (ξ, φ, η) with the property mentioned above is determined by the group $K^0_{comp}(X)$ of virtual vector bundles with compact supports. The elements of the group K^0_{comp} can be represented by triples (ξ, φ, η) such that the vector bundles ξ and η are Hilbert bundles and φ is a Fredholm homomorphism, such that for any point $x \in X \backslash K$ the fiber homomorphism φ_x is an invertible bounded operator. The group $K^0_{comp}(X)$ extends to a $\mathbf{Z}_2$-graded cohomology theory $K^*_{comp}(X)$ and there exists a pairing

$$K^*(X) \otimes K^*_{comp}(X) \to K^*_{comp}(X)$$

which is generated by the tensor product of bundles. Then the Chern character homomorphism is defined as a multiplicative map

$$ch : K^*_{comp}(X) \to H^*_{comp}(X; \mathbf{Q}).$$

3.2

Let M be a connected non-simply connected manifold with fundamental group π. Let $B\pi = K(\pi, 1)$ be the Eilenberg-Maclane complex, that is a CW-complex whose all homotopy groups are trivial with the exception of $\pi_1(B\pi) = \pi$. Then there exists a map, unique up to homotopy

$$f_M : M \to B\pi$$

inducing the given isomorphism of the fundamental groups

$$(f_M)_* : \pi_1(M) \to \pi_1(B\pi) = \pi.$$

It will be convenient to consider the pairs (M, f_M), considering $h : M_1 \to M_2$ to be a homotopy equivalence of the non-simply connected manifolds (more explicitly of the pairs (M_1, f_{M_1}) and (M_2, f_{M_2})) if h is a homotopy equivalence and the diagram

$$\begin{array}{ccc} M_1 & \xrightarrow{f_{M_1}} & B\pi \\ \downarrow h & \nearrow f_{M_2} & \\ M_2 & & \end{array}$$

is homotopy-commutative. When M_1 and M_2 are oriented we require h to preserve orientations.

Similarly two (oriented) non-simply connected manifolds

$$(M_1, f_{M_1}) \text{ and } (M_2, f_{M_2})$$

are called *bordant* if there exists an (oriented) manifold with boundary (W, f_W) with the same fundamental group such that the boundary $(\partial W, f_W|_{\partial W}) = (\partial W, f_{\partial W})$ is diffeomorphic to disjoint union $(M_1, f_{M_1}) \cup (-M_2, f_{M_2})$, where the sign $-M_2$ denotes the exchange of orientation. Then the family of bordism classes forms the oriented bordism ring $\Omega_*(B\pi) = \sum_{n\geq 0} \Omega_n(B\pi)$, which is a module over the bordism ring of a point $\Omega_* = \Omega_*(\mathrm{pt})$. Multiplication by elements of the ring Ω_* is defined by the Cartesian product of the corresponding representatives:

$$\{(M, f_M), N\} \mapsto (M \times N, f_M \circ \mathrm{pr})$$

where $\mathrm{pr} : M \times N \to M$ is the projection, the manifold N being simply connected.

Remark. One can of course allow the manifolds representing the elements of the bordism group $\Omega_*(B\pi)$ to have fundamental groups which differ from π. But the restriction imposed above does not restrict generality (except in dimensions < 4) and it is useful for us for reasons of geometric clarity.

3.3

Let $\mathbf{C}\pi$ denote the involutive group algebra of the group π over the complex number field, and let $L_2(\pi)$ denote the Hilbert space of functions f on the group π for which

$$\sum_{g\in\pi} |f(g)|^2 < \infty.$$

Then left translation operation generates the symmetric representation

$$\varphi_\pi : \mathbf{C}\pi \to B(L_2(\pi))$$

where $B(L_2(\pi))$ is the algebra of bounded operators. The completion of the subalgebra $\varphi_\pi(\mathbf{C}\pi)$ under the operator norm we shall denote by $C^*[\pi]$. Then $C^*[\pi]$ is a C^*-algebra and the homomorphism of the algebra $\mathbf{C}\pi$ into $C^*[\pi]$ generated by the representation φ_π we shall denote with the same symbol φ_π (as well its restriction $\varphi_\pi : \pi \to C^*[\pi]$).

Let Λ be an involutive algebra over the complex number field. Denote by $K^*(\Lambda) = \sum_{i=0}^{3} K^i(\Lambda)$ the 4-periodic Hermitian K-theory which is isomorphic to the Wall groups $L_*(\mathbf{C}\pi)$ (see [18]). It is shown in [18] that one can associate with the manifold (M, f_M) the *symmetric signature* defined as the element $\sigma(M) \in K^*(\mathbf{C}\pi)$.

The construction of the element $\sigma(M)$ is based on the simplicial (co)chain group for the universal covering of the manifold M. Namely let $C^*(M, \mathbf{C}\pi)$ be the group of the cochains with value in the local coefficient system $\mathbf{C}\pi$ corresponding to the natural representation of π in $\mathbf{C}\pi$. Let

$$D : C^*(M, \mathbf{C}\pi) \longrightarrow C_*(M, \mathbf{C}\pi) = (C^*(M, \mathbf{C}\pi))^*$$

be the intersection operator with the open fundamental cycle, that is,

$$Dx = x \cap [\tilde{M}]$$

(more accurately, by its self-conjugate part $\frac{1}{2}(D + D^*)$). It is evident that $Dd + d^*D = 0$ under suitable choice of signs. If $\dim M = n$ is even one can obtain the commutative diagram

$$\begin{array}{ccc} C^{ev} & \xrightarrow{d} & C^{odd} \\ \downarrow D & & \downarrow D \\ (C^{ev})^* & \xrightarrow{d^*} & (C^{odd})^* \end{array}$$

which generates the isomorphism

$$C^{ev} \oplus (C^{odd})^* \xrightarrow{d+d^*+D+D^*} (C^{ev} \oplus (C^{odd})^*)^*$$

which is a quadratic form, that is an element of $L_{ev}(\mathbf{C}\pi)$. If $\dim M = n$ is odd one can obtain an element of $L_{odd}(\mathbf{C}\pi)$ in a similar way.

The element $\sigma(M)$ is both a bordism invariant and a homotopy equivalence invariant. The first means that the correspondence $(M, f) \mapsto \sigma(M)$ gives a homomorphism

$$\sigma : \Omega_*(B\pi) \to K^*(\mathbf{C}\pi). \tag{3.1}$$

Moreover, if $x \in \Omega_*(B\pi)$, $y \in \Omega_*$ then

$$\sigma(x \otimes y) = \sigma(x)\tau(y), \tag{3.2}$$

where $\tau(y)$ is the classical signature of the oriented manifold that represents the element $y \in \Omega_*$. Consider (3.1) after tensor multiplication by the rational number field $\mathbf{Q}$:

$$\sigma : \Omega_*(B\pi) \otimes \mathbf{Q} \to K^*(\mathbf{C}\pi) \otimes \mathbf{Q}. \tag{3.3}$$

The group $\Omega_*(B\pi) \otimes \mathbf{Q}$ can be expressed in terms of the usual homology groups of the space $B\pi$. In fact let $\Omega_*^f(X)$ denote the framed bordism ring. Let h be the natural Hurewicz homomorphism

$$h : \Omega_*^f(X) \to H_*(X; \mathbf{Z})$$

that for any singular (framed) manifold (M, f):

$$f : M \to X$$

associates the homology class

$$h(M, f) = f_*([M]) \in H_*(X; \mathbf{Z})$$

where $[M]$ is the fundamental cycle. Since $\Omega_*^f \otimes \mathbf{Q} = \Omega_0^f \otimes \mathbf{Q} = \mathbf{Q}$, the homomorphism

$$h : \Omega_*^f(X) \otimes \mathbf{Q} \to H_*(X; \mathbf{Q})$$

is an isomorphism. Consider the natural 'forgetful map' of the bordism groups

$$j : \Omega_*^f(X) \to \Omega_*(X)$$

and extend it to the homomorphism

$$j_* : \Omega_*^f(X) \otimes \Omega_* \otimes \mathbf{Q} \to \Omega_*(X) \otimes \mathbf{Q}$$

according to the formula

$$j_*((M, f) \otimes N) = (M \times N, f \circ \mathrm{pr}) \in \Omega_*(X),$$

$$(M, f) \in \Omega_*^f(X), \ N \in \Omega_*.$$

It is evident that j_* is an isomorphism. Hence the following homomorphism is an isomorphism as well:

$$j_*(h^{-1} \otimes 1) : H_*(X; \mathbf{Q}) \otimes \Omega_* \xrightarrow{h^{-1} \otimes 1} \Omega_*^f(X) \otimes \mathbf{Q} \otimes \Omega_* \xrightarrow{j_*} \Omega_*(X) \otimes \mathbf{Q}.$$

Let L be the multiplicative Hirzebruch genus, that is the invertible characteristic class such that

a) $L(\xi \oplus \eta) = L(\xi)L(\eta)$ for any oriented vector bundles ξ, η;

b) for any oriented manifold X, its signature is expressed by the formula

$$\tau(X) = 2^{2k}\langle L(X), [X]\rangle.$$

Then the homomorphism (3.3) can be interpreted as the following. Consider the composition mapping

$$\sigma \circ (j_* \circ (h^{-1} \otimes 1)) : H_*(B\pi; \mathbf{Q}) \otimes \Omega_* \to \Omega_*(B\pi) \otimes \mathbf{Q} \to K^*(\mathbf{C}\pi) \otimes \mathbf{Q}. \quad (3.4)$$

According to (3.2) if $x \in H_*(B\pi; \mathbf{Q})$, $y \in \Omega_*$ then

$$\sigma \circ j_* \circ (h^{-1} \otimes 1)(x \otimes y) = (\sigma \circ j_* \circ (h^{-1} \otimes 1)(x \otimes 1))\tau(y), \qquad (3.5)$$

hence

$$\sigma \circ j_* \circ (h^{-1} \otimes 1) = \overline{\sigma} \otimes \tau, \qquad (3.6)$$

where

$$\overline{\sigma} = (\sigma \circ j_* \circ (h^{-1} \otimes 1))|_{H_*(X;\mathbf{Q})\otimes 1} : H_*(B\pi; \mathbf{Q}) \to K^*(\mathbf{C}\pi) \otimes \mathbf{Q}. \qquad (3.7)$$

Therefore the homomorphism $\overline{\sigma}$ can be considered as the cohomology class

$$[\overline{\sigma}] \in H^*(B\pi; K^*(\mathbf{C}\pi) \otimes \mathbf{Q}). \qquad (3.8)$$

Then the following formula is true:

$$\sigma(M, f_M) = \langle L(M) f_M^*([\overline{\sigma}]), [M]\rangle \in K^*(\mathbf{C}\pi) \otimes \mathbf{Q}. \qquad (3.9)$$

In fact the element $[M, f_M] \in \Omega_*(X) \otimes \mathbf{Q}$ can be decomposed as the sum

$$[M, f_M] = \sum_\alpha [M_\alpha, f_{M_\alpha}] \otimes [N_\alpha],$$

where $[M_\alpha, f_{M_\alpha}] \in \Omega_*^f(B\pi)$, $[N_\alpha] \in \Omega_*$. Then the lefthand part of (3.9) has the form

$$\sigma(M, f_M) = \sum_\alpha \sigma(M_\alpha f_{M_\alpha}) \tau(N_\alpha).$$

The righthand part of (3.9) can be calculated in the following way:

$$\begin{array}{rl} \langle L(M) f_M^*([\overline{\sigma}]), [M]\rangle & = \\ = \sum_\alpha \langle L(M_\alpha \times N_\alpha) f_{M_\alpha \times N_\alpha}^*([\overline{\sigma}]), [M_\alpha \times N_\alpha]\rangle & = \\ = \sum_\alpha \langle (1 \otimes L(N_\alpha))(f_{M_\alpha}^*([\overline{\sigma}] \otimes 1), [M_\alpha \times N_\alpha]\rangle & = \\ = \sum_\alpha \langle f_{M_\alpha}^*([\overline{\sigma}]), [M_\alpha]\rangle \langle L(N_\alpha), [N_\alpha]\rangle & = \\ = \sum_\alpha \langle [\overline{\sigma}], (f_{M_\alpha})_*[M_\alpha]\rangle \tau[N_\alpha] & = \\ = \sum_\alpha \langle [\overline{\sigma}], h(M_\alpha, f_{M_\alpha})\rangle \tau[N_\alpha] & = \\ = \sum_\alpha \sigma(M_\alpha, f_{M_\alpha}) \tau[N_\alpha] & . \end{array}$$

The formula (3.9) gives the whole family of higher signatures which are homotopy invariant.

4 Fredholm representations controlled at infinity

Let $H_1, H_2, \cdots, H_n$ be a sequence of Hilbert spaces connected by bounded operators $F_1, \cdots, F_{n-1}$ giving the Fredholm complex

$$0 \longrightarrow H_1 \xrightarrow{F_1} H_2 \xrightarrow{F_2} H_3 \longrightarrow \cdots \longrightarrow H_{n-1} \xrightarrow{F_{n-1}} H_n \longrightarrow 0. \qquad (4.1)$$

This means that the composition of the operators $F_{i+1} \circ F_i$ is a compact operator:

$$F_{i+1} \circ F_i \in C(H_i, H_{i+2}) \tag{4.2}$$

and there exist 'adjoint' operators

$$0 \longleftarrow H_1 \xleftarrow{G_2} H_2 \xleftarrow{G_3} H_3 \longleftarrow \cdots \longleftarrow H_{n-1} \xleftarrow{G_n} H_n \longleftarrow 0 \tag{4.3}$$

such that

$$F_{j-1}G_j + G_{j+1}F_j - 1 \in C(H_j). \tag{4.4}$$

It is convenient to represent the definition (4.1)–(4.4) of the Fredholm complex in graded form. Let $H = \oplus H_i$ be a graded Hilbert space, let

$$F : H \longrightarrow H \tag{4.5}$$

be a homogeneous (of degree 1) bounded operator such that

$$F \circ F \in C(H), \tag{4.6}$$

there exists a homogeneous bounded operator (of degree -1)

$$G : H \longrightarrow H \tag{4.7}$$

such that

$$F \circ G + G \circ F - 1 \in C(H). \tag{4.8}$$

The operator G can be chosen in different ways, but it can always be chosen so that

$$G \circ G \in C(H). \tag{4.9}$$

From conditions (4.5)–(4.9) it follows that the operator

$$\tilde{F} = (F + G) : H \to H \tag{4.10}$$

is a Fredholm operator homogeneous of degree one under the $\mathbf{Z}_2$-grading of the space H. In other words

$$(F + G)(H^{ev}) \subset H^{odd},$$

$$(F + G)(H^{odd}) \subset H^{ev},$$

where

$$H^{ev} = \oplus H_{2i}, \quad H^{odd} = \oplus H_{2i+1}.$$

Denote

$$\tilde{F}^{ev} = (F + G)|_{H^{ev}}, \quad \tilde{F}^{odd} = (F + G)|_{H^{odd}}. \tag{4.11}$$

Let T be a unitary graded representation (of degree 0) of the group π in the Hilbert space H. Then the pair (F, T) is called a *Fredholm representation* if the operator F satisfies the conditions (4.5)–(4.8) and additionally if the following condition holds

$$FT(g) - T(g)F \in C(H) \tag{4.12}$$

for any element $g \in \pi$.

The condition (4.12) implies that

$$\tilde{F}T(g) - T(g)\tilde{F} \in C(H). \tag{4.13}$$

The pair $(\tilde{F}^{ev}, T)$ is a Fredholm representation (on a short exact sequence of Hilbert spaces), and as well it satisfies

$$\tilde{F}^{ev}T(g) - T(g)\tilde{F}^{ev} \in C(H^{ev}, H^{odd}). \tag{4.14}$$

Therefore we shall focus our attention on the operators (4.10) and (4.11).

We need to consider the notion of a trivial Fredholm representation. At first, it seems sufficient to substitute zero for the algebra $C(H)$ in the conditions (4.8) and (4.12):

$$F \circ G + G \circ F - 1 = 0. \tag{4.15}$$

$$\tilde{F}T(g) - T(g)\tilde{F} = 0. \tag{4.16}$$

Conditions (4.15, 4.16) were used to define trivial elements in the KK-theory of G. Kasparov. But really this modification will not be sufficient.

Indeed, in [9] there is an attempt to use some weaker conditions than (4.15), (4.16). There the conditions (4.8, 4.12) were replaced by the stronger condition (4.15). But the class of Fredholm representations which satisfy the condition (4.15) cannot be considered as trivial elements in KK-theory, because there are no reasons which lead to the triviality of the homomorphism

$$\rho^* : K_A^*(X) \longrightarrow K_{comp}^*(X \times Y).$$

In particular, in [9] on page 86 the following is written: "...if $\tau = \tau(Y, B)$ then A is exact $(\operatorname{mod} C)$ and it is easily seen that $\bar{A}^{(j)}$ must also be exact over B." This statement seems not to be true. As a counterexample consider the splitting ([9], p. 85)

$$Y \times (H_0^{(1)} \oplus H_0^{(2)}) \xrightarrow{\bar{A}} Y \times (H_1^{(1)} \oplus H_1^{(2)})$$

with $\operatorname{index} \bar{A} = 0$, $\operatorname{index} A^{(1)} \neq 0$, $\operatorname{index} A^{(2)} \neq 0$. Then there exists a compact operator $K \in C$ such that $\bar{A} + K$ is invertible but $A^{(1)} + K^{(1)}$ will never be invertible.

The right way is to change not only condition (4.8) but also condition (4.12) as it was formulated in the Definition 1.

Now we shall give a proof of Theorem 1 by induction on the cardinality of the covering family $\{U_\alpha\}$ of X. Without loss of generality one can assume that the covering $\{U_\alpha\}$ consists of the stars of vertices of some simplicial division on X. Let $\Phi_\alpha(x,y)$ be constructed for all $\alpha < \alpha_0$ and satisfy the conditions (2.5), (2.6), (2.7). The set $\overline{U_{\alpha_0}}\backslash(\cup_{\alpha<\alpha_0}U_\alpha \cup X_0)$ is covered by some simplices whose some faces lie in $\overline{\cup_{\alpha<\alpha_0}U_\alpha}$ and there, due to (2.5), Φ_{α_0} is defined. Therefore there exists an extension of Φ_{α_0} from $U_{\alpha_0} \cap \overline{\cup_{\alpha<\alpha_0}U_\alpha}$ to U_{α_0} satisfying (2.6). Moreover for any $\varepsilon > 0$ there exists a compact K such that for any $x \in (\cup_{\alpha<\alpha_0}U_\alpha \cup X_0)$, $y \in Y\backslash K$ the condition (2.7) is holds. Hence there exists an extension Φ_{α_0} to U_{α_0} such that condition (2.7) is true (for some sufficiently small $\varepsilon' > 0$).

Remarks.

1. If $\varphi'_{\alpha\beta}$ are other transition functions, that is if

$$\varphi'_{\alpha\beta}(x) = h_\beta(x)\varphi_{\alpha\beta}(x)h_\alpha^{-1},$$

where

$$h_\alpha : U_\alpha \longrightarrow \mathrm{Aut}_A(V),$$

then the operators

$$\Phi'_\alpha(x,y) = T_y^*(h_\alpha(x))\Phi_\alpha(x,y)T_y^*(h_y^{-1}(x))$$

are a realization of the same representation ρ^* on the same bundle ξ.

2. Let G be a discrete group acting on $\tilde{X}$ and let $X = \tilde{X}/G$. Let $\tilde{\xi} = (\tilde{E}\longrightarrow\tilde{X})$ be an equivariant bundle and

$$\tilde{\Phi} : \tilde{E}_A \otimes H \longrightarrow \tilde{E}_A \otimes H$$

be an equivariant homomorphism satisfying (2.5), (2.6), (2.7). Then

$$\Phi/G : E_A \otimes H \longrightarrow E_A \otimes H,$$

where $E = \tilde{E}/G$ satisfies the conditions (2.5), (2.6), (2.7) as well.

5 Remarks

1. The notion of a family of Fredholm representations with compact supports one can interpret as a continuous family of Fredholm representations in some new topology on the space of Fredholm representations. Namely a neighborhood of the trivial element consists of not only of the Fredholm representations which are near the trivial element with

respect to the uniform norm but also the elements (F, T) which satisfy the conditions

$$\|F\| \leq C, \ \|F^{-1}\| \leq C, \ \ \|FT(g) - T(g)F\| < \varepsilon, \ g \in \pi_0 \subset \pi \quad (5.1)$$

where π_0 is a finite subset of π. The family $O(C, \varepsilon, \pi_0)$ of all Fredholm representations which satisfy the conditions (5.1) is a neighborhood of the trivial element in the space of all Fredholm representations.

Therefore the family of the Fredholm representations with compact supports is a continuous family from the one point compactification Y^+ to the space of the Fredholm representations with new topology.

2. The topology defined above is not homogeneous. It would be interesting to give similar definition of neighborhoods for arbitrary point of the space of Fredholm representations.

3. One can extend the notion of special manifold which was considered in [27] to the class of non-compact manifolds which satisfy similar conditions with slight modifications. Namely, call a complete Riemannian manifold M special if:

   (1) The universal cover $\tilde{M}$ is diffeomorphic to $\mathbf{R}^n$ and

   (2) There is a point $\tilde{x}_0 \in \tilde{M}$ and a vector field ω such that ω is nontrivial away from some compact $K \subset \tilde{M}$ and index $\omega = +1$, and there exist constants C_1, C_2 such that

$$0 < C_1 < \|\omega(x)\| < C_2, \ x \in (\tilde{M} \backslash K),$$

   and for all $g \in \pi_1(M)$

$$\lim_{x \longrightarrow \infty} \|g_* \omega(x) - \omega(gx)\| = 0.$$

   It is clear that for this larger class of manifolds the theorem on homotopy invariance of the higher signatures holds.

4. The construction of a new topology on the space of Fredholm representations shows that in a similar way one can introduce a notion of quasi-representation of a discrete group π. In [4] the notion of almost flat bundles on a manifold M was introduced.

   Namely, let $\alpha \in K^0(M)$ be an element of the K-theory on a (compact) Riemannian manifold M, and let $\alpha = (E^+, \nabla^+) - (E^-, \nabla^-)$, where E^+, E^- are hermitian complex vector bundles with connections ∇^+, ∇^-. One can say that the element α is almost flat if for any $\varepsilon > 0$ there exists a representation $\alpha = (E^+, \nabla^+) - (E^-, \nabla^-)$ such that

$$\|(E^+, \nabla^+)\| \leq \varepsilon, \ \|(E^-, \nabla^-)\| \leq \varepsilon,$$

where

$$\|(E,\nabla)\| = \mathrm{Sup}_{x\in M}\{\|\theta_x(X\wedge Y)\| : \|X\wedge Y\| \leq 1\},$$

and $\theta = \nabla^2$ is the curvature.

Then for an almost flat element $\alpha \in K^0(M)$ and any $\varepsilon > 0$ and a finite subset $F \subset \pi$ there exists a representation $\alpha = (E^+,\nabla^+) - (E^-,\nabla^-)$ with

$$\|(E^+,\nabla^+)\| \leq \varepsilon,\ \|(E^-,\nabla^-)\| \leq \varepsilon,$$

such that corresponding quasi-representations

$$\sigma^+ : \pi \longrightarrow U(N^+),\ \sigma^- : \pi \longrightarrow U(N^-)$$

have the following property:

$$\|\sigma^+\|_F \leq \varepsilon,\ \|\sigma^-\|_F \leq \varepsilon,$$

where

$$\|\sigma\|_F = \mathrm{Sup}\{\|\sigma(ab) - \sigma(a)\sigma(b)\| : a,b \in F\}.$$

This notion has a shortcoming because the choice of quasi-representation depends on the property of almost flatness, which in its turn depends on the smooth structure on the manifold M. It would be interesting to construct a natural inverse correspondence from the family of quasi-representations of a discrete group to the family of almost flat bundles. I suggest a new definition of quasi-representation of a discrete group π, such that there is a natural map from the family $\mathcal{R}(\pi)$ of such quasi-representations to $K(B\pi)$.

Definition. Let

$$\sigma = \{\sigma_n : \pi \longrightarrow U(N_n) \subset U(\infty)\}$$

be a sequence of maps (here we think of all the finite unitary groups as embedded in in the inductive limit $U(\infty) = \lim_{\rightarrow} U(N)$) such that for any $\varepsilon > 0$ and finite subset $F \subset \pi$ there exists a number N_0 such that if $n_1, n_2 > N_0$ then

$$\|\sigma_{n_1}(a) - \sigma_{n_2}(a)\| \leq \varepsilon,\ a \in F,$$

$$\|\sigma_{n_1}\|_F \leq \varepsilon.$$

Then one can construct a natural homomorphism

$$\phi : \mathcal{R}(\pi) \longrightarrow K(B\pi)$$

and a pairing

$$L_*(\mathbf{C}\pi) \otimes \mathcal{R}(\pi) \longrightarrow \mathbf{Z},$$

such that one has the following formula

$$\rho(\sigma(M)) = \langle L(M)\mathrm{ch}(\phi(\rho)), [M]\rangle = \sigma_x(M),$$

where $x = \mathrm{ch}(\phi(\rho))$. Therefore the numbers $\sigma_x(M)$ are homotopy invariant for all $x \in \mathrm{Im}\ \mathrm{ch} \circ \phi \subset H^*(B\pi)$. It seems that the family of quasi-representations gives more homotopy invariant higher signatures.

One can extend the notion of quasi-representations to Fredholm quasi-representations and construct a corresponding Kasparov KK-theory. It is interesting to compare the Kasparov KK-theory with this generalization.

References

[1] M. F. Atiyah, *K-Theory*, Benjamin, New York, 1967.

[2] W. Browder, *Homotopy type of differential manifolds*, Colloquium on Algebraic Topology (1962, Århus Universitet), Århus, Matematisk Institut, 1962, pp. 42–46; reprinted in these proceedings.

[3] S. E. Cappell, *Manifolds with fundamental group a generalized free product, I*, Bull. Amer. Math. Soc. **80** (1974), 1193–1198.

[4] A. Connes, M. Gromov et H. Moscovici, *Conjecture de Novikov et fibrés presque plats*, C. R. Acad. Sci. Paris **310**, Série I, (1990), 273–277.

[5] A. Connes, M. Gromov and H. Moscovici, *Group Cohomology with Lipschitz Control and Higher Signatures*, Geom. Funct. Anal. **3** (1993), 1–78.

[6] A. Connes and H. Moscovici, *Cyclic Cohomology, The Novikov Conjecture and Hyperbolic Groups*, Topology **29** (1990), No. 3, 345–388.

[7] F. T. Farrell and W.-C. Hsiang, *On Novikov's conjecture for non-positively curved manifolds, I*, Annals of Math. **113** (1981), 199–209.

[8] F. Hirzebruch, *On Steenrod's reduced powers, the index of inertia, and the Todd genus*, Proc. Nat. Acad. Sci. U.S.A. **39** (1953), 951–956.

[9] W. C. Hsiang and H. D. Rees, *Mishchenko's work on Novikov's conjecture*, Contemporary Mathematics, v. **10** (1982), 77–98.

[10] G. G. Kasparov, *The homotopy invariance of rational Pontrjagin numbers*, Dokl. Akad. Nauk SSSR, **190** (1970), 1022–1025 = Soviet Math. Dokl. **11** (1970), 235–238.

[11] G. G. Kasparov, *Equivariant KK-theory and the Novikov conjecture*, Invent. math. **91** (1988), 147–201.

[12] G. G. Kasparov, *The generalized index of elliptic operators*, Funct. Analysis and its Applications **7** (1973), No. 3, 238–240.

[13] G. G. Kasparov, *K-theory, Group C^*-algebras and Higher Signatures*, Preprint, I, II, Chernogolovka (1981); reprinted in these proceedings.

[14] G. G. Kasparov, *Topological invariants of elliptical operators, I*, Izv. Akad. Nauk SSSR, Ser. Mat. **39** (1975), 796–838 = Math. USSR–Izv. **9** (1975), 751–792.

[15] G. Lusztig, *Novikov's higher signature and families of elliptic operators*, J. Diff. Geometry **7** (1972), 229–256.

[16] A. S. Mishchenko, *Homotopy invariants of non simply connected manifolds. Rational invariants, I*, Izv. Akad. Nauk SSSR, Ser. Mat. **34** (1970) No. 3, 501–514 = Math. USSR–Izv. **4** (1970), 506–519.

[17] A. S. Mishchenko, *Infinite-dimensional representations of discrete groups and homotopy invariants of multiply-connected manifolds*, Uspechi Mat. Nauk **28** (1973) No. 2, 239–240.

[18] A. S. Mishchenko, *Infinite-dimensional representations of discrete groups and higher signatures*, Izv. Akad. Nauk SSSR, Ser. Mat. **38** (1974), 81–106 = Math. USSR–Izv. **8** (1974), 85–111.

[19] A. S. Mishchenko, *Hermitian K-theory, The Theory of Characteristic Classes and Methods of Functional Analysis*, Russian Math. Surveys **31** (1976), No. 2, 71–138.

[20] A. S. Mishchenko, *C^*-algebras and K-theory*, Algebraic Topology, Århus, 1978, Lecture Notes in Math., No. 763, Springer-Verlag, Berlin–Heidelberg, 1979, pp. 262–274.

[21] A. S. Mishchenko, *The theory of elliptic operators over C^*-algebras*, Dokl. Akad. Nauk SSSR **239** (1978), No. 6 = Soviet Math. Dokl. **19** (1978), No. 2, 512–515.

[22] A. S. Mishchenko and A. T. Fomenko, *The index of elliptic operators over C^*-algebras*, Izv. Akad. Nauk SSSR **43** (1979), 831–859 = Math. USSR–Izv. **15** (1980), 87–112.

[23] A. S. Mishchenko and Ju. P. Solov'ev, *Representations of Banach algebras and formulas of Hirzebruch type*, Mat. Sb. **111** (1980) No. 2 = Math. USSR Sbornik **39** (1981), No. 2, 189–205.

[24] S. P. Novikov, *Homotopy equivalent smooth manifolds, I*, Izv. Akad. Nauk SSSR, Ser. Mat. **28** (1964), 365–474.

[25] S. P. Novikov, *The homotopy and topological invariance of certain rational Pontrjagin classes*, Dokl. Akad. Nauk SSSR **162** (1965), 1248–1251 = Soviet Math. Dokl. **6** (1965), 854–857.

[26] S. P. Novikov, *Rational Pontrjagin classes. Homeomorphism and homotopy type of closed manifolds, I*, Izv. Akad. Nauk SSSR, Ser. Mat. **29** (1965), No. 6, 1373–1388.

[27] H. D. Rees, *Special manifolds and Novikov's conjecture*, Topology **22** (1983), 365–378.

[28] V. A. Rokhlin, *New results in the theory of four-dimensional manifolds*, Dokl. Akad. Nauk SSSR **84** (1952), 221–224.

[29] V. A. Rokhlin, *Pontrjagin-Hirzebruch class of codimension 2*, Izv. Akad. Nauk SSSR, Ser. Mat. **30** (1966), 705–718.

Department of Mathematics and Mechanics, Moscow State University, 119899 Moscow, Russia

email: `asmish@mech.math.msu.su`

Assembly maps in bordism-type theories

Frank Quinn

Preface

This paper is designed to give a careful treatment of some ideas which have been in use in casual and imprecise ways for quite some time, particularly some introduced in my thesis. The paper was written in the period 1984–1990, so does not refer to recent applications of these ideas.

The basic point is that a simple property of manifolds gives rise to an elaborate and rich structure including bordism, homology, and "assembly maps." The essential property holds in many constructs with a bordism flavor, so these all immediately receive versions of this rich structure. Not everything works this way. In particular, while bundle-type theories (including algebraic K-theory) also have assembly maps and similar structures, they have them for somewhat different reasons.

One key idea is the use of spaces instead of sequences of groups to organize invariants and obstructions. I first saw this idea in 1968 lecture notes by Colin Rourke on Dennis Sullivan's work on the Hauptvermutung ([21]). The idea was expanded in my thesis [14] and article [15], where "assembly maps" were introduced to study the question of when PL maps are homotopic to block bundle projections. This question was first considered by Andrew Casson, in the special case of bundles over a sphere. The use of obstruction spaces instead of groups was the major ingredient of the extension to more general base spaces. The space ideas were expanded in a different direction by Buoncristiano, Rourke, and Sanderson [4], to provide a setting for generalized cohomology theories.

Another application of these ideas was a "homological" description of the surgery sequence. The classical formulation of this sequence describes "normal maps" as a cohomology group—in particular contravariant—while the surgery obstruction is covariant. Applying duality in generalized homology describes the normal map set as a homology group and relates the classical surgery obstruction to an assembly map. This idea was made precise and useful by Andrew Ranicki [19].

The careful development of the material in the generality given here was largely motivated by the work of Lowell Jones [10], [11]. He developed an approach to the classification of piecewise linear actions of cyclic groups as a profound application of surgery theory. This material allows direct recognition of one of Jones' obstructions as a generalized homology class with coefficients in the "fiber of the transfer." The relation to Jones' work is sketched in section 6.4.

I would like to thank Andrew Ranicki for his encouragement over the years to bring this work into the light.

0: Introduction

Let J_n represent a group-valued functor of spaces, for example bordism groups, or Wall surgery groups, or algebraic K-groups (of the fundamental group). In these examples there is an associated generalized homology theory $H_*(X; \mathbf{J})$ and a natural homomorphism

$$H_*(X; \mathbf{J}) \to J_*(X)$$

called the "assembly." These homomorphisms are important for two reasons; they offer a first step in the computation of the functors $J_*(X)$, and some of them arise in geometric situations. For example the assembly map for surgery groups is closely related to surgery obstructions, and the "Novikov conjecture" is equivalent to rational injectivity of the assembly when X is a $K(\pi, 1)$.

The objective is to give two descriptions of these homomorphisms. The first description is very general, in the context of homology with coefficients in a spectrum-valued functor. This yields a wealth of naturality properties and useful elaborations. However it is difficult to see specific elements, particularly homology classes, from this point of view.

The second description is complementary to this. For certain types of theories homology classes can be described explicitly in terms of "cycles." Assembly maps are directly defined by "glueing" (assembling) the pieces in a cycle. This gives an explicit element-by-element view which is good for specific calculations and recognizing homology classes when they occur as obstructions. But the naturality properties become obscure.

The main result is that these two constructions do in fact describe the same groups, spaces, maps, etc. Special cases been used in calculations of surgery groups and obstructions [14], [15], [5], [28], [8]. With this description the assembly map is seen to describe obstructions for certain block bundle problems [15], and constructions of PL regular neighborhoods [10], [11]. The cycle description is well adapted to constructions divided into blocks, like PL regular neighborhoods of polyhedra.

This paper begins with definitions of various types of homology; generalized, twisted, Čech, and spectral sheaf. This logically comes first, but the reader may find it more interesting to begin with the bordism material of section 3.

The second section defines homology with coefficients in a spectrum-valued functor. Assembly maps are part of the functorial structure of these homology theories. The idea is to begin with a map $p\colon E \to X$, apply the functor fiberwise to point inverses of p to get a "spectral sheaf" over X, and take the homology of this. In this setting the usual assembly appears as a morphism induced by a map of data: the constant coefficient homology $H_*(X;\mathbf{J})$ is the $\mathbf{J}$-coefficient homology of the identity map $X \to X$, the groups $J_*(X)$ are the homology of the point map $X \to pt$, and the assembly is induced by the diagram

$$\begin{array}{ccc} X & \xrightarrow{=} & X \\ \downarrow{=} & & \downarrow \\ X & \longrightarrow & pt \end{array}$$

regarded as a morphism from the identity to the point map.

From this point of view the usual assembly is a small part of a rich structure: there are lots of maps more interesting than the identity and the point map. Special cases were defined by D. W. Anderson [2], and in algebraic K-theory by Loday [12] and Waldhausen [24].

Bordism-type theories are described in section 3. This is a fairly primitive notion, designed so the conditions can be easily verified in examples. These have associated bordism groups, and bordism spectra. The spectrum construction is used to define functors which satisfy the conditions of the second section, so homology with coefficients in these spectra are defined.

This description applies naturally to surgery groups, and bordism groups defined using manifolds, Poincaré spaces, normal spaces, or chain complexes. There is an existence theorem in 3.7 which asserts that one can contrive to obtain any homology theory from a bordism-type theory. However we regard it as a conceptual error to use this result: the theory is designed to take advantage of special structure in a class of examples, and has no special benefits as an approach to general theory.

Roughly speaking the approach applies to theories with classifying spaces which are simplicial complexes satisfying the Kan condition. There is a general-nonsense construction which replaces a space with a Kan complex of the same weak homotopy type, and this gives the existence theorem. However, for example, the natural classifying spaces for algebraic K-theory do not satisfy the Kan condition, so the approach does not naturally or usefully apply to K-theory.

In section 4 "cycles" are introduced as representatives for homology classes in bordism-type theories. These are defined on covers of the space, and associate to an element of the cover a "fragment" of an object, with various "faces." The prototypical example of such a fragment is a manifold, with its boundary subdivided into submanifolds (this example leads to bordism, hence the title of the paper). The pieces of a cycle fit together; over an intersection of two elements of the cover corresponding faces of the fragments are equal. Over three-fold intersections certain "edges" agree, etc. The assembly map simply glues (assembles) the pieces together using these identifications to get a single object.

Another way to view cycles is in terms of transversality. Suppose X is a finite simplicial complex. The dual cone (or cell) decomposition of X describes X as being assembled from pieces, each a cone on a union of smaller pieces. The boundaries of largest cones are bicollared in X; boundaries of smaller cones are bicollared as subsets of the boundary of the next larger cones. A manifold could therefore be made transverse to all these cones. This breaks the manifold into pieces over each maximal cone, intersecting in faces over the next smaller cones, etc. A cycle is an abstraction of this pattern. Thus a $\mathbf{J}$-cycle may be thought of as a $\mathbf{J}$-object which is transverse to a dual cone decomposition.

There is an associated description of cocycles, representing cohomology classes, given in 4.7. These associate to each simplex of a complex an object with the same pattern of faces as the simplex. Since there is a correspondence between simplices and dual cones, a cycle also associates an object to each simplex. But the objects in a cycle have faces corresponding to faces of the cone dual to the simplex, rather than the simplex itself. So for example in a cocycle dimensions of associated objects increase with dimension of the simplex, while in a cycle the dimension decreases.

Section 5 contains the proof of the main theorem, that cycles represent homology classes.

The final sections presents examples of bordism-type theories, and applications of the representation theorem. In section 6.1 manifolds are shown to form a bordism-type theory. Details are included as a model for verifications in other contexts. In 6.2 this is extended to manifolds with a map to a space. This construction defines a manifold-type theory depending functorially on a space, so leads to a full array of functor-coefficient homology groups, assembly maps, etc.

Transversality is used to show that the assembly maps in the manifold theories are isomorphisms. This is an analog of the classical Pontrjagin-Thom theorem that the bordism groups form a homology theory represented by the Thom spectrum.

This suggests thinking of assemblies in general bordism-type theories in terms of transversality. The fiber of the assembly map, which measures the deviation from isomorphism, then classifies obstructions to transversality.

Poincaré chain complexes are considered in 6.3, and the relation of this development to the work of Ranicki [20] and Weiss [26] is briefly described.

Finally in section 6.4 we sketch a sophisticated application. This begins with the observation that in some circumstances PL regular neighborhoods are equivalent to PL manifold cycles. Then a formulation of surgery in these terms gives a classification of manifold structures on Poincaré cycles. Putting these observations together gives a way to construct PL regular neighborhoods. In particular an obstruction encountered by Jones [10], [11] in the construction of PL group actions is reformulated as a generalized homology class.

Important topics not covered here are applications to surgery classification problems, product structures and duality, and computational aids like spectral sequences.

We mention that there is another class of theories with a description of homology classes and the assembly. These are the controlled theories, which deal with objects with a naturally associated "size," over a metric space. The representation theorem asserts that objects with sufficiently small size represent homology classes. The assembly map simply forgets the size restriction. These theories are well adapted to problems where things cannot be broken into blocks, for example in the study of purely topological neighborhoods [18]. The methods are more those of sheaf theory with things given on overlapping open sets, rather than the articulated fragments of the bordism-type theories.

This paper can be considered a completed version of the author's thesis, where some limited assembly maps for surgery were described, and the term "assembly" was introduced.

1: Homology

Generalized homology spectra (with coefficients in a spectrum) are defined in 1.1, and extended in 1.2 to homology with spectral sheaf coefficients. Twisted homology is discussed in 1.3 as a special case. Finally in 1.4 there is a description of Čech (or "shape") homology, which will be the setting for the general theory.

1.1 Spectra

A "spectrum" is a sequence of based spaces J_n together with based maps $j_n\colon J_n \wedge S^1 \to J_{n+1}$; see Whitehead [27] The spaces in a spectrum will be understood to be compactly generated space with the homotopy type of a CW complex.

We usually require these to be Ω-spectra in the sense that the adjoint of the structure map $J_n \to \Omega J_{n+1}$ is a homotopy equivalence. An arbitrary spectrum has a canonically associated Ω-spectrum with n^{th} space $\operatorname{holim}_{i\to\infty} \Omega^i J_{n+i}$. Generally the homotopy limits used here will be those defined by Bousfield and Kan [3]. In this particular case (a countable ordered direct system) it is just the mapping telescope (union of the mapping cylinders). An Ω-spectrum will be denoted by a boldface character; $\mathbf{J}$.

Given a spectrum $\mathbf{J}$ and a pair (X, Y), the *homology spectrum* $\mathbf{H}_\bullet(X, Y; \mathbf{J})$ is defined to be the Ω-spectrum associated to the spectrum $(X/Y) \wedge J_*$.

Referring to the definition just above of "associated Ω-spectrum" we see that the nth space in the homology spectrum is given by

$$\operatorname*{holim}_{i\to\infty} \Omega^{i-n}(X/Y \wedge J_i).$$

Homology *groups* are defined (by Whitehead [27]) to be the homotopy groups of the homology spectrum.

We can at this point describe the simplest example of an assembly (see also [2]). Suppose $\mathbf{J}$ is a functor from spaces to spectra. Then there is a natural transformation

$$X = \operatorname{maps}(\mathrm{pt}, X) \to \operatorname{maps}(\mathbf{J}(\mathrm{pt}), \mathbf{J}(X)).$$

The adjoint of this is a map $X_+ \wedge \mathbf{J}(\mathrm{pt}) \to \mathbf{J}(X)$. But the left side of this gives the homology spectrum, so this is a map from the homology of X to $\mathbf{J}(X)$.

1.2 Spectral sheaf homology

We think of the homology $\mathbf{H}_\bullet(X, Y; \mathbf{J})$ as homology with coefficients in the constant coefficient system given by $\mathbf{J}$ over each point in X. The twisted coefficient construction extends this to coefficient systems which are "locally constant"; fibered over X. The next step is to generalize to coefficient systems which vary almost arbitrarily. This construction is based on Quinn [16, §8].

The description takes place in the category of "spaces over X" described by James [9]. Fix a base space X, then a space over X is E together with maps $i\colon X \to E$ and $p\colon E \to X$ whose composition is the identity. Maps in the category are continuous maps $E \to F$ which commute with the inclusion of, and projection to, X.

The "suspension" of a space over X is given by $S^k_X E = S^k \times E/\sim$, where the equivalence relation $\sim$ identifies each $S^k \times p^{-1}(x)$ to a point, and $i\colon X \to S^k_X E$ takes x to this identification point. A spectrum in this category is therefore a sequence E_n of spaces over X together with maps $E_n \to S^1_X E_{n+1}$. Note that over each point the sequence of spaces $p_n^{-1}(x)$ form an ordinary spectrum (except they might violate our convention about having the homotopy type of CW complexes).

We refer to spectra in the category of spaces over X as *spectral sheaves* over X. There are technical connection with ordinary sheaves, but at this point the name is primarily intended to be suggestive.

The simplest examples of these spectral sheaves are products $\mathbf{J} \times X$. Then come the twisted products $\mathbf{J} \times_G \hat{X}$ described in the next section. More elaborate examples will be constructed in section 2.3.

We now define homology with coefficients in a spectral sheaf. As motivation note that in the constant coefficient case we begin with the total space of the product sheaf $\mathbf{J} \times X$, divide out X to get an ordinary spectrum, and pass to the associated Ω-spectrum. More generally, note that identifying the image of X to a point in a suspension over X gives the ordinary suspension; $S^1_X E/i(X) = S^1 E$. Therefore if $\{E_n\}$ is a spectral sheaf over X then an ordinary spectrum is obtained by dividing out X. The homology is the Ω-spectrum associated to this ordinary spectrum:

$$\mathbf{H}_\bullet(X; E) = \operatorname*{holim}_{n \to \infty} \Omega^n(E_n/i(X)).$$

Similarly if $Y \subset X$ then the relative homology is defined by dividing out both X and the inverse image of Y; $E_n \;/\; (i(X) \cup p^{-1}(Y))$.

We caution that we have not included the hypothesis that these "spectra" should have the homotopy type of CW complexes. To ensure the smooth functioning of the machinery of homotopy theory it is important to restrict to cases where this can be verified.

1.3 Twisted homology

This is defined to give a class of examples of the general theory. It will not be used here, so can be skipped by the purposeful reader. This construction does occur in spectral sequences describing general spectral sheaf homology in terms of simpler objects.

Suppose that G is a discrete group (see below for a non-discrete version) which acts on the spectrum $\mathbf{J}$, and $\omega\colon \pi_1 X \to G$ is a homomorphism. We use this data to define a spectrum denoted $\mathbf{H}_\bullet(X; \mathbf{J}, \omega)$. Let $\hat{X} \to X$ denote the covering space with G action associated to ω. Then define twisted homology to be the Ω-spectrum associated to the spectrum $(\hat{X}/\hat{Y}) \wedge_G \mathbf{J}$. More explicitly this means take $(\hat{X}/\hat{Y}) \times \mathbf{J}_n$, divide by the diagonal G action, and identify the invariant subset $(\hat{X}/\hat{Y}) \vee \mathbf{J}_n$ to a point.

In the terms of the previous section, $\hat{X} \times_G \mathbf{J}$ is a spectral sheaf over X, and the twisted homology is the homology with coefficients in this sheaf; $\mathbf{H}_\bullet(X, Y; \hat{X} \times_G \mathbf{J})$.

We give another description of this which has better space-level functoriality properties. First, the G action on $\mathbf{J}$ determines fibrations over the classifying space B_G by $\mathbf{J}_n \times_G E_G \to B_G$, where E_G denotes the universal cover of B_G. (This is a fibered spectral sheaf over B_G.) The homomorphism $\pi_1 X \to G$ determines, up to homotopy, a map $\nu\colon X \to B_G$. The G-product $X \times_G \mathbf{J}_n$ is then obtained from the pullback of these two maps to B_G. The G-smash $X \wedge_G \mathbf{J}_n$ is obtained from this by identifying to a point the 0-section and the inverse image of the basepoint in X. The twisted homology is therefore the Ω-spectrum associated to these quotiented pullbacks.

The difference between a map $X \to B_G$ and a homomorphism $\pi_1 X \to G$ is that the first specifies a particular covering space (by pulling back the universal cover of B_G) whereas the second only specifies a cover up to isomorphism. There are also problems with basepoints and disconnected spaces. These are not important for single spaces since changes in basepoints, covers, etc. only change the homology spectrum by homotopy equivalence. The differences become more significant when we consider families of spaces, in section 2.

This point of view is also more general, since B_G can be replaced by the classifying space of a topological monoid (or anything else). We describe an interesting example which can be expressed in these terms. Suppose ν is an oriented vector bundle over X, and let $\Omega_n(X, \nu)$ denote the bordism group of smooth n-manifolds together with a bundle map from the stable normal bundle to ν. These groups occur in the study of intersections and singularities. They have also been used to study surgery normal maps.

To describe this as a twisted theory, let Ω^{fr} be the spectrum classifying framed bordism. The infinite orthogonal group SO acts on this by changing the framing, so defines a bundle over B_{SO} with fiber Ω^{fr}. The oriented vector bundle determines a map $X \to B_{SO}$. The bordism groups defined above are then the SO-twisted homology groups defined by this data;

$$\Omega_n(X,\nu) \simeq H_n(X;\Omega^{\text{fr}},\nu).$$

1.4 Čech homology

Finally we define Čech, or "shape" homology spectra. This is usually thought of as a way to extend homology in a reasonable way to pathological spaces (eg. not locally connected). This is not the motivation here; Čech homology coincides with the usual notion for all the spaces we really care about. Instead, both the definition and the the transversality view of the assembly naturally take place in Čech homology. It can be avoided, but only at the cost of some technical awkwardness.

The discussion here is for constant coefficients. In section 2.3 we will use a Čech version of spectral sheaf homology, which is a straightforward mixture of this section and 1.2.

Suppose $\mathcal{U}$ is a collection of subsets of a space X. We usually think of $\mathcal{U}$ as an open covering, though it is technically convenient to work with more general collections. Also suppose $\mathcal{U}$ has a partial ordering such that any finite number of elements with nonempty common intersection is totally ordered. Then the nerve of the collection, denoted nerve($\mathcal{U}$), is a simplicial complex with k-simplices the sets of $k+1$ elements from $\mathcal{U}$ with nonempty intersection. We do not (at this point) require these elements to be distinct, so these sets are partially ordered, and fail to be totally ordered only as a result of duplications. The face operator ∂_j is defined by omission of the jth element, and the degeneracy s_j duplicates the jth entry. (Note that the results are well defined even though the "j^{th} entry" may not be well defined because of the duplications.)

If $Y \subset X$ and $\mathcal{U}$ is a collection of subsets of X, then $\mathcal{U} \cap Y$ is a collection of subsets of Y. The nerve of this is a subcomplex; nerve($\mathcal{U} \cap Y$) $\subset$ nerve($\mathcal{U}$).

Given a spectrum $\mathbf{J}$ and a collection $\mathcal{U}$, we can form the homology of the nerve; $\mathbf{H}_\bullet(\text{nerve}(\mathcal{U});\mathbf{J})$. Strictly speaking this should be the homology of the realization, and denoted $\mathbf{H}_\bullet(\|\text{nerve}(\mathcal{U})\|;\mathbf{J})$, but the simpler notation seems to be clear. More generally if (X,Y) is a pair then we can form the relative homology $\mathbf{H}_\bullet(\text{nerve}(\mathcal{U}),\text{nerve}(\mathcal{U}\cap Y);\mathbf{J})$, as an approximation to the homology of (X,Y).

A *morphism* of collections of subsets $\theta\colon \mathcal{U} \to \mathcal{V}$ is a function compatible with the partial orders, and such that $U \subset \theta(U)$. (So $\mathcal{U}$ is a refinement of $\mathcal{V}$.) A morphism induces a simplicial map of nerves nerve($\mathcal{U}$) $\to$ nerve(V), which in turn induces a map of geometric realizations and a map of homology spectra. Note that if $\mathcal{U}$ and $\mathcal{V}$ are two partially ordered collections then the collection obtained from intersections $\mathcal{U} \cap \mathcal{V}$ has natural morphisms to both $\mathcal{U}$ and $\mathcal{V}$.

The partially ordered open covers of X form an inverse system. We define the Čech, or "shape" homology spectrum of (X, Y) to be the homotopy inverse limit of homologies of nerves of this inverse system:

$$\check{\mathbf{H}}_\bullet(X, Y; \mathbf{J}) = \underset{\leftarrow \mathcal{U}}{\operatorname{holim}} \mathbf{H}_\bullet(\operatorname{nerve}(\mathcal{U}), \operatorname{nerve}(\mathcal{U} \cap Y); \mathbf{J}).$$

We are primarily interested in this as an alternative description of the definition of 1.1, so we show

1.5 Lemma. *If (X, Y) is a metric pair with the homotopy type of a CW pair (K, L), then there is a natural equivalence $\check{\mathbf{H}}_\bullet(X, Y; \mathbf{J}) \simeq \mathbf{H}_\bullet(K, L; \mathbf{J})$.*

Proof. The realization $\|K\|$ of a simplicial complex K has a canonical open collection of subsets consisting of stars of vertices. If v is a vertex the *star*, denoted star(v), is the union of all open simplices whose closures contain v. An ordered simplicial complex comes equipped with a partial ordering of its vertices so that the vertices of every simplex are totally ordered. This induces a partial ordering of the covering. Further, a finite collection of sets in the cover intersect if and only if the corresponding vertices span a simplex, so this ordering satisfies the hypotheses above. This partially ordered covering of $\|K\|$ is denoted stars(K).

Now suppose X is a metric space, and $\mathcal{U}$ is a partially ordered open cover. A partition of unity subordinate to $\mathcal{U}$ can be used to construct a map $f\colon X \to \|\operatorname{nerve}(\mathcal{U})\|$. Specifically, suppose $h_U\colon X \to [0,1]$ are functions with locally finite support, and that the support of h_U is contained in U. Let $x \in X$, and let h_{U_i} for $i = 0, \dots, n$ denote the functions which are nonzero on x. Then $x \in \bigcap_{i=0}^n U_i$, so $(U_0, \dots, U_n)$ defines an n-simplex in the nerve. Represent the simplex Δ^n as the points in real $(n+1)$-space with nonnegative entries with sum 1, then f takes the point x is to the point $(U_0, \dots, U_n) \times \{h_{U_0}(x), \dots, h_{U_n}(x))\} \in \operatorname{nerve}(\mathcal{U})^n \times \Delta^n \subset \|\operatorname{nerve}(\mathcal{U})\|$.

This map induces a morphism from the inverse image of the open star cover of the nerve to the original cover: $\theta_f\colon f^{-1}(\text{stars}(\operatorname{nerve}(\mathcal{U}))) \to \mathcal{U}$. It follows that the inverse system of inverse images of star covers of complexes is cofinal in the system of all covers. Therefore the homotopy inverse limit over maps to complexes is homotopy equivalent to the limit over covers. Explicitly, if X is metric then

$$\check{\mathbf{H}}_\bullet(X, Y; \mathbf{J}) \simeq \underset{(K,L) \to (X,Y)}{\operatorname{holim}} \mathbf{H}_\bullet(K, L; \mathbf{J})$$

where (K, L) is a complex pair.

Note there is an analogous definition of "singular" homology obtained by taking the homotopy *direct* limit of homology of complexes mapping *to* (X, Y).

If (X, Y) has the homotopy type of a CW pair then there is a homotopy equivalence to the realization of a pair of simplicial complexes, $(X, Y) \to (K, L)$. The simplicial approximation theorem implies that the subdivisions of (K, L) are cofinal up to homotopy in the inverse system of complexes to which (X, Y) map. Therefore the homotopy inverse limit over the subsystem is homotopy equivalent to the limit over the full system. But homology of the realization of a complex is independent of subdivisions, so the homology is constant on this subsystem. Therefore the inverse limit of the whole system (the Čech homology) is equivalent to $\mathbf{H}_\bullet(K, L; \mathbf{J})$ as required. □

2: Functor coefficient homology

In this section we define the homology of a map with coefficients in a spectrum-valued functor. The functors are discussed in 2.1. A simple special case of the construction, which gives the constant coefficient assembly maps, is described in 2.2. The full construction is then given in detail in 2.3. This construction takes place in Čech homology, which involves a homotopy inverse limit. Proposition 2.4 shows that these limits are unnecessary in some cases.

2.1 Spectrum-valued functors

Suppose that $\mathbf{J}(X)$ is a covariant functor which assigns an Ω-spectrum to a space. In detail this means each space is functorially assigned a sequence of pointed spaces $\mathbf{J}_n(X)$, with natural maps $\mathbf{J}_n(X) \wedge S^1 \to \mathbf{J}_{n+1}(X)$.

A functor is *homotopy invariant* if a homotopy equivalence $X \to Y$ induces a homotopy equivalence $\mathbf{J}(X) \to \mathbf{J}(Y)$. Alternative descriptions of this property are that homotopic maps induce homotopic morphisms of spectra, or that the inclusion $X \times \{0\} \to X \times I$ induces a homotopy equivalence of $\mathbf{J}$ spectra.

A homotopy invariant functor induces a functor on the associated homotopy categories, but we will not use this. For our purposes it is quite important that $\mathbf{J}$ be a functor on maps, and take values in morphisms of spectra, not just homotopy classes.

A slight extension will be required in the applications. Consider the category of pairs (X, ω) where $\omega \colon X \to B$, for some fixed B ($= B_{Z/2}$ in the applications). Morphisms in this category are $X \to Y$ which commute with the maps of B. Then our functors will be defined on this category; $\mathbf{J}(X,\omega)$. In this context "homotopy invariant" means $\mathbf{J}(X,\omega) \to \mathbf{J}(Y,\nu)$ is a homotopy equivalence if $X \to Y$ is a homotopy equivalence. Note this requirement on $X \to Y$ is weaker than homotopy equivalence in the category of spaces over B, since the homotopies are not required to commute with maps to B.

The results of this chapter can be extended to this setting simply by including ω in the notation. Since it plays no essential role we have simplified the notation by omitting it.

2.2 Constant coefficient assembly

Now suppose that $\mathbf{J}$ is a homotopy invariant spectrum-valued functor of pairs. We will define a natural (up to homotopy) morphism of spectra $\mathbf{H}_\bullet(X, \mathbf{J}(pt)) \to \mathbf{J}(X)$.

Heuristically the construction is described as follows: Think of $X \times \mathbf{J}(pt)$ as obtained by applying $\mathbf{J}$ fiberwise to the identity map $X \to X$. Similarly we can obtain $\mathbf{J}(X)$ by applying $\mathbf{J}$ fiberwise to the projection $X \to pt$. Then the commutative diagram

$$\begin{array}{ccc} X & \xrightarrow{=} & X \\ \downarrow{=} & & \downarrow \\ X & \longrightarrow & pt \end{array}$$

maps the first construction into the second. Dividing out X in the first construction and passing to associated Ω-spectra gives $\mathbf{H}_\bullet(X; \mathbf{J}(pt)) \to \mathbf{J}(X)$, as desired.

Rather than literally applying $\mathbf{J}$ fiberwise we do an analogous simplicial construction.

Suppose $X = \|K\|$ is the realization of a simplicial complex K. Regard K as a category, with one objects for each simplex σ and morphisms generated by the face and degeneracy maps ∂_j, s_j. If F is a covariant functor from K to spaces, then $\coprod_{\sigma \in K} F(\sigma)$ is a simplicial space. In particular it has a "geometric realization," defined by:

$$\|F\| = \Big(\coprod_{k, \sigma \in K^k} F(\sigma) \times \Delta^k \Big) / \sim$$

where $\sim$ is the equivalence relation generated by: if $x \in F(\partial\sigma)$, $t \in \Delta^{k-1}$, and $u \in \Delta^{k+1}$ then $(x, \partial_j^* t) \in F(\sigma) \times \Delta^k$ is equivalent to $(\partial_j x, t) \in F(\partial_j \sigma) \times \Delta^{k-1}$, and $(x, s_j^* u)$ is equivalent to $(s_j x, u)$. The following properties of the realization are immediate:

2.2A Lemma.

(1) *Realization is natural in K and F,*
(2) *the constant functor $F(\sigma) = X$ has realization $\|K\| \times X$, and*
(3) *suppose $\|F\| \to \|K\|$ is defined using the natural transformation $F \to pt$ and statements* (1), (2). *If σ is a nondegenerate simplex of K then the inverse image of $\operatorname{int}\|\sigma\| \subset \|K\|$ is $F(\sigma) \times \operatorname{int}\|\sigma\|$.* □

We use this to construct a simplicial version of the assembly.

Consider the covariant functor from K to spaces which takes a simplex to its open star; $\sigma^k \mapsto \operatorname{star}(\sigma)$. Compose this functor with the nth space functor in $\mathbf{J}_n$, to get a functor $\sigma \mapsto \mathbf{J}_n(\operatorname{star}(\sigma))$. Denote the realization of this functor by $\mathbf{J}_n(I_{\|K\|})$. Substituting in the definition above this is

$$\mathbf{J}_n(I_{\|K\|}) = \Big(\coprod_{k,\sigma \in K^k} \mathbf{J}_n(\operatorname{star}(\sigma)) \times \Delta^k \Big) / \sim \ .$$

Next define maps by realizing natural transformations, as in the lemma. The natural transformation $\mathbf{J}_n(\operatorname{star}(\sigma)) \to pt$ gives a projection $\mathbf{J}_n(I_{\|K\|}) \to \|K\|$. Next, begin with the transformation from star to the point functor. Apply $\mathbf{J}_n$ to this and realize to get a map $\mathbf{J}_n(I_{\|K\|}) \to \|K\| \times \mathbf{J}_n(pt)$. This fits with the previous construction to give a commutative diagram

$$\begin{array}{ccc} \mathbf{J}_n(I_{\|K\|}) & \longrightarrow & \|K\| \times \mathbf{J}_n(pt) \\ \downarrow & & \downarrow \\ \|K\| & \xrightarrow{=} & \|K\|. \end{array}$$

It follows from the homotopy invariance of $\mathbf{J}$ that this is a fiber homotopy equivalence over $\|K\|$; according to (3) of the lemma the inverse image of the interior of a nondegenerate simplex σ is $\mathbf{J}_n(\operatorname{star}(\sigma)) \times \operatorname{int}\|\sigma\|$ on the left, and $\mathbf{J}_n(pt) \times \operatorname{int}\|\sigma\|$ on the right. But $\operatorname{star}(\sigma)$ is contractible, so $\mathbf{J}_n(\operatorname{star}(\sigma)) \simeq \mathbf{J}_n(pt)$. Thus $\mathbf{J}_n(I_{\|K\|}) \to \|K\| \times \mathbf{J}_n(pt)$ is a homotopy equivalence.

In the other direction, the inclusion $\operatorname{star}(\sigma) \subset \|K\|$ gives a natural transformation from the star functor to the constant functor with value $\|K\|$. Applying $\mathbf{J}_n$ and realizing gives $\mathbf{J}_n(I_{\|K\|}) \to \|K\| \times \mathbf{J}_n(\|K\|)$. Compose this with the projection to $\mathbf{J}_n(\|K\|)$.

Now consider the spectrum structure of $\mathbf{J}_n$. The spectrum maps give maps $\mathbf{J}_n(I_{\|K\|}) \times S^1 \to \mathbf{J}_{n+1}(I_{\|K\|})$, which in fact give $\mathbf{J}(I_{\|K\|})$the structure of a spectral sheaf over $\|K\|$. Divide by the 0-section $i\colon \|K\| \to \mathbf{J}_n(I_{\|K\|})$ to get a spectrum, and pass to the associated Ω-spectrum, to get the homology with coefficients in the spectral sheaf. The analogous construction on $\|K\| \times$

$\mathbf{J}_n(pt)$ gives the constant coefficient homology. $\mathbf{J}(\|K\|)$ is already an Ω-spectrum so this construction gives

$$\mathbf{H}_\bullet(\, \|K\|; \mathbf{J}(pt)) \leftarrow \mathbf{H}_\bullet(\, \|K\|; \mathbf{J}(I_{\|K\|})) \rightarrow \mathbf{J}(\|K\ |).$$

We have shown that the left of these maps comes from a sequence of homotopy equivalences, so is a homotopy equivalence. Composing with a homotopy inverse gives the desired map $\mathbf{H}_\bullet(\, \|K\|; \mathbf{J}(pt)) \rightarrow \mathbf{J}(\|K\|)$.

2.3 Functor coefficient homology

In this section we begin with a spectrum-valued functor $\mathbf{J}$, and a map $p\colon E \rightarrow X$. Roughly, a spectral sheaf $\mathbf{J}(p) \rightarrow X$ is constructed by applying $\mathbf{J}$ fiberwise to p, generalizing the construction of the previous section. Homology with coefficients in this sheaf is then defined. As pointed out in the introduction, when the construction is done in this generality the assembly does not have to be treated separately; it is functorially induced by the morphism from p to the map which projects E to a point.

The definition takes place in Čech homology. This gives a definition for arbitrary maps, which is useful in naturality arguments. The maps encountered in applications are essentially simplicial, and the definition is shown (in 2.4) to simplify in this case.

Now suppose $p\colon E \rightarrow X$ is given. If $\mathcal{U}$ is a partially ordered open cover we define a covariant functor from nerve($\mathcal{U}$) to spaces, by $\sigma \mapsto p^{-1}(\cap\sigma)$. Recall that a simplex of the nerve is given by a monotone sequence of elements of $\mathcal{U}$, $\sigma = (U_0, \dots, U_k)$, and $\cap\sigma$ denotes the intersection $\cap\sigma = \cap_i U_i$. Compose this functor with $\mathbf{J}$ to obtain a functor from nerve($\mathcal{U}$) to Ω-spectra. Geometric realization, as in the previous section, defines a spectral sheaf $\mathbf{J}(p, \mathcal{U}) \rightarrow \|\text{nerve}(\mathcal{U})\|$. The spectra associated with this spectral sheaf have the homotopy type of CW complexes since the total space of the sheaf is defined by geometric realization. We therefore get a homology spectrum $\mathbf{H}_\bullet(\,\text{nerve}(\mathcal{U}); \mathbf{J}(p, \mathcal{U}))$.

Next suppose $\theta\colon \mathcal{U} \rightarrow \mathcal{V}$ is a morphism of partially ordered covers, as considered in 1.4 ($\mathcal{U}$ "refines" $\mathcal{V}$). This induces a simplicial map nerve($\mathcal{U}$) $\rightarrow$ nerve($\mathcal{V}$), and a natural transformation of inverse image functors $\sigma \mapsto p^{-1}(\cap\sigma)$. Composing with $\mathbf{J}$ gives a natural transformation of spectrum-valued functors. Realizing defines a morphism of spectral sheaves $\mathbf{J}(p, \mathcal{U}) \rightarrow \mathbf{J}(p, \mathcal{V})$ covering the map $\|\text{nerve}(\mathcal{U})\| \rightarrow \|\text{nerve}(\mathcal{V})\|$. This in turn induces a morphism of homology spectra;

$$\mathbf{H}_\bullet(\,\text{nerve}(\mathcal{U}); \mathbf{J}(p, \mathcal{U})) \rightarrow \mathbf{H}_\bullet(\,\text{nerve}(\mathcal{V}); \mathbf{J}(p, \mathcal{V})).$$

These maps give the homology spectra the structure of an inverse system indexed by the partially ordered covers. We define the (**J** coefficient, Čech) homology to be the homotopy inverse limit:

$$\check{\mathbf{H}}_{\bullet}(X; \mathbf{J}(p)) = \operatorname*{holim}_{\leftarrow \mathcal{U}} \mathbf{H}_{\bullet}(\operatorname{nerve}(\mathcal{U}); \mathbf{J}(p, \mathcal{U})).$$

We caution that this homology may not actually be obtained from some spectral sheaf $\mathbf{J}(p)$ on X itself, as the notation suggests. Proposition 2.4 below does imply this when the map p is simplicial.

As an aside we remark that the construction can be simplified to involve only sheaves over simplices, rather than over nerves. If the non-empty intersection requirement is dropped in the definition of the nerve we get a simplex with vertices $\mathcal{U}$. The spectral sheaf $\mathbf{J}(p, \mathcal{U})$ extends to a sheaf over this by the same formula: $\sigma \mapsto \mathbf{J}(p^{-1}(\cap\sigma)) = \mathbf{J}(\phi)$ if σ is not in the nerve. If $\mathbf{J}(\phi)$ is contractible—and $\mathbf{J}$ can always be redefined so this is the case—then the spectral sheaf over the simplex has the same homology as the sheaf over the nerve. This point will be developed further in section 5.3.

2.4 Naturality

Naturality for this definition follows from the naturality of all the ingredients. Specifically, suppose

$$\begin{array}{ccc} F & \xrightarrow{\hat{f}} & E \\ {\scriptstyle q}\downarrow & & \downarrow{\scriptstyle p} \\ Y & \xrightarrow{f} & X \end{array}$$

commutes. If $\mathcal{U}$ is a partially ordered cover of X then $f^{-1}(\mathcal{U})$ is a cover of Y. There is an induced simplicial map (an inclusion in fact) $\operatorname{nerve}(f^{-1}\mathcal{U}) \to \operatorname{nerve}(\mathcal{U})$. Covering this is a natural transformation of inverse image functors; $\hat{f}\colon q^{-1}(\cap f^{-1}(\sigma)) \to p^{-1}(\cap\sigma)$. Compose with $\mathbf{J}$ to get a natural transformation of spectrum-valued functors, and realize to get a morphism of spectral sheaves $\mathbf{J}(q, f^{-1}\mathcal{U}) \to \mathbf{J}(p, \mathcal{U})$ covering the map of realizations of nerves. This induces a morphism of homology spectra,

$$\mathbf{H}_{\bullet}(\operatorname{nerve}(f^{-1}\mathcal{U}); \mathbf{J}(q, f^{-1}\mathcal{U})) \to \mathbf{H}_{\bullet}(\operatorname{nerve}(\mathcal{U}); \mathbf{J}(p, \mathcal{U})).$$

Now take homotopy inverse limits. These are both indexed by the inverse system of covers of X, so there is a natural induced map between the limits. On the right we get homology of X. The homology of Y is obtained by taking the limit of spectra on the left over the larger inverse system of all covers of Y. But there is a natural map from the inverse limit over the larger system to the inverse limit over the subsystem. Composition with the map above gives

$$\check{\mathbf{H}}_\bullet(Y;\mathbf{J}(q)) \to \check{\mathbf{H}}_\bullet(X;\mathbf{J}(p)).$$

We define this to be the morphism functorially associated to $(f,\hat{f})$. The Čech homology is thus a functor of X and p (and $\mathbf{J}$).

Definition

Suppose $\mathbf{J}$ and $p\colon E \to X$ are as above. Then the *total assembly map* is defined to be the map $\check{\mathbf{H}}_\bullet(X;\mathbf{J}(p)) \to \mathbf{J}(E)$ induced by the commutative diagram

$$\begin{array}{ccc} E & \xrightarrow{=} & E \\ {\scriptstyle p}\downarrow & & \downarrow \\ X & \longrightarrow & pt. \end{array}$$

2.5 The long exact sequence of a pair

We can define the relative homology spectrum $\check{\mathbf{H}}_\bullet(X,Y;\mathbf{J}(p))$ to be the cofiber (in the category of spectra) of the natural map $\check{\mathbf{H}}_\bullet(Y;\mathbf{J}(q)) \to \check{\mathbf{H}}_\bullet(X;\mathbf{J}(p))$ induced by the inclusion $Y \subset X$. Applying π_* then gives the usual long exact sequence of homology groups.

The same spectrum can be obtained less trivially by taking the homotopy inverse limit of relative homology spectra of nerves:

$$\check{\mathbf{H}}_\bullet(X,Y;\mathbf{J}(p)) \simeq \operatorname*{holim}_{\leftarrow \mathcal{U}} \mathbf{H}_\bullet(\,\|\mathrm{nerve}(\mathcal{U})\|, \|\mathrm{nerve}(\mathcal{U}\cap Y)\|; \mathbf{J}(p,\mathcal{U})).$$

The reason these two constructions agree is that the relative homology spectra for nerves was also defined by taking the cofiber, and homotopy inverse limits preserve cofibers, up to homotopy.

To see this last point, note that cofibers in the category of spectra can also be described as deloopings of homotopy fibers of maps of spaces. But it follows from Bousfield and Kan [3, XI 5.5] that homotopy inverse limits preserve homotopy fibrations.

As an application of the long exact sequence we get a description of the cofiber of the total assembly map defined just above. Let $\hat{p}\colon E\times I \to \mathrm{cone}X$ denote the map obtained from $p\times 1$ by dividing out $X\times\{0\}\subset X\times I$. The map $\hat{p}$ fiberwise deformation retracts to $E \to pt$, so the homology of the cone is just $\mathbf{J}(E)$. The homotopy fibration for the pair $(\mathrm{cone}X, X)$ therefore gives a homotopy fibration

$$\check{\mathbf{H}}_\bullet(X,\mathbf{J}(p)) \to \mathbf{J}(E) \to \check{\mathbf{H}}_\bullet(\mathrm{cone}X, X;\mathbf{J}(\hat{p})).$$

The long exact sequence of homotopy groups of this homotopy fibration therefore give

$$\cdots \to \check{H}_n(X, \mathbf{J}(p)) \to J_n(E) \to \check{H}_n(\text{cone}X, X; \mathbf{J}(\hat{p})) \\ \to \check{H}_{n-1}(X, \mathbf{J}(p)) \to \cdots .$$

2.6 Simplicial maps

The point here is that when the coefficient map is simplicial, the functor coefficient Čech homology is equivalent to one of the terms in the inverse limit which defines it. This will be used to avoid homotopy inverse limits.

Proposition. *Suppose $p\colon E \to X$, Y is relatively fiber homotopy equivalent to the realization of a simplicial map $q\colon A \to K$, L, with $L \subset K$ a subcomplex. Then the Čech homology is equivalent to the homology of the open cover of $(\|K\|, \|L\|)$ by stars:*

$$\check{\mathbf{H}}_\bullet(X, Y; \mathbf{J}(p)) \xrightarrow{\simeq} \mathbf{H}_\bullet(K, L; \mathbf{J}(q, stars(K))).$$

A "fiber homotopy" of a map between maps p and $\|q\|$ is a commutative diagram

$$\begin{array}{ccc} E \times I & \longrightarrow & \|A\| \\ {\scriptstyle P\times 1}\downarrow & & \downarrow{\scriptstyle \|q\|} \\ X \times I & \longrightarrow & \|K\| \end{array}$$

such that P restricts to p on $E \times \{0\}$ and $\|q\|$ on $E \times \{1\}$. Such a homotopy is "relative" if the image of $Y \times I$ is contained in $\|L\|$. Accordingly two maps are fiber homotopy *equivalent* if there are maps both ways between them and fiber homotopies of the compositions to the identities. Note that this notion of homotopy equivalence preserves the homotopy type of point inverses.

We recall the definition of the star cover (see the proof of 1.5). The star of a vertex $v \in K$ is the union of all open simplices in the realization $\|K\|$ whose closures contain v. More generally, if σ is a simplex of K, then star(σ) is the intersection of the stars of the vertices. Note such intersections also define simplices in the nerve, and in fact the function $\sigma \mapsto \text{star}(\sigma)$ defines an isomorphism of simplicial complexes $K \to \text{nerve}(\text{stars}(K))$.

Proof. First, the homology is homotopy invariant so

$$\check{\mathbf{H}}_\bullet(X; \mathbf{J}(p)) \to \check{\mathbf{H}}_\bullet(K; \mathbf{J}(q)))$$

is a homotopy equivalence. Next, the star covers of subdivisions of K are cofinal in the system of all covers of $\|K\|$ so it is sufficient to consider the inverse limit over this subsystem. We show that $\mathbf{H}_\bullet$ is constant (up to homotopy) on this subsystem, so they are all homotopy equivalent to the limit.

It is sufficient to consider a subdivision obtained by adding a single vertex; let K' be obtained by adding v'. The choice of ordering for the new vertex determines a simplicial map $K' \to K$. This is covered by a map of spectral sheaves $\mathbf{J}(p, \text{stars}(K')) \to \mathbf{J}(p, \text{stars}(K))$. We will show that this map is a homology equivalence in each degree. This implies that the associated map of homology spectra (obtained by dividing by the bases and passing to associated Ω-spectra) is a homotopy equivalence.

Next suppose $\sigma \in K$ is a nondegenerate simplex, and consider the interior of the realization $\text{int}(\|\sigma\|) \subset \|K\|$. It is sufficient to show that the inverse image of this in $\mathbf{J}(p, \text{stars}(K'))$ maps by homology equivalence to the inverse image in $\mathbf{J}(p, \text{stars}(K))$. To see this is sufficient, consider the filtration of the realization of K by skeleta, and show by induction that the restriction of the spectral sheaves to the skeleta are homotopy equivalent. The induction step follows from the long exact sequence relating skeleta of adjacent dimensions, and the homology equivalence fact for individual simplices.

Now consider inverses of $\text{int}(\|\sigma\|)$, first in the subdivision K'. If σ does not contain the new vertex v' then the inverse is again the simplex σ. If σ contains both v' and its image v under the simplicial map, then the inverse image of the interior is an open simplex in the interior of $\|\sigma\|$. Finally suppose σ contains v but not v'. Then the inverse image is a $(k+1)$-simplex γ of K' with σ as a face and v' as additional vertex. Let σ' denote the face of γ with the same vertices as σ except with v' substituted for v. The map $\|\gamma\| \to \|\sigma\|$ is the linear projection which is the identity on $\|\sigma\|$ and takes v' to v. The inverse image of the interior is therefore the interior of $\|\gamma\|$ union with the interiors of the two faces $\|\sigma'\|$ and $\|\sigma\|$.

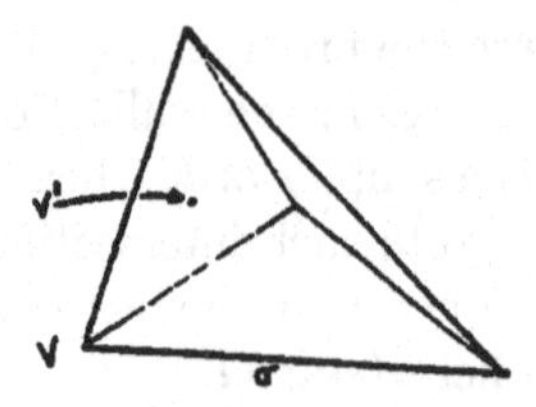

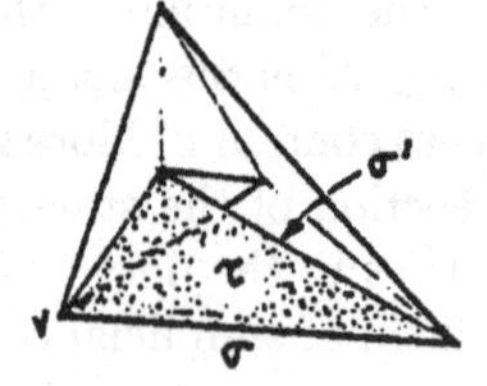

To complete the proof we consider inverses in the spectral sheaves. Since

σ is nondegenerate the inverse image of its interior in $\mathbf{J}(p, \text{stars}(K))$ is $\mathbf{J}(p^{-1}(\text{star}(\sigma))) \times \text{int}\|\sigma\|$. (This follows directly from the definition of the spectral sheaf as the realization of a functor; see the lemma in 2.2.) If σ does not contain the new vertex, then the inverse image in $\mathbf{J}(p, \text{stars}(K'))$ is the same, so the condition is satisfied.

Suppose next that σ contains both v' and v. Then the inverse image in K' is a simplex τ in the subdivision of σ, whose realization is taken homeomorphically to the realization of σ. Further, the map $\text{star}(\tau) \to \text{star}(\sigma)$ is a homeomorphism. In the spectral sheaves the inverses are given by this homeomorphism times the morphism of spectra induced by the inclusion $q^{-1}(\text{star}(\tau)) \subset q^{-1}(\text{star}(\sigma))$. For any simplex α the set $q^{-1}(\text{star}(\alpha))$ deformation retracts to $q^{-1}(t)$ for any $t \in \text{int}\|\alpha\|$. This implies that the inverses of the stars of both σ and τ deformation retract to the inverse of any point in $\text{int}(\|\tau\|)$, so the inclusion is a homotopy equivalence. According to the homotopy invariance the morphism induced on $\mathbf{J}$ is also a homotopy equivalence, so inverses of σ satisfy the homology equivalence property.

Finally suppose σ contains v but not v'. According to the above, the inverse image of the interior in $\|K'\|$ is the union $\text{int}\|\gamma\| \cup \text{int}\|\sigma\| \cup \|\sigma'\|$. The inverse image in $\mathbf{J}(q, \text{stars}(K'))$ thus has the homotopy type of the union of the mapping cylinders of the morphisms of $\mathbf{J}$ induced by the inclusions

$$q^{-1}(\text{star}(\sigma)) \leftarrow q^{-1}(\text{star}(\gamma)) \to q^{-1}(\text{star}(\sigma')).$$

The rightmost map is a homotopy equivalence: both inverses deformation retract to point inverses in the interior of the respective simplices, but these point inverses are homotopy equivalent since $\text{int}\|\sigma\| \cup \text{int}\|\gamma\|$ lies in the interior of one of the simplices of K. From this we conclude that the union of mapping cylinders deformation retracts to the left end, $\mathbf{J}(q^{-1}(\text{star}(\sigma)))$. Thus the preimage of $\text{int}\|\sigma\|$ in $\mathbf{J}(q, \text{stars}(K'))$ has the homotopy type of the preimage in $\mathbf{J}(q, \text{stars}(K))$, as required. □

3: Bordism-type theories

A "bordism-type theory" consists of a class of objects with faces, indexed by arbitrary sets. The prototype example of oriented manifolds, with faces codimension 0 submanifolds of the boundary, is presented in 3.1. The definition itself is given in 3.2. Bordism groups and spectra are defined in 3.3 and 3.4. Morphisms of these theories are defined in 3.5; these are necessary to define functors taking values in the category of bordism-type theories. The relative theory associated to a morphism is defined in 3.6. Finally in 3.7 the existence theorem is given, which asserts that up to weak homotopy any spectrum can be obtained as a bordism spectrum of some bordism-type theory.

3.1 An example

Before giving the abstraction we describe an example which displays the essential features. This example leads to the bordism theory of oriented manifolds; further examples are given in section 6.

Suppose A is a set. A *manifold A-ad* is a manifold with subsets $\partial_a M \subset \partial M$ for each $a \in A$. We require the union to be ∂M, and allow only finitely many of these to be nonempty. Finally we require each $\partial_a M$ together with the subsets $\partial_b M \cap \partial_a M$ to be a manifold $A - \{a\}$-ad.

Logically speaking this is an inductive definition: we should define A-ads with at most k nonempty faces inductively in k, so that the requirement that the faces be "-ads" is well defined. In any case the faces are codimension 0 submanifolds of ∂M, which intersect in codimension 0 submanifolds of their boundaries, etc.

For example the n-simplex Δ^n with its faces $\partial_i \Delta^n$ is a manifold $[n]$-ad of dimension n. Here we use the notation $[n]$ for the set $\{0, 1, \ldots, n\}$.

The "bordism theory" consists of the collection of all manifold -ads, together with some operations on them. More specifically,

(1) for all sets A and integers n, the collections of compact oriented manifold A-ads of dimension n;
(2) face operations ∂_a which take n-dimensional A-ads to $(n-1)$-dimensional $(A - \{a\})$-ads;
(3) reindexing operations which change the labels on the faces and add empty faces;
(4) an involution obtained by reversing the orientations; and
(5) a "Kan" condition wherein a collection of -ads with appropriate incidence relations are assembled to form a single manifold.

The only odd thing which occurs is a sign change in iterated boundaries: when $a \neq b$ then $\partial_a \partial_b M = -\partial_b \partial_a M$, due to the way boundaries work in homology.

Note that the finiteness condition on faces and the reindexing of (3) imply that an arbitrary A-ad is obtained by reindexing a $[k]$-ad, for some $[k]$ and injection $[k] \to A$. It follows that it is logically sufficient to define $[k]$-ads, for each k. However direct definition of general A-ads is no more difficult, and saves a lot of trouble with reindexing.

The thing which gives these theories their characteristic flavor is the addition of empty boundaries in reindexing, i.e. reindexing using injections rather than only bijections.

We now abstract this. The symbol "$\mathcal{J}$" is a script "J", representing a generic theory just as **J** represented a generic functor in the previous section.

3.2 Definition

A *bordism-type theory*, $\mathcal{J}$, consists of :

(1) For every set A a collection $\mathcal{J}_A^n$ of "A-ads of dimension n," with a basepoint denoted $\phi \in \mathcal{J}_A^n$;
(2) for each $a \in A$ a function $\partial_a : \mathcal{J}_A^n \to \mathcal{J}_{A-a}^{n-1}$ such that $\partial_a \phi = \phi$, and if $M \in \mathcal{J}_A^n$ then $\partial_a M = \phi$ for all but finitely many a;
(3) corresponding to each injection $\theta \colon A \to B$ a basepoint-preserving function $\ell_\theta \colon \mathcal{J}_A^n \to \mathcal{J}_B^n$ which is natural in θ. Further this satisfies $\partial_{\theta(a)}(\ell_\theta M) = \ell_\theta(\partial_a M)$ and ℓ_θ is a bijection onto $\{M \in \mathcal{J}_B^n \mid \partial_b M = \phi \text{ for all } b \in B - \theta(A)\}$;
(4) there is an involution (-1) on each $\mathcal{J}_A^n$ which commutes with ℓ_θ and ∂_a, and leaves ϕ fixed. Further, if $a \neq b$ in A then $\partial_a \partial_b M = -\partial_b \partial_a M$; and
(5) these satisfy the Kan condition described below.

The most restrictive aspects of this are the bijection hypothesis in (3), and the Kan condition. In manifolds the bijection hypothesis is obvious, and the Kan condition follows from glueing together manifolds along faces in their boundaries. To see these axioms verified in a non-geometric situation look at the proof in 3.7.

Suppose A is a set and $a \in A$ is a fixed element. Then define an n-dimensional Kan (A, a)-cycle in $\mathcal{J}$ to be a function $N \colon A - \{a\} \to \mathcal{J}$ so that $N(b) \in \mathcal{J}_{A-\{b\}}^n$, and if $b \neq c$ are in $A - \{a\}$ then $\partial_b N(c) = -\partial_c N(b)$. Also assume only finitely many of the $N(b)$ are different from ϕ,=. Note no object is assigned to a. The principal example is: if M is an object in $\mathcal{J}_A^{n+1}$ then the function $b \mapsto \partial_b M$ for $b \neq a$ is an n-dimensional Kan (A, a)-cycle.

The Kan condition asserts that all Kan cycles arise in this way: if N is an n-dimensional (A, a)-cycle then there is an $(n+1)$-dimensional object M so that $N(b) = \partial_b M$, for all $b \neq a$.

The name is by analogy with the Kan condition for simplicial sets, which requires that a simplicial map defined on all but one of the faces of a simplex, extends to a simplicial map of the whole simplex.

3.3 Bordism groups

Suppose $\mathcal{J}$ is a bordism-type theory. Define $\Omega_n^{\mathcal{J}}$ to be the set of equivalence classes of n-dimensional ϕ-ads (no faces), where the equivalence is defined by $M \sim N$ if there is an $(n+1)$-dimensional $[1]$-ad W with $\partial_0 W = M$, and $\partial_1 W = -N$. (The symbol "$\mathit{\Omega}$" is a slanted version of Ω, used to try to distinguish between bordism groups and loop spaces.)

Proposition. *$\sim$ is an equivalence relation, and the set of equivalence classes $\mathit{\Omega}_n^{\mathcal{J}}$ has a natural abelian group structure.*

"Naturality" will not make sense until we have defined morphisms of bordism-type theories in 3.5.

Proof. The direct proof is elementary but long; we sketch a few pieces of it. The most efficient proof comes from the recognition as homotopy groups of a spectrum, in the next proposition.

We show that $\sim$ is transitive. Suppose W_0 expresses the equivalence $M \sim N$ and W_1 expresses the equivalence $N \sim P$. Then the function $i \mapsto W_i$ for $i = 0, 1$ defines a $([2], \{2\})$-cycle, in the sense defined in the Kan condition. The Kan condition therefore asserts there is an $(n+2)$-dimensional $[2]$-ad V such that $\partial_0 V = W_0$ and $\partial_1 V = W_1$. Then $\partial_2 V$ is an $(n+1)$-dimensional $[1]$-ad. Calculations using the axioms reveal that $\partial_0 \partial_2 V = -M$ and $\partial_1 \partial_2 V = P$. Thus $-\partial_2 V$ expresses an equivalence $M \sim P$, and $\sim$ is transitive.

The group structure is defined by: if W is a $[2]$-ad with $\partial_0 W = M$ and $\partial_1 W = N$, then $[M] + [N] = [-\partial_2 W]$. The Kan condition can be used to show that given M and N such a W exists, and that the equivalence class of $\partial_2 W$ is independent of the choice. This implies the operation is well defined.

The identity element is the equivalence class of the basepoint, $[\phi]$, and inverses come from the involution: $-[M] = [-M]$. To see the inverses, consider M as a $([1], \{1\})$-cycle and apply the Kan condition to get a $[1]$-ad V with $\partial_0 V = M$. Now define a $([2], \{0\})$-cycle by $2 \mapsto V$ and 1 goes to $-V$ reindexed so $\partial_1 = -\partial_1 V$ and $\partial_2 = -M$. Then apply the Kan condition to get a $[2]$-ad W with these as faces. The new face, $\partial_1 W$, has faces M and $-M$. Reindex to introduce ϕ as a third face, then it expresses the relation $[M] + [-M] = [\phi]$.

The fact that the group is abelian comes directly from reindexing: let W be as above, expressing $[M] + [N] = [\partial_2 W]$. Let θ be the bijection $[2] \to [2]$ which interchanges 0 and 1. Then $\partial_0 \ell_\theta W = M$ and $\partial_1 \ell_\theta W = N$, showing $[N] + [M] = [\partial_2 W]$ and therefore $[M] + [N] = [N] + [M]$. □

For example if SDiff is the theory of oriented smooth manifolds defined as in example 3.1, the bordism groups $\mathit{\Omega}_n^{\mathrm{SDiff}}$ are the classical smooth oriented bordism groups.

3.4 Bordism spectra

The next step is the construction of spectra which serve as classifying spaces for these theories.

We work with Δ-sets, in the sense of Rourke and Sanderson [22]. A Δ-set K is like a simplicial set in having sets $K^{(i)}$ of "i-simplices," the 0-simplices are appropriately partially ordered, and face operators $b_j\colon K^{(i)} \to K^{(i-1)}$ are given for $0 \leq j \leq i$. (We sometimes denote face operators in Δ-sets by ∂_i, but this conflicts somewhat with face operators in the bordism-type theory.) Δ-sets do not have the degeneracy operators of a simplicial set. Geometric realizations are defined for Δ-sets in essentially the same way as for simplicial sets.

Define the Δ-set $\Omega_n^{\mathcal{J}}$ to have k-simplices the $\mathcal{J}$-$[k]$-ads of dimension $k+n$. We also require that the "total intersection" of all faces $\partial_0\partial_1\cdots\partial_n M$ is the basepoint ϕ. The face operator $b_i M$ is defined by reindexing the $(n-\{i\})$-ad $(-1)^i\partial_i M$ using the order-preserving bijection $[n-1] \to [n]-\{i\}$.

The notation for this Δ-set is the same as for the bordism group. This doubling up of notation seems to be relatively harmless since the group is π_0 of the Δ-set. At any rate it is less harmful than introducing yet another notation. The boldface analog $\mathbf{\Omega}^{\mathcal{J}}$ is reserved for the associated Ω-spectrum.

We define the Ω-spectrum $\mathbf{\Omega}^{\mathcal{J}}$ by geometrically realizing the Δ-set: $\mathbf{\Omega}_n^{\mathcal{J}} = \|\Omega_{-n}^{\mathcal{J}}\|$. Note the minus sign in the index on the Δ-set: this results from an incompatibility in the indexing conventions for bordism and spectra.

Proposition. *The spaces $\mathbf{\Omega}_*^{\mathcal{J}}$ have a natural Ω-spectrum structure with homotopy groups the bordism groups defined above; $\pi_n\mathbf{\Omega}^{\mathcal{J}} = \Omega_n^{\mathcal{J}}$.*

As with bordism groups, the naturality will be considered after morphisms of bordism-type theories are defined.

Proof. First we verify that the simplices described above do in fact give a Δ-set, namely that the face identities $b_j b_i = b_i b_{j+1}$ (if $j \geq i$) hold. The only interesting thing about this is is the role of the sign $(-1)^i$. The point is that when $\partial_i M$ is reindexed to define $b_i M$ the previous faces with index i or higher are all shifted down by one. This shift, with the $(-1)^*$, gives a net change of -1 on these faces. This cancels the -1 in the iterated boundary formula in definition 3.2(3).

Next we observe that the Δ-sets $\Omega_n^{\mathcal{J}}$ satisfy the Kan condition. Let Λ_j^k denote the subcomplex of the k-simplex which consists of all but the jth face. The Kan condition asserts that any Δ-map $\Lambda_j^k \to \Omega_n^{\mathcal{J}}$ extends to a Δ-map $\Delta^k \to \Omega_n^{\mathcal{J}}$. A Δ-map $\Lambda_j^k \to \Omega_n^{\mathcal{J}}$ defines a $([k],\{j\})$-cycle in the sense of 3.2, so the Kan condition in 3.2(4) implies the Kan condition for the Δ-set.

The Kan condition implies a simplicial approximation theorem ([22, §5]) which in turn implies that the homotopy groups are the bordism groups: elements in the homotopy group π_k are represented by maps $\Delta^k \to \|\Omega_n^{\mathcal{J}}\|$ which take $\partial\Delta^k$ to the basepoint. Simplicial approximation asserts that this is homotopic to the realization of a Δ-map, which is exactly a k-simplex with all faces are ϕ. This in turn is obtained by reindexing a $k+n$ dimensional ϕ-ad to get a $[k]$-ad. This ϕ-ad defines an element in the bordism group $\Omega_{n+k}^{\mathcal{J}}$. Similarly homotopies can be interpreted as maps of Δ^{k+1} which take all but two faces to the basepoint. These can be approximated by Δ-maps which can be interpreted as bordisms.

Now we describe the spectrum structure. The cone on a Δ-set K can be described as a Δ-set with k-simplices $K^k \cup \{\mathrm{cone}\,\sigma \mid \sigma \in K^{k-1}\}$. The cone point is put last in the partial ordering, so if σ is a k-simplex $\partial_i\mathrm{cone}(\sigma) = \mathrm{cone}(\partial_i\sigma)$ if $i \leq k$, and $\partial_{k+1}\mathrm{cone}(\sigma) = \sigma$.

Now define a Δ-map $\mathrm{cone}(\Omega_n^{\mathcal{J}}) \to \Omega_{n-1}^{\mathcal{J}}$ by taking both $\Omega_n^{\mathcal{J}}$ and the cone point to the basepoints, and $\mathrm{cone}(M) \mapsto \ell_k(M)$. Here ℓ_θ reindexes the $[k]$-ad M to be a $[k+1]$-ad using the inclusion $[k] \subset [k+1]$. Taking geometric realizations, and dividing out the end $\|\Omega_n^{\mathcal{J}}\|$ of the cone gives a map $\|\Omega_n^{\mathcal{J}}\| \wedge S^1 \to \|\Omega_{n-1}^{\mathcal{J}}\|$. This defines a spectrum structure.

To see this is an Ω-spectrum we show that the adjoint $\|\Omega_n^{\mathcal{J}}\| \to \Omega(\|\Omega_{n-1}^{\mathcal{J}}\|)$ is a homotopy equivalence. For this we use the model of the loop space of a based Kan Δ-set given in [4, p. 36]; the k-simplices of ΩK are defined to be $(k+1)$-simplices $\sigma \in K$ with $b_{k+1}\sigma = \phi = v_{k+1}\sigma$. Here v_{k+1} denotes the $k+1$ vertex, and is obtained by applying all the face maps except b_{k+1}. It is shown in [4] that there is a natural homotopy equivalence $\|\Omega K\| \simeq \Omega\|K\|$. The adjoint of the spectrum structure maps defined above are homotopic to the realizations of $\Omega_n^{\mathcal{J}} \to \Omega(\Omega_{n-1}^{\mathcal{J}})$ defined by ℓ_k on k-simplices. But according to condition (3) of the definition, this is an isomorphism of Δ-sets. It therefore induces a homotopy equivalence on realizations. □

3.5 Morphisms and naturality

Suppose $\mathcal{J}$ and $\mathcal{K}$ are bordism-type theories. A *morphism* $\mathcal{J} \to \mathcal{K}$ is a collection of basepoint-preserving functions $\mathcal{J}_A^n \to \mathcal{K}_A^n$ for all n and sets A, which commute with face functions, the involutions -1, and the reindexing functions. Clearly these can be composed, and form a category.

3.5A Example

If X is a space, define the oriented manifold bordism theory of X to have A-

ads (M, f), where M is an oriented manifold A-ad as in 3.1, and $f\colon M \to X$. The same operations as defined in 3.1 give this the structure of a bordism-type theory. A map of spaces $X \to Y$ induces, by composition, a morphism from the bordism theory of X to that of Y.

3.5B Lemma. *The bordism groups of 3.3 and the bordism spectrum of 3.4 are natural with respect to morphisms of bordism-type theories. Further, for a morphism $F\colon \mathcal{J} \to \mathcal{K}$ the following are equivalent:*

(1) *F induces isomorphisms of bordism groups,*
(2) *F induces homotopy equivalence of bordism spectra, or*
(3) *for every [0]-ad M in $\mathcal{K}$ such that $\partial_0 M = F(N_1)$ for $N_1 \in \mathcal{J}$, there is a [1]-ad W in $\mathcal{K}$ such that $\partial_1 W = M$ and a {1}-ad $N \in \mathcal{J}$ with $\partial_1 N = N_1$ and $\partial_0 W = F(N)$.*

We say that a morphism which satisfies the conditions of the lemma is a "homotopy equivalence" of theories. The last condition is the one which will be checked in practice; it can be paraphrased as saying a pair in $\mathcal{K}$ with boundary from $\mathcal{J}$ deforms rel boundary into $\mathcal{J}$. It is also equivalent to the vanishing of the relative bordism groups defined in the next section.

Define a functor from spaces to bordism-type theories to be "homotopy invariant" if homotopy equivalences of spaces induce homotopy equivalences of theories. Then the following is immediate from the lemma.

3.5C Corollary. *If $\mathcal{J}$ is a homotopy invariant functor from spaces to bordism-type theories then the associated bordism spectra $\Omega^{\mathcal{J}}$ define a homotopy invariant spectrum-valued functor in the sense of 2.1.*

Proof of the lemma. The naturality is evident from the definitions. The equivalence of (1) and (2) follows from the fact that the bordism groups are the homotopy groups of the spectrum. It remains to show that conditions (1) and (3) are equivalent. We give a direct proof here; a much slicker one comes from the relative bordism groups of 3.6.

Suppose (3) holds. Let $[M]$ represent an element in the group $\Omega_n^{\mathcal{K}}$, and reindex M as a [0]-ad with $\partial_0 M = \phi$. Since $\phi = F(\phi)$ we can apply (3) to find a [1]-ad W with $\partial_1 W = M$, $\partial_0 W = F(N)$, and $\partial_1 N = \phi$. This is a bordism showing $[M] = F_*(-[N])$, so $F_*\colon \Omega_n^{\mathcal{J}} \to \Omega_n^{\mathcal{K}}$ is onto.

Similarly we show F_* is injective by showing $F_*([N_1]) = 0$ implies $[N_1] = 0$, for $[N] \in \Omega_n^{\mathcal{J}}$. The hypothesis implies there is a [0]-ad in $\mathcal{K}$ with $\partial_0 M = F(N_1)$. Condition (3) implies there is a bordism from M to $F(N)$. But $\partial_1 N = N_1$, so $[N] = 0$. Thus (3) implies (1). The other direction is similar. □

4: Cycles

In this section we describe "cycles" which represent functor-coefficient homology classes, when the coefficient functor is obtained from bordism-type theories. Fix a homotopy invariant functor (in the sense of 3.5) $\mathcal{J}$ from spaces to bordism-type theories. The associated spectrum-valued functor is denoted $\Omega^{\mathcal{J}}(X)$. The homology to be described is $H_n(\text{nerve}(\mathcal{U}); \Omega^{\mathcal{J}}(p, \mathcal{U}))$, where $p: E \to X$ is a map and $\mathcal{U}$ is a cover of X.

Cycles are described in 4.2, and induced morphisms in 4.3. Groundwork for the main theorem is laid in 4.4 with the development of a bordism-type theory whose objects are themselves cycles. The proof is given in section 5, where the bordism spectrum of this theory is shown to be equivalent to the homology spectrum.

4.1 Δ-nerves

Cycles will be defined using the nerve of a covering. For ease and efficiency we use a more compact model for the nerve than the simplicial complex described in 1.4.

Suppose $\mathcal{U}$ is a set of subsets of a space X, partially ordered as in 1.4. The Δ-nerve, denoted $\text{nerve}_\Delta(\mathcal{U})$, is defined to be the Δ-set with k-simplices the collections of $k+1$ *distinct* elements of $\mathcal{U}$ with nonempty intersection. (We caution that this distinctness is in $\mathcal{U}$, and does not imply that the corresponding subsets of X are distinct.) The face operator b_j is defined by omitting the jth set; this is well defined since a collection with nonempty intersection is totally ordered.

The Δ-nerve is exactly the set of nondegenerate simplices in the simplicial nerve (the degenerate simplices are ones in which some set is repeated). It follows (see [22]) that the geometric realizations of the two nerves are equal.

Some notations are needed involving a k-simplex $\sigma = (U_0, \ldots, U_k)$ of the nerve. $\cap\sigma$ denotes, as before, the intersection $\cap_{i=0}^{k} U_i$. The complement of σ in $\mathcal{U}$ is denoted $\mathcal{U} - \sigma$. And as indicated above $\partial_j \sigma = \{U_i \mid i \neq j\}$.

4.2 Cycles

Suppose $\mathcal{U}$ is a partially ordered cover of X, and $p: E \to X$ is given. Then a $\mathcal{J}$-n-cycle in $(X, p; \mathcal{U})$ is a function $N: \text{nerve}_\Delta(\mathcal{U}) \to \mathcal{J}$, specifically

(1) if σ is a k-simplex of $\text{nerve}_\Delta(\mathcal{U})$ then $N(\sigma)$ is an $(n-k)$-dimensional $(\mathcal{U} - \sigma)$-ad in $\mathcal{J}(p^{-1}(\cap\sigma))$,
(2) let $\text{incl}_*: \mathcal{J}(p^{-1}(\cap\sigma)) \to \mathcal{J}(p^{-1}(\cap b_j\sigma))$ denote the morphism induced by the inclusion, then $\text{incl}_*(N(\sigma)) = (-1)^j \partial_{U_j} N(b_j\sigma)$, and
(3) all but finitely many of the $N(\sigma)$ are ϕ.

For a source of geometric examples, suppose X is the realization of a simplicial complex and $\mathcal{U}$ is the covering by stars, as in 1.5. The dual cones provide a refinement of this cover in which the faces of the cones have collars. Suppose $M \to X$ is a map from a manifold, then M can be made transverse to the cones. This breaks M into pieces which are manifolds with boundary faces indexed by the cones. The function (simplex) $\mapsto$ (inverse image of cone dual to the simplex) defines a cycle in the manifold bordism-type theory. The incidence relations in (2) record how these pieces fit together inside M.

Next define a "homology" between two cycles. This is a function

$$H\colon \text{nerve}_\Delta(\mathcal{U}) \to \mathcal{J}$$

which takes σ to an $(n-k+1)$-dimensional $(\mathcal{U}-\sigma)\amalg[1]$-ad in $\mathcal{J}(p^{-1}(\cap\sigma))$. The disjoint union means $H(\sigma)$ is an -ad with faces ∂_U for U not in σ, and in addition faces ∂_0 and ∂_1. These are required to satisfy the cycle conditions above on the ∂_U faces, and also $\partial_0\partial_1 H = \phi$. It follows that $\partial_0 H$ and $\partial_1 H$ are n-dimensional cycles in the sense above. We then say that $\partial_0 H$ is homologous to $-\partial_1 H$.

In these terms the main theorem can be stated as:

4.2A Theorem. *Suppose $\mathcal{U}$ is a partially ordered cover of X, and $p\colon E \to X$ is given. Then there is a canonical isomorphism from the group of homology classes of $\mathcal{J}$-n-cycles in $(X, p; \mathcal{U})$ and the homology group*

$$H_n(\text{nerve}_\Delta(\mathcal{U}); \Omega^{\mathcal{J}}(p, \mathcal{U}))$$

defined in 2.3.

This isomorphism is also natural with respect to the functorially induced functions of cycles defined in section 4.5. This statement will follow from Theorem 5.1.

There is also a relative version of the theorem, and for that we define a relative version of cycles. The idea is that a cycle as defined above is "closed" in the sense that all $(n-1)$-dimensional pieces correspond to intersections $U \cap V$, and consequently occur as faces of the two n-dimensional pieces lying over U and V. We obtain "free boundary" which occurs only once as a face simply by failing to define pieces corresponding to certain subsets. Specifically, if $Y \in \mathcal{U}$ then a relative cycle over $(\mathcal{U}, Y)$ is defined to be a function $(\text{nerve}_\Delta(\mathcal{U}) - \{Y\}) \to \mathcal{J}$ satisfying the conditions (1)–(3) above.

The "Kan cycles" used in the definition of the Kan condition in 3.2 are relative cycles in this sense, taking values in a constant functor. To see these as cycles in a covering, embed the set A as a linearly independent set in a real vector space, and let X be the convex hull. Then X is covered by sets U_a consisting of all points with nonzero a coordinate when expressed as a convex linear combination. An (A, a)-cycle in the sense of the Kan condition is a relative cycle over $(\{U_b \mid b \in A\}, U_a)$.

If N is a relative $\mathcal{J}$-n-cycle in $(X, p; \mathcal{U}, Y)$, then there is a boundary $\partial_Y N$ defined. This is the $(n-1)$-cycle in $(Y, p; (\mathcal{U} - \{Y\}) \cap Y)$, specified by $(\partial_Y N)(\sigma) = \partial_Y(N(\sigma \cup \{Y\}))$.

4.2B Proposition. *There is a canonical isomorphism from the group of homology classes of relative $\mathcal{J}$-n-cycles in $(X, Y, p; \mathcal{U} \cup \{Y\})$ to the relative homology $H_n(X, Y; \Omega^{\mathcal{J}}(p, \mathcal{U}))$. The homomorphism of cycles induced by the boundary operation ∂_Y agrees with the boundary homomorphism in homology.*

4.3 Naturality of cycles

In this section we construct functions of cycles induced by morphisms of data. According to the representation theorem, cycles represent homology classes. According to the general construction in section 2, morphisms of data induce homomorphisms of homology groups, including assembly maps. The objective here is to give cycle-level descriptions of these natural homomorphisms.

The simplest case is the total assembly corresponding to the map $X \to pt$, and this is considered first. For this the construction is a simple application of the Kan condition. In general the functoriality of $\mathcal{J}$ is applied to get from a cycle on a cover of X a "multivalued" cycle on a cover of Y. The Kan condition is used to assemble the multiple pieces to get an honest cycle on Y.

Suppose $p\colon E \to X$ and $\mathcal{U}$ is a cover of X, as usual, and M is a $\mathcal{J}$-n-cycle in $(X, p; \mathcal{U})$. Let $\mathrm{incl}_*\colon \mathcal{J}(p^{-1}(U)) \to \mathcal{J}(E)$ denote the morphism induced by the inclusion. Then $\mathrm{incl}_*(M)$ is a function from $\mathcal{U}$ into $\mathcal{J}(E)$. (Note that once we are in a single theory $\mathcal{J}(E)$ the values of M on higher simplices of the nerve are determined by values on $\mathcal{U}$, so are unnecessary.)

If we add a disjoint element a and reindex the $M(*)$ to add $\partial_a M = \phi$, then this defines a $(\mathcal{U} \cup \{a\}, a)$-cycle in the sense of the Kan condition. Apply the Kan condition to obtain a $(\mathcal{U} \cup \{a\})$-ad N with $\partial_U N = M(U)$ for $U \in \mathcal{U}$. Then define $A(M) = \partial_a N$. Since $\partial_U \partial_a N = \phi$ this is a reindexed n-dimensional ϕ-ad in $\mathcal{J}(E)$.

We can now state the following, which is a special case of 4.3B.

4.3A Proposition. *This construction induces a homomorphism from homology classes of $\mathcal{J}$-n-cycles in $(X, p, \mathcal{U})$ to $\Omega_n^{\mathcal{J}}(E)$. Under the canonical isomorphism with homology this corresponds to the total assembly*

$$H_n(\mathrm{nerve}_\Delta(\mathcal{U}); \Omega^{\mathcal{J}}(p, \mathcal{U})) \to H_n(pt; \Omega^{\mathcal{J}}(E)) \simeq \Omega_n^{\mathcal{J}}(E) .$$

Now begin the general construction. Suppose $(f, \hat{f})$ is a map between maps p, q, ie. there is a commutative diagram

$$\begin{array}{ccc} E & \xrightarrow{\hat{f}} & F \\ \downarrow p & & \downarrow q \\ X & \xrightarrow{f} & Y. \end{array}$$

Suppose $\mathcal{U}$ is a cover of X, $\mathcal{V}$ a cover of Y, and $\theta\colon \mathcal{U} \to \mathcal{V}$ is a morphism. By this last we mean an order-preserving function so that $f(U) \subset \theta(U)$ for every $U \in \mathcal{U}$.

The construction will be in two parts; first a function from cycles in $(X, p, \mathcal{U})$ to cycles in $(Y, q, \mathcal{V}_\theta)$, where $\mathcal{V}_\theta$ is a cover of Y obtained by introducing multiple copies of the elements of $\mathcal{V}$ indexed by $\mathcal{U}$. The second part goes from cycles in $(Y, q, \mathcal{V}_\theta)$ to cycles in $(Y, q, \mathcal{V})$, by assembling pieces over multiple copies using the Kan condition as above.

Define $\mathcal{V}_\theta$ to be the collection of subsets of Y isomorphic as a partially ordered set with $\mathcal{U}$, by the correspondence $U \mapsto V_U = \theta(U)$. This may not be a cover of Y. There are morphisms of covers (or perhaps just collections of subsets) $\mathcal{U} \to \mathcal{V}_\theta \to \mathcal{V}$, the first defined by $U \mapsto V_U$ and the second by $V_U \mapsto \theta(U)$. Since the first is an isomorphism of partially ordered sets it induces an injective Δ-map $\mathrm{nerve}_\Delta(\mathcal{U}) \to \mathrm{nerve}_\Delta(\mathcal{V}_\theta)$. The second induces a simplicial map of simplicial nerves, as does any morphism, but usually not a Δ-map.

Next suppose M is an n-cycle in $(X, p, \mathcal{U})$, so for σ a k-simplex in $\mathrm{nerve}_\Delta(\mathcal{U})$ we get an $(n-k)$-dimensional $\mathcal{U} - \sigma$-ad $M(\sigma) \in \mathcal{J}(p^{-1}(\cap\sigma)$. Let V_σ denote the image of σ in $\mathrm{nerve}_\Delta(\mathcal{V}_\theta)$. Then $\hat{f}$ induces a function $\hat{f}_*\colon \mathcal{J}(p^{-1}(\cap\sigma)) \to \mathcal{J}(q^{-1}(\cap V_\sigma))$. Therefore we can define a function

$$\hat{f}_* M\colon \mathrm{nerve}_\Delta(\mathcal{V}_\theta) \to \mathcal{J}$$

by: if τ is the image of a simplex in $\mathrm{nerve}_\Delta(\mathcal{U})$, so $\tau = V_\sigma$, then $\hat{f}_* M(\tau) = \hat{f}_*(M(\sigma))$. If τ is not in the image define $\hat{f}_* M(\tau) = \phi$.

It is immediate that this $\hat{f}_*M$ is an n-cycle in $(Y, q, \mathcal{V}_\theta)$, and further that the same procedure defines a function on homologies. This therefore defines a function on homology classes of cycles.

Now we begin the second step; the passage from $\mathcal{V}_\theta$ to $\mathcal{V}$. This is broken into simpler pieces by a relative version of the θ-construction: suppose $\mathcal{V}' \subset \mathcal{V}$, then define $(\mathcal{V}, \mathcal{V}')_\theta$ to be the collection $\mathcal{V}' \cup \{V_U \mid \theta(U) \notin \mathcal{V}'\}$. There are morphisms $\mathcal{V}_\theta \to (\mathcal{V}, \mathcal{V}')_\theta \to \mathcal{V}$. By using a sequence of $\mathcal{V}'$ which differ by single sets we get a factorization of $\mathcal{V}_\theta \to \mathcal{V}$ into a sequence of morphisms which are bijections except over a single set. It is therefore sufficient to do the construction for such morphisms. Note that the finiteness requirement on cycles implies that for a given cycle only a finite number of such special morphisms are required, even if $\mathcal{V}$ is infinite.

Suppose then that $\mathcal{V}$ is a cover of Y, $V_1 \in \mathcal{V}$, and $V_2, \ldots, V_r$ are additional copies of V_1. Given an n-cycle in $(Y, q, \mathcal{V} \cup \{V_2, \ldots, V_r\})$ we construct an n-cycle in $(Y, q, \mathcal{V})$, by constructing a type of "homology" in which the cover changes. For a given simplex the construction depends on whether or not the simplex contains V_1.

Let M be the n-cycle over $\mathcal{V} \cup \{V_2, \ldots, V_r\}$, and let a, b, c be elements (to be used for indices) not in the cover. On simplices of $\text{nerve}_\Delta(\mathcal{V})$ containing V_1, say $\sigma \cup \{V_1\}$, we think of $M(\sigma \cup \{V_*\})$ as a $(\{V_*, a\}, a)$-cycle and fill in using the Kan condition. Specifically we want a function N so that $N(\sigma \cup \{V_1\})$ is an $(n-k+1)$-dimensional $(\mathcal{V} - \sigma \cup \{V_*\} \cup \{a\})$-ad in $\mathcal{J}(q^{-1}(\cap(\sigma \cup \{V_1\})))$ which is finite, satisfies the face relations as in (2) of the definition of cycles, and $\partial_{V_i} N(\sigma \cup \{V_1\}) = M(\sigma \cup \{V_i\})$ for $1 \le i \le r$.

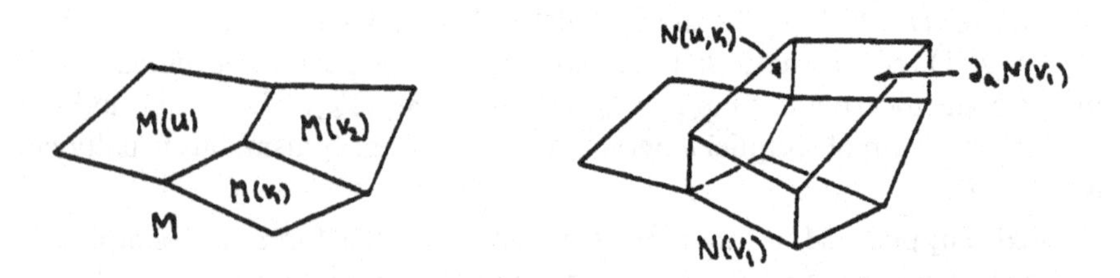

The function N is constructed by induction on dimension, beginning with large dimensions and working down. $M(\tau) = \phi$ for all but finitely many $\tau \in \text{nerve}_\Delta(\mathcal{V} \cup \{V_i\})$, so there is a dimension above which $M = \phi$ and above this dimension we can set $N = \phi$. Now suppose N is defined for simplices of dimension greater than k, and suppose $\sigma \cup \{V_1\}$ has dimension k. Then a Kan-type cycle over $(\mathcal{V} - \sigma \cup \{V_*\} \cup \{a\}, a)$ in $\mathcal{J}(q^{-1}(\cap(\sigma \cup \{V_1\}))$

is defined by $U \mapsto N(\sigma \cup \{V_1, U\})$ and $V_i \mapsto N(\sigma \cup \{V_i\})$ for $1 \leq i \leq r$. Apply the Kan condition to obtain $N(\sigma \cup \{V_1\})$.

We would like to continue applying the Kan condition to extend N over simplices which do not contain V_1. However the pieces $M(\sigma)$ and $N(\sigma \cup \{V_1\})$ do not fit together correctly; the intersection is the cycle $V_i \mapsto M(\sigma \cup \{V_i\})$ rather than a single face. To fix this we introduce some new pieces, which should be thought of as subdividing a collar neighborhood of $\cup M(\sigma \cup \{V_*\})$ in $M(\sigma)$.

Let C be a homology from $\partial_a C = M$ to some other cycle $\partial_b C$. (We think of C as a collar $M \times I$, but construct it by inductive application of the Kan condition as in the construction of N above.) Next we construct a function W on simplices of $\text{nerve}_\Delta(\mathcal{V})$ which do not contain V_1. We want $W(\sigma)$ to be a $(\mathcal{V} - \sigma \cup \{V_*\} \cup \{a, b, c\})$-ad in the same theory as $M(\sigma)$ satisfying

(1) $\partial_U W(\sigma) = W(\sigma \cup \{U\})$ if $U \neq V_1$, and $U \notin \sigma$;
(2) $\partial_{V_i} W(\sigma) = C(\sigma \cup \{V_i\})$; and
(3) $\partial_b W(\sigma) = \partial_b C(\sigma)$ and $\partial_c W(\sigma) = N(\sigma \cup \{V_1\})$.

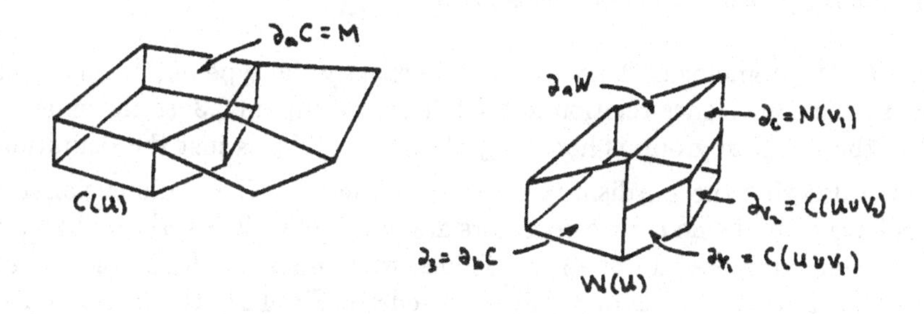

As with N we define W to be ϕ on high-dimensional simplices and work down by induction. If σ has dimension k and W is defined on higher dimensional simplices then all the faces specified above are defined and form a $(\mathcal{V} - \sigma \cup \{V_*\} \cup \{a, b, c\}, a)$-cycle. Applying the Kan condition yields an -ad which we define to be $W(\sigma)$.

Now define a function on $\text{nerve}_\Delta(\mathcal{V})$ by $\sigma \mapsto \partial_a N(\sigma)$ if $V_1 \in \sigma$, and $\sigma \mapsto \partial_a W(\sigma)$ if $V_1 \notin \sigma$. This defines a $\mathcal{J}$-n-cycle in $(Y, q, \mathcal{V})$, which is defined to be the functorial image of M.

4.3B Proposition. *This construction induces a homomorphism from homology classes of $\mathcal{J}$-n-cycles in $(X, p, \mathcal{U})$ to $\mathcal{J}$-n-cycles in $(Y, q, \mathcal{V})$. Under the canonical isomorphism with homology this corresponds to the induced homomorphism*

$$H_n(\|\text{nerve}(\mathcal{U})\|; \Omega^{\mathcal{J}}(p, \mathcal{U})) \to H_n(\|\text{nerve}(\mathcal{V})\|; \Omega^{\mathcal{J}}(q, \mathcal{V})).$$

4.4 The bordism-type theory of cycles

Fix $p\colon E \to X$ and a cover $\mathcal{U}$ of X. In this section we define a bordism-type theory denoted $\text{Cycles}^{\mathcal{J}}(X,p,\mathcal{U})$ whose ϕ-ads are $\mathcal{J}$-cycles in $(X,p,\mathcal{U})$.

Suppose A is a set. An A-ad of dimension n in $\text{Cycles}^{\mathcal{J}}(X,p,\mathcal{U})$ is a function $N\colon \text{nerve}_{\Delta}(\mathcal{U}) \to \mathcal{J}$ satisfying exactly the definition of cycles given above, except that it takes values in A-ads. Explicitly,

(1) if σ is a k-simplex of $\text{nerve}_{\Delta}(\mathcal{U})$ then $N(\sigma)$ is an $(n-k)$-dimensional $(\mathcal{U} - \sigma \cup A)$-ad in $\mathcal{J}(p^{-1}(\cap\sigma))$,
(2) let $\text{incl}_*\colon \mathcal{J}(p^{-1}(\cap\sigma)) \to \mathcal{J}(p^{-1}(\cap b_j\sigma))$ denote the morphism induced by the inclusion, then $\text{incl}_*(N(\sigma)) = (-1)^j \partial_{U_j} N(b_j\sigma)$, and
(3) all but finitely many of the $N(\sigma)$ are ϕ.

In (2), U_j is the jth element of σ with respect to the partial ordering, as in 4.2.

4.4A Proposition. *$\text{Cycles}^{\mathcal{J}}(X,p,\mathcal{U})$ has the structure of a bordism-type theory. The n-dimensional bordism group of this theory is exactly the set of homology classes of $\mathcal{J}$-n-cycles in $(X,p,\mathcal{U})$.*

Proof. The homologies defined in 4.2 are exactly the type of [1]-ads used to define the equivalence relation in the bordism group, in 2.3, so the assertion that the bordism group is homology classes of cycles is just the definition.

We describe the bordism-type theory structure. The n-dimensional A-ads have been defined. Face operators are defined by $(\partial_a N)(\sigma)) = \partial_a(N(\sigma)$. The reindexing operations ℓ_θ are defined by reindexing all the pieces, and the bijectivity condition in 3.2(3) is immediate. Similarly the involution -1 is defined by applying the involution in $\mathcal{J}$ to each piece.

The remaining ingredient is to see that the Kan condition holds. This construction is similar to a step in the construction of induced maps of cycles in 4.3.

Suppose M is an n-dimensional (A,a)-cycle. This is a function from $A - \{a\}$ to cycles so that $M(b)$ is an n-dimensional $(A - \{b\})$-ad in

$$\text{Cycles}^{\mathcal{J}}(X,p,\mathcal{U}).$$

Unraveling a little further, this is a function $M\colon (A-\{a\}) \times \text{nerve}_{\Delta}(\mathcal{U}) \to \mathcal{J}$ so that $M(b,\sigma)$ is a $(A-\{b\}) \cup (\mathcal{U}-\sigma)$-ad of dimension $n-k$ in $\mathcal{J}(p^{-1}(\cap\sigma))$, where k is the dimension of σ. We can think of this as a cycle of (A,a)-cycles, over $\mathcal{U}$.

What we want is a cycle of A-ads over $\mathcal{U}$, so that for each σ the (A,a)-cycle $N(*,\sigma)$ is obtained from it by taking faces. In more detail this is a function N: $\text{nerve}_\Delta(\mathcal{U}) \to \mathcal{J}$ so that $N(\sigma)$ is a $A \cup (\mathcal{U} - \sigma)$-ad of dimension $n - k + 1$ in $\mathcal{J}(p^{-1}(\cap\sigma))$. This should satisfy $M(b, \sigma) = \partial_b N(\sigma)$, and the cycle face relation 4.4(2) relating $N(\sigma)$ and $N(b_j\sigma)$.

We construct N by induction downward on dimensions of simplices in $\text{nerve}_\Delta(\mathcal{U})$. Since $M(b, \sigma) = \phi$ for all but finitely many (b, σ), there is a dimension for σ above which $M = \phi$ and we can set $N(\sigma) = \phi$.

Now suppose M is defined on simplices of dimensions greater than k, and consider a k-simplex σ. Define a Kan-type $(A \cup \mathcal{U} - \sigma, a)$-cycle in $\mathcal{J}(p^{-1}(\cap\sigma))$ by: take $b \in A - \{a\}$ to $M(b, \sigma)$. Take $V \in \mathcal{U} - \sigma$ to $\text{incl}_* N(\sigma \cup V)$, if $\sigma \cup V$ is a simplex in the nerve (ie. $\cap\sigma \cap V \neq 0$), and take it to ϕ otherwise. Note $N(\sigma \cup V)$ is defined since $\sigma \cup V$ is a simplex of dimension greater than k. This formula does in fact form a cycle, so the Kan condition in $\mathcal{J}(p^{-1}(\cap\sigma))$ implies there is a $(A \cup \mathcal{U} - \sigma)$-ad which has this cycle as faces. Select one of these to be $N(\sigma)$, then this satisfies the conditions required for the induction step. □

4.5 Naturality of theories

The constructions of 4.3 give homomorphisms of homology groups of cycles, corresponding to appropriate morphisms of data. The definitions of 4.4 extends the homology groups to entire bordism-type theories. It would be nice to similarly extend the homomorphism construction to give morphisms of bordism-type theories. Such morphisms would by naturality induce maps of the associated bordism spectra. The constructions of 4.3 are not canonical enough to give morphisms of theories, but they do extend directly to the bordism spectra.

Proposition. *Suppose $(f, \hat{f})\colon p \to q$ is a morphism of maps and $\theta\colon \mathcal{U} \to \mathcal{V}$ is a morphism of covers, as in 4.3. Then there is an associated morphism of bordism spectra*

$$(f, \hat{f}, \theta)_*\colon \Omega(\mathit{Cycles}(X, p, \mathcal{U})) \to \Omega(\mathit{Cycles}(Y, q, \mathcal{V}))$$

which is natural up to homotopy, and on homotopy groups induces the homomorphism defined in 4.3.

Proof. We will not give the proof in detail. It is primarily an elaboration on the construction in 4.3, and although the idea behind it can be described easily, the indexing on the various -ads considered gets too complex to be informative. Also relatively little use is made of it here; it is primarily used to replace the word "canonical" in Theorem 4.2A with the word "natural" in the final result (however, see 4.6).

The spaces $\Omega(\mathrm{Cycles}(X, p, \mathcal{U}))$ and $\Omega(\mathrm{Cycles}(Y, q, \mathcal{V}))$ are geometric realizations of Δ-sets (see 3.4), so we get a map between them by realizing Δ-maps.

A k-simplex of $\Omega_n(\mathrm{Cycles}(X, p, \mathcal{U}))$ is by definition a $[k]$-ad in the bordism-type theory $\mathrm{Cycles}(X, p, \mathcal{U})$, which in turn is a function $\mathrm{nerve}_\Delta(\mathcal{U}) \to$ ($[k]$-ads in $\mathcal{J}$). If θ is an injection then it induces a Δ-injection of nerves $\mathrm{nerve}_\Delta(\mathcal{U}) \to \mathrm{nerve}_\Delta(\mathcal{V})$. In this case we get a k-simplex of

$$\Omega_n(\mathrm{Cycles}(Y, q, \mathcal{V}))$$

by applying morphisms induced in $\mathcal{J}$ by $(f, \hat{f})$, and then extending the function to $\mathrm{nerve}_\Delta(\mathcal{V})$ simply by defining it to be ϕ on the complement of $\mathrm{nerve}_\Delta(\mathcal{U})$.

As for single cycles this reduces the construction to $X = Y$ and θ a morphism which eliminates duplicate copies of a single set V_1. In this case we define the map by induction on skeleta of $\Omega_n(\mathrm{Cycles}(X, p, \mathcal{U}))$. If M is a 0-simplex then it is a cycle, and we define $(f, \hat{f}, \theta)(M)$ as in 4.3A. Suppose the map is defined on the $(k-1)$-skeleton and M is a k-simplex. The construction from this point is essentially the same as that of 4.3A, except there are more faces. As in 4.3A we proceed by induction downwards on dimension of simplices of $\mathrm{nerve}_\Delta(\mathcal{V})$. The induction is started by setting the values to be ϕ for simplices of sufficiently high dimension.

On simplices containing V_1 we basically want a cycle of solutions to Kan extension problems. For a given simplex we get a Kan cycle with three types of pieces: from the construction on higher simplices, ones of the form $M(\sigma \cup \{V_i\})$, as before, and also pieces from the construction on $b_j M$. According to the induction hypotheses all these are already defined, so the Kan condition can be used to extend the construction to $(\sigma \cup \{V_1\})$.

On simplices not containing V_1 the construction involves first constructing a "collar" to introduce more faces, then finding a cycle of solutions to the resulting Kan problems. Again if this has been done for higher simplices of $\mathrm{nerve}_\Delta(\mathcal{V})$ and also for faces $b_j M$ then we get extension problems whose solutions extend the construction over the simplex.

Finally we indicate why this is well-defined and natural up to homotopy. To see it is well-defined suppose there are two such constructions, and think of these as defined on $\mathrm{nerve}_\Delta(\mathcal{V}) \times \{0, 1\}$. Then use the same proceedure to fill in between these to get a homotopy defined on $\mathrm{nerve}_\Delta(\mathcal{V}) \times [0, 1]$. The only difference is that there are yet more faces, coming from the ends of the homotopy where the construction is already given.

This, together with the Δ-set model for the loop space used in 3.4, also shows that these maps of spaces fit together (up to homotopy) to give maps of spectra.

The proof of naturality is straightforward except for changing covers. For this we need to see that if $\mathcal{V}$ is obtained from $\mathcal{U}$ by eliminating duplicates of two different sets, then the compositions are independent (up to homotopy) of the order in which this is done. Filling in a homotopy between the two compositions proceeds by double induction as before, but the proceedure for each piece is substantially more complicated. It separates into four cases, depending on whether or not the sets being changed are contained in the simplex. The worst case, when the simplex contains neither, seems to require four different applications of the Kan condition. It is not too difficult to guess what to do, but it does seem to be a lot of trouble to verify that the resulting formulae do in fact give Kan-type cycles. □

4.6 Nonsimplicial situations

Cycles are associated with a covering, and describe homology of the nerve of that cover. A homotopy inverse limit is used to define general homology, and although this can be avoided if the situation is simplicial (by 2.6) it is necessary in general. The maps in the inverse system are induced by morphisms of the data, so the proposition gives a description of the system in terms of cycles. This leads to a cycle description of general homology classes.

When the space is reasonable, for example metric, the description of such homotopy inverse limits in terms of arcs can be employed. In these terms a homology class is represented by a half-open arc of cycles: a triangulation of $[0,\infty)$ is given, the vertices are mapped to $(X, p, \mathcal{U})$-cycles, where $\mathcal{U}$ depends on the vertex. Edges are mapped to homologies of the type considered in 4.3 which change the covering. Finally we require that the covers have diameters which go to 0 as we go toward ∞.

Carrying out this algorithm using the constructions as given results in an unpleasantly complicated mess, so we will not do it here. For this to work better it would be helpful to have a direct description of homologies which change covers arbitrarily, not just by duplicating a single subset. Also better naturality constructions would be needed. For this it might be useful to consider more elaborate forms of the Kan condition, for example allowing pieces to intersect in cycles instead of just faces.

4.7 Cocycles and cohomology

There is a representation result for cohomology which is easier than cycles and homology, and we describe this here. They are particularly useful in descriptions of products and duality, but these will not be discussed here. We restrict to the constant coefficient case. The reader who needs the functor coefficient analog should not have trouble working it out.

These representative are basically the same as the unfortunately named "mock bundles" of [4].

Cohomology with coefficients in a spectrum $\mathbf{J}$ is defined, dually to homology, by maps:

$$H^n(X; \mathbf{J}) = \pi_0(\text{maps}(X, \mathbf{J}_n)).$$

When $\mathbf{J}$ is an Ω-spectrum this involves the single space $\mathbf{J}_n$, as written. For general spectra it has to be interpreted in terms of maps of spectra.

Now suppose $X = \|K\|$ is the realization of a Δ-set, and $J_n = \Omega_n^{\mathcal{J}}$ is the bordism spectrum of a bordism-type theory $\mathcal{J}$. According to the simplicial approximation theorem a map of realizations is homotopic to the realization of a Δ-map $f\colon K \to \Omega_{-n}^{\mathcal{J}}$. (Note, as in 3.4, the sign difference on the subscripts on the spectrum and Δ-set.) Thus to obtain representatives for cohomology classes we have only to refer to the definition of $\Omega_{-n}^{\mathcal{J}}$ and spell out what such a Δ-map looks like. The result is:

Proposition. *Classes in $H_n(\|K\|; \Omega^{\mathcal{J}})$ are represented by functions f from simplices of K to -ads in $\mathcal{J}$ such that*

(1) *if σ is an k-simplex then $f(\sigma)$ is an $[k]$-ad of dimension $k-n$, and*
(2) *$f(b_j\sigma)$ is obtained from $(-1)^j\partial_j f(\sigma)$ by reindexing by the order-preserving bijection $[k-1] \to [k] - \{j\}$.* □

To contrast this with the definition of a cycle we make explicit some of the differences. For this consider an n-cycle M in the star cover associated to the triangulation of $\|K\|$. Since the simplices of the nerve of this cover are indexed by simplices in K, the n-cycle M is also a function from simplices of K to -ads in $\mathcal{J}$. However

(1) the dimension of $M(\sigma)$ is the negative of the dimension of $f(\sigma)$. In particular the dimension of $f(\sigma)$ increases with the dimension of σ while the dimension of $M(\sigma)$ decreases.
(2) the face structures are also dual: $f(\sigma)$ has faces corresponding to the faces of σ, while roughly speaking $M(\sigma)$ has faces corresponding to simplices disjoint from σ.

In its face and dimension structure a cocycle behaves like a product $K \times F$, or more generally like a bundle over K. It can be thought of as a sort of block bundle over K in which the fibers are allowed to change from point to point (hence the term "mock bundle" in [4]).

5: Proof of the representation theorem

We now show that cycles represent homology. This is done on the spectrum level; the bordism spectrum of cycles is equivalent to the homology spectrum.

5.1 Theorem. *Suppose $\mathcal{U}$ is a cover of X, and $p\colon E \to X$. Then there is a homotopy equivalence of spectra,*

$$\Omega(\mathit{Cycles}^{\mathcal{J}}(X,p,\mathcal{U})) \to \mathbf{H}_{\bullet}(\mathit{nerve}(\mathcal{U}); \Omega^{\mathcal{J}}(p,\mathcal{U}))$$

which is natural up to homotopy.

Before giving the formal proof we describe the idea. Given a cycle we want to construct a point in one of the loop spaces in the direct limit defining the homology space. So we seek a map $S^k \to \Omega(p,\mathcal{U})/\mathrm{nerve}_\Delta(\mathcal{U})$, where $\Omega(p,\mathcal{U}) \to \mathrm{nerve}_\Delta(\mathcal{U})$ is the spectral sheaf constructed as in 2.3, and the quotient indicates dividing out the 0-section.

A cycle is a function $M\colon \mathrm{nerve}_\Delta(\mathcal{U}) \to \mathcal{J}$, but not any sort of Δ-map; $M(\sigma)$ has the wrong face structure to be a simplex which is an image of σ. However, assume $\mathcal{U}$ is finite with $n+1$ elements. There is a natural simplicial embedding $\mathrm{nerve}_\Delta(\mathcal{U}) \subset \Delta^{n+1}$, and we can associate to each simplex a dual simplex $D^n(\sigma)$. The cycle has the correct face structure to define a Δ-map $M(\sigma)\colon D^n(\sigma) \to \Omega^{\mathcal{J}}(p^{-1}(\cap\sigma))$.

Geometric realization of this Δ-map gives a map of spaces $\|D^n(\sigma)\| \to \Omega^{\mathcal{J}}(p^{-1}(\cap\sigma)$. This is natural in σ, so defines a natural transformation of functors. Here both $\|D^n(\sigma)\|$ and $\Omega^{\mathcal{J}}$ are regarded as functors from

$$\mathrm{nerve}_\Delta(\mathcal{U})$$

into spaces. The geometric realization of the functor $\sigma \mapsto \Omega^{\mathcal{J}}(p^{-1}(\cap\sigma)$ gives the spectral sheaf $\Omega^{\mathcal{J}}(p,\mathcal{U})$. The central geometric fact in the argument is that the geometric realization of the dual simplex functor $\sigma \mapsto \|D^n(\sigma)\|$ can be canonically identified with S^n. The realization of the natural transformation gives a map between these spaces, and therefore $S^n \to \Omega(p,\mathcal{U})/\mathrm{nerve}_\Delta(\mathcal{U})$, as required.

To reverse the process begin with a map $f\colon S^j \to \Omega(p,\mathcal{U})/\mathrm{nerve}_\Delta(\mathcal{U})$. First f is deformed to be transverse with respect to the simplex coordinates of $\Omega(p,\mathcal{U})$ coming from the construction as a realization. This describes f as coming from the realization of a natural transformation of functors, but defined on some manifold-valued functor usually different from the dual simplex functor. After stabilizing to get into the stable range (for embeddings) we can embed this functor in the dual simplex functor. Extension of the natural transformation to the whole dual simplex functor corresponds to suspension in the spectrum structure. Therefore after suspension the map becomes homotopic to a realization of the desired kind.

5.2 Functors on Δ-sets

Functors on simplicial nerves, regarded as categories, are used in 2.3 to define functor-coefficient homology. Since we are now using the Δ-nerve, we need a Δ-set version of this.

Suppose K is a Δ-set, and regard it as a category with morphisms generated by identity maps and the face operators ∂_j. If F is a covariant functor from K to spaces, then $\coprod_{\sigma\in K} F(\sigma)$ is a Δ-space. The geometric realization of this is defined by:

$$\|F\| = \Big(\coprod_{k,\sigma\in K^k} F(\sigma)\times\Delta^k\Big)/\sim$$

where $\sim$ is the equivalence relation generated by: if $x \in F(\partial\sigma)$, $t\in\Delta^{k-1}$, and $u\in\Delta^{k+1}$ then $(x,\partial_j^* t)\in F(\sigma)\times\Delta^k$ is equivalent to $(\partial_j x, t)\in F(\partial_j\sigma)\times\Delta^{k-1}$.

This construction is natural, with the same properties as the simplicial version in 2.2A. The reason, by the way, that we did not use the Δ version from the beginning is that morphisms of covers defines simplicial maps of simplicial nerves, but not Δ-maps of Δ-nerves. This makes changing covers awkward in the Δ version, and is part of the difficulty encountered in the naturality constructions in section 4.

Now we relate this to the simplicial version. If K is a simplicial set then $\operatorname{core} K \subset K$ is the Δ-set of nondegenerate simplices; the example of particular concern is $\operatorname{nerve}_\Delta(\mathcal{U}) \subset \operatorname{nerve}(\mathcal{U})$.) Then $\operatorname{core} K$ is a subcategory of K. If F is a functor from K to spaces, in the sense of 2.2 then the restriction to $\operatorname{core} K$ is a functor in the sense above. The inclusion $\operatorname{core} K \subset K$ induces a map of realizations.

5.2A Lemma. *Suppose K is a simplicial set and F is a functor from K to spaces. The natural map, from the Δ-realization of the restriction of F to $\operatorname{core} K$, to the simplicial realization of F, is a homeomorphism.*

This is straightforward, basically the same argument as the proof that the Δ-realization of the core is the same as the simplicial realization of K. □

5.3 Extensions to simplices

For convenience we extend both cycles and the spectral sheaves over a simplex containing the nerve. When the cover is finite we will actually work with the extension over $K = \partial\Delta^{\mathcal{U}}$, rather than over the whole simplex.

Suppose $\mathcal{U}$ is a totally ordered collection of subsets of X. Define a Δ-set $\Delta^{\mathcal{U}}$ with k-simplices all collections of $k+1$ distinct elements of $\mathcal{U}$. If $\mathcal{U}$ is finite this is a simplex. With this notation the simplex Δ^n is $\Delta^{[n]}$. The nerve is a sub-Δ-set of this; $\text{nerve}_\Delta(\mathcal{U}) \subset \Delta^{\mathcal{U}}$.

Now suppose $\mathcal{J}$ is a bordism-theory valued functor of spaces, which on the empty set consists only of basepoints. A $\mathcal{J}$-n-cycle over $(X, p, \mathcal{U})$ is defined to be a function from $\text{nerve}_\Delta(\mathcal{U})$ into $\mathcal{J}$ which takes a simplex σ to a $\mathcal{U} - \sigma$-ad in $\mathcal{J}(p^{-1}(\cap\sigma))$. We can consider cycles defined on all of $\Delta^{\mathcal{U}}$ satisfying the same conditions. The value on the additional simplices is determined: if $\sigma \notin \text{nerve}_\Delta(\mathcal{U})$ then $\cap\sigma = \phi$ so $\mathcal{J}(p^{-1}(\cap\sigma)) = \mathcal{J}(\phi)$, which consists only of basepoints. We formalize this as a lemma.

5.3A Lemma. *Suppose $\mathcal{J}(\phi)$ consists only of basepoints, and K is a Δ-set, $\text{nerve}_\Delta(\mathcal{U}) \subset K \subset \Delta^{\mathcal{U}}$. Then restriction defines a bijection from $\mathcal{J}$ cycles defined on K to cycles defined on $\text{nerve}_\Delta(\mathcal{U})$.* □

Spectral sheaves extend in a similar fashion. Suppose $\mathbf{J}$ is a spectrum-valued functor of spaces. The spectral sheaf over $\|\text{nerve}_\Delta(\mathcal{U})\|$ is defined by realizing the functor $\sigma \mapsto \mathbf{J}(p^{-1}(\cap\sigma))$. This functor is evidently defined on all of $\Delta^{\mathcal{U}}$, not just $\text{nerve}_\Delta(\mathcal{U})$. Realization of the extended functor defines a sheaf over $\|\Delta^{\mathcal{U}}\|$, which we continue to denote by $\mathbf{J}(p, \mathcal{U})$.

5.3B Lemma. *Suppose $\mathbf{J}$ is a spectrum-valued functor such that $\mathbf{J}(\phi)$ is contractible, and K is a Δ-set $\text{nerve}_\Delta(\mathcal{U}) \subset K \subset \Delta^{\mathcal{U}}$. Then the induced inclusion of homology spectra*

$$\mathbf{H}_\bullet(\textit{nerve}_\Delta(\mathcal{U}); \mathbf{J}(p, \mathcal{U})) \to \mathbf{H}_\bullet(K; \mathbf{J}(p, \mathcal{U}))$$

is a homotopy equivalence.

Proof. The homology of $\text{nerve}_\Delta(\mathcal{U})$ comes from the sheaf over $\text{nerve}_\Delta(\mathcal{U})$, divided by $\text{nerve}_\Delta(\mathcal{U})$. Similarly the homology of K comes from the sheaf over K, divided by K. The cofiber of the inclusion is the sheaf over K, divided by K and the sheaf over $\text{nerve}_\Delta(\mathcal{U})$. This is homeomorphic to $(K \times \mathbf{J}(\phi))/(\text{nerve}_\Delta(\mathcal{U}) \times \mathbf{J}(\phi) \cup K \times *)$, which is contractible because $\mathbf{J}(\phi)$ is. Since the cofiber is contractible, the inclusion is a homotopy equivalence. □

Note that if $\mathcal{J}$ is a bordism-theory valued functor as in 5.4A then the associated bordism spectrum is a spectrum-valued functor which satisfies the hypotheses of 5.4B. The bordism spectrum of the empty set is the realization of the Δ-set $\{\phi\}$ with a single simplex in each dimension. This realization is easily seen to be contractible (the fundamental group and homology groups are trivial).

5.4 Dual simplices

Fix an integer n. The "dual simplex functor" is a function $D^n: \partial\Delta^{n+1} \to \partial\Delta^{n+1}$, where $\partial\Delta^{n+1}$ denotes the Δ-set. If σ is a simplex of $\partial\Delta^{n+1}$ then $D^n(\sigma)$ is defined to be the simplex spanned by the vertices of Δ^{n+1} not in σ.

So for example Δ^{n+1} is the join $\sigma * D^n(\sigma)$.

Lemma.

(1) *The function $\sigma \mapsto \|D^n\|$ defines a functor from $\partial\Delta^{n+1}$ to spaces, and*

(2) *there is a canonical homeomorphism from the realization of this functor to S^n.*

Proof. D^n is contravariant in σ, in the sense that if $\tau \subset \sigma$ then $D^n(\tau) \supset D^n(\sigma)$. The face maps are contravariant with respect to inclusion, so D^n is a covariant functor on the category with morphisms generated by the ∂_*.

The realization of this functor is built of pieces $\|\sigma\| \times \|D^n(\sigma)\|$. Geometrically we think of this as a tubular neighborhood of σ. The homeomorphism to S^n gives the handlebody structure obtained by thickening up the cell decomposition of $\partial\Delta^{n+1}$.

The proof uses some spherical geometry, so we understand S^n to mean the sphere with its usual Riemannian metric. Suppose $X \subset S^n$ is contained in the interior of some hemisphere. Then we define the *convex hull* of X, denoted hull(X), to be the smallest set containing X, contained in the hemisphere, and intersecting each geodesic in a connected set. The hull is also the intersection of all hemispheres containing X, so in particular is independent of any particular hemisphere.

The fact we will use is that certain convex hulls are naturally homeomorphic to simplices. There is, for each set of points $V = (v_0, \dots, v_j)$ in S^n which are equidistant from each other, and lie in the interior of a hemisphere, a homeomorphism $f_V: \Delta^j \to \operatorname{hull} V$. This is continuous in V and natural with respect to faces and isometries; the restriction to the face obtained by omitting the ith vertex is the function associated to the set obtained by omitting v_i, and if $g: S^n \to S^n$ is an isometry then $f_V g$ is $f_{g(V)}$.

Choose points $x_0, \dots, x_n$ in S^n which are equidistant and the maximal distance apart; these are the vertices of an inscribed regular simplex. To make the construction canonical these points should be chosen in some canonical way. The point $-x_k$ is the barycenter of the simplex determined by $\{x_i \mid i \neq k\}$. Define the point $y_{i,k}$ for $i \neq k$ to be the midpoint of the shortest geodesic between x_i and $-x_k$.

Note that given i, j there is an isometry of S^n (in fact a reflection) which interchanges x_i and x_j and leaves the other x_* fixed. Since isometries preserve geodesics this reflection also interchanges $y_{i,k}$ and $y_{j,k}$, and interchanges $y_{k,i}$ and $y_{k,j}$

Next we define, for each $\sigma \in \partial\Delta^{n+1}$, a map $F_\sigma\colon \|\sigma\| \times \|D^n(\sigma)\| \to S^n$. This will be natural in σ and a homeomorphism onto the convex hull of the points $\{y_{i,k} \mid i \in \sigma, \text{ and } k \notin \sigma\}$ (here we have written $i \in \sigma$ if the ith vertex of Δ^{n+1} is in σ). These facts will imply the lemma: the naturality implies these fit together to define a map from the realization of the functor $\|D^n(*)\|$ to S^n, and the homeomorphism statement implies this map is a homeomorphism.

Suppose $(s,t) \in \|\sigma\| \times \|D^n(\sigma)\|$. For any fixed $k \notin \sigma$ the points in the set $y_{\sigma,k} = \{y_{i,k} \mid i \in \sigma\}$ are equidistant from each other. This is because the reflections which interchange the $\{x_i \mid i \in \sigma\}$ also interchange these points, and reflections preserve distances. Therefore the functions $f_{y_{\sigma,k}}$ are defined. Let $V = \{v_k\}$ denote the set obtained by applying these functions to the point s.

The points in the set V are also equidistant from one another. This is because the reflections which interchange x_k with $i \notin \sigma$ interchange the sets $y_{\sigma,k}$ and therefore—by naturality with respect to isometry—the functions $f_{y_{\sigma,k}}$. This implies the reflections also interchange images of a specific point, in this case s. From this we conclude the function f_V is defined, and we define $F_\sigma(s,t) = f_V(t)$.

The function F_σ is continuous because the f_* are continuous in $*$. It is natural in σ because the f_* are. It therefore remains to verify that it is a homeomorphism.

Consider the function f_V again. The image of this intersects σ in a single point, and is perpendicular to σ at that point. This is because the reflections which interchange points of V leave σ invariant; if w is a vector in σ it makes the same angle with each of the geodesics from the intersection point to a vertex of the image of f_V. This angle must therefore be 0. This identifies f_V as the intersection of the linear sphere perpendicular to σ, and the hull of $\{y_{i,k} \mid i \in \sigma, \text{ and } k \notin \sigma\}$.

We can now reverse the construction. If z is a point in the hull then it lies in some sphere perpendicular to σ. Let $V = \{V_k\}$ denote the intersection of this sphere with the hull of $y_{\sigma,k}$, where $k \notin \sigma$. The sphere is invariant under isometries which fix σ, so invariant under the reflections used above. Since these interchange elements of V the same argument used above implies these are equidistant. The point z is therefore $f_V(t)$ for some $t \in \|D^n(\sigma)\|$. Further for each $k \notin \sigma$ the point z_k is $f_{y_{\sigma,k}}(s_k)$, for some $s_k \in \|\sigma\|$. Using the symmetry again we see that all the s_k are equal. This identifies z as $F_\sigma(s_k, t)$.

Explicitly we have explained why F_σ is onto. But the points s and t are easily seen to be uniquely defined, so it is also injective. Therefore it is a homeomorphism. □

5.5 From cycles to homology

Suppose $\mathcal{U}$ is a finite collection of subsets of X. In this section we describe the map, from the bordism spaces of cycles to the homology spaces, both over $\mathcal{U}$. We verify this is a map of spectra in the next section, and consider infinite collections in the section 5.7.

Denote the elements of $\mathcal{U}$ by $U_0, \ldots, U_{n+1}$, so the simplex spanned by this is canonically Δ^{n+1}. Suppose that U_0 is empty. This assumption implies that $\text{nerve}_\Delta(\mathcal{U}) \subset \partial_0 \Delta^{n+1}$, and in particular $\text{nerve}_\Delta(\mathcal{U}) \subset \partial \Delta^{n+1}$.

Suppose M is a $\mathcal{J}$-r-cycle in $(X, p, \mathcal{U})$, extended trivially to $\partial\Delta^{n+1}$ as in the previous section. Then $M(\sigma)$ is a $(\mathcal{U} - \sigma)$-ad of dimension $r - j$ in $\mathcal{J}(p^{-1}(\cap\sigma))$, where j is the dimension of σ. Using the canonical bijection $[n-j] \to \mathcal{U} - \sigma$ we can regard $M(\sigma)$ as an $(n-j)$-dimensional simplex of the bordism Δ-set $\Omega^{\mathcal{J}}_{r-n}(p^{-1}(\cap\sigma))$. Or, since $D^n(\sigma)$ is the simplex spanned by $\mathcal{U} - \sigma$, this can be regarded as a Δ-map $D^n(\sigma) \to \Omega^{\mathcal{J}}_{r-n}(p^{-1}(\cap\sigma))$

The geometric realization of this Δ-map defines a map of spaces

$$\|D^n(\sigma)\| \to \Omega^{\mathcal{J}}_{n-r}(p^{-1}(\cap\sigma)).$$

(Remember that indices are reversed in forming the bordism spectrum Ω.) This is a natural transformation of functors defined on $\partial\Delta^{n+1}$, in other words

$$\begin{array}{ccc}
\|D^n(\partial_i\sigma)\| & \longrightarrow & \Omega^{\mathcal{J}}_{n-r}(p^{-1}(\cap\partial_i\sigma)) \\
\downarrow & & \downarrow \\
\|D^n(\sigma)\| & \longrightarrow & \Omega^{\mathcal{J}}_{n-r}(p^{-1}(\cap\sigma))
\end{array}$$

commutes. This is condition (2) in the definition of cycles in 4.2. This natural transformation induces a map of realizations of these functors. According to 5.4 the realization of $\sigma \mapsto \|D^n(\sigma)\|$ is S^n. The other realization is the spectral sheaf (over $\partial\Delta^{n+1}$) so we get a map $S^n \to \Omega^{\mathcal{J}}_{n-r}(p, \mathcal{U})$.

The initial hypothesis that $U_0 = \phi$ implies that the basepoint of S^n (the 0th vertex of $\partial\Delta^{n+1}$) maps to the basepoint. Thus this map defines a point in the loop space $\Omega^n(\Omega^{\mathcal{J}}_{n-r}(p,\mathcal{U}))$. Divide by $\partial\Delta^{n+1}$ and include into the homology spectrum to get a point in $\mathbf{H}_{-r}(\partial\Delta^{n+1};\Omega^{\mathcal{J}}(p,\mathcal{U}))$.

This defines a function from the vertices of the bordism spectrum of $\mathcal{J}$-cycles on $(X,p,\mathcal{U})$ to the homology spectrum of $\partial\Delta^{n+1}$. Next we extend this to a map of the whole space of cycles, essentially by adding a Δ^i coordinate to the construction above.

Let M be an i-simplex of the bordism space of r-dimensional cycles. This takes a j-simplex σ of $\partial\Delta^{n+1}$ to a $((\mathcal{U}-\sigma)\cup[i])$-ad of dimension $r-j+i$ in $\mathcal{J}(p^{-1}(\cap\sigma))$. Regard $M(\sigma)$ as an $(n-j+i+1)$-simplex of the associated bordism space. More precisely regard it as a Δ-map of the join simplex $D^n(\sigma) * \Delta^i$ into $\Omega^{\mathcal{J}}_{r-n-1}(p^{-1}(\cap\sigma))$. Realize this to get $\|D^n(\sigma)\| * \|\Delta^i\| \to \Omega^{\mathcal{J}}_{n-r+1}(p^{-1}(\cap\sigma))$.

Again this is a natural transformation of functors of σ, so induces a map of realizations of functors. The realization of the left side is $S^n * \|\Delta^i\|$, and that of the right side is the spectral sheaf, so this gives a map $S^n * \|\Delta^i\| \to \Omega^{\mathcal{J}}_{n-r+1}(p,\mathcal{U})$. Note $\|D^n(\sigma)\|$ and $\|\Delta^i\|$ are taken to the basepoint: The image of $D^n(\sigma)$ corresponds to $\partial_{[i]}M(\sigma)$, which is ϕ by definition of simplices of the bordism space. The image of Δ^i corresponds to $\partial_{(\mathcal{U}-\sigma)}M(\sigma)$ which is ϕ by condition (3) in the definition of cycles.

Regard the join as $S^n \times \|\Delta^i\| \times I$ with identifications at the ends of the I coordinate. Then the realization of $M(\sigma)$ defines a map $S^n \times \|\Delta^i\| \times I \to \Omega^{\mathcal{J}}_{n-r+1}(p,\mathcal{U})$. Let $S^n \times I \subset S^{n+1}$ denote the standard embedding, then the map extends by the point map on the complement to give $S^{n+1} \times \|\Delta^i\| \to \Omega^{\mathcal{J}}_{n-r+1}(p,\mathcal{U})$.

Regard this map as a simplex in the Δ-set of maps from S^{n+1} to $\Omega^{\mathcal{J}}_{n-r+1}(p,\mathcal{U})$. These also preserve basepoints, so this is a simplex in the loop space. We denote the loop space by maps$(S^{n+1},\Omega^{\mathcal{J}}_{n-r+1}(p,\mathcal{U}))$, using "maps" to avoid another Ω. We get one of these for each i-simplex M of the bordism Δ-set of cycles. The naturality of the construction implies these fit together to define a Δ-map

$$R_0 \colon \Omega_r(\text{Cycles}(X,p,\mathcal{U})) \to \text{maps}(S^{n+1},\Omega^{\mathcal{J}}_{n-r+1}(p,\mathcal{U})).$$

Homology is obtained from the geometric realization of the right side of this by dividing by the image of the 0-section, including into

$$\text{maps}(S^{n+j},\Omega_{n-r+j}),$$

and taking the limit as $j \to \infty$. Therefore realizing R_0 and including the right side in this limit defines maps

$$R\colon \Omega_{-r}(\text{Cycles}(X,p,\mathcal{U})) \to \mathbf{H}_{-r}(\partial\Delta^{n+1};\Omega^{\mathcal{J}}(p,\mathcal{U})).$$

(Again we note the reversal of the index on the left upon realization of the bordism Δ-set.) The right side of this is equivalent to the homology of $\text{nerve}_\Delta(\mathcal{U})$ by lemma 5.3B, so this gives the desired maps from cycles to homology.

5.6 The spectrum structure

For each $r \in Z$ we have a map from the space of r-cycles to the rth homology space. The next step is to show these form a map of spectra, ie. commute up to homotopy with the spectrum structure maps.

It is sufficient to show that the Δ-map R_0 homotopy commutes with appropriate spectrum structure maps. The appropriate diagram is

$$\begin{array}{ccc}
\Omega_r(\text{Cycles}(X,p,\mathcal{U})) & \xrightarrow{R_0} & \text{maps}(S^{n+1},\Omega^{\mathcal{J}}_{n-r+1}(p,\mathcal{U})) \\
\Big\downarrow{\scriptstyle \ell} & & \Big\downarrow \\
\Omega\Big(\Omega_{r-1}(\text{Cycles}(X,p,\mathcal{U}))\Big) & \xrightarrow{\Omega R_0} & \Omega\Big(\text{maps}(S^{n+1},\Omega^{\mathcal{J}}_{n-r+2}(p,\mathcal{U}))\Big).
\end{array}$$

The outer Ωs in the bottom row denote loop spaces. The right vertical map is induced by the structure map in the spectral sheaf, which roughly speaking is the fiberwise union of the structure maps over points in $\|\partial\Delta^{n+1}\|$. We refine the diagram so we can use a Δ-model for this.

The map R_0 is defined by realizing functors, so we factor it through a Δ-set of functors. Suppose F and G are functors from $\partial\Delta^{n+1}$ to Δ-sets. Then $\text{nat}(F,G)$ will be a space of natural transformations between these (actually we define something closer to the natural transformations from F to the loopspace ΩG). An i-simplex of this space associates to each $\sigma \in \partial\Delta^{n+1}$ a Δ-map $F(\sigma) * \Delta^i \to G(\sigma)$, compatible with the maps induced from the face maps in $\partial\Delta^{n+1}$. We also require that $F(\sigma) * \{\phi\}$ and $\{\phi\} * \Delta^i$ are taken to the basepoint of $G(\sigma)$. Here $K * \Delta^i$ is the Δ-set with simplices $\tau * \sigma$ where $\tau \in K$ and $\sigma \in \Delta^i$. $\tau * \sigma$ denotes a simplex with vertices the union of the vertices of τ and σ, ordered so that those of σ are last.

The construction of the map R_0 proceeds by constructing such natural transformations, from $D^n(\sigma)$ to $\Omega^{\mathcal{J}}_{r-n-1}(p^{-1}(\cap\sigma))$, then geometrically realizing. R_0 therefore factors as

$$\Omega_r(\text{Cycles}(X,p,\mathcal{U})) \xrightarrow{R_1} \text{nat}(D^n(*),\Omega^{\mathcal{J}}_{r-n-1}(p^{-1}(\cap *)))$$
$$\to \text{maps}(S^n_{n+1}\Omega^{\mathcal{J}}_{n-r+1}(p,\mathcal{U})).$$

The second map is obtained by realization (twice; the Δ-sets to get space-valued functors, and then the functors). In fact unraveling all the definitions will show that R_1 is an isomorphism.

Realization preserves spectrum structures, so the part we are concerned with is R_1. It is sufficient to show the following diagram commutes:

$$\begin{array}{ccc} \Omega_r(\mathrm{Cycles}(X,p,\mathcal{U})) & \xrightarrow{R_1} & \mathrm{nat}(D^n(*), \Omega^{\mathcal{J}}_{r-n}(p^{-1}(\cap *))) \\ {\scriptstyle \ell}\downarrow & & \downarrow \\ \Omega\Big(\Omega_{r-1}(\mathrm{Cycles}(X,p,\mathcal{U}))\Big) & \longrightarrow & \mathrm{nat}(D^n(*), \Omega\Omega^{\mathcal{J}}_{r-n-2}(p^{-1}(\cap *))). \end{array}$$

In these terms we can be more explicit about the right vertical map; this is induced by composition with the natural transformation

$$\ell\colon \Omega^{\mathcal{J}}_{r-n-1}(p^{-1}(\cap\sigma)) \to \Omega\Omega^{\mathcal{J}}_{r-n-2}(p^{-1}(\cap *)).$$

As in the definition of the spectrum structure in 3.4 we will use the Δ-set model for the loop space: the k-simplices of ΩK are the $(k+1)$-simplices of K with ∂_{k+1} and the opposite vertex both equal to basepoints. (And we denote this opposite vertex by $\partial_0^{k+1}\sigma$.) In this model the map ℓ is defined by reindexing a $[k]$-ad (= a k-simplex) to a $[k+1]$-ad using the natural inclusion $[k] \subset [k+1]$.

We now describe the lower horizontal map. This is a composition of two maps, first

$$\Omega R_1\colon \Omega(\Omega_{r-1}(\mathrm{Cycles}(X,p,\mathcal{U}))) \to \Omega(\mathrm{nat}(D^n(*), \Omega^{\mathcal{J}}_{r-n-1}(p^{-1}(\cap *)))$$

obtained by restricting R_1 to the models of loopspaces as subsets with face restrictions. Then there is the identification

$$\Omega(\mathrm{nat}(D^n(*), \Omega^{\mathcal{J}}_{r-n-1}(p^{-1}(\cap *))) \simeq \mathrm{nat}(D^n(*), \Omega\Omega^{\mathcal{J}}_{r-n-2}(p^{-1}(\cap *))).$$

We discuss this identification.

A k-simplex of $\Omega(\mathrm{nat}(D^n(*), \Omega^{\mathcal{J}}_{r-n-1}(p^{-1}(\cap *))))$ is a $(k+1)$-simplex of $\mathrm{nat}(D^n(*), \Omega^{\mathcal{J}}_{r-n-1}(p^{-1}(\cap *)))$ with $\partial_{k+1} = \phi = \partial_0^{k+1}$. This is a natural transformation $D^n(\sigma) * \Delta^{k+1} \to \Omega^{\mathcal{J}}_{r-n-1}(p^{-1}(\cap *))$, so associates to an $(n-j)$-simplex $\sigma \in \partial\Delta^{n+1}$ a $(j+k+2)$-simplex of $\Omega^{\mathcal{J}}_{r-n-1}(p^{-1}(\cap *))$. By definition of "nat" the restrictions to $\Delta^j * \{\phi\} = \partial_{j+1}^{k+2}$, and to $\{\phi\} * \Delta^{k+1} = \partial_0^{j+1}$ are both ϕ. Further, the restriction imposed to get the Δ loopspace is $\partial_{j+k+2} = \phi = \partial_{j+1}^{k+1}$.

Similarly a k-simplex of $\mathrm{nat}(D^n(*), \Omega\Omega^{\mathcal{J}}_{r-n-2}(p^{-1}(\cap *)))$ is a natural transformation which takes an $(n-j)$-simplex $\sigma \in \partial\Delta^{n+1}$ to a $(j+k+1)$-simplex of $\Omega\Omega^{\mathcal{J}}_{r-n-2}(p^{-1}(\cap *))$, again with $\partial^{k+1}_{j+1} = \phi = \partial^{j+1}_0$. This is a $j+k+2$-simplex of $\Omega^{\mathcal{J}}_{r-n-2}(p^{-1}(\cap *))$ with face restrictions $\partial_{j+k+2} = \phi = \partial^{j+k+2}_0$. These are the same conditions as in the previous paragraph, so the two sets are equal.

It is now straightforward to verify that the diagram commutes. Let M be a k-simplex of bordisms of cycles, so it assigns to $\sigma \in \partial\Delta^{n+1}$ a $(\mathcal{U}-\sigma\cup[k])$-ad in $\mathcal{J}(p^{-1}(\cap\sigma))$. Both compositions take this to the natural transformation which takes $D^n(\sigma)*\Delta^k$ to the $(j+k+2)$-simplex of $\Omega^{\mathcal{J}}_{r-n-2}(p^{-1}(\cap *))$ defined by $\ell_{j+k+2}M(\sigma)$, where ℓ_{j+k+2} reindexes by the inclusion $(\mathcal{U}-\sigma\cup[k]) \subset (\mathcal{U}-\sigma\cup[k+1])$.

This completes the verification that the maps defined in the previous section give a map of spectra, from the bordism spectrum of cycles, to homology. □

5.7 Reduction to finite collections

Elsewhere in this section we assume the collection $\mathcal{U}$ of subsets of X is finite. In this section we show this is sufficient for most purposes, and indicate the modifications necessary in the others.

The finiteness condition on cycles implies that any finite subcomplex of the bordism space $\Omega(\mathrm{Cycles}^{\mathcal{J}}(X,p,\mathcal{U}))$ is contained in the image of $\Omega(\mathrm{Cycles}^{\mathcal{J}}(X,p,\mathcal{V}))$ for some finite $\mathcal{V} \subset \mathcal{U}$. In particular homology classes of cycles can be defined solely in terms of finite subsets.

Similarly the homology spectrum is obtained from loop spaces of a quotient $\Omega^{\mathcal{J}}(p,\mathcal{U})/\mathrm{nerve}(\mathcal{U})$. A map of a finite complex into this deforms into the inverse image of a finite subcomplex of $\mathrm{nerve}(\mathcal{U})$ under the projection $\Omega^{\mathcal{J}}(p,\mathcal{U}) \to \mathrm{nerve}(\mathcal{U})$. But this corresponds to the homology spectrum of a finite subset of $\mathcal{U}$. The conclusion is that maps of spheres and homotopies between them lie in homology spaces of finite $\mathcal{V} \subset \mathcal{U}$.

It follows from this and naturality that if the passage from cycles to homology is a homotopy isomorphism for finite $\mathcal{U}$ then it is an isomorphism in general. Also, to define this passage on the group level it is sufficient to consider finite collections. The only ingredient for which this is not sufficient is the definition of the map on the spectrum level, when $\mathcal{U}$ is infinite. Therefore we discuss the construction of the map in this case.

Suppose $\mathcal{V}$ is a finite subset of $\mathcal{U}$. Denote by $\Delta^{\mathcal{V}}$ the simplex with vertices $\mathcal{V}$. In these terms the construction of 5.5 defines a map

$$\Omega(\mathrm{Cycles}^{\mathcal{J}}(X,p,\mathcal{V})) \to \mathrm{maps}(\|\partial\Delta^{\mathcal{V}}\|, \Omega^{\mathcal{J}}(p,\mathcal{V})/\|\partial\Delta^{\mathcal{V}}\|).$$

This construction is natural in $\mathcal{V}$, so forms a direct system of maps.

Suppose $\mathcal{V}$ is enlarged by addition of a copy of the empty set. This does not change the cycles and on the mapping space is equivalent to the suspension (since $\partial\Delta^{\mathcal{U}\cup\{\phi\}} \sim S^1 \wedge \partial\Delta^{\mathcal{U}}$). This means the direct limit used to define the homology can also be obtained as the mapping spaces associated to the sequence of inclusions $\cdots \subset \mathcal{V} \cup n\{\phi\} \subset \mathcal{V} \cup (n+1)\{\phi\} \subset \cdots$.

In general expand $\mathcal{U}$ by adding infinitely many copies of the empty set, and take the direct limit over all finite subsets $\mathcal{V}$ of the map above;

$$\lim_{\mathcal{V}\rightarrow} \Omega(\mathrm{Cycles}^{\mathcal{J}}(X,p,\mathcal{V})) \rightarrow \lim_{\mathcal{V}\rightarrow} \mathrm{maps}(\|\partial\Delta^{\mathcal{V}}\|, \Omega^{\mathcal{J}}(p,\mathcal{V})/\|\partial\Delta^{\mathcal{V}}\|).$$

It follows from the discussion above that the left side of this is equivalent to the cycle space for $\mathcal{U}$, and the right side is equivalent to the homology space. This therefore defines the desired map when $\mathcal{U}$ is infinite.

5.8 Completion of the proof

Again assume $\mathcal{U}$ is a collection of $n+2$ subsets of X, with the first one empty. The objective is to show that the map defined above, from the bordism spectrum of cycles over $\mathcal{U}$ to the homology with coefficients in the spectral sheaf $\Omega^{\mathcal{J}}(p,\mathcal{U})$ is a homotopy equivalence.

In the previous part of the proof the cycle spectrum has been identified with the Δ-set of natural transformations $\mathrm{nat}(D^n(\sigma), \Omega^{\mathcal{J}}_{r-n+1}(p^{-1}(\cap\sigma)))$, so it is sufficient to show the map from this to homology is a homotopy equivalence.

The first step is to compare the Δ-natural transformations with topological ones. Suppose F and G are topological functors on $\partial\Delta^{n+1}$. Define $\mathrm{nat}_*(F,G)$ to be the Δ-set with k-simplices the topological natural transformations $F(\sigma) * \Delta^k \rightarrow G(\sigma)$ which take $F(\sigma)$ and Δ^k to the basepoint. The subscript $*$ indicates the use of the join to define the simplex structure (there will be a product version below).

5.8A Lemma. *Realization defines a map of Δ-sets*

$$nat(D^n(\sigma), \Omega^{\mathcal{J}}_{r-n+1}(p^{-1}(\cap\sigma))) \rightarrow nat_*(\|D^n(\sigma)\|, \Omega^{\mathcal{J}}_{-r+n-1}(p^{-1}(\cap\sigma))).$$

As usual note the reversal of the index upon realization; $\Omega^{\mathcal{J}}_{-r+n-1} = \|\Omega^{\mathcal{J}}_{r-n+1}\|$.

Proof. It is sufficient to show the map induces an isomorphism of homotopy groups. These are both Kan Δ-sets so we can work with single simplices.

Specifically suppose M is a k-simplex of the topological version, $\partial_0 M$ is the realization of N_0, a $(k-1)$-simplex of the Δ version, and $\partial_j M = \phi$ for $j > 0$. Then it is sufficient to construct a (k-simplex N of the Δ version with $\partial_0 N = N_0$ and a homotopy rel ∂_0 of $\|N\|$ to M. This N is to be a functor on $\partial\Delta^{n+1}$, and will be constructed by induction downward on the dimension of σ.

Suppose $N(\sigma)$ is defined for $\sigma \in \partial\Delta^{n+1}$ of dimension greater than j, and suppose τ is a j-simplex. Then N is defined on $D^n(\partial\tau) * \Delta^k \cup D^n(\tau) * \partial\Delta^k = \partial\Delta^{n-j+k+1}$. Realize this, then the induction hypothesis provides a homotopy of this realization to the restriction of M to $\|\partial\Delta^{n-j+k+1}\|$. Regard this homotopy as an extension of $\|N\|$ over a collar of $\|\partial\Delta^{n-j+k+1}\|$, then M provides an extension over the rest of $\|\Delta^{n-j+k+1}\|$.

Now apply the simplicial approximation theorem, [22, Theorem 5.3]. This asserts that a map of a realization into the realization of a Kan Δ-set is homotopic to the realization of a Δ-map. Further, it can be held fixed where it is already a realization. Applying this to the map of $\|\Delta^{n-j+k+1}\|$ constructed above gives an $(n-j+k+1)$-simplex which we define to be $N(\tau)$, together with an extension of the previous homotopies to a homotopy of $\|N(\tau)\|$ to $M(\tau)$. This completes the induction step, and therefore the proof of the lemma. □

The next step is a minor modification, replacing the join by the product in the definition of the spaces "nat." If F, G are topological functors as above, define $\text{nat}_\times(F, G)$ to have k-simplices the natural transformations $F(\sigma) \times \Delta^k \to G(\sigma)$.

Recall the definition of the join $F(\sigma) * \Delta^k$ as the product $F(\sigma) \times I \times \Delta^k$ with identification of the subset with I coordinate 0 with $F(\sigma)$, and identification with Δ^k when the I coordinate is 1. A map of the join defines $F(\sigma) \times I \times \Delta^k \to G(\sigma)$. Use adjointness to shift the I coordinate to G, then this gives a map to the loopspace $F(\sigma) \times \Delta^k \to \Omega G(\sigma)$. This construction defines an isomorphism of Δ-sets $\text{nat}_*(F, G) \simeq \text{nat}_\times(F, \Omega G)$.

When we set G to be a bordism spectrum the loopspace is obtained by shifting the index by one. Putting these definitions and remarks together we get

5.8B Lemma. *The natural morphism*

$$\Omega_r(\mathit{Cycles}^{\mathcal{J}}(X, p, \mathcal{U})) \to \mathit{nat}_\times(\|D^n(\sigma)\|, \Omega^{\mathcal{J}}_{r-n}(p^{-1}(\cap\sigma)))$$

is a homotopy equivalence of spectra. □

The task is now to show the $\text{nat}_\times$ spectrum is equivalent to homology. The realization of a natural transformation $D^n(\sigma) \times \Delta^k \to \Omega^{\mathcal{J}}_{r-n}(p^{-1}(\cap\sigma))$

defines a k-simplex of the space of pointed maps $\text{maps}(S^n, \Omega^{\mathcal{J}}_{r-n}(p,\mathcal{U}))$. The mapping space then maps into the homology, which is defined to be the limit $\lim_{j\to\infty} \text{maps}(S^{n+j}, \Omega^{\mathcal{J}}_{r-n-j}(p,\mathcal{U})/S^n)$.

To see this is a homotopy equivalence it is sufficient to see it induces an isomorphism on homotopy. Since these are Kan Δ-set it is sufficient to see that a k-simplex of homology deforms rel boundary to a k-simplex of the natural transformation space. In what follows we do this for 0-simplices, ie. set $k = 0$. The reason this is sufficient is that these are Ω-spectra, so any homotopy group appears as a 0th homotopy group by adjusting the index r. Or we could note that the proof for k-simplices is obtained simply by multiplying everything by Δ^k. In any case this will usefully simplify the notation.

The first step is to deform a point in the homology space to the realization of a natural transformation of functors, but not quite the right functors.

Let $f\colon S^j \to \Omega^{\mathcal{J}}_{r-j}(p,\mathcal{U})/S^n$ represent a point in the homology space $\mathbf{H}_r(X; \Omega^{\mathcal{J}}(p,\mathcal{U}))$. Think of dividing by S^n as adding the cone on S^n, then make f transverse to the $\frac{1}{2}$ level in the cone. This gives a codimension 0 submanifold $W \subset S^j$ and a map $f\colon (W, \partial W) \to (\Omega^{\mathcal{J}}_{r-j}(p,\mathcal{U}), S^n)$. The original f is obtained up to homotopy by dividing $\Omega^{\mathcal{J}}$ by S^n and extending the map to all of S^j by taking $S^j - W$ to the basepoint.

We now will use a transversality construction on W, f to produce

(1) a functor $\sigma \mapsto (W(\sigma), \partial_\alpha W(\sigma))$ from $\partial\Delta^{n+1}$ to pairs of spaces, and a homeomorphism of the realization $\|W(*)\| \simeq W$ taking $\|\partial_\alpha W(*)\|$ to ∂W,
(2) a natural transformation $F\colon (W(\sigma), \partial_\alpha W(\sigma)) \to (\Omega^{\mathcal{J}}_{r-j}(p^{-1}(\cap\sigma), pt)$ of functors of σ, and
(3) a homotopy of maps of pairs from the realization of the transformation $\|F\|$ to the map f.

This construction proceeds by downward induction on dimensions of simplices in $\partial\Delta^{n+1}$. To describe this we need some notation for realization of functors defined on part of $\partial\Delta^{n+1}$.

Suppose $W(\sigma)$ is defined for r-simplices. with $r \geq k$. Define the realization, as in 5.2, to be

$$\|W(*)\|_k = \left(\coprod W(\sigma) \times \|\sigma\|\right)/\sim$$

where the union is over simplices of dimension $\geq k$, and $\sim$ is the equivalence relation generated by: $(x, \partial_r^*(t)) \sim (W(\partial_r)(x), t)$. Here ∂^* denotes the inclusion of the realization of a face in the realization of the whole simplex, and $W(\partial_r)$ denotes the map functorially associated by W to the face map ∂_r. Finally, denote by $\partial_\beta\|W(*)\|_k$ the part of this coming from simplices of dimension less than k; the image of $W(\sigma) \times \|\partial_r\sigma\|$, where σ is a k-simplex.

The induction hypothesis for the construction is, for given k,

(1) a functor $(W(\sigma), \partial_\alpha W(\sigma))$ defined for r-simplices, $r \geq k$, and a homeomorphism $\|W(*)\|_k \to W$ onto a codimension 0 submanifold so that $\|\partial_\alpha W(*)\|_k$ is taken to the intersection of the image with ∂W and $\partial_\beta\|W(*)\|_k$ is taken to the interface (intersection of the image with the closure of its complement),

(2) a natural transformation $F\colon (W(\sigma), \partial_\alpha W(\sigma)) \to (\Omega^{\mathcal{J}}_{r-j}(p^{-1}(\cap\sigma), pt)$, and

(3) a homotopy of maps of pairs from f to f_k, whose restriction to $\|W(*)\|_k$ is the realization of $\|F\|$, and which takes the complement to the part of $\Omega^{\mathcal{J}}_{r-j}(p, \mathcal{U})$ lying over the $(k-1)$-skeleton of $\partial\Delta^{n+1}$.

For the induction step we split off a piece of the complement suitable to be the realization over the $(k-1)$-simplices.

Define W_k to be the closure of the complement of the realization $\|W(*)\|_k$. This has boundary $\partial_\beta\|W(*)\|_k \cup \partial_\alpha W_k$, where the second piece is defined to be $W_k \cap \partial W$. According to (3) the map f_k restricts to a map of this into the part of the spectral sheaf lying over the $(k-1)$-skeleton of $\partial\Delta^{n+1}$, namely $\cup_{\{\sigma^r | r<k\}} \Omega^{\mathcal{J}}_{r-j}(p^{-1}(\cap\sigma)) \times \|\sigma\|$.

Consider the barycenters of the simplices of dimension $k-1$; b_σ. Since the restriction to $\partial_\beta\|W(*)\|_k$ comes from the realization of a natural transformation, this restriction is transverse to $\Omega^{\mathcal{J}}_{r-j}(p^{-1}(\cap\sigma)) \times b_\sigma$. Modify f_k by homotopy fixed on $\partial_\beta\|W(*)\|_k$ to make it transverse to these barycenters on all of W_k. Define the preimage of σ^{k-1} to be $W(\sigma)$.

There is a normal bundle for $W(\sigma^{k-1})$ whose fibers project to concentric copies of $\|\sigma\|$ about b_σ. By a small additional homotopy we may also arrange that on this normal bundle f_k composed with the projection

$$\Omega^{\mathcal{J}}_{r-j}(p^{-1}(\cap\sigma)) \times \|\sigma\| \to \Omega^{\mathcal{J}}_{r-j}(p^{-1}(\cap\sigma))$$

is constant on fibers. Finally since both these conditions are already satisfied on ∂_β (since it is a realization there) we may assume this normal bundle extends the one given in ∂_β by the realization structure. Then use radial expansion in $\|\sigma\|$ to stretch each fiber out to a homeomorphism to $\|\sigma\|$. This gives a homotopy of f_k to f_{k-1} which takes the complement of these normal bundles to the part of the spectral sheaf lying over the $(k-2)$-skeleton, and is a product over each $k-1$ simplex. This map satisfies the induction hypothesis for $k-1$, and so completes the induction step.

This construction produces manifold-valued functors, specifically $W(\sigma)$ is a manifold $(\mathcal{U} - \sigma \cup \{\alpha\})$-ad of dimension $j - k$, when σ is a k-simplex. Strictly speaking this should have been included in the induction hypothesis, since it was used (for transversality) and does not quite follow from the other hypotheses.

Recall that the original goal was to construct a cycle representing a particular homology class. At this point we can describe the cycle corresponding to the class represented by f. We will not actually use this; after a brief description we return to the more technical goal of constructing an appropriate natural transformation.

Corresponding to a simplex $\sigma \in \mathrm{nerve}_\Delta(\mathcal{U})$ we have a map $W(\sigma) \to \Omega^{\mathcal{J}}_{r-j}(p^{-1}(\cap\sigma))$. Triangulate $W(\sigma)$ and approximate this map by a Δ-map into the Δ-set $\Omega^{\mathcal{J}}_{j-r}(p^{-1}(\cap\sigma))$. This can be interpreted as a Kan-type cycle and totally assembled to give a single $\mathcal{J}$ object. Since $W(\sigma)$ is a manifold $(\mathcal{U} - \sigma \cup \{\alpha\})$-ad, with $\partial_\alpha W(\sigma) = \phi$, the assembly can be arranged to yield a $(\mathcal{U} - \sigma)$-ad in $\mathcal{J}$. By doing this inductively downwards with respect to the dimension of σ these can be arranged to fit together. The result is a $\mathcal{J}$-cycle in $(p, \mathcal{U})$.

The final step in the proof is to modify the construction to yield a natural transformation defined on the dual simplex functor, and therefore a cycle.

Enlarge the collection $\mathcal{U}$ by adding m copies of the empty set put at the end in the ordering; we denote the result by $\mathcal{U} \cup m\{\phi\}$. This does not affect the homotopy type of either the homology spectrum or the cycles. There is a natural inclusion $\Omega^{\mathcal{J}}_{r-j}(p, \mathcal{U}) \subset \Omega^{\mathcal{J}}_{r-j}(p, \mathcal{U} \cup m\{\phi\})$ covering the inclusion $\partial\Delta^{n+1} \subset \partial\Delta^{n+1+m}$. The homology class represented by the map f is also represented by the composition with this inclusion, $(W, \partial W) \to (\Omega^{\mathcal{J}}_{r-j}(p, \mathcal{U} \cup m\{\phi\}), \partial\Delta^{n+1+m})$. Further, the functor and natural transformation $W(*)$ and F constructed above for f also gives the composition. The point is that by this construction we can adjust n to be arbitrarily large. In particular we can assume $n > 2j$.

There is a natural transformation of topological functors on $\partial\Delta^{n+1}$, from $W(*)$ to $D^n(*)$. This is constructed by induction downwards on dimensions of simplices, using collars of boundaries in $W(*)$ and the contractibility of $D^n(*)$. Further for each σ this can be arranged to be an embedding with natural trivial normal bundle $W(\sigma) \times D^{n-j} \subset D^n(\sigma)$. Again we proceed downwards on dimension of σ, using the facts that the dimension of $D^n(\sigma)$ is greater than twice that of $W(\sigma)$. The triviality of the normal bundle comes from the fact that $W(\sigma)$ has trivial normal bundle in S^j.

Next we use the Ω-spectrum structure

$$\Omega^{\mathcal{J}}_{r-j}(p^{-1}(\cap\sigma)) \sim \text{maps}_0(D^{n-j}, \Omega^{\mathcal{J}}_{r-n}(p^{-1}(\cap\sigma))).$$

Here we are using maps_0 to indicate maps of the disk which take the boundary to the basepoint; the $(n-j)$-fold loop space. The adjoints of the maps $W(\sigma) \to \text{maps}_0(D^{n-j}, \Omega^{\mathcal{J}}_{r-n}(p^{-1}(\cap\sigma)))$ define natural maps $W(\sigma) \times D^{n-j} \to \Omega^{\mathcal{J}}_{r-n}(p^{-1}(\cap\sigma))$.

These maps take $W(\sigma) \times \partial D^{n-j} \cup \partial_\alpha W(\sigma) \times D^{n-j}$ to the basepoint. But this is the interface between the embedding in $D^n(\sigma)$ and its complement, so these extend by the basepoint on the complement to give maps $\hat{F}\colon D^n(\sigma) \to \Omega^{\mathcal{J}}_{r-n}(p^{-1}(\cap\sigma))$. These form a natural transformation of the type equivalent to a cycle, so if we show these extended maps represent the same homology class as f then the proof is complete.

Let $S^j \times D^{n-j} \subset S^n$ denote the standard embedding. Part of the direct limit used to define homology is the suspension operation: composition with the spectrum structure map

$$S^j \to \Omega^{\mathcal{J}}_{r-j}(p, \mathcal{U})/\partial\Delta^{n+1} \to \text{maps}_0(D^{n-j}, \Omega^{\mathcal{J}}_{r-n}(p, \mathcal{U}))/\partial\Delta^{n+1}$$

followed by adjunction to $S^j \times D^{n-j} \to \Omega^{\mathcal{J}}_{r-n}(p, \mathcal{U})/\partial\Delta^{n+1}$, extended by the point map to give $\tilde{f}\colon S^n \to \Omega^{\mathcal{J}}_{r-n}(p, \mathcal{U})/\partial\Delta^{n+1}$. Since this is part of the limit, the maps f and $\tilde{f}$ represent the same homology class.

Recall that the realization of the functor $W(*)$ is S^j, and the realization of $D^n(*)$ gives S^n. The embeddings $W(\sigma) \times D^{n-j} \subset D^n(\sigma)$ thus realize to give an embedding $S^j \times D^{n-j} \subset S^n$. Since $n > 2j$ this embedding is isotopic to the standard embedding. The adjunction construction on the functor level realizes to give the adjunction of maps. Therefore the isotopy between the embeddings defines a homotopy from $\|\hat{F}\|$ to $\tilde{f}$.

This represents the homology class of f by the realization of a transformation from the functor $D^n(*)$, and therefore completes the proof of the representation theorem. □

6: Examples

A few of the main examples of bordism-type theories are described here, along with their special features. In section 6.1 the prototype examples of manifolds are described. Then in 6.2 these are extended to functors from spaces to bordism-type theories, by including a map in the data. From these the general machinery developed earlier gives bordism spectra, functor-coefficient homology, and assembly maps.

Applying the representation theorem gives manifold cycles which represent homology classes in these theories. Cycles are identified with maps transverse to dual cones, so the total assembly is seen to be just a matter of forgetting that a map is transverse. The fact that maps of manifolds can be made transverse to bicollared subsets is then used to reverse this: an arbitrary map can be made transverse to dual cones, thus can be realized as a cycle. This shows that assembly maps are isomorphisms for these categories. This is an analog of the classical Pontrjagin-Thom theorem which asserts that manifold bordism is a homology theory with coefficients the appropriate Thom spectra.

The second class of examples are constructed from chain complexes. The Poincaré chain complexes developed by Mishchenko, Ranicki, Weiss, and others fit into this framework, giving assembly maps described by glueing cycles. This is related to the papers of Ranicki [20] and Weiss [26] on algebraic assemblies.

6.1 Manifolds

We begin with the definition of manifold A-ads, adding precision to the sketch in 3.1. Let $S\mathcal{M}$ denote one of the categories of oriented manifolds, TOP, DIFF or PL.

The definition is inductive in the number of elements in A, or equivalently after reindexing, the number of nonempty faces. To begin the induction, suppose A is empty and define an A-ad to be a compact oriented $S\mathcal{M}$ manifold without boundary. Define the involution by letting $-M$ be the same manifold with the opposite orientation. The basepoint is the empty manifold ϕ.

An -ad has an underlying manifold (forgetting the face structure) which for the purposes of the definition we denote by $|M|$. If A is empty define $|M| = M$.

Now suppose -ads with k faces have been defined, and A has $k+1$ elements. Then a manifold A-ad is a compact oriented manifold with boundary $|M|$ together with an $(A-a)$-ad $\partial_a M$ for each $a \in A$ such that

(1) $|\partial_a M| \subset \partial M$ as an oriented codimension 0 submanifold, and $\cup_a |\partial_a M| = \partial M$,
(2) if $a \neq b$ then $|\partial_a \partial_b M| = |\partial_a M| \cap |\partial_b M|$, and further
(3) $\partial_a \partial_b M = -\partial_b \partial_b M$ as $(A-a-b)$-ads.

The involution $-M$ is defined to have underlying manifold $|M|$ with opposite orientation, and face structure $\partial_a(-M) = -\partial_a M$.

By induction this defines A-ads for all finite A. If A is infinite then an A-ad is defined to be a B-ad for some finite subset $B \subset A$, and $\partial_a M = \phi$ if

$a \notin B$.

A manifold -ad has dimension n if its underlying manifold has dimension n. In the notation of section 3 this completes the definition of classes $S\mathcal{M}^n_A$, of A-ads of dimension n. These objects can be reindexed via an injection $\theta: A \to B$ simply by defining $\partial_{\theta(a)}\ell_\theta M = \partial_a M$, and $\partial_b \ell_\theta M = \phi$ if b is not in the image of θ.

This definition needs to be refined in the smooth category. Strictly speaking we need manifolds with "corners" so that three or more can fit together around lower-dimensional face to give a smooth structure. The iterated codimension-1 approach used here can be made to work in the smooth category using the "straightening the angle" device to change face angles when necessary. We have chosen this approach because it requires less detail on the structure of cone complexes, and it emphasizes that only the simplest type of transversality—to trivial 1-dimensional bundles —is needed. The more direct approach would be required if we were considering more rigid objects, like manifolds with a Riemannian metric, or a conformal or affine structure.

We briefly describe the more direct approach, which gives a technically better way to approach the topic in any category. The basic idea is to consider -ads as objects modeled on specific examples of -ads, just as manifolds with boundary are modeled on disks.

To get an appropriate model suppose A is a collection of points in $\mathbf{R}^n$ equidistant from each other (so the number of points is no greater than $n+1$). Let R_a denote the points in the space whose distance from a is less than or equal to the distance to the other points. This has faces $R_a \cap R_b$ lying in the $(n-1)$-plane orthogonal to the center of the edges joining a and b. Similarly an iterated intersection $\cap_{a \in S} R_a$ lies in the affine subspace orthogonal to the center of the simplex spanned by S.

A smooth manifold -ad of dimension n should have coordinate charts modeled on open sets in some R_a, so that faces in the -ad correspond to faces of R_a. The model establishes particular angles at which faces meet. This particular model is chosen so that when pieces of a Kan cycle of smooth -ads are glued together the result has an obvious natural smooth structure.

Lemma. *The collections $S\mathcal{M}^n_A$ together with the reindexing operations form a bordism-type theory.*

Proof. The reindexing hypothesis is clear, that reindexing defines a bijection from $S\mathcal{M}^n_A$ to $\{M \in S\mathcal{M}^n_B \mid \partial_b = \phi \text{ if } b \notin \theta(A)\}$.

The other thing to check is the Kan condition. Suppose $N\colon (A-a) \to$

$S\mathcal{M}$ is a Kan cycle, in the sense of 3.2. Let $\cup_b N(b)$ denote the union of these, then this is a manifold (assuming the interiors are disjoint; see the appendix) with boundary $\cup_b \partial_a N(b)$. The fact it is a manifold can be seen by inductively adding one piece at a time to the union, and observing that the union is over codimension 0 submanifolds of the boundary. Or, thinking of the pieces $N(b)$ as locally modeled on convex regions $R_b \subset \mathbf{R}^n$ as above, then the union is a manifold because the union of the model regions is a manifold.

We define an A-ad M of dimension $n+1$ by: the underlying manifold is $\cup_b N(b) \times I$, and the faces are $\partial_b M = N(b) \times \{0\}$ for $b \neq a$. Finally $\partial_a M$ has underlying manifold $(\cup N) \times \{1\} \cup \partial(\cup N) \times I$. The face structure of $\partial_a M$ is specified by $\partial_b \partial_a M = \partial_a N(b) \times \{0\}$.

This A-ad satisfies the conclusion of the Kan condition, so the $S\mathcal{M}$ are bordism-type theories. □

The bordism groups associated to these theories by 3.3 are exactly the classical manifold bordism groups (see [23]). The bordism spectrum of 3.4 is similarly homotopy equivalent to the Thom spectrum, whose homotopy groups are identified with bordism groups by the Pontrjagin-Thom construction.

6.2 Manifolds over spaces

We augment the construction above with a map to a space, to obtain a (bordism-type theory) valued functor. Then we show that assembly maps in the associated homology theory are isomorphisms.

If $S\mathcal{M}$ is a category of oriented manifolds as in the previous section, and X is a topological space, then define $S\mathcal{M}^n_A(X)$ to be the collection of (M, f), where M is an n-dimensional A-ads in $S\mathcal{M}$, and $f\colon M \to X$. Define $\partial_a(M, f)$ to be $(\partial_a M, f|\partial_a M)$, and define the involution and reindexing using these operations on M, without changing f.

6.2A Lemma. *The collections $S\mathcal{M}^n_A(X)$ together with these operations are bordism-type theories, natural in X. The resulting (bordism-type theory)-valued functors of spaces are homotopy invariant, in the sense of 3.5.*

Proof. The only part of the bordism-type theory structure which might need comment is the Kan condition. Since the maps are part of the structure, the maps on pieces of a Kan cycle N fit together to define a map on the union used in the previous proof, $\cup_b N(b) \to X$. A suitable map $M \to X$ for the solution to the problem is obtained by projecting on the first factor $M = \cup_b N(b) \times I \to \cup_b N(b)$, and composing with the map on the union.

To check the homotopy invariance we use the criterion in Lemma 3.5B (3). Also, rather than general homotopy equivalences it is sufficient to check invariance under inclusions $Y \subset X$ which are deformation retracts. (Because both spaces in a homotopy equivalence embed as deformation retracts in a mapping cylinder.)

Suppose, then, that X deformation retracts to Y by a deformation $H: X \times I \to X$. Suppose M is a [0]-ad in $S\mathcal{M}(X)$, so a pair (M, f) with M a manifold with boundary, which is the face $\partial_0 M$, and $f/ : M \to X$. Suppose $\partial_0 M$ comes from $S\mathcal{M}(Y)$, which means $f(\partial M) \subset Y$. We deform M into $S\mathcal{M}(Y)$ rel $\partial_0 M$. Define $W = M \times I$ as a [1]-ad with $\partial_1 W = M \times \{0\}$ and $\partial_0 W$ the rest of the boundary. Define a map to X by $H(f \times \mathrm{id})$. Then since H is a deformation retraction the restriction of this to $M \times \{0\}$ is f, and the rest of the boundary maps into Y. Therefore $\partial_0 W$ is an element of $S\mathcal{M}(Y)$, as required. □

According to Corollary 3.5C the bordism spectra of these theories define homotopy invariant spectrum-valued functors of spaces. The notation established in Section 3 for these spectra is $\Omega^{S\mathcal{M}}(X)$. The constructions of section 2 define homology with coefficients in these functors.

6.2 B Proposition. *Suppose $S\mathcal{M}$ is one of the manifold theory functors defined above, and $p\colon E \to X$ is fiber homotopy equivalent to the realization of a simplicial map. Then the assembly $H_n(X; \Omega^{S\mathcal{M}}(p)) \to H_n(\mathrm{pt}, \Omega^{S\mathcal{M}}(E)) = \Omega_n^{S\mathcal{M}}(E)$ is an isomorphism.*

This is a version of the classical result that bordism groups form a homology theory. It also identifies the coefficient spectrum of the theory as the bordism spectrum of a point, which is therefore equivalent to the appropriate Thom spectrum.

Proof. The proof uses the transversality to dual cones referred to several times, and here we describe the process in some detail. There are three stages to the discussion: first define the dual cone decomposition and transversality to it, second observe that manifold cycles are exactly manifolds transverse to the dual cones, and finally show that any manifold can be made transverse.

Suppose K is a simplicial complex. Take the first barycentric subdivision of the realization. If σ is a simplex of K define the dual $D(\sigma)$ to be the union of all simplices of the subdivision which intersect σ in exactly the barycenter. The dual of a vertex is the closure of the star used in 1.5.

It is not hard to see (eg. in [6]) that

(1) $D(\sigma)$ is the cone on $\cup D(\tau)$, where the union is over τ which contain σ as a face, and

(2) the boundary of $D(\sigma)$ is bicollared in the boundary of $D(\partial_i\sigma)$ (or in $\|K\|$ if σ is a vertex).
(3) the boundary of $D(\sigma)$ is naturally equivalent as a union of cones to the dual cone decomposition of the link of σ.

For example we give a picture of a complex K and its dual cones:

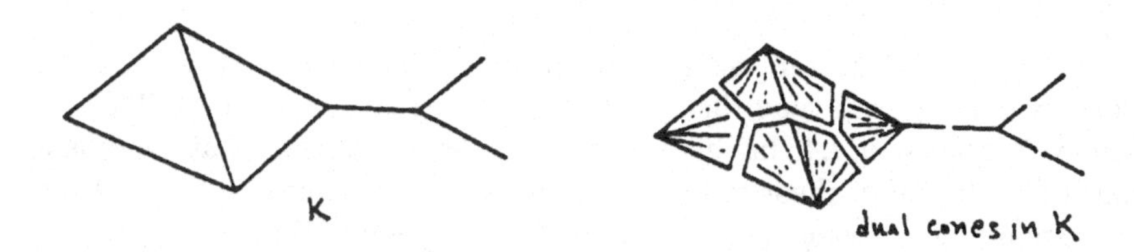

A little more information about the collaring in (2) is necessary. $\partial D(\sigma)$ separates $\partial D(\partial_i\sigma)$ (or $|K|$ if σ is a vertex) into two pieces: the cone and the exterior. There is an obvious collar on the cone side given by the cone parameter. On the outside the collar is also radial. In consequence it respects intersections with other cones: if $\partial D(\sigma) \cap D(\tau) = \partial D(\tau)$ then the intersection of $D(\tau)$ with the collar is a collar on $\partial D(\tau)$. Further, the collar mapping is transverse to the interior of $D(\tau)$.

Now if M is a manifold then we say $f\colon M \to |K|$ is *transverse to the cone structure* (or "trans-simplicial" [6]) if for each σ the restriction of f to $f^{-1}(D(\partial_i\sigma)) \to D(\partial_i\sigma)$ (or $\to |K|$ if σ is a vertex) is transverse to the bicollared subset $\partial D(\sigma)$.

This should be understood inductively: if v is a vertex then f is transverse to the bicollared subset $\partial D(v) \subset K$. Therefore $f^{-1}(\partial D(v)) \to \partial D(v)$ is a manifold. Next, if τ is an edge with vertex v then $f^{-1}(\partial D(v))$ is transverse to the bicollared subset $\partial D(\tau) \subset \partial D(v)$, and so on. Note this is all codimension 1 transversality (to trivial line bundles) so no sophisticated theory of normal bundles is necessary.

Finally some technical adjustments should be made in the smooth case, along the lines of the comments following the definition of -ads. Namely, rather than iterated codimension 1 situations the local structure around $\partial D(\sigma)$ should be recognized as a product with some $\mathbf{R}^s$, which the other cones intersect in the pattern described in the earlier comment. Smooth transversality to this gives -ads with face structure with the correct angles, etc.

The next step identifies cycles and transverse maps as being essentially the same. More precisely we show a transverse map naturally determines a cycle, and that any cycle is homotopic to one obtained this way.

6.2C Lemma. *Suppose $p\colon |E| \to |K|$ is the realization of a simplicial map, and $f\colon M \to |E|$ is a map such that pf is transverse to the dual cones in $|K|$. Then the function $\sigma \mapsto pf^{-1}(D(\sigma))$ defines an $S\mathcal{M}$ cycle in E in the inverse of the star cover of $|K|$. Conversely, any such cycle is homotopic to one obtained in this way.*

There is an important space version of this, namely there is a Δ-set, and even a bordism-type theory, of transverse maps defined similarly to the cycle theory in 4.4. In this language the lemma asserts that there is a natural inclusion of theories from the transverse maps into cycles, and the corresponding inclusion of bordism spectra is a deformation retraction.

Proof. Recall that a cycle is a function on the nerve of the cover, and the nerve of the star cover is K. The cover itself is indexed by the vertices of K, which we denote K^0. Similarly denote the vertices of σ by σ^0, then the vertices of K not in σ are $K^0 - \sigma^0$. With this notation the definition 4.2 becomes: a $\Omega^{S\mathcal{M}}$-cycle of dimension n in $(|K|, p, \text{stars}(K))$ is a function $N\colon K \to \Omega^{S\mathcal{M}}$ such that

(1) if σ is a k-simplex then $N(\sigma)$ is an $(n-k)$-dimensional $(K^0 - \sigma^0)$-ad in $\Omega^{S\mathcal{M}}(p^{-1}(\sigma))$,
(2) let $\text{incl}_*\colon \Omega^{S\mathcal{M}}(p^{-1}(\sigma)) \to \Omega^{S\mathcal{M}}(p^{-1}(b_j\sigma))$ denote the morphism induced by the inclusion, then $\text{incl}_*(N(\sigma)) = (-1)^j \partial_{U_j} N(b_j\sigma)$, and
(3) all but finitely many of the $N(\sigma)$ are empty.

The function $\sigma \mapsto pf^{-1}(D(\sigma))$ does satisfy these conditions. If σ is a k-simplex then $pf^{-1}(D(\sigma))$ is the result of k layers of codimension-1 transversality, so has codimension k in M, therefore dimension $n-k$. The faces of $pf^{-1}(D(\sigma))$ correspond to the faces of $D(\sigma)$, therefore to simplices τ which have σ as a face. Such simplices are determined by their vertices not in σ^0, so $pf^{-1}(D(\sigma))$ is naturally a $(K^0 - \sigma^0)$-ad. Finally, since M is compact only finitely many of these inverse images can be nonempty.

Now for the converse suppose N is a cycle. $N(\sigma)$ is a $S\mathcal{M}$-ad with a map to $p^{-1}(\text{star}\sigma)$, and its faces $N(\tau)$ map to subsets $p^{-1}(\text{star}\,\tau) \subset p^{-1}(\text{star}\,\sigma)$. But the inclusions $p^{-1}(D(\sigma)) \subset p^{-1}(\text{star}\,\sigma)$ are homotopy equivalences (both spaces deformation retract to the inverse image of the barycenter of σ). Therefore the reference maps are (coherently) homotopic to maps $N(\sigma) \to p^{-1}(D(\sigma))$.

Next take the union of the pieces $N(\sigma)$ to get a manifold M with a map $f\colon M \to |E|$. (See the proof of the Kan condition to see that this is a manifold. (M, f) also represents the total assembly of the cycle N, in $\Omega_n^{S\mathcal{M}}(|E|)$.) This map has the property that the inverse images $f^{-1}(D(\sigma)) = N(\sigma)$ are manifolds, but may need a little modification to actually be transverse.

Since $p\colon |E| \to |K|$ is transverse to the dual cones, the inverse image $p^{-1}(\partial D(\sigma))$ is collared in $p^{-1}(D(\sigma))$. But $N(\partial\sigma) = fp^{-1}(\partial D(\sigma))$ is the boundary of the manifold $fp^{-1}(D(\sigma))$ so is also collared. Thus the map from the second to the first can be change by homotopy rel boundary to preserve collars. f is then "transverse on one side" to $p^{-1}(\partial D(\sigma))$.

To arrange transversality use this construction inductively beginning with the largest simplices (smallest dual cones) over which N is nonempty. Suppose S is a collection of cones, so that for each $D(\sigma) \in S$ the map

$$pf^{-1}(\partial D(\sigma)) \to p^{-1}(\partial D(\sigma))$$

is transverse to the inverse images of the cones in $\partial D(\sigma)$. Change f rel all these boundaries so that it preserves collars of boundaries of $D(\tau)$ for $D(\tau) \in S$. Then f is transverse with respect to the larger collection obtained by adding to S the c ones whose boundaries lie in S.

In the smooth category a little more precision is appropriate. The interior of each cone in $|K|$, thus the inverses in $|E|$, have neighborhoods canonically isomorphic to the cone crossed with one of the smooth models described in 6.1. Neighborhoods of pieces of cycles also have such structures. Rather than working with collars inductively one works directly with the models, arranging the maps to be the identity on the model coordinate near the center stratum. □

Since transverse maps are the same as cycles, we can complete the proof of the proposition by showing that any map $f\colon M \to E$ with M a manifold, is homotopic to one such that pf is transverse to the dual cones in $|K|$.

The basic idea is that since the boundaries of duals of vertices are bicollared in $|K|$, ordinary transversality can be used to make f transverse to them. The inverse image $f^{-1}(\partial D(v)) \to \partial D(v)$ is again a map of a manifold to a complex with dual cones, but the dimension (of both the manifold and the complex) is smaller. Therefore this serves as the induction step in obtaining transversality by induction on dimension.

In more detail first note that since $p\colon |E| \to |K|$ is transverse to the dual cones, the inverse images $p^{-1}(\partial D(\sigma))$ have the same collaring properties as the boundaries themselves. Next suppose f is transverse over an open set $U \subset |K|$, and let v be a vertex. Then there is a homotopy fixed over a closed set slightly smaller than U to a new f which is also transverse to

$p^{-1}(\partial D(v))$. Then $pf^{-1}(\partial D(v)) \to p^{-1}(\partial D(v))$ is a map of a manifold to a complex over the dual cone decomposition of the link of v.

By induction on dimension we can assume this is homotopic, fixed over a closed set slightly smaller than $U \cap \partial D(v)$, to a map transverse to the inverse images of the cones. Use this homotopy to modify f to a map which restricts to the new one on the inverse image. Since the collar on $p^{-1}(\partial D(v))$ is transverse to the inverse images of the other cones, this new f is transverse to all the cones over a neighborhood of $\partial D(v)$. Add this to the set U. By induction the links of all vertices of K can be added to U, at which point f is transverse to all cones.

This completes the proof of Proposition 6.2B. □

6.3 Chain complexes

A chain complex together with a chain equivalence with its dual serves as an algebraic analog of a manifold. This idea and elaborations have been developed by Mishchenko, Ranicki, Weiss, and others as a powerful tool for the investigation of surgery theory.

Assembly maps have been important in the algebraic theory: we mention particularly the total surgery obstruction of Ranicki [19], [20] which lies in a fiber of an assembly map, and the visible theory of Weiss [26], for which an assembly map is an isomorphism and provides a calculation.

In this section we describe the constructions of Ranicki and Weiss, roughly and with little detail, and relate them to the approach taken in this paper. Specifically the first subsection describes the theory, and the way in which cycles appear in it. Section 6.3B describes how chain A-ads are defined, thereby giving bordism-type theories to which this paper applies. Then 6.3C extends additive and algebraic bordism categories to be functors of spaces, thereby defining (bordism-type theory)-valued functors. This leads to functor-coefficient assembly maps, etc. resulting from the general theory. Finally in 3.6D there are some remarks about an analog for "bounded" chain complexes over a metric space.

6.3A Ranicki's construction

These constructions take place in an additive category, rather than the usual setting of modules over a ring. There are substantial benefits to working in this generality, as will be pointed out later.

An algebraic bordism category Λ is defined by Ranicki [20, §3] to be a triple $\Lambda = (\mathbb{A}, \mathbb{B}, \mathbb{C})$. In this $\mathbb{A}$ is an additive category with chain duality [20, 1.1]: the model is the category of modules over a commutative ring, with the functor which takes a module to its hom dual. (Make this a "chain" duality by thinking of M^* as a very short chain complex.) $\mathbb{C} \subset \mathbb{B}$ are subcategories of the chain complexes in $\mathbb{A}$. The models for these are: $\mathbb{B}$ is finitely generated free chain complexes, with morphisms chain homotopy equivalences, and $\mathbb{C}$ is the full subcategory of contractible complexes.

We describe the way this data is used. Consider complexes from $\mathbb{B}$ together with a duality structure, roughly a chain map $C \to C^*$, whose mapping cone is in $\mathbb{C}$. Depending on the type of duality structure used one gets symmetric or quadratic Poincaré complexes, with bordism groups denoted by $L^n(\Lambda)$ and $L_n(\Lambda)$. Adjusting $\mathbb{B}$ gives variations: finitely generated projective complexes gives the L_n^p groups, finitely generated free with homotopy equivalences the L_n^h, and free based complexes with simple equivalences gives L_n^s. In all these cases $\mathbb{C}$ consists of the contractible complexes. Other variations are obtained by changing this: contractible over some other ring gives the Cappell-Shaneson Γ-groups, and $\mathbb{C} = \mathbb{B}$ gives the "normal" bordism groups.

Ranicki's next step is to construct new bordism categories $\Lambda_*(K)$ and $\Lambda^*(K)$ depending on the original bordism category and a simplicial complex K ([20, §5]). The two versions, distinguished by the position of the $*$, correspond to cycles and cocycles in K. Applying the previous construction gives symmetric or quadratic Poincaré objects in these categories. These, it turns out, represent homology or cohomology classes of K , with coefficients in an appropriate spectrum $\mathbf{L}(\Lambda)$.

When Λ is the bordism category of modules over a ring R, then a glueing construction called "universal assembly" defines a morphism from the bordism category $\Lambda_*(K)$ to the bordism category of modules over the ring $R[\pi_1 K]$. Naturality then gives morphisms of L-groups,

$$L(\Lambda_*(K)) \to L(R[\pi_1 K]).$$

Using the identification of the L-groups of $\Lambda_*(K)$ as homology then gives an assembly map

$$H_n(|K|; \mathbf{L}(R)) \to L_n(R[\pi_1 |K|]).$$

6.3B Poincaré chain -ads

In order to engage the machinery of this paper in the chain complex context we need -ads, and there are two ways to approach this. The low-tech way is to observe that Poincaré pairs are defined, and appropriate glueings are possible. Thus a definition of Λ-Poincaré A-ads can be pieced together inductively as was done with manifolds in 6.1. The high-tech approach is to use Ranicki's machinery, and obtain n-ads of dimension m as Poincaré objects of dimension $m-n$ in the algebraic bordism category associated to the n-simplex, $\Lambda^*(\Delta^n)$ (see [20, 5.4]). Then reindex the faces to get arbitrary A-ads.

These -ads can be reindexed in obvious ways, and they satisfy the Kan condition (Weiss [26,1.10]), so they form bordism-type theories in the sense of §3. Denote by $\mathcal{L}_*(\Lambda)$ and $\mathcal{L}^*(\Lambda)$ respectively the bordism-type theory of quadratic and symmetric -ads in Λ. There are then bordism spectra, homology, cycles, etc. associated to this theory. We extend this to a functor of spaces to get a full version in the next section, but first explain how this is related to the Ranicki constructions.

The theory of Poincaré chain complexes may be thought of as being obtained in three stages: first one has the category of modules over a ring R, with the duality functor which sends a module to its dual. Next one forms the category of chain complexes over R, again with a duality operation. Finally symmetric, quadratic, etc. Poincaré complexes are obtained as chain complexes together with some sort of elaboration of a chain homotopy equivalence with the dual complex. The L-groups appear as bordism groups of these Poincaré complexes.

We could think of the formation of cycles as a fourth stage in this development, using Poincaré chain A-ads. However the cycle construction "commutes" with the other constructions. If $\mathbb{A}$ is an additive category one can basically think of Ranicki's category $\mathbb{A}_*(K)$ as the category of cycles of $\mathbb{A}$-objects. Chain complexes in this cycle category are cycles of $\mathbb{A}$-chain complexes. The duality operation becomes a little more complex, which is why "chain duality" is introduced [20, 1.1]. Finally given a bordism category the formal approach defines Poincaré objects in the chains-of-cycles category, and these are exactly cycles of Poincaré complexes. Therefore by doing the Poincaré chain constructions in general additive categories with chain duality, cycles are obtained as a special case.

From our point of view the key to being able to see assemblies by this approach is the functoriality of glueing. In the bordism-type theories of section 3, pieces are glued together by application of the Kan condition: the result is known to exist but not naturally or canonically. In the algebra it is given by a natural formula (for manifolds too; see the proof of the lemma in 6.1). Thus it works out that (in a sense) glueings of modules lead to glueings of chain complexes, and glueings of the chain complexes underlying Poincaré complexes lead to glueings of the Poincaré complexes. Because of this, assemblies of Poincaré complexes can be obtained by naturality from module-level assemblies.

6.3C Categories over spaces

Our machinery is set up to produce functor-coefficient homology and assemblies from functors of spaces. Accordingly we extend the algebra along the lines of [17] to incorporate a space.

Suppose $\mathbb{A}$ is an additive category, and X a space. Define a new additive category $\mathbb{A}_X$ with objects (M, S, i), where

(1) S is a set, and $i\colon S \to X$ a function which is locally finite,
(2) $M\colon S \to \text{objects}\mathbb{A}$ is a function.

Morphisms in this category are equivalence classes of paths in X together with morphisms in $\mathbb{A}$. Specifically a morphism $(M, S, i) \to (M', S', i')$ is a collection $(\rho_j, 0_j, 1_j, f_j)$, where

(1) $0_j \in S$, $1_j \in S'$, and ρ_j is a path in X from $i(0_j)$ to $i'(1_j)$,
(2) $f_j\colon M(0_j) \to M'(1_j)$ is a morphism in $\mathbb{A}$, and
(3) j runs over some index set, and each elem ent of S (respectively S') occurs only finitely many times as 0_j, (respectively 1_j).

The equivalence relation on morphisms is generated by:

(1) the paths can be changed by homotopy in X holding the ends fixed,
(2) if for indices j, k the endpoints and paths are the same, then the data for these indices can be replaced in the collection by $(\rho_j, 0_j, 1_j, f_j + f_k)$, and
(3) if $f_j = 0$ then the datum $(\rho_j, 0_j, 1_j, f_j)$ can be deleted from the collection.

For example, if R is a ring and $\mathbb{A}$ is the category of finitely generated free R-modules, and X is compact, then $\mathbb{A}_X$ is equivalent to the category of free finitely generated $R[\pi_1 X]$ modules [17].

If X is a space then at least for the standard choices of subcategories $\mathbb{B}$, $\mathbb{C}$ there are standard ways to lift to subcategories $\mathbb{B}_X$, $\mathbb{C}_X$ of chain complexes in the category $\mathbb{A}_X$. (We will not try to mechanize this in general). Therefore given an appropriate algebraic bordism category Λ there is a functor from the category of spaces to the category of algebraic bordism categories; $X \mapsto \Lambda_X$. The definition of Poincaré -ads in an algebraic bordism category functorially associates bordism-type theories $\mathcal{L}_*(\Lambda_X)$ and $\mathcal{L}^*(\Lambda_X)$, as explained above.

This now makes contact with the earlier development. Applying the bordism spectrum functor gives spectrum-valued functors $X \mapsto \Omega(\mathcal{L}_*(\Lambda_X))$ and $X \mapsto \Omega(\mathcal{L}^*(\Lambda_X))$. Denote these functors more compactly by $\mathbf{L}^{\Lambda}_*(X)$ and $\mathbf{L}^*_{\Lambda}(X)$. Associated to these functors are functor coefficient homology, assembly maps, etc. For example taking $p\colon E \to X$ to the point map gives the total assembly

$$\mathbf{H}_\bullet(X; \mathbf{L}^\Lambda_*(p)) \to \mathbf{L}^\Lambda_*(E)$$

(and similarly for the symmetric case $\mathbf{L}^*$.)

The main theorems of this paper identify the functor-coefficient homology as represented by cycles, and describe assembly maps in terms of glueing cycles together. If p is the identity map of K then unraveling the definitions shows that cycles over the star cover of $|K|$ are the same as Poincaré objects in Ranicki's category $\Lambda_*(K)$. Further the algebraic assembly described in [20, §9] is the same as the glueing via the Kan condition used here, since the algebraic assembly is the mechanism by which the Kan condition is verified. Putting these together, we see that the algebraic assembly of Ranicki coincides with the constant coefficient spectrum assembly. More generally the straightforward generalization of Ranicki's construction to variable coefficients using the algebraic bordism categories $\Lambda_{p^{-1}(*)}$ coincides with the associated spectrum-functor assembly.

Weiss [26] shows that the assembly for "visible hyperquadratic" Poincaré complexes is an isomorphism. These occur as the relative theory relating quadratic and finite, or "visible" symmetric complexes. The isomorphism theorem provides a calculation, particularly as the coefficient spectrum is a product of Eilenberg-MacLane spectra of 8-torsion groups. It also has important theoretical consequences for example in the structure of Ranicki's total surgery obstruction [20,§17].

6.3D Bounded algebra

Ferry and Pedersen [7] have described a bounded version of surgery, expanding on analogous K-theory work by Pedersen and Weibel [13] , and controlled surgery by Yamasaki [2]. The constructions in this section are so formal that much of it can be applied to the bounded theory.

The constructions of categories over spaces in the previous section can easily be modified to give additive categories of bounded homomorphisms over metric spaces, see [17]. Then using the machinery of Ranicki one can consider chain complexes in these categories. Bordism groups of Poincaré quadratic chain complexes in these categories give the obstruction groups for bounded surgery. These Poincaré complexes also define algebraic bordism categories, so bordism-type theories, cycles, assemblies, etc.

In some significant special cases an assembly map from homology into bounded surgery groups is an isomorphism. This occurs for the $L^{-\infty}$ over control space an open cone. On the chain complex level it is proved using a transversality theorem of Yamasaki [28], beginning with a global object and using transversality to divide it up into a cycle in exactly the same way as was done with manifolds in section 6.2. See also the second appendix in [20].

6.4 An application to group actions

In this section we briefly sketch an application to the construction of PL group actions, slightly reformulating work by Lowell Jones. We begin with an outline of Jones' construction, then use the machinery of this paper to formulate the obstruction.

The problem is: given $K \subset U$ a subcomplex of a PL manifold, and a prime p, when is there a PL Z/p action on U with K as fixed set? This breaks into two pieces: construction of an action on a neighborhood of K, and the extension to the rest of U. Our focus is on the first part, so we will assume U is a regular neighborhood of K. For the next step we note (see 6.4D, below) that a regular neighborhood is the "mapping cylinder" of a PL manifold cycle over K. The problem is therefore to construct a free Z/p action on such a cycle. More precisely, let M denote the cycle corresponding to the boundary of the regular neighborhood. Suppose N is another manifold cycle over K with a homomorphism from π_1 of the assembly (glued up total space) to Z/p, and suppose there is an isomorphism of cycles from M to the associated Z/p cover $\hat{N}$ of N. This gives an isomorphism of mapping cylinders. The mapping cylinder of $\hat{N}$ has an action of Z/p with fixed set exactly K. Since the mapping cylinder of M is U this provides the desired action on U.

This reformulates the problem to: construct a free Z/p action on a PL manifold cycle M over K. The next step in Jones' program is to construct a *homotopy* action. This is a cycle of Poincaré spaces, together with a homomorphism from π_1 of the assembly to Z/p, and a *homotopy equivalence* of cycles from M to the associated Z/p cover. There are obstructions to this. The first come from Smith theory: K must be a mod p homology manifold. The solution of the "homotopy fixed point conjecture" shows that the remaining obstructions to finding a homotopy action are rational. Jones avoids them by assuming that a PL action is already given on the cycle over $\partial_0 K \subset K$, and the rational homology $H_*(K, \partial_0 K; Q)$ is trivial. This is a very important special case, and under these conditions there is a unique homotopy action on M extending the PL action given on $\partial_0 M$. Another obstruction argument gives a *normal* structure on this Poincaré cycle. This is a reduction of the stable normal bundle to the structure group PL, whose Z/p cover agrees with the PL normal bundle of M.

6.4A The standard situation

The data is now close to a surgery situation. We have a Poincaré cycle X

over K, together with a manifold structure on the restriction to $\partial_0 K$ and a PL structure on the stable normal bundle. There is a homomorphism from the total fundamental group of X to Z/p, an equivalence of the Z/p cover with a manifold cycle M, and a PL isomorphism of normal bundles covering this equivalence. The problem is to extend the manifold structure over $\partial_0 K$ to a manifold structure on X with the specified normal bundle and Z/p cover.

The objective is to extract an obstruction from this, whose vanishing implies that there is a solution to the "problem" and therefore a Z/p action on the cycle M.

We review some surgery theory. A "surgery problem" is a Poincaré space Y with a manifold structure on part of it's boundary, say $\partial_0 Y$, and an extension of the PL normal bundle of $\partial_0 Y$ to a PL bundle structure on the stable normal bundle of Y. If we are given a homomorphism $\pi' \to \pi$ then the surgery obstruction group $L(\pi, \pi')$ is the bordism group of such surgery problems together with homomorphisms

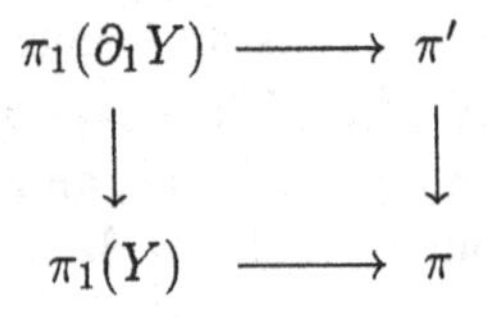

Here $\partial_1 Y$ denotes the (closure of) the complement of $\partial_0 Y$ in ∂Y. The fundamental theorem of surgery states that if the obstruction is trivial in $L(\pi_1 Y, \pi_1 \partial_1 Y)$ (and the dimension is at least 6) then there is a "solution" to the surgery problem: a manifold homotopy equivalent to Y with the given ∂_0 and PL normal bundle.

We modify the definition of "surgery problem" to include the covering information in the standard situation. Suppose (A, B) is a pair with a homomorphism $\rho\colon \pi_1 A \to Z/p$. A "standard problem" over (A, B) is a Poincaré triad $(Y, \partial_0 Y, \partial_1 Y)$ with $\partial_0 Y$ a PL manifold, an extension of the normal bundle of $\partial_0 Y$ to a PL structure on the stable normal bundle of Y, a map $(Y, \partial_1 Y) \to (A, B)$, and a homotopy equivalence of triads $(M, \partial_0 M, \partial_1 M) \to (\hat{Y}, \partial_0 \hat{Y}, \partial_1 \hat{Y})$. Here $\hat{Y}$ is the Z/p cover of Y induced by the homomorphism ρ, M is a PL manifold, the equivalence is a PL isomorphism on ∂_0, and the resulting isomorphism of PL normal bundles over ∂_0 extends to a PL isomorphism over Y refining the natural bundle homotopy equivalence.

This is a lot of data, but it can be managed by recalling where it came from. There is an empty standard problem, and it is straightforward—if tedious—to define ads of standard problems and verify that the Kan condition is satisfied. This therefore forms a bordism-type theory which we denote by $\mathcal{L}$. This theory has bordism groups, spaces, etc. The bordism spaces can be described in terms of traditional surgery problems. Recall that $\Omega^{\mathcal{L}}(A, B, \rho)$ denotes the bordism space of standard problems over (A, B, ρ). Then there is a homotopy fibration

$$\Omega^{\mathcal{L}}(A, B, \rho) \to \mathbb{L}(A, B) \to \mathbb{L}(\hat{A}, \hat{B})$$

where $(\hat{A}, \hat{B})$ is the cover induced by the homomorphism ρ, $\mathbb{L}$ is the surgery space (the bordism space of surgery problems) and the second map is the transfer. (The transfer is defined on the simplex level by taking induced covers). As a consequence of this description the space $\Omega^{\mathcal{L}}$ is often called "the fiber of the transfer."

Now return to the standard situation in 6.4A. Assembling the cycle X over K gives a map $p\colon |X| \to K$. The rest of the data gives a cycle of "standard problems" mapping to $|X|$, subordinate to the cover of K by stars of simplices. This cycle is constructed from the boundary of a regular neighborhood in the original manifold M, so the dimension is $m-1$. Applying the main theorem 4.2A identifies the homology class of this cycle as an element in the functor-coefficient homology group defined in 2.3,

$$H_{m-1}(K, \partial_0 K; \Omega^{\mathcal{L}}(p, \text{ star } (K))).$$

In brief, this is a homology class with coefficients in the the "fiber of the transfer."

Again according to the main theorem this homology class vanishes if and only if the cycle is homologous to the empty cycle. Applying the fundamental theorem of surgery to a nullhomology shows that the original problem can be "solved" and there is a Z/p action.

6.4B Vague Proposition. *In the "standard situation" of 6.4A there is an obstruction in the $(m-1)$-dimensional homology of $(K, \partial K)$ with coefficients in the fiber of the transfer, applied fiberwise to $p\colon |X| \to K$, ie. $\Omega^{\mathcal{L}}(p, \text{ star } (K))$. If the dimensions are sufficiently high then there is a Z/p action on M with K as fixed set and link quotient in the quotient homotopic to X, if and only if this homology class vanishes.*

6.4C More precision

"Dimensions sufficiently high" means that no manifold encountered in the cycle should have dimension less than 5. This happens if the codimension of

K in M at least 6, and can be arranged if $(K, \partial_0 K)$ is 6-connected. We have neglected several issues in the discussion. One is orientation, though that is easily incorporated by making the notation more complicated. A more significant omission is discussion of simple homotopy issues. Since a PL isomorphism is desired at the end, and this has to come from application of the s-cobordism theorem, we want to work with Poincaré spaces, homotopy equivalences, etc. with torsions lying in the kernel of the transfer to the fundamental group of the fragments of the cycle M. This is only really a problem if the embedding is "locally knotted," and this can only happen in codimensions 1 and 2.

In fact it is usual in this problem to assume that the embedding $K \subset M$ has codimension at least 4, so it is locally 1-connected. This simplifies the situation a great deal: the coefficient functor becomes constant, and equal to the fiber of the transfer $\mathbb{L}^h(Z/p) \to \mathbb{L}(1)$. The obstruction therefore lies in an standard constant-coefficient homology group. In Jones' treatment the obstruction is not directly recognized as a homology class. Rather the characteristic variety theorem is used to derive invariants from it, and these derived classes are shown to characterize the obstruction and also define a homology class. Directly recognizing the obstruction as a homology class allows a simpler treatment of parts of the construction.

6.4D Cycles and regular neighborhoods

This section explains the equivalence between PL cycles and PL regular neighborhoods. For more detail on this construction see Akin [1].

Suppose U is a regular neighborhood of a polyhedron K, and suppose it is compact to avoid finiteness and subdivision problems. Then U can be described as the mapping cylinder of a map $\partial U \to K$ which is simplicial with respect to an appropriate triangulation. This map is transverse to the dual cones of the triangulation of K, so it defines a function on the nerve of the covering, as in 4.2. Assume in addition that $U - K$ is a PL n-manifold, then this function defines a PL manifold $(n-1)$-cycle. This gives a construction going from regular neighborhoods with $U - K$ a manifold, to PL manifold cycles over K. More precisely the output is a $(K, \mathrm{id}, \text{ star } (K))$ cycle in the sense of 4.2.

There is a converse to this construction. If we begin with a PL manifold cycle over the star cover of a triangulation of K then there is an associated map from the pieces of the cycle to the dual cones. This is not well-defined, but there is a standard construction which is well-defined up to PL cell-like automorphisms of the cycle. The mapping cylinder of this map gives a regular neighborhood U of K, with $U - K$ a manifold, and this neighborhood is well-defined up to isomorphism rel K by the original cycle.

These constructions are inverses:

6.4E Proposition. *Fix a triangulation of K in which stars of simplices are contractible. These constructions give a bijection between isomorphism classes of PL manifold cycles over the dual cells of the triangulation of K, and isomorphism classes rel K of regular neighborhoods $K \subset U$ with $U - K$ a manifold, and with embedding simplicial with respect to the triangulation.*

A proof can be extracted from Akin [1] in a reasonably straightforward way, though it is not stated explicitly. Here "isomorphism" of cycles means the following: cycles are functions from the nerve of the star cover to PL manifolds. Two such are isomorphic if for each simplex in the nerve there is a PL isomorphism of the corresponding manifolds, and all these isomorphisms commute with the boundary relations in the definition of a cycle. Note that isomorphism is a much stronger relation than homology, and the associated regular neighborhood is definitely not an invariant of the homology class of the cycle.

The notion of "isomorphism" of cycles over K can be elaborated to allow for subdivision of the triangulation. This gives a statement that isomorphism classes of cycles correspond to isomorphism classes of regular neighborhoods, with no reference to a particular triangulation. This refinement is not needed here.

References

[1] E. Akin, *Transverse cellular mappings of polyhedra*, Trans. Amer. Math Society **169** (1972), 401–438.

[2] D. W. Anderson, *Chain functors and homology groups*, Proceedings 1971 Seattle Algebraic Topology Symposium, Springer Lecture Notes in Math. **249**, vol. 249, 1971, pp. 1–12.

[3] A. K. Bousfield, D. M. Kan, *Homotopy limits, completions, and localizations*, Springer Lecture Notes in Math. **304** (1972).

[4] S. Buoncristiano, C. P. Rourke, and B. J. Sanderson, *A geometric approach to homology theory*, London Math. Soc. Notes Vol. **18**, Cambridge U. Press, 1976.

[5] S. Cappell, *A splitting theorem for manifolds*, Invent. Math. **33** (1976), 69–170.

[6] M. Cohen, *Simplicial structures and transverse cellularity*, Ann. Maths. **85** (1967), 218–245.

[7] S. Ferry and E. Pedersen, *Epsilon surgery theory*, (these proceedings).

[8] I. Hambleton, R.J. Milgram, L. Taylor, and B. Williams, *Surgery with finite fundamental group*, Proc. London Math Soc. **56** (1988), 349–379.

[9] I. M. James, *Ex-homotopy theory*, Ill. J. Math. **15** (1971), 324–337.

[10] L. E. Jones, *Construction of surgery problems*, Geometric Topology, J. Cantrell, ed., Academic Press, 1979, pp. 367–391.

[11] L. E. Jones, *Combinatorial symmetries of the m-dimensional ball*, Memoir of the Amer. Math. Soc. **352** (1986).

[12] J. L. Loday, *K-théorie algébrique et représentations des groupes*, Ann. Sc. Ec. Norm. Sup. **9** (1976), 309–377.

[13] E. Pedersen and C. Weibel, *K-theory homology of spaces*, Proceedings 1986 Arcata Conference on Algebraic Topology, Springer Lecture Notes in Math. **1370**, 1989, pp. 346–361.

[14] F. Quinn, *Thesis*, Princeton University (1970).

[15] F. Quinn, *A geometric formulation of surgery*, Topology of manifolds, J. C. Cantrell and C. H. Edwards, ed., Markham, 1970, pp. 500–511.

[16] F. Quinn, *Ends of maps, II*, Invent. Math. **68** (1982), 353–424.

[17] F. Quinn, *Geometric algebra*, Proceedings 1983 Rutgers Conference on Algebraic Topology, Springer Lecture Notes in Math. **1126**, 1985, pp. 182–198.

[18] F. Quinn, *Homotopically stratified sets*, J. Amer. Math. Soc. **1** (1988), 441–499.

[19] A. A. Ranicki, *The total surgery obstruction*, Proceedings 1978 Århus Conference on Algebraic Topology, Springer Lecture Notes in Math. **763**, 1979, pp. 275–316.

[20] A. A. Ranicki, *Algebraic L-theory and Topological Manifolds*, Cambridge Tracts in Mathematics **102**, Cambridge U. Press, 1992.

[21] A. A. Ranicki (ed.), *The Hauptvermutung book. Papers by M. A. Armstrong, A. J. Casson, G. E. Cooke, A. A. Ranicki, C. P. Rourke and D. P. Sullivan*, K-theory journal book series (to appear).

[22] C. P. Rourke, and B. J. Sanderson, *Δ-sets I: Homotopy theory*, Quart. J. Math. Oxford **22** (1971), 321–338.

[23] R. Stong, *Notes on cobordism theory*, Princeton Mathematical Notes **7**, Princeton U. Press, 1968.

[24] F. Waldhausen, *Algebraic K-theory of generalized free products*, Ann. Maths. **108** (1978), 135–256.

[25] C. T. C. Wall, *Surgery on Compact Manifolds*, Academic Press, New York, 1970.

[26] M. Weiss, *Visible L-theory*, Forum Math. **4** (1992), 465–498.

[27] G. W. Whitehead, *Generalized homology theories*, Transactions of the A. M. S. **102** (1962), 227–283.

[28] M. Yamasaki, *L-groups of crystallographic groups*, Invent. Math. **88** (1987), 571–602.

Department of Mathematics, Virginia Polytech Institute & State University, Blacksburg, VA 24061-0123, USA

email: quinn@math.vt.edu

On the Novikov conjecture

Andrew Ranicki

Introduction

Signatures of quadratic forms play a central role in the classification theory of manifolds. The Hirzebruch theorem expresses the signature $\sigma(N) \in \mathbb{Z}$ of a $4k$-dimensional manifold N^{4k} in terms of the $\mathcal{L}$-genus $\mathcal{L}(N) \in H^{4*}(N;\mathbb{Q})$. The 'higher signatures' of a manifold M with fundamental group $\pi_1(M) = \pi$ are the signatures of the submanifolds $N^{4k} \subset M$ which are determined by the cohomology $H^*(B\pi;\mathbb{Q})$. The Novikov conjecture on the homotopy invariance of the higher signatures is of great importance in understanding the connection between the algebraic and geometric topology of high-dimensional manifolds. Progress in the field is measured by the class of groups π for which the conjecture has been verified. A wide variety of methods has been used to attack the conjecture, such as surgery theory, elliptic operators, C^*-algebras, differential geometry, hyperbolic geometry, bounded/controlled topology, and algebra.

The diffeomorphism class of a closed differentiable m-dimensional manifold M^m is distinguished in its homotopy type up to a finite number of possibilities by the rational Pontrjagin classes $p_*(M) \in H^{4*}(M;\mathbb{Q})$. Thom and Rokhlin-Shvartz proved that the rational Pontrjagin classes $p_*(M)$ are

This is an expanded version of talks delivered at the Oberwolfach meetings 'Algebraic K-theory', 28 June, 1993 and 'Novikov conjectures, index theory and rigidity', 6 September, 1993.

combinatorial invariants by showing that they determine and are determined by the signatures of closed $4k$-dimensional submanifolds $N^{4k} \subset M \times \mathbb{R}^j$ (j large) with trivial normal bundle. A homotopy equivalence of manifolds only preserves the global algebraic topology, and so need not preserve the local algebraic topology given by the Pontrjagin classes. The Browder-Novikov-Sullivan-Wall surgery theory shows that modulo torsion invariants for $m \geq 5$ a homotopy equivalence of closed differentiable m-dimensional manifolds is homotopic to a diffeomorphism if and only if it preserves the signatures of submanifolds and the non-simply-connected surgery obstruction is in the image of the assembly map; this map is onto in the simply-connected case. (Here, torsion means both Whitehead groups and finite groups). Novikov proved the topological invariance of the rational Pontrjagin classes by showing that a homeomorphism preserves signatures of submanifolds with trivial normal bundles, using the fundamental group and non-compact manifold topology.

The object of this largely expository paper is to outline the relationship between the Novikov conjecture, the exotic spheres, the topological invariance of the rational Pontrjagin classes, surgery theory, codimension 1 splitting obstructions, the bounded/controlled topology of non-compact manifolds, the algebraic theory of Ranicki [47], [50], [51], and the method used by by Carlsson and Pedersen [14] to prove the conjecture for a geometrically defined class of infinite torsion-free groups π with $B\pi$ a finite complex and $E\pi$ a non-compact space with a sufficiently nice compactification. See Ferry, Ranicki and Rosenberg [19] for a wider historical survey of the Novikov conjecture.

The surgery obstruction groups $L_m(\mathbb{Z}[\pi])$ of Wall [58] are defined for any group π and $m(\text{mod } 4)$, to be the Witt group of $(-)^k$-quadratic forms over the group ring $\mathbb{Z}[\pi]$ for $m = 2k$, and a stable automorphism group of such forms for $m = 2k+1$. In [58] the groups $L_*(\mathbb{Z}[\pi])$ were understood to be the simple quadratic L-groups $L^s_*(\mathbb{Z}[\pi])$, the obstruction groups for surgery to simple homotopy equivalence, involving based f.g. free $\mathbb{Z}[\pi]$-modules and simple isomorphisms. Here, $L_*(\mathbb{Z}[\pi])$ are understood to be the free quadratic L-groups $L^h_*(\mathbb{Z}[\pi])$, the obstruction groups for surgery to homotopy equivalence, involving unbased f.g. free $\mathbb{Z}[\pi]$-modules and all isomorphisms. The simple and free L-groups differ in 2-torsion only, being related by the Rothenberg exact sequence

$$\ldots \longrightarrow L^s_m(\mathbb{Z}[\pi]) \longrightarrow L_m(\mathbb{Z}[\pi]) \longrightarrow \widehat{H}^m(\mathbb{Z}_2; Wh(\pi)) \longrightarrow L^s_{m-1}(\mathbb{Z}[\pi]) \longrightarrow \ldots$$

with $\widehat{H}^*(\mathbb{Z}_2; Wh(\pi))$ the (2-torsion) Tate $\mathbb{Z}_2$-cohomology groups of the duality involution on the Whitehead group $Wh(\pi)$. A normal map $(f,b): M \longrightarrow N$ from an m-dimensional manifold M to an m-dimensional geometric Poincaré complex N with $\pi_1(N) = \pi$ has a surgery obstruction $\sigma_*(f,b) \in L_m(\mathbb{Z}[\pi])$ such that $\sigma_*(f,b) = 0$ if (and for $m \geq 5$ only if) (f,b) is normal bordant to a homotopy equivalence. The original treatment in [58] using forms and automorphisms was extended in Ranicki [47] to quadratic Poincaré complexes (= chain complexes with Poincaré duality). The surgery obstruction groups $L_*(\mathbb{Z}[\pi])$ were expressed in [47] as the cobordism groups of quadratic Poincaré complexes over $\mathbb{Z}[\pi]$.

The assembly maps in quadratic L-theory

$$A \ : \ H_*(X;\mathbb{L}_\bullet(\mathbb{Z})) \longrightarrow L_*(\mathbb{Z}[\pi_1(X)])$$

are defined in Ranicki [51] for any topological space X, abstracting a geometric construction of Quinn. The generalized homology groups $H_*(X;\mathbb{L}_\bullet(\mathbb{Z}))$ with coefficients in the simply-connected surgery spectrum $\mathbb{L}_\bullet(\mathbb{Z})$ are the cobordism groups of sheaves Γ over X of quadratic Poincaré complexes over $\mathbb{Z}$. Here, X is taken to be a simplicial complex, and the 'sheaf' Γ is taken to be a quadratic Poincaré cycle in the sense of [51], i.e. a contravariant functor on the category with objects the simplices of X and morphisms the face inclusions*. The assembly map A sends a quadratic Poincaré cycle Γ over X to the quadratic Poincaré complex over $\mathbb{Z}[\pi_1(X)]$

$$A(\Gamma) \ = \ q_! p^! \Gamma$$

with $p^!$ the pullback along the universal covering projection $p : \widetilde{X} \longrightarrow X$ and $q_!$ the pushforward along the unique map $q : \widetilde{X} \longrightarrow \{\text{pt.}\}$.

Novikov conjecture for a group π
The assembly maps for the classifying space $B\pi$

$$A \ : \ H_*(B\pi;\mathbb{L}_\bullet(\mathbb{Z})) \longrightarrow L_*(\mathbb{Z}[\pi])$$

are rational split injections.

This will be called the **rational Novikov conjecture**, to distinguish it from:

Integral Novikov conjecture for a group π
The assembly maps $A : H_*(B\pi;\mathbb{L}_\bullet(\mathbb{Z})) \longrightarrow L_*(\mathbb{Z}[\pi])$ are split injections.

The rational Novikov conjecture is trivially true for finite groups π; it has been verified for infinite groups which have strong geometric properties. In principle, it is possible that the conjecture is true for *all* groups, although Gromov [21] suggests there may be a counterexample.

* The simplicial method applies to an arbitrary space by considering algebraic Poincaré cycles over the simplicial complexes defined by the nerves of open covers. Hutt [26] has developed the actual sheaf theory of algebraic Poincaré complexes over an arbitrary space.

The integral Novikov conjecture is known to be false for finite groups π; it has been verified for many torsion-free infinite groups which have strong geometric properties.

The verification of the integral Novikov conjecture π requires the construction of a 'disassembly' map

$$B : L_m(\mathbb{Z}[\pi]) \longrightarrow H_m(B\pi;\mathbb{L}_\bullet(\mathbb{Z})) ; C \longrightarrow B(C)$$

such that $BA = 1$. Such a map B has to send a quadratic Poincaré complex C over $\mathbb{Z}[\pi]$ to a sheaf $B(C)$ over $B\pi$ of quadratic Poincaré complexes over $\mathbb{Z}$, with $BA(\Gamma)$ cobordant to Γ for any sheaf Γ over $B\pi$ of quadratic Poincaré complexes over $\mathbb{Z}$. It is possible to construct such B for any group π which has sufficient geometry that manifolds with fundamental group π have rigidity, meaning that homotopy equivalences can be deformed to homeomorphisms. Novikov [39] constructed B algebraically in the case of a free abelian group $\pi = \mathbb{Z}^n$, when $B\pi = T^n$ and A is an isomorphism. See Farrell and Jones [17] for a geometric construction of B in the case when $B\pi$ is realized by a compact aspherical Riemannian manifold all of whose sectional curvatures are nonpositive (when A is also an isomorphism), and the connection with the original Mostow rigidity theorem for hyperbolic manifolds.

The locally finite assembly maps in quadratic L-theory

$$A^{lf} : H_*^{lf}(X;\mathbb{L}_\bullet(\mathbb{Z})) \longrightarrow L_*(\mathbb{C}_X(\mathbb{Z}))$$

are defined in Ranicki [51] for any metric space X, using the X-graded $\mathbb{Z}$-module category $\mathbb{C}_X(\mathbb{Z})$ of Pedersen and Weibel [43]. The locally finite generalized homology groups $H_*^{lf}(X;\mathbb{L}_\bullet(\mathbb{Z}))$ are the cobordism groups of locally finite sheaves Γ over X of quadratic Poincaré complexes over $\mathbb{Z}$. It was shown in Ranicki [50] that A^{lf} is an isomorphism for $X = O(K) \subseteq \mathbb{R}^{N+1}$ the open cone of a compact polyhedron $K \subseteq S^N$, which can be used to prove the topological invariance of the rational Pontrjagin classes (see 9.13 below). It is easier to establish that the locally finite assembly maps A^{lf} are isomorphisms than the ordinary assembly maps A. This is an algebraic reflection of the observed fact that rigidity theorems deforming homotopy equivalences to homeomorphisms are easier to prove for non-compact manifolds than for compact manifolds.

Carlsson and Pedersen [14] prove the integral Novikov conjecture for groups π with $B\pi$ a finite complex realized by a compact metric space such that the universal cover $E = E\pi$ admits a contractible π-equivariant compactification $\overline{E}$ with a metric such that compact sets in E become small when translated under π near the boundary $\partial E = \overline{E}\backslash E$. Bounded/controlled algebra is used to prove that A^{lf} is an isomorphism for $X = E$, and equivariant topology is used to construct an algebraic disassembly map B by means of $(A^{lf})^{-1}$. The conditions on the compactification allow E-bounded algebra/topology to be deformed to ∂E-controlled algebra/topology, i.e. to pass from homotopy equivalences to homeomorphisms. See Ferry and Weinberger [20] for a more geometric approach. The computation $Wh_{-*}(\{1\}) = 0$ of Bass, Heller and Swan [4] is an essential ingredient of both [14] and [20], since the lower K-groups of $\mathbb{Z}$ are potential obstructions to the disassembly of quadratic Poincaré complexes over $\mathbb{Z}$ in bounded algebra, or equivalently to compactifying simply-connected open manifolds in bounded topology.

In dealing with vector bundles, manifolds, homotopy equivalences, etc., only the oriented and orientation-preserving cases are considered. Manifolds are understood to be compact and differentiable, unless specified otherwise. Also, except for classifying spaces, only topological spaces which are finite-dimensional locally finite polyhedra or topological manifolds are considered.

§1. Pontrjagin classes and the $\mathcal{L}$-genus

The **Pontrjagin classes** of an m-plane bundle $\eta : X \longrightarrow BO(m)$ over a space X are integral characteristic classes

$$p_*(\eta) \in H^{4*}(X) .$$

The rational Pontrjagin character defines an isomorphism

$$\mathrm{ph} : KO(X) \otimes \mathbb{Q} = [X, \mathbb{Z} \times BO] \otimes \mathbb{Q} \xrightarrow{\cong} H^{4*}(X; \mathbb{Q}) .$$

The **$\mathcal{L}$-genus** of an m-plane bundle $\eta : X \longrightarrow BO(m)$ is a rational cohomology class

$$\mathcal{L}(\eta) \in H^{4*}(X; \mathbb{Q})$$

whose components $\mathcal{L}_k(\eta) \in H^{4k}(X; \mathbb{Q})$ can be expressed as polynomials in the Pontrjagin classes $p_1, p_2, \ldots$ with rational coefficients. The $\mathcal{L}$-genus determines and is determined by the rational Pontrjagin classes $p_k(\eta) \in H^{4k}(X; \mathbb{Q})$. The first two $\mathcal{L}$-polynomials are given by

$$\mathcal{L}_1 = \frac{1}{3} p_1 \ , \ \mathcal{L}_2 = \frac{1}{45}(7p_2 - (p_1)^2) .$$

See Hirzebruch [24] and Milnor and Stasheff [34] for the textbook accounts of the Pontrjagin classes and the $\mathcal{L}$-genus.

The **Pontrjagin classes** and the **$\mathcal{L}$-genus** of an m-dimensional differentiable manifold M are the Pontrjagin classes and the $\mathcal{L}$-genus of the tangent m-plane bundle $\tau_M : M \longrightarrow BO(m)$

$$p_*(M) = p_*(\tau_M) \in H^{4*}(M) ,$$

$$\mathcal{L}(M) = \mathcal{L}(\tau_M) \in H^{4*}(M; \mathbb{Q}) .$$

By construction, the Pontrjagin classes and $\mathcal{L}$-genus are invariants of the differentiable structure of M: if $h : M' \longrightarrow M$ is a diffeomorphism then

$$\begin{aligned} \tau_{M'} &= h^*\tau_M : M' \longrightarrow BO(m) , \\ p_*(M') &= h^*p_*(M) \in H^{4*}(M') , \\ \mathcal{L}(M') &= h^*\mathcal{L}(M) \in H^{4*}(M';\mathbb{Q}) . \end{aligned}$$

§2. Signature

Definition 2.1 The **intersection form** of a closed $4k$-dimensional manifold N^{4k} is the nondegenerate symmetric form

$$\phi : H^{2k}(N;\mathbb{Q}) \times H^{2k}(N;\mathbb{Q}) \longrightarrow \mathbb{Q} \; ; \; (x,y) \longrightarrow \langle x \cup y, [N] \rangle$$

on the finite-dimensional $\mathbb{Q}$-vector space $H^{2k}(N;\mathbb{Q})$. The **signature** of N^{4k} is

$$\sigma(N) = \text{signature}(H^{2k}(N;\mathbb{Q}),\phi) \in \mathbb{Z} .$$

□

Remarks 2.2 (i) An m-dimensional geometric Poincaré complex X is a finite CW complex with a fundamental class $[X] \in H_m(X)$ inducing isomorphisms

$$[X] \cap - : H^*(X) \xrightarrow{\simeq} H_{m-*}(X) .$$

Closed topological manifolds are the prime examples of geometric Poincaré complexes. The intersection form $(H^{2k}(X;\mathbb{Q}),\phi)$ and the signature $\sigma(X) \in \mathbb{Z}$ are defined for any $4k$-dimensional geometric Poincaré complex X, and are homotopy invariants of X.
(ii) The intersection form and signature are also defined for any $4k$-dimensional geometric Poincaré pair $(X,\partial X)$, such as a manifold with boundary $(M,\partial M)$.

□

Signature Theorem 2.3 (Hirzebruch) *The signature of a closed differentiable manifold N^{4k} is the evaluation of the $\mathcal{L}$-genus $\mathcal{L}(N) \in H^{4*}(N;\mathbb{Q})$ on* $[N] \in H_{4k}(N;\mathbb{Q})$

$$\sigma(N) = \langle \mathcal{L}(N),[N] \rangle \in \mathbb{Z} .$$

□

Transversality Theorem 2.4 *A continuous map $h : M'^m \longrightarrow M^m$ of differentiable m-dimensional manifolds is homotopic to a differentiable map. Given an n-dimensional submanifold $N^n \subset M^m$ it is possible to choose the* *homotopy in such a way that the differentiable map (also denoted by h) is transverse regular at N, with the restriction*

$$f = h| : N'^n = h^{-1}(N) \longrightarrow N^n$$

a degree 1 map of n-dimensional manifolds which is covered by a map of the normal $(m-n)$-plane bundles $b : \nu_{N' \subset M'} \longrightarrow \nu_{N \subset M}$.

□

Definition 2.5 A submanifold $N^n \subset M^m \times \mathbb{R}^j$ is **special** if it is closed, $n = 4k$ and the normal bundle is trivial

$$\nu_{N \subset M} = \epsilon^i : N \longrightarrow BSO(i) \quad (i = m + j - 4k) .$$

□

Proposition 2.6 (Thom) *The rational Pontrjagin classes and the $\mathcal{L}$-genus of a manifold M are determined by the signatures of the special submanifolds $N^{4k} \subset M \times \mathbb{R}^j$.*

Proof The Pontrjagin classes and the $\mathcal{L}$-genus of a special submanifold $N^{4k} \subset M^m \times \mathbb{R}^j$ are the images in $H^{4*}(N;\mathbb{Q})$ of the Pontrjagin classes and the $\mathcal{L}$-genus of M, that is

$$p_*(N) = e^* p_*(M) \ , \ \mathcal{L}(N) = e^* \mathcal{L}(M)$$

with

$$e : N \longrightarrow M \times \mathbb{R}^j \longrightarrow M .$$

The signature of N thus depends only on the homology class $e_*[N] \in H_{4k}(M;\mathbb{Q})$ represented by N

$$\begin{aligned} \sigma(N) &= \langle \mathcal{L}(N), [N] \rangle \\ &= \langle e^* \mathcal{L}(M), [N] \rangle = \langle \mathcal{L}(M), e_*[N] \rangle \in \mathbb{Z} . \end{aligned}$$

From now on, we shall write $e_*[N] \in H_{4k}(M;\mathbb{Q})$ as $[N]$. The cobordism classes of special submanifolds $N^{4k} \subset M^m \times \mathbb{R}^j$ are in one-one correspondence with the proper homotopy classes of proper maps

$$f : M \times \mathbb{R}^j \longrightarrow \mathbb{R}^i \quad (i = m + j - 4k)$$

with $N = f^{-1}(0)$ (assuming transverse regularity at $0 \in \mathbb{R}^i$). The set of proper homotopy classes is in one-one correspondence with the cohomotopy group $\pi^i(\Sigma^j M_+)$ of homotopy classes of maps $\Sigma^j M_+ \longrightarrow S^i$, with $\Sigma^j M_+$ the

j-fold suspension of $M_+ = M \cup \{\text{pt.}\}$. By the Serre finiteness of the stable homotopy groups of spheres and Poincaré duality

$$\pi^i(\Sigma^j M_+) \otimes \mathbb{Q} = H^{m-4k}(M;\mathbb{Q}) = H_{4k}(M;\mathbb{Q}) .$$

The $\mathbb{Q}$-vector space $H_{4k}(M;\mathbb{Q})$ is thus spanned by the homology classes $[N]$ of special submanifolds $N^{4k} \subset M \times \mathbb{R}^j$, and

$$\mathcal{L}(M) \in H^{4k}(M;\mathbb{Q}) = \mathrm{Hom}_{\mathbb{Q}}(H_{4k}(M;\mathbb{Q}),\mathbb{Q})$$

is given by

$$\begin{aligned} \mathcal{L}(M) \ &: H_{4k}(M;\mathbb{Q}) \longrightarrow \mathbb{Q} ; \\ &[N] \longrightarrow \langle \mathcal{L}(M),[N]\rangle = \langle \mathcal{L}(N),[N]\rangle = \sigma(N) . \end{aligned}$$

□

A PL homeomorphism of differentiable manifolds cannot in general be approximated by a diffeomorphism, by virtue of the exotic spheres of Milnor [33].

Theorem 2.7 (Thom, Rokhlin-Shvarts) *The rational Pontrjagin classes and the $\mathcal{L}$-genus are combinatorial invariants.*
Proof Transversality also works in the PL category, so that the characterization (2.6) of the $\mathcal{L}$-genus in terms of signatures of special submanifolds $N^{4k} \subset M \times \mathbb{R}^j$ can be carried out in the PL category. In particular, if $h : M' \longrightarrow M$ is a PL homeomorphism then

$$p_*(M') = h^* p_*(M) \ , \ \mathcal{L}(M') = h^*\mathcal{L}(M) .$$

□

Remark 2.8 Thom used PL transversality and the Hirzebruch signature theorem to define rational Pontrjagin classes $p_*(M)$ and the $\mathcal{L}$-genus $\mathcal{L}(M) \in H^{4*}(M;\mathbb{Q})$ for a PL manifold M. It is not possible to prove the topological invariance of the rational Pontrjagin classes by a mimicry of Thom's PL transversality argument: on the contrary, topological invariance is required for topological transversality.

□

Proposition 2.9 (Dold, Milnor) *The rational Pontrjagin classes and the $\mathcal{L}$-genus are not homotopy invariants.*
Proof The stable classifying space G/O for fibre homotopy trivialized vector bundles is such that there is defined a fibration

$$G/O \longrightarrow BO \longrightarrow BG$$

with an exact sequence

$$\ldots \longrightarrow \pi_{n+1}(BG) \longrightarrow \pi_n(G/O) \longrightarrow \pi_n(BO) \longrightarrow \pi_n(BG) \longrightarrow \ldots .$$

The homotopy groups of the stable classifying space BG for spherical fibrations are the stable homotopy groups of spheres

$$\pi_*(BG) = \pi^S_{*-1} ,$$

so that by Serre's finiteness theorem

$$\pi_*(BG) \otimes \mathbb{Q} = \pi^S_{*-1} \otimes \mathbb{Q} = 0 \quad (* > 1) .$$

By Bott periodicity $\pi_{4k}(BO) = \mathbb{Z}$, detected by the kth Pontrjagin class p_k. For any $k \geq 1$ there exists a fibre homotopy trivial $(j+1)$-plane bundle $\eta : S^{4k} \longrightarrow BO(j+1)$ (j large) over S^{4k} with

$$p_k(\eta) \neq 0 \in H^{4k}(S^{4k}) = \mathbb{Z} .$$

The sphere bundle $S(\eta)$ is a closed $(4k+j)$-dimensional manifold which is homotopy equivalent to $S(\epsilon^{j+1}) = S^{4k} \times S^j$, such that

$$\begin{aligned} p_k(S(\eta)) &= -p_k(\eta) \neq p_k(\epsilon^{j+1}) = 0 , \\ \mathcal{L}_k(S(\eta)) &= s_k p_k(S(\eta)) \neq \mathcal{L}_k(S(\epsilon^{j+1})) = 0 \\ &\in H^{4k}(S^{4k} \times S^j) = \mathbb{Z} \end{aligned}$$

with $s_k \neq 0 \in \mathbb{Z}$ the coefficient of p_k in $\mathcal{L}_k$. See 2.10 for a more detailed account.

□

Remark 2.10 Let Θ^m be the group of m-dimensional exotic differentiable spheres, and let $bP_{m+1} \subseteq \Theta^m$ be the subgroup of the exotic spheres Σ^m which occur as the boundary ∂W of a framed $(m+1)$-dimensional manifold W, as in Kervaire and Milnor [27]. For $m \geq 5$

$$\Theta^m = \pi_m(PL/O)$$

is a finite group. The classifying space PL/O for PL trivialized vector bundles fits into a fibration

$$PL/O \longrightarrow TOP/O \longrightarrow TOP/PL \simeq K(\mathbb{Z}_2, 3) .$$

(See Ranicki [52] for information on $TOP/PL \simeq K(\mathbb{Z}_2, 3)$.) Thus for $m \geq 5$

$$\Theta^m = \pi_m(PL/O) = \pi_m(TOP/O) ,$$

and the subgroup

$$\begin{aligned} bP_{m+1} &= \mathrm{im}(\pi_{m+1}(G/TOP) \longrightarrow \pi_m(TOP/O)) \\ &= \mathrm{im}(L_{m+1}(\mathbb{Z}) \longrightarrow \Theta^m) \subseteq \Theta^m \end{aligned}$$

is cyclic if m is odd, and is zero if m is even. The class $[\Sigma^m] \in bP_{m+1}$ of an exotic sphere Σ^m such that $\Sigma^m = \partial W$ for a framed $(m+1)$-dimensional manifold W is the image of the surgery obstruction

$$\sigma_*(f,b) \in \pi_{m+1}(G/TOP) = L_{m+1}(\mathbb{Z})$$

of the corresponding normal map $(f,b) : (W, \partial W) \longrightarrow (D^{m+1}, S^m)$ with $\partial f : \partial W \longrightarrow S^m$ a homotopy equivalence. We only consider the case $m = 4k-1$ here, with $k \geq 2$; the subgroup $bP_{4k} \subseteq \Theta^{4k-1}$ is cyclic of order

$$t_k = a_k 2^{2k-2}(2^{2k-1}-1)\,\mathrm{num}(B_k/4k)$$

with B_k the kth Bernoulli number and $a_k = 1$ (resp. 2) if k is even (resp. odd). Let (W^{4k}, Σ^{4k-1}) be the framed $(2k-1)$-connected $4k$-dimensional manifold with homotopy $(4k-1)$-sphere boundary obtained by the E_8-plumbing of 8 copies of $\tau_{S^{2k}} : S^{2k} \longrightarrow BSO(2k)$, so that $[\Sigma^{4k-1}] \in bP_{4k}$ is a generator. Let

$$Q^{4k} = W^{4k} \cup c\Sigma^{4k-1}$$

be the framed $(2k-1)$-connected $4k$-dimensional PL manifold with signature $\sigma(Q) = 8$ obtained from (W^{4k}, Σ^{4k-1}) by coning off the boundary. The t_k-fold connected sum $\#_{t_k}\Sigma^{4k-1}$ is diffeomorphic to the standard $(4k-1)$-sphere S^{4k-1}, so that $\#_{t_k}Q^{4k}$ has a differentiable structure. The topological K-group of isomorphism classes of stable vector bundles over S^{4k}

$$\widetilde{KO}(S^{4k}) = \pi_{4k}(BO) = \pi_{4k}(BO(j+1)) \quad (j \text{ large})$$

is such that there is defined an isomorphism

$$\pi_{4k}(BO) \xrightarrow{\cong} \mathbb{Z} \; ; \; \eta \longrightarrow \langle p_k(\eta), [S^{4k}] \rangle / a_k(2k-1)! \; ,$$

by the Bott integrality theorem. The subgroup of fibre homotopy trivial bundles

$$\mathrm{im}(\pi_{4k}(G/O) \longrightarrow \pi_{4k}(BO)) = \ker(J : \pi_{4k}(BO) \longrightarrow \pi_{4k}(BG)) \subseteq \pi_{4k}(BO)$$

is the infinite cyclic subgroup of index

$$j_k = \mathrm{den}(B_k/4k)$$

with the generator $\eta : S^{4k} \longrightarrow BO(j+1)$ such that

$$p_k(\eta) = a_k j_k (2k-1)! \in H^{4k}(S^{4k}) = \mathbb{Z} .$$

For any fibre homotopy trivialization

$$h : J\eta \simeq J\epsilon^{j+1} : S^{4k} \longrightarrow BG(j+1)$$

the corresponding homotopy equivalence

$$S(h) : S(\eta) \xrightarrow{\simeq} S(\epsilon^{j+1}) = S^{4k} \times S^j$$

is such that the inverse image of $S^{4k} \times \{*\} \subset S^{4k} \times S^j$ is a submanifold of the type

$$N^{4k} = \#_{t_k} Q^{4k} \subset S(\eta) ,$$

and $S(h)$ restricts to a normal map

$$(f, b) = S(h)| : N^{4k} \longrightarrow S^{4k}$$

with $b : \nu_N \longrightarrow -\eta$. Moreover,

$$\begin{aligned} \tau_N &= f^*(\eta) : N \longrightarrow BO(4k) , \\ p_k(N) &= f^* p_k(\eta) = a_k j_k (2k-1)! \in H^{4k}(N) = \mathbb{Z} , \\ \sigma(N) &= s_k p_k(N) = s_k a_k j_k (2k-1)! = 8t_k \in \mathbb{Z} , \end{aligned}$$

with

$$s_k = \frac{8t_k}{a_k j_k (2k-1)!} = \frac{2^{2k}(2^{2k-1}-1)B_k}{(2k)!}$$

the coefficient of p_k in $\mathcal{L}_k$. The homotopy equivalence $S(h) : S(\eta) \longrightarrow S^{4k} \times S^j$ does not preserve the $\mathcal{L}$-genus, since

$$\begin{aligned} \langle \mathcal{L}_k(S(\eta)), [N] \rangle &= \sigma(N) = 8t_k \\ &\neq \langle \mathcal{L}_k(S^{4k} \times S^j), [S^{4k}] \rangle = \sigma(S^{4k}) = 0 \in \mathbb{Z} . \end{aligned}$$

(See 3.3 for more details in the special case $k = 2$.) The homotopy equivalence $S(h) : S(\eta) \longrightarrow S^{4k} \times S^j$ is not homotopic to a diffeomorphism since the surgery obstruction of (f, b) is

$$\begin{aligned} \sigma_*(f, b) &= \frac{1}{8}(\sigma(N) - \sigma(S^{4k})) \\ &= t_k \neq 0 \in L_{4k}(\mathbb{Z}) = \mathbb{Z} . \end{aligned}$$

These were the original examples due to Novikov [36] of homotopy equivalences of high-dimensional simply-connected manifolds which are not homotopic to diffeomorphisms. By the topological invariance of the rational Pontrjagin classes these homotopy equivalences are not homotopic to homeomorphisms.

□

§3. Splitting homotopy equivalences

Let M^m be an m-dimensional manifold, and let $N^n \subset M^m$ be an n-dimensional submanifold. Every map of m-dimensional manifolds $h : M' \longrightarrow M$ is homotopic to a map (also denoted by h) which is transverse regular at $N \subset M$, with the restriction

$$f = h| : N' = h^{-1}(N) \longrightarrow N$$

a degree 1 map of n-dimensional manifolds such that the normal $(m-n)$-plane bundle of N' in M' is the pullback along f of the normal $(m-n)$-plane bundle of N in M

$$\nu_{N' \subset M'} : N' \xrightarrow{f} N \xrightarrow{\nu_{N \subset M}} BSO(m-n) .$$

Let $i : N \longrightarrow M$, $i' : N' \longrightarrow M'$ be the inclusions. For any embedding $M' \subset S^{m+k}$ (k large) define a map of $(m-n+k)$-plane bundles covering f

$$\begin{aligned} b : \nu_{N' \subset S^{m+k}} &= \nu_{N' \subset M'} \oplus i'^*(\nu_{M' \subset S^{m+k}}) \\ &\longrightarrow \eta = \nu_{N \subset M} \oplus (h^{-1}i)^*(\nu_{M' \subset S^{m+k}}) , \end{aligned}$$

so that $(f, b) : N' \longrightarrow N$ is a normal map. If $h : M' \longrightarrow M$ is a homotopy equivalence it need not be the case that (f, b) is a homotopy equivalence.

Definition 3.1 (i) A homotopy equivalence $h : M' \longrightarrow M$ of manifolds **splits** along a submanifold $N \subset M$ if h is homotopic to a map (also denoted h) which is transverse regular along $N \subset M$, and such that the restriction $h| : N' = h^{-1}(N) \longrightarrow N$ is a homotopy equivalence.
(ii) A homotopy equivalence $h : M' \longrightarrow M$ of manifolds h-**splits** along a submanifold $N \subset M$ if there exists an extension of $h : M' \longrightarrow M$ to a homotopy equivalence

$$(g; h, h') : (W; M', M'') \longrightarrow M \times ([0,1]; \{0\}, \{1\})$$

with $(W; M', M'')$ an h-cobordism and $h' : M'' \longrightarrow M$ split along $N \subset M$.

□

For $m \geq 5$ a homotopy equivalence $h : M' \longrightarrow M$ of m-dimensional manifolds splits along a submanifold $N^n \subset M^m$ if and only if $h : M' \longrightarrow M$ h-splits with $\tau(M' \longrightarrow W) = 0 \in Wh(\pi_1(M))$, by the s-cobordism theorem.

See Chapter 23 of Ranicki [51] for an account of the Browder-Wall surgery obstruction theory for splitting homotopy equivalences along submanifolds. Here is a brief summary:

Proposition 3.2 (i) *If a homotopy equivalence of manifolds $h : M' \longrightarrow M$ is homotopic to a diffeomorphism then h splits along every submanifold $N \subset M$ and $\tau(h) = 0 \in Wh(\pi_1(M))$. A homotopy equivalence which does not split along a submanifold or is such that $\tau(h) \neq 0$ cannot be homotopic to a diffeomorphism.*
(ii) *The (free) LS-groups LS_* of Wall [58, §11] are defined for a manifold M^m and a submanifold $N^n \subset M^m$ with normal bundle*

$$\xi \;=\; \nu_{N\subset M} \;:\; N \longrightarrow BO(q) \;\;(q \;=\; m-n)$$

to fit into a commutative braid of exact sequences

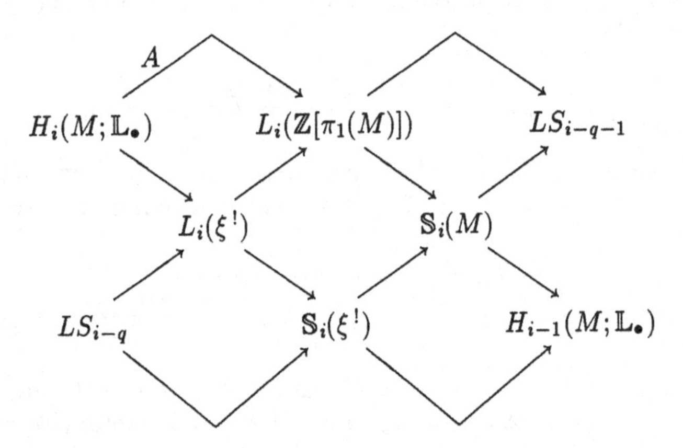

with A the algebraic L-theory assembly map (§7), $L_*(\xi^!)$ the relative L-groups in the transfer exact sequence

$$\ldots \longrightarrow L_i(\mathbb{Z}[\pi_1(M\backslash N)]) \longrightarrow L_i(\xi^!) \longrightarrow L_{i-q}(\mathbb{Z}[\pi_1(N)]) \longrightarrow L_{i-1}(\mathbb{Z}[\pi_1(M\backslash N)]) \longrightarrow L_{i-1}(\xi^!) \longrightarrow$$

and similarly for $\mathbb{S}_*(\xi^!)$. The structure invariant $s(h) \in \mathbb{S}_{m+1}(M)$ (7.1) of a homotopy equivalence $h : M' \longrightarrow M$ of m-dimensional manifolds has image $[s(h)] \in LS_n$, which has image $\sigma_*(f,b) \in L_n(\mathbb{Z}[\pi_1(N)])$ the surgery obstruction of the n-dimensional normal map given by transversality

$$(f,b) \;=\; h| \;:\; N' \;=\; h^{-1}(N) \longrightarrow N\;.$$

For $n \geq 5$, $q \geq 1$ $h : M' \longrightarrow M$ h-splits along $N \subset M$ if and only if $[s(h)] = 0 \in LS_n$. For $q \geq 3$

$$\begin{aligned} \pi_1(M) &= \pi_1(N) = \pi_1(M \backslash N) , \\ L_*(\xi^!) &= L_*(\mathbb{Z}[\pi_1(M)]) \oplus L_{*-q}(\mathbb{Z}[\pi_1(M)]) , \\ LS_* &= L_*(\mathbb{Z}[\pi_1(M)]) \end{aligned}$$

and

$$[s(h)] = \sigma_*(f,b) \in LS_n = L_n(\mathbb{Z}[\pi_1(M)])$$

so that for $n \geq 5$ $h : M' \longrightarrow M$ h-splits if and only if $\sigma_*(f,b) = 0 \in L_n(\mathbb{Z}[\pi_1(M)])$.

□

The lens spaces give rise to homotopy equivalences $h : M' \longrightarrow M$ of manifolds in dimensions ≥ 3 with $\tau(h) \neq 0 \in Wh(\pi_1(M))$.

The exotic spheres give rise to homotopy equivalences $h : M' \longrightarrow M$ of manifolds which do not split along submanifolds. The following example gives an explicit homotopy equivalence $h : M'^m \longrightarrow M^m$ which does not split along a special submanifold $N^{4k} \subset M^m$ in the simply-connected case $\pi_1(N) = \pi_1(M) = \{1\}$.

Example 3.3 Take $k = 2$ in 2.10, with

$$a_2 = 1 , \; B_2 = \frac{1}{30} , \; j_2 = 240 , \; s_2 = \frac{7}{45} , \; t_2 = 56 .$$

Let (W^8, Σ^7) be the framed 3-connected 8-dimensional differentiable manifold with signature $\sigma(W) = 8$ obtained by the E_8-plumbing of 8 copies of $\tau_{S^4} : S^4 \longrightarrow BO(4)$, with boundary $\partial W = \Sigma^7$ the homotopy 7-sphere generating the exotic sphere group $\Theta^7 = \mathbb{Z}_{28}$. The 28-fold connected sum $\#_{28}\Sigma^7$ is diffeomorphic to the standard 7-sphere S^7. Let $\eta : S^8 \longrightarrow BO(q+1)$ (q large) be a fibre homotopy trivial $(q+1)$-plane bundle over S^8 such that

$$\eta \in \ker(J : \pi_8(BO) \longrightarrow \pi_S^7) = 240\mathbb{Z} \subset \pi_8(BO) = \mathbb{Z}$$

is the generator with

$$p_2(\eta) = -1440 \in H^8(S^8) = \mathbb{Z} .$$

The sphere bundle is a closed $(8+q)$-dimensional differentiable manifold $M' = S(\eta)$ with a homotopy equivalence

$$h : M' = S(\eta) \longrightarrow M = S(\epsilon^{q+1}) = S^8 \times S^q$$

which does not split along the special submanifold

$$N^8 = S^8 \times \{\text{pt.}\} \subset M^{8+q} = S^8 \times S^q .$$

The inverse image of N is the special submanifold

$$N'^8 = h^{-1}(N) = \#_{28} W \cup D^8 \subset M'^{8+j}$$

with

$$\begin{aligned} \mathcal{L}_2(M') &= \sigma(N') = \frac{7}{45}\langle -p_2(\eta), [S^8]\rangle = 28 \cdot \sigma(W) = 224 \\ &\neq h^*\mathcal{L}_2(M) = \sigma(N) = 0 \in H^8(M';\mathbb{Q}) = \mathbb{Q} , \\ p_2(M') &= 1440 \neq h^*p_2(M) = 0 \in H^8(M') = \mathbb{Z} . \end{aligned}$$

The codimension q splitting obstruction of h along $N \subset M$ is the surgery obstruction of the 8-dimensional normal map

$$(f, b) = h| : N' \longrightarrow N ,$$

which is

$$\begin{aligned} [s(h)] &= \sigma_*(f,b) = \frac{1}{8}(\sigma(N') - \sigma(N)) \\ &= 28 \in LS_8 = L_8(\mathbb{Z}) = \mathbb{Z} . \end{aligned}$$

□

Example 3.4 If $m-4k \geq 3$ and $k \geq 2$ a homotopy equivalence $h : M' \longrightarrow M$ of simply-connected m-dimensional manifolds splits along a simply-connected $4k$-dimensional submanifold $N^{4k} \subset M$ if and only if the surgery obstruction

$$\sigma_*(f,b) = \frac{1}{8}(\sigma(N') - \sigma(N)) \in LS_{4k} = L_{4k}(\mathbb{Z}) = \mathbb{Z}$$

is 0, which for special $N \subset M$ is equivalent to

$$\langle (h^{-1})^*\mathcal{L}_k(M') - \mathcal{L}_k(M), [N]\rangle = 0 \in \mathbb{Q} .$$

□

Codimension 1 splitting obstruction theory is particularly significant for the topological invariance of the rational Pontrjagin classes and the Novikov conjectures. See §8 below for an account of the codimension 1 theory for homotopy equivalences of compact manifolds. In §10 there is a corresponding account for proper homotopy equivalences of open manifolds, making use of the evident modification of Definition 3.1 :

Definition 3.5 (i) A proper homotopy equivalence $h : W' \longrightarrow W$ of open manifolds **splits** along a closed submanifold $N \subset W$ if h is proper homotopic to a map (also denoted h) which is transverse regular along $N \subset W$, and such that the restriction $h| : N' = h^{-1}(N) \longrightarrow N$ is a homotopy equivalence.
(ii) A proper homotopy equivalence $h : W' \longrightarrow W$ of open manifolds h-**splits** along a closed submanifold $N \subset W$ if there exists an extension of $h : W' \longrightarrow W$ to a proper homotopy equivalence

$$(g; h, h') \ : \ (V; W', W'') \longrightarrow W \times ([0,1]; \{0\}, \{1\})$$

with $(V; W', W'')$ a proper h-cobordism and $h' : W'' \longrightarrow W$ split along $N \subset W$.

□

See Ranicki [44] for an algebraic development of the projective L-groups $L^p_*(\mathbb{Z}[\pi])$, which are related to the free L-groups $L_*(\mathbb{Z}[\pi])$ by a Rothenberg-type exact sequence

$$\begin{aligned} \dots \longrightarrow L_m(\mathbb{Z}[\pi]) \longrightarrow L^p_m(\mathbb{Z}[\pi]) \longrightarrow \widehat{H}^m(\mathbb{Z}_2; \widetilde{K}_0(\mathbb{Z}[\pi])) \\ \longrightarrow L_{m-1}(\mathbb{Z}[\pi]) \longrightarrow \dots . \end{aligned}$$

See Pedersen and Ranicki [42] for a geometric interpretation of projective L-theory in terms of normal maps from compact manifolds to finitely dominated geometric Poincaré complexes.

Proposition 3.6 *Let $h : W' \longrightarrow W = N \times \mathbb{R}$ be a proper homotopy equivalence of open m-dimensional manifolds, with N a closed $(m-1)$-dimensional manifold. Let*

$$(f, b) \ = \ h| \ : \ N' \ = \ h^{-1}(N \times \{0\}) \longrightarrow N$$

be the normal map of closed $(m-1)$-dimensional manifolds obtained by transversality, with

$$W'^{+} \ = \ h^{-1}(N \times \mathbb{R}^{+}) \ , \ W'^{-} \ = \ h^{-1}(N \times \mathbb{R}^{-}) \subset W'$$

such that

$$h \ = \ h^{+} \cup_f h^{-} \ : \ W' \ = \ W'^{+} \cup_{N'} W'^{-} \longrightarrow W \ = \ (N \times \mathbb{R}^{+}) \cup_{N \times \{0\}} (N \times \mathbb{R}^{-})$$

and

$$\pi_1(N) \ = \ \pi_1(N') \ = \ \pi_1(W'^{+}) \ = \ \pi_1(W'^{-}) \ (= \ \pi \ say) \ .$$

(i) *The spaces W'^{+}, W'^{-} are finitely dominated, and the Wall finiteness obstruction*

$$[W'^{+}] \ = \ -[W'^{-}] \ = \ (-)^m [W'^{+}]^* \in \widetilde{K}_0(\mathbb{Z}[\pi])$$

is the splitting obstruction, such that $[W'^+] = 0$ *if (and for* $m \geq 6$*) only if* h *splits along* $N \times \{0\} \subset W = N \times \mathbb{R}$.
(ii) *The Tate* $\mathbb{Z}_2$*-cohomology class*

$$[W'^+] \in \widehat{H}^m(\mathbb{Z}_2; \widetilde{K}_0(\mathbb{Z}[\pi]))$$

is the proper h-splitting obstruction, such that $[W'^+] = 0$ *if (and for* $m \geq 6$*) only if* $h : W' \longrightarrow W$ *h-splits along* $N \times \{0\} \subset W = N \times \mathbb{R}$.
(iii) *The surgery obstruction of* (f, b) *is the image of the Tate* $\mathbb{Z}_2$*-cohomology class of* $[W'^+]$

$$\begin{aligned} \sigma_*(f,b) &= [W'^+] \in \mathrm{im}(\widehat{H}^m(\mathbb{Z}_2; \widetilde{K}_0(\mathbb{Z}[\pi])) \longrightarrow L_{m-1}(\mathbb{Z}[\pi])) \\ &= \ker(L_{m-1}(\mathbb{Z}[\pi]) \longrightarrow L^p_{m-1}(\mathbb{Z}[\pi])) \ . \end{aligned}$$

In particular, the projective surgery obstruction of (f, b) *is*

$$\sigma^p_*(f,b) = 0 \in L^p_{m-1}(\mathbb{Z}[\pi]) \ .$$

Proof (i)+(ii) The finiteness obstruction for arbitrary $\pi_1(N)$ is just the end invariant of Siebenmann [55], and is the obstruction to killing $\pi_*(W'^+, N')$ by handle exchanges (= ambient surgeries).
(iii) Let $\widetilde{W}'^+, \widetilde{W}'^-, \widetilde{N}, \widetilde{N}'$ be the universal covers of W'^+, W'^-, N, N' respectively. The homology $\mathbb{Z}[\pi]$-modules are such that

$$H_*(\widetilde{N}') = H_*(\widetilde{N}) \oplus H_{*+1}(\widetilde{W}'^+, \widetilde{N}') \oplus H_{*+1}(\widetilde{W}'^-, \widetilde{N}')$$

and the quadratic Poincaré kernel of (f, b) (Ranicki [47]) is the hyperbolic $(m-1)$-dimensional quadratic Poincaré complex on

$$C(\widetilde{W}'^+, \widetilde{N}')_{*+1} \oplus C(\widetilde{W}'^-, \widetilde{N}')_{*+1} \simeq C(\widetilde{W}'^+, \widetilde{N}')_{*+1} \oplus C(\widetilde{W}'^+, \widetilde{N}')^{m-*}$$

which is equipped with a projective null-cobordism.

□

Remarks 3.7 (i) 3.6 (i) is a special case of the codimension 1 bounded splitting Theorem 10.1. The unobstructed case $\pi = \{1\}$ is the splitting result of Browder [6].
(ii) The projective L-groups are such that

$$L_m(\mathbb{Z}[\pi][z, z^{-1}]) = L_m(\mathbb{Z}[\pi]) \oplus L^p_{m-1}(\mathbb{Z}[\pi])$$

with

$$\begin{aligned} \sigma_*((f,b) \times 1_{S^1}) &= (0, \sigma^p_*(f,b)) \\ &\in L_m(\mathbb{Z}[\pi][z, z^{-1}]) = L_m(\mathbb{Z}[\pi]) \oplus L^p_{m-1}(\mathbb{Z}[\pi]) \end{aligned}$$

for any normal map (f, b) of finitely dominated $(m-1)$-dimensional geometric Poincaré complexes with fundamental group π (Ranicki [45]).
(iii) The vanishing of the projective surgery obstruction $\sigma_*^p(f, b) = 0$ in 3.6 (iii) corresponds to the vanishing of the free surgery obstruction

$$\sigma_*((f, b) \times 1_{S^1}) \ = \ 0 \in L_m(\mathbb{Z}[\pi][z, z^{-1}]) \ .$$

For $m \geq 5$ this is realized by the geometric wrapping up construction (Hughes and Ranicki [25]) of an $(m+1)$-dimensional normal bordism

$$(F, B) \ : \ (L; N' \times S^1, \partial_+ L) \longrightarrow N \times S^1 \times ([0,1]; \{0\}, \{1\})$$

with $\partial_+ F = F| : \partial_+ L \longrightarrow N \times S^1$ a homotopy equivalence and

$$\begin{aligned}(F, B)| \ &= \ (f, b) \times 1_{S^1} \ : \ N' \times S^1 \longrightarrow N \times S^1 \ , \\ (L \backslash \partial_+ L, N' \times S^1) \ &= \ (W'^+, N') \times S^1 \ , \\ \tau(\partial_+ F) \ &= \ [W'^+] \in \mathrm{im}(\widetilde{K}_0(\mathbb{Z}[\pi]) \longrightarrow Wh(\pi \times \mathbb{Z}))\end{aligned}$$

with

$$\widetilde{K}_0(\mathbb{Z}[\pi]) \longrightarrow Wh(\pi \times \mathbb{Z}) \ ; \ [P] \longrightarrow \tau(-z : P[z, z^{-1}] \longrightarrow P[z, z^{-1}])$$

the geometrically significant variant of the injection of Bass [3, XII]. The infinite cyclic covering of (F, B) induced from the universal covering $\mathbb{R} \longrightarrow S^1$

$$(\overline{F}, \overline{B}) \ : \ (\overline{L}; N' \times \mathbb{R}, \overline{\partial_+ L}) \longrightarrow N \times \mathbb{R} \times ([0,1]; \{0\}, \{1\})$$

is homotopy equivalent to an extension of (f, b) to a finitely dominated m-dimensional geometric Poincaré bordism

$$(F_1, B_1) \ : \ (W'^+; N', N) \longrightarrow N \times ([0,1]; \{0\}, \{1\})$$

with $(F_1, B_1)| = 1 : N \longrightarrow N$.
(iv) By the codimension 1 splitting theorem of Farrell and Hsiang [15] (8.1) the Whitehead torsion $\tau(h) \in Wh(\pi \times \mathbb{Z})$ of a homotopy equivalence $h : M' \longrightarrow M = N \times S^1$ of closed m-dimensional manifolds is such that

$$\tau(h) \in \mathrm{im}(Wh(\pi) \longrightarrow Wh(\pi \times \mathbb{Z})) \ \ (\pi \ = \ \pi_1(N))$$

if (and for $m \geq 6$ only if) h splits along $N \times \{*\} \subset N \times S^1$. The projection $Wh(\pi \times \mathbb{Z}) \longrightarrow \widetilde{K}_0(\mathbb{Z}[\pi])$ of Bass [3, XII] sends the Whitehead torsion $\tau(h) \in Wh(\pi \times \mathbb{Z})$ to the splitting obstruction of 3.6

$$[\tau(h)] \ = \ [\overline{M}'^+] \in \widetilde{K}_0(\mathbb{Z}[\pi])$$

for the proper homotopy equivalence $\overline{h} : \overline{M}' \longrightarrow \overline{M} = N \times \mathbb{R}$ obtained from h by pullback from the universal cover $\mathbb{R} \longrightarrow S^1$. The h-splitting obstruction of $h : M' \longrightarrow M$ is the Tate $\mathbb{Z}_2$-cohomology class

$$[\tau(h)] \;=\; [\overline{M}'^{+}] \in LS_{m-1} \;=\; \widehat{H}^m(\mathbb{Z}_2; \widetilde{K}_0(\mathbb{Z}[\pi]))\ .$$

(The identification $LS_{m-1} = \widehat{H}^m(\mathbb{Z}_2; \widetilde{K}_0(\mathbb{Z}[\pi]))$ is the h-version of the identification $LS^s_{m-1} = \widehat{H}^m(\mathbb{Z}_2; Wh(\pi))$ obtained by Wall [58, Thm. 12.5] for the corresponding codimension 1 s-splitting obstruction group).

□

§4. Topological invariance

A homeomorphism of differentiable manifolds cannot in general be approximated by a diffeomorphism, by virtue of the exotic spheres. Thus it is not at all obvious that the $\mathcal{L}$-genus $\mathcal{L}(M)$ and the rational Pontrjagin classes $p_*(M)$ are topological invariants of a differentiable manifold M. Surgery theory for simply-connected compact manifolds is adequate for the construction and classification of exotic spheres. The topological invariance of the rational Pontrjagin classes requires surgery on non-simply-connected compact manifolds and/or simply-connected non-compact manifolds. The original proof due to Novikov [25] made use of the torus, as subsequently formalized in Ranicki [39, Appendix C16] using bounded L-theory: if $h : M' \longrightarrow M$ is a homeomorphism of manifolds then for any $j \geq 1$ the homeomorphism $h \times 1 : M' \times \mathbb{R}^j \longrightarrow M \times \mathbb{R}^j$ can be approximated by a differentiable $\mathbb{R}^j$-bounded homotopy equivalence, and the signatures of special submanifolds are $\mathbb{R}^j$-bounded homotopy invariants. (See §9 for a brief account of bounded surgery theory). Recently, Gromov [22] obtained a new proof of the topological invariance using the non-multiplicativity of the signature on surface bundles instead of torus geometry and the algebraic K- and L-theory of the group rings of the free abelian groups. I am grateful to Gromov for sending me a copy of [22]. In 4.1 the two methods are related to each other using algebraic surgery theory.

Theorem 4.1 (Novikov [38]) *The rational Pontrjagin classes and the $\mathcal{L}$-genus are topological invariants.*

Proof By 2.6 it suffices to prove that the signatures of special submanifolds are homeomorphism invariant, i.e. that if $h : M'^m \longrightarrow M^m$ is a homeomorphism of differentiable (or PL) manifolds then

$$\sigma(N) \;=\; \sigma(N') \in L^{4k}(\mathbb{Z}) \;=\; \mathbb{Z}$$

for any special submanifold $N^{4k} \subset M^m \times \mathbb{R}^j$, with

$$N' \;=\; h'^{-1}(N) \subset M' \times \mathbb{R}^j$$

the transverse inverse image of any differentiable (or PL) approximation $h' : M' \times \mathbb{R}^j \longrightarrow M \times \mathbb{R}^j$ to $h \times 1_{\mathbb{R}^j}$. Every special submanifold is (ambient) cobordant to a simply-connected one, so it may be assumed that N is simply-connected, $\pi_1(N) = \{1\}$. The surgery obstruction of the $4k$-dimensional normal map

$$(f, b) \;=\; h'| \;:\; N' \longrightarrow N$$

is

$$\sigma_*(f, b) \;=\; \frac{1}{8}(\sigma(N') - \sigma(N)) \in L_{4k}(\mathbb{Z}) \;=\; \mathbb{Z}$$

which is a codimension i splitting obstruction, with $i = m + j - 4k$. There are at least four distinct ways of showing that $\sigma_*(f, b) = 0$:

1. use $\mathbb{R}^i$-bounded L-theory as in Ranicki [50], [51], and the computation $L_*(\mathbb{C}_{\mathbb{R}^i}(\mathbb{Z})) = L_{*-i}(\mathbb{Z})$,
2. as in the original proof of Novikov [38] use $T^{i-1} \subset \mathbb{R}^i$ and the computation $\widetilde{K}_0(\mathbb{Z}[\mathbb{Z}^{i-1}]) = 0$,
3. use $T^{i-1} \subset \mathbb{R}^i$ and the computation $L_*(\mathbb{Z}[\mathbb{Z}^{i-1}]) = H_*(T^{i-1}; \mathbb{L}_\bullet(\mathbb{Z}))$ of Novikov [39] and Ranicki [45]*,
4. follow Gromov [22] and use a hypersurface $B^{i-1} \subset \mathbb{R}^i$ (assuming i is odd) with a fibre bundle $F^{i-1} \longrightarrow E \longrightarrow B^{i-1}$ such that the total space has signature $\sigma(E) \neq 0$.

For 1. note that the homeomorphism

$$h_0 \;=\; (h \times 1_{\mathbb{R}^j})| \;:\; (h \times 1_{\mathbb{R}^j})^{-1}(N \times \mathbb{R}^i) \longrightarrow N \times \mathbb{R}^i$$

can be approximated by a differentiable $\mathbb{R}^i$-bounded homotopy equivalence which is normal bordant to the $\mathbb{R}^i$-bounded normal map

$$(f, b) \times 1_{\mathbb{R}^i} \;:\; N' \times \mathbb{R}^i \longrightarrow N \times \mathbb{R}^i\,,$$

so that

$$\begin{aligned}\sigma_*(f, b) \;=\; \sigma_*((f, b) \times 1_{\mathbb{R}^i}) \;&=\; \sigma_*(h_0) \;=\; 0 \\ &\in L_{4k}(\mathbb{Z}) \;=\; L_{4k+i}(\mathbb{C}_{\mathbb{R}^i}(\mathbb{Z})) \;=\; \mathbb{Z}\end{aligned}$$

---

* The full L-theoretic computation $L_*(\mathbb{Z}[\mathbb{Z}^{i-1}]) = H_*(T^{i-1}; \mathbb{L}_\bullet(\mathbb{Z}))$ requires the K-theoretic computation $\widetilde{K}_0(\mathbb{Z}[\mathbb{Z}^{i-1}]) = 0$, but for the topological invariance of the rational Pontrjagin classes it suffices to know that the map $-\otimes\sigma^*(T^{i-1}) : L_{4*}(\mathbb{Z}) = \mathbb{Z} \longrightarrow L_{4*+i-1}(\mathbb{Z}[\mathbb{Z}^{i-1}])$ is a rational injection – this is a formal consequence of the splitting theorem $L^h_*(A[z, z^{-1}]) = L^h_*(A) \oplus L^p_{*-1}(A)$ of [39] and [45] applied inductively to $\mathbb{Z}[\mathbb{Z}^{i-1}] = \mathbb{Z}[\mathbb{Z}^{i-2}][z, z^{-1}]$, and the identity $L^h_*(A)[1/2] = L^p_*(A)[1/2]$ given by the Rothenberg-type sequence relating the projective and free L-groups of a ring with involution A.

– see 9.14 for a (somewhat) more detailed account.
For 2. proceed as in [38], making repeated use of codimension 1 splitting (3.6). To start with, approximate the homeomorphism

$$h_1 = (h \times 1_{\mathbb{R}^j})| : W_1 = (h \times 1_{\mathbb{R}^j})^{-1}(N \times T^{i-1} \times \mathbb{R}) \longrightarrow N \times T^{i-1} \times \mathbb{R}$$

by a differentiable proper homotopy equivalence $h'_1 : W_1 \longrightarrow N \times T^{i-1} \times \mathbb{R}$. Since $\widetilde{K}_0(\mathbb{Z}[\mathbb{Z}^{i-1}]) = 0$ it is possible to split h'_1 along $N \times T^{i-1} \times \{0\} \subset N \times T^{i-1} \times \mathbb{R}$, with the restriction

$$f_1 = h'_1| : N_1 = h'^{-1}_1(N \times T^{i-1} \times \{0\}) \longrightarrow N \times T^{i-1}$$

a homotopy equivalence normal bordant to

$$(f,b) \times 1_{T^{i-1}} : N' \times T^{i-1} \longrightarrow N \times T^{i-1} .$$

Pass to the infinite cyclic cover $\overline{T}^{i-1} = T^{i-2} \times \mathbb{R}$ of $T^{i-1} = T^{i-2} \times S^1$ and apply the same procedure to the proper homotopy equivalence

$$h_2 = \overline{f}_1 : W_2 = \overline{N}_1 \longrightarrow N \times \overline{T}^{i-1} = N \times T^{i-2} \times \mathbb{R} .$$

After $i-1$ applications of 3.6 there is obtained a homotopy equivalence of $4k$-dimensional manifolds $f_i : N_i \longrightarrow N$ normal bordant to $(f,b) : N' \longrightarrow N$, so that

$$\sigma_*(f,b) = \sigma_*(f_i) = 0 \in L_{4k}(\mathbb{Z}) .$$

(Alternatively, apply the splitting theorem of Farrell and Hsiang [15] i times – cf. 3.7 (iii)).
For 3. and 4. suppose given a closed hypersurface $U^{i-1} \subset \mathbb{R}^i$ with a neighbourhood $U \times \mathbb{R} \subset \mathbb{R}^i$, regard $N \times U \times \mathbb{R}$ as a codimension 0 submanifold of $M \times \mathbb{R}^j$ by

$$N \times U \times \mathbb{R} \subset N \times \mathbb{R}^i \subset M \times \mathbb{R}^j ,$$

and define the codimension 0 submanifold of $M' \times \mathbb{R}^j$

$$W^{4k+i} = (h \times 1_{\mathbb{R}^j})^{-1}(N \times U \times \mathbb{R}) \subset M' \times \mathbb{R}^j .$$

The restriction

$$(h \times 1_{\mathbb{R}^j})| : W \longrightarrow N \times U \times \mathbb{R}$$

is a homeomorphism. Let

$$V^{4k+i-1} = h''^{-1}(N \times U \times \{0\}) \subset W^{4k+i}$$

be the codimension 1 transverse inverse image of any differentiable (or PL) approximation $h'' : W \longrightarrow N \times U \times \mathbb{R}$ to $(h \times 1_{\mathbb{R}^j})|$. The $(4k+i-1)$-dimensional normal map

$$(g,c) = h''| : V \longrightarrow N \times U$$

is normal bordant to $(f,b)\times 1_U : N'\times U \longrightarrow N\times U$. The surgery obstruction of (g,c) is thus given by the surgery product formula of Ranicki [47]

$$\begin{aligned}\sigma_*(g,c) &= \sigma_*((f,b)\times 1_U) = \sigma_*(f,b)\otimes\sigma^*(U) \\ &\in \mathrm{im}(L_{4k}(\mathbb{Z})\otimes L^{i-1}(\mathbb{Z}[\pi_1(U)])\longrightarrow L_{4k+i-1}(\mathbb{Z}[\pi_1(U)]))\ ,\end{aligned}$$

with $\sigma^*(U)\in L^{i-1}(\mathbb{Z}[\pi_1(U)])$ the symmetric signature of U – see §6 for a brief account of the symmetric signature. Also, by 3.6 (ii)

$$\begin{aligned}\sigma_*(g,c) &= [W^+]\in \mathrm{im}(\widehat{H}^{4k+i}(\mathbb{Z}_2;\widetilde{K}_0(\mathbb{Z}[\pi_1(U)]))\longrightarrow L_{4k+i-1}(\mathbb{Z}[\pi_1(U)])) \\ &= \ker(L_{4k+i-1}(\mathbb{Z}[\pi_1(U)])\longrightarrow L^p_{4k+i-1}(\mathbb{Z}[\pi_1(U)]))\ .\end{aligned}$$

For 3. take $U=T^{i-1}\subset\mathbb{R}^i$. It follows from $\widetilde{K}_0(\mathbb{Z}[\mathbb{Z}^{i-1}])=0$ that $\sigma_*(g,c)=0$. The map

$$-\otimes\sigma^*(T^{i-1})\ :\ L_{4k}(\mathbb{Z})\longrightarrow L_{4k+i-1}(\mathbb{Z}[\mathbb{Z}^{i-1}])$$

is a (split) injection which sends $\sigma_*(f,b)$ to

$$\sigma_*(f,b)\otimes\sigma^*(T^{i-1}) = \sigma_*(g,c) = 0\in L_{4k+i-1}(\mathbb{Z}[\mathbb{Z}^{i-1}])$$

so that $\sigma_*(f,b)=0$.
For 4. assume that i is odd, say $i=2n+1$, and let $U=B^{2n}\subset\mathbb{R}^{2n+1}$ be a hypersurface for which there exists a fibre bundle

$$F^{2n}\longrightarrow E^{4n}\xrightarrow{\ p\ } B^{2n}$$

such that the total space E is a $4n$-dimensional manifold with signature

$$\sigma(E)\neq 0\in L^{4n}(\mathbb{Z}) = \mathbb{Z}\ .$$

(Any such B bounds a simply-connected manifold, so that the simply-connected symmetric signature is $\sigma(B)=0\in L^{2n}(\mathbb{Z})$, but $\sigma^*(B)\neq 0\in L^{2n}(\mathbb{Z}[\pi_1(B)])$.) For example, take the n-fold cartesian product

$$F = F_1^{(n)}\longrightarrow E = E_1^{(n)}\xrightarrow{p=p_1^{(n)}} B = B_1^{(n)}$$

of one of the surface bundles over a surface

$$F_1^2\longrightarrow E_1^4\xrightarrow{\ p_1\ } B_1^2$$

with $\sigma(E_1)\neq 0\in L^4(\mathbb{Z})=\mathbb{Z}$ constructed by Atiyah [2], using an embedding $B_1\times\mathbb{R}\subset\mathbb{R}^3$ to define an embedding

$$B_1^{(n)} = \prod_1^n B_1\subset\mathbb{R}^{2n+1}$$

by

$$B_1^{(n)} \subset B_1^{(n-1)} \times \mathbb{R}^3 = B_1^{(n-2)} \times (B_1 \times \mathbb{R}) \times \mathbb{R}^2 \subset B_1^{(n-2)} \times \mathbb{R}^5 \subset \ldots \subset \mathbb{R}^{2n+1}.$$

Let

$$F^{2n} \longrightarrow Q^{4k+4n+1} \longrightarrow W^{4k+2n+1}$$

be the induced fibre bundle over W. The algebraic surgery transfer map induced by p

$$p^! : L_{4k+2n}(\mathbb{Z}[\pi_1(B)]) \longrightarrow L_{4k+4n}(\mathbb{Z}[\pi_1(E)])$$

sends $\sigma_*(g,c)$ to

$$\begin{aligned} p^!\sigma_*(g,c) &= [Q^+] \\ &\in \operatorname{im}(\widehat{H}^{4k+4n+1}(\mathbb{Z}_2; \widetilde{K}_0(\mathbb{Z}[\pi_1(E)])) \longrightarrow L_{4k+4n}(\mathbb{Z}[\pi_1(E)])) \\ &= \ker(L_{4k+4n}(\mathbb{Z}[\pi_1(E)]) \longrightarrow L^p_{4k+4n}(\mathbb{Z}[\pi_1(E)])) \end{aligned}$$

with signature

$$\sigma_*(f,b)\sigma(E) = 0 \in L_{4k+4n}(\mathbb{Z}) = \mathbb{Z}$$

(Lück and Ranicki [29]), so that $\sigma_*(f,b) = 0$.

□

Remarks 4.2 (i) Novikov's proof of the topological invariance of the rational Pontrjagin classes was in the differentiable category, but it applies equally well in the PL category. In fact, the proof led to the disproof of the manifold Hauptvermutung by Casson and Sullivan — see Ranicki [52]. A homeomorphism of PL manifolds cannot in general be approximated by a PL homeomorphism. The proof also led to the subsequent development by Kirby and Siebenmann [28] of the classification theory of high-dimensional topological manifolds. It is now possible to define the $\mathcal{L}$-genus and the rational Pontrjagin classes for a topological manifold, and the Hirzebruch signature theorem $\sigma(N) = \langle \mathcal{L}(N), [N] \rangle$ also holds for topological manifolds N^{4k}.

(ii) Let W be an open $(4k+1)$-dimensional manifold with a proper map $g : W \longrightarrow \mathbb{R}$ transverse regular at $0 \in \mathbb{R}$, so that

$$V^{4k} = g^{-1}(0) \subset W^{4k+1}$$

is a closed $4k$-dimensional submanifold. Novikov [37] defined the signature of (W,g) by

$$\sigma(W,g) = \text{signature}(H^{2k}(W)/H^{2k}(W)^{\perp}, [\phi]) \in \mathbb{Z}$$

with

$$\begin{aligned}\phi\ :\ H^{2k}(W)\times H^{2k}(W) &\longrightarrow \mathbb{Z}\ ;\ (x,y)\longrightarrow \langle x\cup y,[V]\rangle\ ,\\ H^{2k}(W)^{\perp} &= \{x\in H^{2k}(W)\,|\,\phi(x,y)=0 \text{ for all } y\in H^{2k}(W)\}\ ,\end{aligned}$$

and identified

$$\sigma(W,g)\ =\ \sigma(V)\in L^{4k}(\mathbb{Z})\ =\ \mathbb{Z}\ ,$$

thus proving that $\sigma(V)$ is a proper homotopy invariant of (W,g). (In the context of the bounded L-theory of Ranicki [50] this is immediate from the computation $L^{4k+1}(\mathbb{C}_{\mathbb{R}}(\mathbb{Z}))=L^{4k}(\mathbb{Z})$.) This signature invariant was used in [37] to prove that $\mathcal{L}_k(M)\in H^{4k}(M;\mathbb{Q})$ is a homotopy invariant for any closed $(4k+1)$-dimensional manifold M, as follows. $\mathcal{L}_k(M)$ is detected by the signatures of special $4k$-dimensional submanifolds $N^{4k}\subset M^{4k+1}\times\mathbb{R}^j$ with

$$\begin{aligned}\sigma(N)\ &=\ \langle\mathcal{L}(N),[N]\rangle\\ &=\ \langle\mathcal{L}_k(M),[N]\rangle\in L^{4k}(\mathbb{Z})\ =\ \mathbb{Z}\ .\end{aligned}$$

The Poincaré dual $[N]^*\in H^1(M)$ of $[N]\in H_{4k}(M)$ is represented by a map $f:M\longrightarrow S^1$ with a lift to a proper map $\overline{f}:\overline{M}\longrightarrow\mathbb{R}$ such that the transverse inverse image

$$V_N^{4k}\ =\ f^{-1}(1)\subset M$$

is diffeomorphic to $\overline{f}^{-1}(0)\subset\overline{M}$ and cobordant to N, so that

$$\sigma(N)\ =\ \sigma(V_N)\ =\ \sigma(\overline{M},\overline{f})\in L^{4k}(\mathbb{Z})\ =\ \mathbb{Z}$$

is a homotopy invariant of (M,f). A homotopy equivalence $h:M'\longrightarrow M$ induces a proper homotopy equivalence $\overline{h}:\overline{M}'\longrightarrow\overline{M}$, so that

$$\mathcal{L}_k(M')\ =\ h^*\mathcal{L}_k(M)\in H^{4k}(M';\mathbb{Q})\ .$$

For any map $f:M\longrightarrow S^1$ with transverse inverse image

$$V^{4k}\ =\ f^{-1}(1)\subset M^{4k+1}$$

the 'higher signature' of (M,f)

$$\begin{aligned}f_*(\mathcal{L}_k(M)\cap[M])\ &=\ \langle\mathcal{L}(V),[V]\rangle\\ &=\ \sigma(V)\in H_1(S^1)\ =\ \mathbb{Z}\ \subset\ H_1(S^1;\mathbb{Q})\ =\ \mathbb{Q}\end{aligned}$$

is thus a homotopy invariant of (M,f), verifying the Novikov conjecture for $\pi=\mathbb{Z}$ (5.2). The proof of topological invariance of the rational Pontrjagin classes in Novikov [38] grew out of this, leading on to the formulation of the

general conjecture and the verification for free abelian π in Novikov [39].
(iii) Gromov's proof of topological invariance does not use surgery theory: the actual method of [22] extends the symmetric form defined by Novikov [37] for open $(4k+1)$-dimensional manifolds to the context of cohomology with coefficients in a flat hermitian bundle (as used by Lusztig [30] and Meyer [31]).
(iv) The topological invariance of Whitehead torsion (originally proved by Chapman) was proved in Ranicki and Yamasaki [54] using controlled K-theory. The parallel development of controlled L-theory will give yet another proof of the topological invariance of the rational Pontrjagin classes.

□

§5. Homotopy invariance

Definition 5.1 The **higher $\mathcal{L}$-genus** of an m-dimensional manifold M with fundamental group $\pi_1(M) = \pi$ is

$$\mathcal{L}_\pi(M) \;=\; f_*(\mathcal{L}(M) \cap [M]) \in H_{m-4*}(B\pi; \mathbb{Q}) \ ,$$

with $f : M \longrightarrow B\pi$ classifying the universal cover $\widetilde{M}$ and $\mathcal{L}(M) \cap [M] \in H_{m-4*}(M; \mathbb{Q})$ the Poincaré dual of the $\mathcal{L}$-genus $\mathcal{L}(M) \in H^{4*}(M; \mathbb{Q})$.

□

Conjecture 5.2 (Novikov [39, §11]) *The higher $\mathcal{L}$-genus is a homotopy invariant: if $h : M'^m \longrightarrow M^m$ is a homotopy equivalence of m-dimensional manifolds then*

$$\mathcal{L}_\pi(M) \;=\; \mathcal{L}_\pi(M') \in H_{m-4*}(B\pi; \mathbb{Q}) \ .$$

□

Definition 5.3 A submanifold $N^{4k} \subset M^m \times \mathbb{R}^j$ is **π-special** if it is special and the Poincaré dual $[N]^* \in H^{m-4k}(M; \mathbb{Q})$ of $[N] \in H_{4k}(M; \mathbb{Q})$ is such that

$$[N]^* \in \mathrm{im}(f^* : H^{m-4k}(B\pi; \mathbb{Q}) \longrightarrow H^{m-4k}(M; \mathbb{Q})) \ .$$

The **higher signatures** of M are the signatures $\sigma(N) \in \mathbb{Z}$ of the π-special manifolds $N \subset M \times \mathbb{R}^j$.

□

Remarks 5.4 (i) The higher $\mathcal{L}$-genus of an m-dimensional manifold M with $\pi_1(M) = \pi$ is detected by the higher signatures. As before, let $f : M \longrightarrow B\pi$ classify the universal cover $\widetilde{M}$ of M. The $\mathbb{Q}$-vector space $H^{m-4k}(B\pi; \mathbb{Q})$ is spanned by the elements of type $x = e^*(1)$ for a proper map

$$e \;:\; B\pi \times \mathbb{R}^j \longrightarrow \mathbb{R}^i \quad (i = m + j - 4k)$$

with large j. (It is convenient to assume here that $B\pi$ is compact). For any such x, e the composite

$$e(f\times 1)\ :\ M\times\mathbb{R}^j \xrightarrow{f\times 1} B\pi\times\mathbb{R}^j \xrightarrow{e} \mathbb{R}^i$$

can be made transverse regular at $0\in\mathbb{R}^i$, with

$$N^{4k} \ =\ (e(f\times 1))^{-1}(0)\subset M\times\mathbb{R}^j$$

a π-special submanifold. The higher $\mathcal{L}$-genus of M is such that

$$\begin{aligned}&\mathcal{L}_\pi(M)\ :\ H^{m-4k}(B\pi;\mathbb{Q})\longrightarrow\mathbb{Q}\ ;\\ &x\longrightarrow\langle x,\mathcal{L}_\pi(M)\rangle\ =\ \langle\mathcal{L}(M)\cup f^*(x),[M]\rangle\ =\ \langle\mathcal{L}(N),[N]\rangle\ =\ \sigma(N)\ .\end{aligned}$$

(ii) The Novikov conjecture is equivalent to the homotopy invariance of the higher signatures: if $h: M'^m\longrightarrow M^m$ is a homotopy equivalence then

$$\sigma(N)\ =\ \sigma(N')\in\mathbb{Z}$$

for any π-special submanifold $N^{4k}\subset M^m\times\mathbb{R}^j$, with the inverse image

$$N'\ =\ (h\times 1)^{-1}(N)\subset M'\times\mathbb{R}^j$$

also a π-special submanifold.

□

See Chapter 24 of Ranicki [51] for a more detailed account of the higher signatures.

§6. Cobordism invariance

Very early on in the history of the Novikov conjecture (essentially already in [39]) it was recognized that the conjecture is equivalent to the algebraic Poincaré cobordism invariance of the higher $\mathcal{L}$-genus, and also to the injectivity of the rational assembly map $A_\pi: H_{m-4*}(B\pi;\mathbb{Q})\longrightarrow L_m(\mathbb{Z}[\pi])\otimes\mathbb{Q}$.

See Ranicki [47] for the **symmetric** (resp. **quadratic**) **L-groups** $L^m(R)$ (resp. $L_m(R)$) of a ring with involution R, which are the cobordism groups of m-dimensional symmetric (resp. quadratic) Poincaré complexes (C,ϕ) consisting of an m-dimensional f.g. free R-module chain complex C with a symmetric (resp. quadratic) Poincaré duality $\phi: C^{m-*}\simeq C$. The symmetrization maps $1+T: L_m(R)\longrightarrow L^m(R)$ are isomorphisms modulo 8-torsion. The quadratic L-groups $L_*(R)$ are the Wall surgery obstruction groups, and depend only on $m(\bmod 4)$. The symmetric L-groups are not 4-periodic in general. The L-groups of $\mathbb{Z}$ are given by

$$L^m(\mathbb{Z}) = \begin{cases} \mathbb{Z} & \text{if } m \equiv 0(\bmod 4) \\ \mathbb{Z}_2 & \text{if } m \equiv 1(\bmod 4) \\ 0 & \text{otherwise} \end{cases}, \quad L_m(\mathbb{Z}) = \begin{cases} \mathbb{Z} & \text{if } m \equiv 0(\bmod 4) \\ \mathbb{Z}_2 & \text{if } m \equiv 2(\bmod 4) \\ 0 & \text{otherwise} . \end{cases}$$

The symmetric L-groups $L^*(R)$ and the symmetric signature were introduced by Mishchenko [35].

Definition 6.1 The **symmetric signature** of an m-dimensional geometric Poincaré complex X with universal cover $\widetilde{X}$ is the symmetric Poincaré cobordism class

$$\sigma^*(X) = (C(\widetilde{X}), \phi) \in L^m(\mathbb{Z}[\pi_1(X)])$$

with ϕ the m-dimensional symmetric structure of the Poincaré duality chain equivalence $[X] \cap - : C(\widetilde{X})^{m-*} \longrightarrow C(\widetilde{X})$.

□

The standard algebraic mapping cylinder argument shows:

Proposition 6.2 *The symmetric signature is both a cobordism and a homotopy invariant of a geometric Poincaré complex.*

□

The symmetric signature is a non-simply-connected generalization of the signature; for $m = 4k$ the natural map $L^m(\mathbb{Z}[\pi_1(X)]) \longrightarrow L^m(\mathbb{Z}) = \mathbb{Z}$ sends $\sigma^*(X)$ to the signature $\sigma(X)$.

Definition 6.3 The **quadratic signature** of a normal map of m-dimensional manifolds with boundary $(f, b) : (M', \partial M') \longrightarrow (M, \partial M)$ and with $\partial f : \partial M' \longrightarrow \partial M$ a homotopy equivalence is the cobordism class of the quadratic Poincaré complex kernel

$$\sigma_*(f, b) = (C(f^!), \psi) \in L_m(\mathbb{Z}[\pi_1(M)]) ,$$

with ψ the quadratic structure on the algebraic mapping cone $C(f^!)$ of the Umkehr $\mathbb{Z}[\pi_1(M)]$-module chain map

$$f^! : C(\widetilde{M}) \simeq C(\widetilde{M}, \partial\widetilde{M})^{m-*} \xrightarrow{\widetilde{f}^*} C(\widetilde{M}', \partial\widetilde{M}')^{m-*} \simeq C(\widetilde{M}') .$$

□

Proposition 6.4 (Ranicki [47])
(i) *The quadratic signature $\sigma_*(f, b) \in L_m(\mathbb{Z}[\pi_1(M)])$ of an m-dimensional*

normal map $(f,b) : (M', \partial M') \longrightarrow (M, \partial M)$ *is the surgery obstruction of Wall [58], such that* $\sigma_*(f,b) = 0$ *if (and for* $m \geq 5$ *only if)* (f,b) *is normal bordant to a homotopy equivalence.*
(ii) *The symmetrization of the quadratic signature is the symmetric signature of Mishchenko [35]*

$$(1+T)\sigma_*(f,b) \;=\; \sigma^*(M' \cup_{\partial f} -M) \in L^m(\mathbb{Z}[\pi_1(M)])$$

where $-M$ *refers to* M *with the opposite orientation* $[-M] = -[M]$.

□

The rational surgery obstruction of a normal map $(f,b) : M' \longrightarrow M$ of closed m-dimensional manifolds with fundamental group π

$$\sigma_*(f,b) \otimes \mathbb{Q} \in L_m(\mathbb{Z}[\pi]) \otimes \mathbb{Q}$$

depends only on the difference of the higher $\mathcal{L}$-genera

$$\mathcal{L}_\pi(M') - \mathcal{L}_\pi(M) \in H_{m-4*}(B\pi;\mathbb{Q}) \ .$$

For any finitely presented group π the $\mathbb{Q}$-vector space $H_{m-4*}(B\pi;\mathbb{Q})$ is spanned by the differences $\mathcal{L}_\pi(M') - \mathcal{L}_\pi(M)$ for normal maps $(f,b) : M' \longrightarrow M$ of closed m-dimensional manifolds with fundamental group π.

Definition 6.5 The **rational assembly map** in quadratic L-theory is

$$\begin{aligned} A_\pi \;&: H_{m-4*}(B\pi;\mathbb{Q}) \longrightarrow L_m(\mathbb{Z}[\pi]) \otimes \mathbb{Q} \;; \\ &\mathcal{L}_\pi(M') - \mathcal{L}_\pi(M) \longrightarrow \sigma_*(f,b) \otimes \mathbb{Q} \;=\; \frac{1}{8}(\sigma^*(M') - \sigma^*(M)) \ , \end{aligned}$$

with

$$A_\pi \mathcal{L}_\pi(M) \;=\; \frac{1}{8}\sigma^*(M) \in L_m(\mathbb{Z}[\pi]) \otimes \mathbb{Q} \;=\; L^m(\mathbb{Z}[\pi]) \otimes \mathbb{Q} \ .$$

□

The $\mathcal{L}$-genus $\mathcal{L}(M) \in H^{4*}(M;\mathbb{Q})$ is not in general a homotopy invariant of an m-dimensional manifold M, except for the $4k$-dimensional component $\mathcal{L}_k(M) \in H^{4k}(M;\mathbb{Q})$ in the case $m = 4k$ – this is a homotopy invariant by virtue of the signature theorem

$$\sigma(M) \;=\; \langle \mathcal{L}_k(M), [M] \rangle \in \mathbb{Z} \ .$$

The simply-connected surgery exact sequence (§7) shows that if M is a simply-connected m-dimensional manifold and $m - 4k \geq 1$, $m \geq 5$ then the $\mathbb{Q}$-vector space $H^{4k}(M;\mathbb{Q})$ is spanned by the differences

$$(h^{-1})^*\mathcal{L}_k(M') - \mathcal{L}_k(M) \in H^{4k}(M;\mathbb{Q})$$

for homotopy equivalences $h : M' \longrightarrow M$.

Proposition 6.6 *The Novikov conjecture holds for* π *if and only if the rational assembly map*

$$A_\pi \ : \ H_{m-4*}(B\pi;\mathbb{Q}) \longrightarrow L_m(\mathbb{Z}[\pi]) \otimes \mathbb{Q}$$

is injective for each $m(\mathrm{mod}\,4)$.

Proof The rational assembly map

$$A \ : \ H^{4*}(M;\mathbb{Q}) \ = \ H_{m-4*}(M;\mathbb{Q}) \xrightarrow{f_*} H_{m-4*}(B\pi;\mathbb{Q}) \xrightarrow{A_\pi} L_m(\mathbb{Z}[\pi]) \otimes \mathbb{Q}$$

is such that

$$A((h^{-1})^*\mathcal{L}(M') - \mathcal{L}(M)) \ = \ A_\pi(\mathcal{L}_\pi(M') - \mathcal{L}_\pi(M)) \in L_m(\mathbb{Z}[\pi]) \otimes \mathbb{Q}$$

for any homotopy equivalence $h : M' \longrightarrow M$ of m-dimensional manifolds with $\pi_1(M) = \pi$, and $f : M \longrightarrow B\pi$ the classifying map. The $\mathbb{Q}$-vector space $H_{m-4*}(B\pi;\mathbb{Q})$ is spanned by the differences $\mathcal{L}_\pi(M') - \mathcal{L}_\pi(M)$. The non-simply-connected surgery exact sequence (§7) identifies the subspace of $H^{4*}(M;\mathbb{Q})$ spanned by the differences

$$(h^{-1})^*\mathcal{L}(M') - \mathcal{L}(M) \in H^{4*}(M;\mathbb{Q})$$

with the kernel of A, and

$$\ker(f_* : H_{m-4*}(M;\mathbb{Q}) \longrightarrow H_{m-4*}(B\pi;\mathbb{Q})) \subseteq \ker(A) \ .$$

The Novikov conjecture predicts that for all m-dimensional manifolds M with $\pi_1(M) = \pi$

$$f_*(\ker(A)) \ = \ \{0\} \subseteq H_{m-4*}(B\pi;\mathbb{Q}) \ ,$$

or equivalently that $\ker(f_*) = \ker(A)$. In turn, this is equivalent to the injectivity of $A_\pi : H_{m-4*}(B\pi;\mathbb{Q}) \longrightarrow L_m(\mathbb{Z}[\pi]) \otimes \mathbb{Q}$.

□

§7. The algebraic L-theory assembly map

The integral versions of the topological invariance of the rational Pontrjagin classes and of the Novikov conjecture on the homotopy invariance of the higher signatures involve the algebraic L-spectra and the algebraic L-theory assembly map defined in Ranicki [51].

The **symmetric L-spectrum $\mathbb{L}^\bullet(R)$** of a ring with involution R is defined in [51] using n-ads of symmetric forms over R, with homotopy groups

$$\pi_*(\mathbb{L}^\bullet(R)) = L^*(R) .$$

The **generalized homology spectrum $\mathbb{H}_\bullet(X;\mathbb{L}^\bullet(R))$** of a topological space X is defined in [51] using sheaves over X of symmetric Poincaré complexes over R, with homotopy groups

$$\pi_*(\mathbb{H}_\bullet(X;\mathbb{L}^\bullet(R))) = H_*(X;\mathbb{L}^\bullet(R))$$

the cobordism groups of such sheaves. The assembly map

$$A : \mathbb{H}_\bullet(X;\mathbb{L}^\bullet(R)) \longrightarrow \mathbb{L}^\bullet(R[\pi_1(X)])$$

is defined by pulling back a symmetric Poincaré sheaf over X to the universal cover $\widetilde{X}$, and then assembling the stalks to obtain a symmetric Poincaré complex over $R[\pi_1(X)]$. Similarly for the **quadratic L-spectrum $\mathbb{L}_\bullet(R)$**.

The 0th space of the quadratic L-spectrum $\mathbb{L}_\bullet(\mathbb{Z})$ is such that

$$\mathbb{L}_0(\mathbb{Z}) \simeq L_0(\mathbb{Z}) \times G/TOP .$$

As usual, G/TOP is the classifying space for fibre homotopy trivialized topological bundles, with a fibration sequence

$$G/TOP \longrightarrow BTOP \longrightarrow BG .$$

Let $\mathbb{L}_\bullet = \mathbb{L}_\bullet\langle 1\rangle(\mathbb{Z})$ be the 1-connective cover of $\mathbb{L}_\bullet(\mathbb{Z})$, with 0th space such that

$$\mathbb{L}_0 \simeq G/TOP .$$

For any space X define the **structure spectrum**

$$\mathbb{S}_\bullet(X) = \text{homotopy cofibre}(A : \mathbb{H}_\bullet(X;\mathbb{L}_\bullet) \longrightarrow \mathbb{L}_\bullet(\mathbb{Z}[\pi_1(X)])) ,$$

to fit into a cofibration sequence of spectra

$$\mathbb{H}_\bullet(X;\mathbb{L}_\bullet) \xrightarrow{A} \mathbb{L}_\bullet(\mathbb{Z}[\pi_1(X)]) \longrightarrow \mathbb{S}_\bullet(X) ,$$

with A the spectrum level assembly map. The **structure groups**

$$\mathbb{S}_*(X) = \pi_*(\mathbb{S}_\bullet(X))$$

are the cobordism groups of sheaves over X of quadratic Poincaré complexes over $\mathbb{Z}$ such that the assembly quadratic Poincaré complex over $\mathbb{Z}[\pi_1(X)]$ is contractible. The structure groups are the relative groups in the **algebraic surgery exact sequence**

$$\ldots \longrightarrow \mathbb{S}_{m+1}(X) \longrightarrow H_m(X;\mathbb{L}_\bullet) \xrightarrow{A} L_m(\mathbb{Z}[\pi_1(X)]) \longrightarrow \mathbb{S}_m(X) \longrightarrow \ldots .$$

If X is an m-dimensional CW complex then $H_*(X;\mathbb{L}_\bullet) = H_*(X;\mathbb{L}_\bullet(\mathbb{Z}))$ for $* > m$ and $\mathbb{S}_*(X) = \mathbb{S}_{*+4}(X)$ for $* > m+1$.

Proposition 7.1 (Ranicki [51]) (i) *An m-dimensional geometric Poincaré complex X has a* **total surgery obstruction**

$$s(X) \in \mathbb{S}_m(X)$$

such that $s(X) = 0$ if (and for $m \geq 5$ only if) X is homotopy equivalent to a closed m-dimensional topological manifold.
(ii) *A closed m-dimensional topological manifold M has a* **symmetric L-theory orientation**

$$[M]_{\mathbb{L}} \in H_m(M;\mathbb{L}^\bullet(\mathbb{Z}))$$

which is represented by the symmetric Poincaré orientation sheaf, with assembly the symmetric signature

$$A([M]_{\mathbb{L}}) = \sigma^*(M) \in L^m(\mathbb{Z}[\pi_1(M)]) .$$

(iii) *A normal map $(f,b) : M' \longrightarrow M$ of closed m-dimensional topological manifolds has a* **normal invariant**

$$[f,b]_{\mathbb{L}} \in H_m(M;\mathbb{L}_\bullet) = [M, G/TOP]$$

which is represented by the sheaf over M of the quadratic Poincaré complex kernels over $\mathbb{Z}$ of the normal maps

$$(f,b) = h| : N' = h^{-1}(N) \longrightarrow N \quad (N^n \subset M^m) ,$$

with assembly the surgery obstruction

$$A([f,b]_{\mathbb{L}}) = \sigma_*(f,b) \in L_m(\mathbb{Z}[\pi_1(M)]) .$$

The symmetrization of the normal invariant is the difference of the symmetric L-theory orientations

$$(1+T)[f,b]_{\mathbb{L}} = f_*[M']_{\mathbb{L}} - [M]_{\mathbb{L}} \in H_m(M;\mathbb{L}^\bullet(\mathbb{Z})) .$$

(iv) *A homotopy equivalence $h : M' \longrightarrow M$ of closed m-dimensional topological manifolds has a* **structure invariant**

$$s(h) \in \mathbb{S}_{m+1}(M)$$

which is represented by the $\mathbb{Z}[\pi_1(M)]$-contractible quadratic Poincaré kernel sheaf of (iii) *and is such that $s(h) = 0$ if (and for $m \geq 5$ only if) h is h-cobordant to a homeomorphism. Moreover, for $m \geq 5$ every element $x \in \mathbb{S}_{m+1}(M)$ is the structure invariant $x = s(h)$ of a homotopy equivalence $h : M' \longrightarrow M$. The structure group $\mathbb{S}_{m+1}(M)$ is thus the topological manifold structure set of the Browder-Novikov-Sullivan-Wall surgery theory*

$$\mathbb{S}_{m+1}(M) = \mathcal{S}^{TOP}(M) ,$$

with a surgery exact sequence

$$\ldots \longrightarrow L_{m+1}(\mathbb{Z}[\pi_1(M)]) \longrightarrow \mathcal{S}^{TOP}(M) \longrightarrow [M, G/TOP] \longrightarrow L_m(\mathbb{Z}[\pi_1(M)]) .$$

□

Remarks 7.2 (i) The symmetric and quadratic L-spectra of $\mathbb{Z}$ are given rationally by

$$\mathbb{L}^\bullet(\mathbb{Z}) \otimes \mathbb{Q} = \mathbb{L}_\bullet(\mathbb{Z}) \otimes \mathbb{Q} = \bigvee_k K(\mathbb{Q}, 4k) ,$$

so that for any space X

$$H_m(X; \mathbb{L}^\bullet(\mathbb{Z})) \otimes \mathbb{Q} = H_m(X; \mathbb{L}_\bullet(\mathbb{Z})) \otimes \mathbb{Q} = H_{m-4*}(X; \mathbb{Q}) .$$

(ii) The symmetric L-theory orientation $[M]_{\mathbb{L}} \in H_m(M; \mathbb{L}^\bullet(\mathbb{Z}))$ of a closed m-dimensional topological manifold M is an integral refinement of the $\mathcal{L}$-genus, with

$$[M]_{\mathbb{L}} \otimes \mathbb{Q} = [M] \cap \mathcal{L}(M) \in H_m(M; \mathbb{L}^\bullet(\mathbb{Z})) \otimes \mathbb{Q} = H_{m-4*}(M; \mathbb{Q})$$

detected by the signatures $\sigma(N)$ of special submanifolds $N^{4k} \subset M \times \mathbb{R}^j$. As before, let $\pi_1(M) = \pi$ and let $f : M \longrightarrow B\pi$ be the classifying map of the universal cover $\widetilde{M}$. The image $f_*[M]_{\mathbb{L}} \in H_m(B\pi; \mathbb{L}^\bullet(\mathbb{Z}))$ is an integral refinement of the higher $\mathcal{L}$-genus, with

$$f_*[M]_{\mathbb{L}} \otimes \mathbb{Q} = \mathcal{L}_\pi(M) \in H_m(B\pi; \mathbb{L}^\bullet(\mathbb{Z})) \otimes \mathbb{Q} = H_{m-4*}(B\pi; \mathbb{Q})$$

detected by the signatures $\sigma(N)$ of π-special submanifolds $N^{4k} \subset M \times \mathbb{R}^j$.
(iii) The normal invariant $[f, b]_{\mathbb{L}} \in H_m(M; \mathbb{L}_\bullet)$ of an m-dimensional normal map $(f, b) : M' \longrightarrow M$ is given rationally by the difference of the Poincaré duals of the $\mathcal{L}$-genera

$$\begin{aligned}[f, b]_{\mathbb{L}} \otimes \mathbb{Q} &= f_*(\mathcal{L}(M') \cap [M']) - (\mathcal{L}(M) \cap [M]) \\ &\in H_m(M; \mathbb{L}_\bullet(\mathbb{Z})) \otimes \mathbb{Q} = H_{m-4*}(M; \mathbb{Q}) .\end{aligned}$$

(iv) The construction and the verification of the combinatorial invariance of the symmetric L-theory orientation $[M]_{\mathbb{L}} \in H_m(M; \mathbb{L}^\bullet(\mathbb{Z}))$ is quite straightforward for a PL manifold M. The construction and topological invariance of $[M]_{\mathbb{L}}$ for a topological manifold M is much more complicated — see §9 below.
(v) For $m \geq 5$ an m-dimensional geometric Poincaré complex X is homotopy equivalent to a closed m-dimensional manifold M if and only if there exists a symmetric L-theory orientation $[X]_{\mathbb{L}} \in H_m(X; \mathbb{L}^\bullet(\mathbb{Z}))$ such that $A([X]_{\mathbb{L}}) = \sigma^*(X) \in L^m(\mathbb{Z}[\pi_1(X)])$, modulo 2-primary torsion invariants.

□

Integral Novikov conjecture 7.3 *The assembly map in quadratic L-theory*

$$A \ : \ H_*(B\pi; \mathbb{L}_\bullet(\mathbb{Z})) \longrightarrow L_*(\mathbb{Z}[\pi])$$

is a split injection.

□

The algebraic surgery exact sequence for the classifying space $B\pi$ of a group π

$$\ldots \longrightarrow \mathbb{S}_{m+1}(B\pi) \longrightarrow H_m(B\pi; \mathbb{L}_\bullet) \xrightarrow{A} L_m(\mathbb{Z}[\pi]) \longrightarrow \mathbb{S}_m(B\pi) \longrightarrow \ldots$$

is such that

$$\begin{aligned} & \operatorname{im}(\mathbb{S}_{m+1}(B\pi) \longrightarrow H_m(B\pi; \mathbb{L}_\bullet)) \\ & \quad = \ker(A : H_m(B\pi; \mathbb{L}_\bullet) \longrightarrow L_m(\mathbb{Z}[\pi])) \subseteq L_m(\mathbb{Z}[\pi]) \end{aligned}$$

consists of the images of the normal invariants

$$f_*[s(h)] \ = \ f_*[h, b]_{\mathbb{L}} \in H_m(B\pi; \mathbb{L}_\bullet)$$

of all homotopy equivalences $h : M' \longrightarrow M$ of m-dimensional topological manifolds with $\pi_1(M) = \pi$, and with $f : M \longrightarrow B\pi$ classifying the universal cover.

Remarks 7.4 (i) The integral Novikov conjecture for π implies the original Novikov conjecture for π, since the integral assembly map A induces the rational assembly map

$$A \otimes 1 \ : \ H_m(B\pi; \mathbb{L}_\bullet(\mathbb{Z})) \otimes \mathbb{Q} \ = \ H_{m-4*}(B\pi; \mathbb{Q}) \longrightarrow L_m(\mathbb{Z}[\pi]) \otimes \mathbb{Q} \ .$$

(ii) The integral Novikov conjecture is generally false if π has torsion, e.g. if $\pi = \mathbb{Z}_2$.
(iii) The integral Novikov conjecture has been verified for many torsion-free groups π using codimension 1 splitting methods, starting with the free

abelian case $\pi = \mathbb{Z}^i$ (when A is an isomorphism) — see §§8,10 below for further discussion.
(iv) The integral Novikov conjecture has been verified geometrically for many groups π such that the classifying space $B\pi$ is realized by an aspherical Riemannian manifold with sufficient symmetry to ensure geometric rigidity, so that homotopy equivalences of manifolds with fundamental group π can be deformed to homeomorphisms — see Farrell and Hsiang [16], Farrell and Jones [17] for example.

□

Here is how algebraic surgery theory translates rigidity results in geometry into verifications of the integral Novikov conjecture:

Proposition 7.5 *If π is a finitely presented group such that (at least for $m \geq 5$) there is a systematic procedure for deforming every homotopy equivalence $h_0 : M_0 \longrightarrow N$ of closed m-dimensional manifolds with $\pi_1(N) = \pi$ to a homeomorphism $h_1 : M_1 \longrightarrow N$, via an $(m+1)$-dimensional normal bordism*

$$(g, c) \ : \ (W; M_0, M_1) \longrightarrow N \times ([0,1]; \{0\}, \{1\})$$

with $g|_{M_i} = h_i : M_i \longrightarrow N$ $(i = 0, 1)$, then the integral Novikov conjecture holds for π.

Proof The realization theorem of Wall [58] identifies $L_{m+1}(\mathbb{Z}[\pi])$ with the bordism group of normal maps $(f, b) : (K, \partial K) \longrightarrow (L, \partial L)$ of compact $(m+1)$-dimensional manifolds with boundary which restrict to a homotopy equivalence $\partial f = f| : \partial K \longrightarrow \partial L$ on the boundary, with $\pi_1(L) = \pi$. The generalized homology group $H_{m+1}(B\pi; \mathbb{L}_\bullet)$ has a similar description, with the added condition that $\partial f : \partial K \longrightarrow \partial L$ be a homeomorphism (including $\partial K = \partial L = \emptyset$ as a special case). Given systematic deformations of homotopy equivalences to homeomorphisms as in the statement there is defined a direct sum system

$$H_{m+1}(B\pi; \mathbb{L}_\bullet) \underset{B}{\overset{A}{\rightleftarrows}} L_{m+1}(\mathbb{Z}[\pi]) \underset{D}{\overset{C}{\rightleftarrows}} \mathbb{S}_{m+1}(B\pi)$$

verifying the integral Novikov conjecture for π, with

$$B \ : \ L_{m+1}(\mathbb{Z}[\pi]) \longrightarrow H_{m+1}(B\pi; \mathbb{L}_\bullet) \ ; \ (f, b) \longrightarrow (f, b) \cup (g, c)$$
$$(h_0 = \partial f : M = \partial K \longrightarrow N = \partial L) \ ,$$
$$C \ : \ L_{m+1}(\mathbb{Z}[\pi]) \longrightarrow \mathbb{S}_{m+1}(B\pi) \ ;$$
$$\sigma_*((f, b) : (K, \partial K) \longrightarrow (L, \partial L)) \longrightarrow s(\partial f : \partial K \longrightarrow \partial L) \ ,$$
$$D \ : \ \mathbb{S}_{m+1}(B\pi) \longrightarrow L_{m+1}(\mathbb{Z}[\pi]) \ ; \ s(h_0) \longrightarrow \sigma_*(g, c) \ .$$

□

The chain complex treatment in Ranicki [47], [51] of the surgery obstruction of Wall [58] associates a quadratic Poincaré complex $\sigma_*(f,b)$ over $\mathbb{Z}[\pi]$ (resp. a sheaf over $B\pi$ of quadratic Poincaré complexes over $\mathbb{Z}$) to any normal map $(f,b):(M,\partial M)\longrightarrow(N,\partial N)$ with ∂f a homotopy equivalence (resp. a homeomorphism) and $\pi_1(N)=\pi$. In principle, this allows the translation into algebra of any geometric construction of a disassembly map B.

§8. Codimension 1 splitting for compact manifolds

The primary obstructions to deforming a homotopy equivalence of high-dimensional manifolds $h: M'\longrightarrow M$ with $\pi_1(M)$ torsion-free to a homeomorphism are the splitting obstructions along codimension 1 submanifolds $N\subset M$. The method was initiated by Browder [7], where manifolds with fundamental group $\pi_1=\mathbb{Z}$ were studied by considering surgery on codimension 1 simply-connected manifolds.

Codimension 1 Splitting Theorem 8.1 (Farrell and Hsiang [15]) *Let M^m be an m-dimensional manifold, and let $N^{m-1}\subset M^m$ be a codimension 1 submanifold with trivial normal bundle, such that*

$$\pi_1(M) = \pi\times\mathbb{Z}\ ,\quad \pi_1(N) = \pi\ .$$

The Whitehead torsion of a homotopy equivalence $h: M'\longrightarrow M$ of m-dimensional manifolds is such that $\tau(h)\in \mathrm{im}(Wh(\pi)\longrightarrow Wh(\pi\times\mathbb{Z}))$ if (and for $m\geq 6$ only if) h splits along $N\subset M$.

K-theoretic proof. This was the original proof in [15]. Let $\widetilde{M}$ be the universal cover of M. The infinite cyclic cover $\overline{M}=\widetilde{M}/\pi$ of M can be constructed from M by cutting along N, with

$$\overline{M} = \overline{M}^+\cup_N\overline{M}^-$$

for two ends $\overline{M}^+,\overline{M}^-$ with

$$\pi_1(\overline{M}^+) = \pi_1(\overline{M}^-) = \pi_1(N) = \pi$$

and similarly for $M', N'=h^{-1}(N)\subset M'$. The $\mathbb{Z}$-equivariant homotopy equivalence $\overline{h}:\overline{M}'\longrightarrow\overline{M}$ has a decomposition

$$\overline{h} = \overline{h}^+\cup_g\overline{h}^-\ :\ \overline{M}' = \overline{M}'^+\cup_{N'}\overline{M}'^-\longrightarrow\overline{M} = \overline{M}^+\cup_N\overline{M}^-$$

with the restriction

$$(g,c) = h| = \overline{h}|\ :\ N' = h^{-1}(N) = \overline{h}^{-1}(N)\longrightarrow N$$

a normal map. Since h is a homotopy equivalence the natural $\mathbb{Z}[\pi]$-module chain map of the relative cellular $\mathbb{Z}[\pi]$-module chain complexes

$$C(\widetilde{N}',\widetilde{N}) \simeq C(\widetilde{M}'^{+},\widetilde{M}^{+})\oplus C(\widetilde{M}'^{-},\widetilde{M}^{-})$$

is a chain equivalence. Now $C(\widetilde{N}',\widetilde{N})$ is a finite f.g. free $\mathbb{Z}[\pi]$-module chain complex, so that $C(\widetilde{M}'^{+},\widetilde{M}^{+})$ and $C(\widetilde{M}'^{-},\widetilde{M}^{-})$ are finitely dominated (i.e. chain equivalent to a finite f.g. projective $\mathbb{Z}[\pi]$-module chain complex). The reduced projective class

$$[C(\widetilde{M}'^{+},\widetilde{M}^{+})] = -[C(\widetilde{M}'^{-},\widetilde{M}^{-})] \in \widetilde{K}_0(\mathbb{Z}[\pi])$$

is the $\widetilde{K}_0$-component of $\tau(h)\in Wh(\pi\times\mathbb{Z})$ in the decomposition

$$Wh(\pi\times\mathbb{Z}) = Wh(\pi)\oplus\widetilde{K}_0(\mathbb{Z}[\pi])\oplus\widetilde{\mathrm{Nil}}_0(\mathbb{Z}[\pi])\oplus\widetilde{\mathrm{Nil}}_0(\mathbb{Z}[\pi])$$

of Bass [3, XII]. The $\mathbb{Z}[\pi]$-module Poincaré duality chain equivalence

$$C(\widetilde{N}',\widetilde{N})^{m-1-*} \simeq C(\widetilde{N}',\widetilde{N})$$

on the chain complex kernel

$$C(\widetilde{N}',\widetilde{N}) = C(g^! : C(\widetilde{N})\longrightarrow C(\widetilde{N}'))$$

restricts to a chain equivalence

$$C(\widetilde{M}'^{+},\widetilde{M}^{+})^{m-1-*} \simeq C(\widetilde{M}'^{-},\widetilde{M}^{-})\ .$$

Thus $h : M'\longrightarrow M$ splits along $N\subset M$ (i.e. $(g,c) : N'\longrightarrow N$ is a homotopy equivalence) if and only if $C(\widetilde{M}'^{+},\widetilde{M}^{+})$ is chain contractible. For $m\geq 6$ the $\widetilde{K}_0$-component is 0 if and only if it is possible to modify N' by handle exchanges inside $\overline{M}'$ in the style of Browder [6] and Siebenmann [55] until $(g,c) : N'\longrightarrow N$ is a homotopy equivalence, if and only if the $\mathbb{R}$-bounded homotopy equivalence $\overline{f} : \overline{M}'\longrightarrow\overline{M}$ splits along $N\subset\overline{M}$. The $\widetilde{\mathrm{Nil}}_0$-components (which are Poincaré dual to each other) are the obstructions to such modifications inside a fundamental domain of the infinite cyclic cover $\overline{M}'$ of M'.

L-theoretic proof. The surgery obstruction theory of Wall [58] can be used to give an L-theoretic proof of the splitting theorem, at least in the unobstructed case $\tau(h)\in \mathrm{im}(Wh(\pi))$. The surgery exact sequence for the appropriately decorated topological manifold structure set of M

$$\begin{aligned}\dots \longrightarrow L^{\mathrm{im}(Wh(\pi))}_{m+1}(\mathbb{Z}[\pi\times\mathbb{Z}]) \longrightarrow \mathbb{S}^{\mathrm{im}(Wh(\pi))}_{m+1}(M)\\ \longrightarrow H_m(M;\mathbb{L}_\bullet) \longrightarrow L^{\mathrm{im}(Wh(\pi))}_m(\mathbb{Z}[\pi\times\mathbb{Z}]) \longrightarrow \dots\end{aligned}$$

combined with the algebraic computation of Ranicki [45]

$$L_m^{\mathrm{im}(Wh(\pi))}(\mathbb{Z}[\pi \times \mathbb{Z}]) \;=\; L_m(\mathbb{Z}[\pi]) \oplus L_{m-1}(\mathbb{Z}[\pi])$$

(or the geometric winding tricks of [58, 12.9] or Cappell [10]) give an exact sequence

$$\begin{aligned} \ldots \longrightarrow \mathbb{S}_{m+1}(M\backslash N) \longrightarrow \mathbb{S}_{m+1}^{\mathrm{im}(Wh(\pi))}(M) \longrightarrow \mathbb{S}_m(N) \\ \longrightarrow \mathbb{S}_m(M\backslash N) \longrightarrow \ldots . \end{aligned}$$

The codimension 1 h-splitting obstruction of [58, §11] is the Tate $\mathbb{Z}_2$-cohomology class

$$[s(h)] \;=\; [C(\widetilde{M}'^{+}, \widetilde{M})] \in LS_{m-1} \;=\; \widehat{H}^m(\mathbb{Z}_2; \widetilde{K}_0(\mathbb{Z}[\pi])) .$$

The structure set $\mathbb{S}_{m+1}^{\mathrm{im}(Wh(\pi))}(M)$ of homotopy equivalences of m-dimensional manifolds $h : M' \longrightarrow M$ such that $\tau(h) \in \mathrm{im}(Wh(\pi))$ is thus identified with the structure set $\mathbb{S}_{m+1}(N \longrightarrow M\backslash N)$ of homotopy equivalences $h : M' \longrightarrow M$ which split along $N \subset M$

$$\mathbb{S}_{m+1}^{\mathrm{im}(Wh(\pi))}(M) \;=\; \mathbb{S}_{m+1}(N \longrightarrow M\backslash N) .$$

(The relative $\mathbb{S}$-group $\mathbb{S}_{m+1}(N \longrightarrow M\backslash N)$ is denoted $\mathbb{S}_{m+1}(\xi^!)$ in the terminology of 3.2).

□

Example 8.2 For $m \geq 6$ a homotopy equivalence of m-dimensional manifolds of the type

$$h \;:\; M'^m \longrightarrow M^m \;=\; N^{m-1} \times S^1$$

is homotopic to

$$g \times \mathrm{id}_{S^1} \;:\; M' \;=\; N' \times S^1 \longrightarrow M \;=\; N \times S^1$$

for a homotopy equivalence of $(m-1)$-dimensional manifolds $g : N' \longrightarrow N$ if and only if $\tau(h) \in \mathrm{im}(Wh(\pi) \longrightarrow Wh(\pi \times \mathbb{Z}))$ ($\pi = \pi_1(N)$). The structure set of homotopy equivalences $h : M' \longrightarrow M$ which split along $N \subset M$ is given in this case by

$$\mathbb{S}_{m+1}^{\mathrm{im}(Wh(\pi))}(M) \;=\; \mathbb{S}_{m+1}(N) \oplus \mathbb{S}_m(N) .$$

□

For the remainder of §8 we shall assume that M is a connected manifold and that $N \subset M$ is a connected codimension 1 submanifold with trivial normal bundle

$$\nu_{N \subset M} = \epsilon : N \longrightarrow BO(1)$$

and such that $\pi_1(N) \longrightarrow \pi_1(M)$ is injective. As in the general theory of Wall [58, §12] there are two cases to consider:
(A) $N \subset M$ separates M, so that $M \backslash N$ has two components M_1, M_2, with

$$\pi_1(M) = \pi_1(M_1) *_{\pi_1(N)} \pi_1(M_2)$$

the amalgamated free product determined by the injections $\pi_1(N) \longrightarrow \pi_1(M_1)$, $\pi_1(N) \longrightarrow \pi_1(M_2)$, by the Seifert-Van Kampen theorem,
(B) $N \subset M$ does not separate M, so that $M_1 = M \backslash N$ is connected, with

$$\pi_1(M) = \pi_1(M_1) *_{\pi_1(N)} \{z\}$$

the HNN extension determined by the two injections $\pi_1(N) \longrightarrow \pi_1(M_1)$. For example, if M is a genus 2 surface and $N = S^1 \subset M$ separates M with $M \backslash N = M_1 \sqcup M_2$ the disjoint union of punctured tori then (N, M) is of type (A), with $\pi_1(M_1) = \pi_1(M_2) = \mathbb{Z} * \mathbb{Z}$, $\pi_1(N) = \mathbb{Z}$, while $(M', N') = (S^1, \{\text{pt.}\})$ is of type (B).

Waldhausen [57] obtained a splitting theorem for the algebraic K-theory of amalgamated free products and HNN extensions along injections, involving the K-groups $\widetilde{\text{Nil}}_*$ of nilpotent objects, generalizing the splitting theorem of Bass [3, XII] for the Whitehead group of a polynomial extension. The Mayer-Vietoris exact sequence of [57] is

$$\begin{aligned} \ldots \longrightarrow & Wh(\pi_1(N)) \oplus \widetilde{\text{Nil}}_1 \longrightarrow Wh(\pi_1(M_1)) \oplus Wh(\pi_1(M_2)) \\ \longrightarrow & Wh(\pi_1(M)) \longrightarrow \widetilde{K}_0(\mathbb{Z}[\pi_1(N)]) \oplus \widetilde{\text{Nil}}_0 \\ \longrightarrow & \widetilde{K}_0(\mathbb{Z}[\pi_1(M_1)]) \oplus \widetilde{K}_0(\mathbb{Z}[\pi_1(M_2)]) \longrightarrow \widetilde{K}_0(\mathbb{Z}[\pi_1(M)]) \longrightarrow \ldots \end{aligned}$$

with $Wh(\pi_1(M)) \longrightarrow \widetilde{\text{Nil}}_0$ a split surjection, setting $M_2 = \emptyset$ in case (B). See Remark 8.7 below for a brief account of the algebraic transversality used in [57], and its extension to algebraic L-theory.

Codimension 1 Splitting Theorem 8.3 (Cappell [9]) *Let M^m be an m-dimensional manifold, and let $N^{m-1} \subset M^m$ be a codimension 1 submanifold with trivial normal bundle, such that $\pi_1(N) \longrightarrow \pi_1(M)$ is injective. The algebraic L-theory of $\mathbb{Z}[\pi_1(M)]$ is such that there is defined a Mayer-Vietoris exact sequence*

$$\begin{aligned} \ldots \longrightarrow & L_n^I(\mathbb{Z}[\pi_1(N)]) \oplus \text{UNil}_{n+1} \longrightarrow L_n(\mathbb{Z}[\pi_1(M_1)]) \oplus L_n(\mathbb{Z}[\pi_1(M_2)]) \\ \longrightarrow & L_n(\mathbb{Z}[\pi_1(M)]) \longrightarrow L_{n-1}^I(\mathbb{Z}[\pi_1(N)]) \oplus \text{UNil}_n \\ \longrightarrow & L_{n-1}(\mathbb{Z}[\pi_1(M_1)]) \oplus L_{n-1}(\mathbb{Z}[\pi_1(M_2)]) \longrightarrow L_{n-1}(\mathbb{Z}[\pi_1(M)]) \longrightarrow \ldots \end{aligned}$$

with

$$\begin{aligned} I = & \ \mathrm{im}(Wh(\pi_1(M)) \longrightarrow \widetilde{K}_0(\mathbb{Z}[\pi_1(N)])) \\ = & \ \ker(\widetilde{K}_0(\mathbb{Z}[\pi_1(N)]) \longrightarrow \widetilde{K}_0(\mathbb{Z}[\pi_1(M_1)]) \oplus \widetilde{K}_0(\mathbb{Z}[\pi_1(M_2)])) \\ \subseteq & \ \widetilde{K}_0(\mathbb{Z}[\pi_1(N)]) \end{aligned}$$

and $L_n(\mathbb{Z}[\pi_1(M)]) \longrightarrow \mathrm{UNil}_n$ *a split surjection onto an* L*-group of unitary nilpotent objects, setting* $M_2 = \emptyset$ *in case (B). The codimension 1* h*-splitting obstruction (3.2) of a homotopy equivalence* $h : M' \longrightarrow M$ *of* m*-dimensional manifolds along* $N \subset M$ *is given by*

$$[s(h)] \ = \ ([\tau(h)], \sigma_*(f,b)) \in LS_{m-1} \ = \ \widehat{H}^m(\mathbb{Z}_2; I) \oplus \mathrm{UNil}_{m+1} \ .$$

The first component is the obstruction to the existence of a normal bordism to a split homotopy equivalence, the image $[\tau(h)] \in \widehat{H}^m(\mathbb{Z}_2; I)$ *of the Tate* $\mathbb{Z}_2$*-cohomology class of the Whitehead torsion*

$$\tau(h) \ = \ (-)^{m+1}\tau(h)^* \in \widehat{H}^{m+1}(\mathbb{Z}_2; Wh(\pi_1(M))) \ .$$

The second component is the surgery obstruction

$$\sigma_*(f,b) \in \mathrm{UNil}_{m+1} \subseteq L_{m+1}(\mathbb{Z}[\pi_1(M)])$$

of a normal bordism

$$(f,b) \ : \ (W; M', M'') \longrightarrow M \times ([0,1]; \{0\}, \{1\})$$

from $h : M' \longrightarrow M$ *to a split homotopy equivalence* $h' : M'' \longrightarrow M$ *given by the nilpotent normal cobordism construction of Cappell [10] in the case* $[\tau(h)] = 0 \in \widehat{H}^m(\mathbb{Z}_2; I)$*.*

□

The I-intermediate quadratic L-groups $L^I_*(\mathbb{Z}[\rho])$ in 8.3 are such that there is defined a Rothenberg-type exact sequence

$$\ldots \longrightarrow L_m(\mathbb{Z}[\rho]) \longrightarrow L^I_m(\mathbb{Z}[\rho]) \longrightarrow \widehat{H}^m(\mathbb{Z}_2; I) \longrightarrow L_{m-1}(\mathbb{Z}[\rho]) \longrightarrow \ldots$$

with $\widehat{H}^*(\mathbb{Z}_2; I)$ the Tate $\mathbb{Z}_2$-cohomology groups of the duality involution $* : I \longrightarrow I$.

See Ranicki [48, §7.6] for a chain complex interpretation of the nilpotent normal cobordism construction and the identification

$$LS_{m-1} \ = \ \widehat{H}^m(\mathbb{Z}_2; I) \oplus \mathrm{UNil}_{m+1} \ .$$

Example 8.4 In the situation of 8.1

$$\pi_1(M) = \pi \times \mathbb{Z} \ , \ \pi_1(N) = \pi$$

it is the case that

$$Wh(\pi_1(M)) = Wh(\pi) \oplus \widetilde{K}_0(\mathbb{Z}[\pi]) \oplus \widetilde{\mathrm{Nil}}_0(\mathbb{Z}[\pi]) \oplus \widetilde{\mathrm{Nil}}_0(\mathbb{Z}[\pi]) \ ,$$

$$\mathrm{UNil}_* = 0 \ , \ \widehat{H}^*(\mathbb{Z}_2; \widetilde{\mathrm{Nil}}_0(\mathbb{Z}[\pi]) \oplus \widetilde{\mathrm{Nil}}_0(\mathbb{Z}[\pi])) = 0 \ , \ I = \widetilde{K}_0(\mathbb{Z}[\pi])$$

with the $\mathbb{Z}_2$-action interchanging the two copies of $\widetilde{\mathrm{Nil}}_0(\mathbb{Z}[\pi])$. Moreover, the codimension 1 h-splitting obstruction of 8.3

$$[s(h)] = [\tau(h)] = [C(\widetilde{M}'^+, \widetilde{M}^+)] \in LS_{m-1} = \widehat{H}^m(\mathbb{Z}_2; \widetilde{K}_0(\mathbb{Z}[\pi]))$$

is the Tate $\mathbb{Z}_2$-cohomology class of the codimension 1 splitting obstruction

$$\begin{aligned} [\tau(h)] &= (-)^{m+1}[\tau(h)]^* \in \mathrm{coker}(Wh(\pi) \longrightarrow Wh(\pi \times \mathbb{Z})) \\ &= \widetilde{K}_0(\mathbb{Z}[\pi]) \oplus \widetilde{\mathrm{Nil}}_0(\mathbb{Z}[\pi]) \oplus \widetilde{\mathrm{Nil}}_0(\mathbb{Z}[\pi]) \ . \end{aligned}$$

□

Corollary 8.5 *Let π be a finitely presented group such that:*
either (A) $\pi = \pi_1 *_\rho \pi_2$ is an amalgamated free product, with π_1, π_2, ρ finitely presented,
or (B) $\pi = \pi_1 *_\rho \{z\}$ is an HNN extension, with π_1, ρ finitely presented, and let

$$I = \mathrm{im}(Wh(\pi) \longrightarrow \widetilde{K}_0(\mathbb{Z}[\rho])) \subseteq \widetilde{K}_0(\mathbb{Z}[\rho]) \ .$$

The algebraic L-theory Mayer-Vietoris exact sequence

$$\begin{aligned} \ldots \longrightarrow L^I_n(\mathbb{Z}[\rho]) \oplus \mathrm{UNil}_{n+1} &\longrightarrow L_n(\mathbb{Z}[\pi_1]) \oplus L_n(\mathbb{Z}[\pi_2]) \longrightarrow L_n(\mathbb{Z}[\pi]) \\ \longrightarrow L^I_{n-1}(\mathbb{Z}[\rho]) \oplus \mathrm{UNil}_n &\longrightarrow L_{n-1}(\mathbb{Z}[\pi_1]) \oplus L_{n-1}(\mathbb{Z}[\pi_2]) \longrightarrow \ldots \end{aligned}$$

extends to a Mayer-Vietoris exact sequence of $\mathbb{S}$-groups

$$\begin{aligned} \ldots \longrightarrow \mathbb{S}^I_n(B\rho) \oplus \mathrm{UNil}_{n+1} &\longrightarrow \mathbb{S}_n(B\pi_1) \oplus \mathbb{S}_n(B\pi_2) \longrightarrow \mathbb{S}_n(B\pi) \\ \longrightarrow \mathbb{S}^I_{n-1}(B\rho) \oplus \mathrm{UNil}_n &\longrightarrow \mathbb{S}_{n-1}(B\pi_1) \oplus \mathbb{S}_{n-1}(B\pi_2) \longrightarrow \ldots \end{aligned}$$

interpreting $\pi_2 = \emptyset$ in case (B).

Proof This a formal consequence of 8.3 and the Mayer-Vietoris exact sequence of generalized homology theory

$$\begin{aligned} \ldots \longrightarrow h_n(B\rho) &\longrightarrow h_n(B\pi_1) \oplus h_n(B\pi_2) \longrightarrow h_n(B\pi) \\ \longrightarrow h_{n-1}(B\rho) &\longrightarrow h_{n-1}(B\pi_1) \oplus h_{n-1}(B\pi_2) \longrightarrow \ldots \ , \end{aligned}$$

with $h_*(\ ) = H_*(\ ;\mathbb{L}_\bullet)$.

□

Theorem 8.6 (Cappell [11]) *The Novikov conjecture holds for the class of finitely presented groups π obtained from $\{1\}$ by amalgamated free products and HNN extensions along injections.*

Proof If the groups $G = \pi_1, \pi_2, \rho$ in 8.5 are such that the algebraic L-theory assembly maps $A : H_*(BG; \mathbb{L}_\bullet(\mathbb{Z})) \longrightarrow L_*(\mathbb{Z}[G])$ are rational isomorphisms then so is the algebraic L-theory assembly map $A : H_*(B\pi; \mathbb{L}_\bullet(\mathbb{Z})) \longrightarrow L_*(\mathbb{Z}[\pi])$, and the Novikov conjecture holds for π. Apply the 5-lemma to the $\mathbb{S}$-group Mayer-Vietoris exact sequence of 8.5, noting that the UNil-groups have exponent ≤ 8 and so make no rational contribution.

□

(The actual inductively defined class of groups for which the Novikov conjecture was verified in [11] is somewhat larger.)

Remark 8.7 The finite presentation conditions in 8.5, 8.6 are necessary because the L-theory Mayer-Vietoris exact sequence of Cappell [9] was only stated for the group rings of finitely presented groups, since the proof used geometric methods. In fact, it is possible to state and prove an L-theory Mayer-Vietoris exact sequence for amalgamated free products and HNN extensions along injections of any rings with involution, allowing the hypothesis of finite presentation to be dropped. Pending the definitive account of Ranicki [53], here is a brief account of the algebraic proof of the L-theory sequence, extending the method used by Waldhausen [57] to prove the algebraic K-theory Mayer-Vietoris exact sequence.

A ring morphism $f : R \longrightarrow R'$ determines induction and restriction functors

$$\begin{aligned} f_! \ &: \ \{R\text{-modules}\} \longrightarrow \{R'\text{-modules}\} \ ; \\ &\qquad M \longrightarrow f_!M \ = \ R' \otimes_R M \text{ with } r'(1 \otimes x) = r' \otimes x \ , \\ f^! \ &: \ \{R'\text{-modules}\} \longrightarrow \{R\text{-modules}\} \ ; \\ &\qquad M' \longrightarrow f^!M' \ = \ M' \text{ with } rx' = f(r)x' \ . \end{aligned}$$

Let R be a ring such that:

either (A) $R = R_1 *_S R_2$ is the amalgamated free product determined by injections of rings $i_1 : S \longrightarrow R_1$, $i_2 : S \longrightarrow R_2$ such that R_1, R_2 are free as (S,S)-bimodules,

or (B) $R = R_1 *_S [z, z^{-1}]$ is the HNN extension determined by two injections $i_1, i_1' : S \longrightarrow R_1$, with respect to both of which R_1 is a free (S,S)-bimodule.

As in the Serre-Bass theory there is an infinite tree T with augmented simplicial R-module chain complex

$$\Delta(T;R) \ : \ 0 \longrightarrow k_!k^!R \longrightarrow (j_1)_!j_1^!R \oplus (j_2)_!j_2^!R \longrightarrow R \longrightarrow 0$$

with

$$j_1 : R_1 \longrightarrow R \ , \ j_2 : R_2 \longrightarrow R \ , \ k = j_1 i_1 = j_2 i_2 : S \longrightarrow R$$

the inclusions, and

$$j_1^! R = \sum_{T_1^{(0)}} R_1 \ , \ j_2^! R = \sum_{T_2^{(0)}} R_2 \ , \ k^! R = \sum_{T^{(1)}} S \ ,$$

setting $R_2 = 0$ in case (B). Thus for any finite f.g. free R-module chain complex C there is defined a Mayer-Vietoris presentation

$$(*) \ \ C \otimes_R \Delta(T;R) \ : \ 0 \longrightarrow k_! k^! C \longrightarrow (j_1)_! j_1^! C \oplus (j_2)_! j_2^! C \longrightarrow C \longrightarrow 0$$

with $j_1^! C$ an infinitely generated free R_1-module chain complex, $j_2^! C$ an infinitely generated free R_2-module chain complex, and $k^! C$ an infinitely generated free S-module chain complex. For any subtree $U \subset T$ the augmented simplicial R-module chain complex $\Delta(U;R)$ defines a Mayer-Vietoris presentation of R

$$C(U;R) \ : \ 0 \longrightarrow k_! \sum_{U^{(1)}} S \longrightarrow (j_1)_! \sum_{U_1^{(0)}} R_1 \oplus (j_2)_! \sum_{U_2^{(0)}} R_2 \longrightarrow R \longrightarrow 0 \ ,$$

such that if U is finite then $\sum_{U_1^{(0)}} R_1$ is a f.g. free R_1-module, $\sum_{U_2^{(0)}} R_2$ is a f.g. free R_2-module, and $\sum_{U^{(1)}} S$ is a f.g. free S-module. Let C be n-dimensional, with

$$C_r = R^{c_r} \ \ (0 \leq r \leq n)$$

a f.g. free R-module of rank c_r. There exist finite subtrees

$$U_r \subset T \ \ (0 \leq r \leq n)$$

such that the f.g. free submodules

$$(D_1)_r = \sum_{U_{r,1}^{(0)}} R_1^{c_r} \subset j_1^! C_r = \sum_{T_1^{(0)}} R_1^{c_r} \ ,$$

$$(D_2)_r = \sum_{U_{r,2}^{(0)}} R_2^{c_r} \subset j_2^! C_r = \sum_{T_2^{(0)}} R_2^{c_r} \ ,$$

$$E_r = \sum_{U_r^{(1)}} S^{c_r} \subset k^! C_r = \sum_{T^{(1)}} S^{c_r}$$

define f.g. free subcomplexes

$$D_1 \subset j_1^! C \ , \quad D_2 \subset j_2^! C \ , \quad E \ = \ D_1 \cap D_2 \subset k^! C$$

with a Mayer-Vietoris presentation

$$0 \longrightarrow k_! E \longrightarrow (j_1)_! D_1 \oplus (j_2)_! D_2 \longrightarrow C \longrightarrow 0 \ .$$

This type of algebraic transversality (a generalization of the linearization trick of Higman [23] for matrices over a Laurent polynomial extension) was used in [57] to obtain the Mayer-Vietoris exact sequence in algebraic K-theory

$$\begin{aligned} \dots \longrightarrow K_n(S) \oplus \widetilde{\mathrm{Nil}}_{n+1} \longrightarrow K_n(R_1) \oplus K_n(R_2) \longrightarrow K_n(R) \\ \longrightarrow K_{n-1}(S) \oplus \widetilde{\mathrm{Nil}}_n \longrightarrow K_{n-1}(R_1) \oplus K_{n-1}(R_2) \longrightarrow K_{n-1}(R) \longrightarrow \dots \end{aligned}$$

with $K_n(R) \longrightarrow \widetilde{\mathrm{Nil}}_n$ split surjections.

Now suppose that R, R_1, R_2, S are rings with involution. Given a finite f.g. free R-module chain complex C apply $C \otimes_R -$ to $(*)$ above, to obtain an exact sequence of $\mathbb{Z}[\mathbb{Z}_2]$-module chain complexes

$$(**) \ \ 0 \longrightarrow k^! C \otimes_S k^! C \longrightarrow (j_1^! C \otimes_{R_1} j_1^! C) \oplus (j_2^! C \otimes_{R_2} j_2^! C) \longrightarrow C \otimes_R C \longrightarrow 0$$

with $\mathbb{Z}[\mathbb{Z}_2]$ acting by $x \otimes y \longrightarrow \pm\, y \otimes x$. In the terminology of Ranicki [47] the following algebraic transversality holds: for any n-dimensional quadratic complex over R

$$(C, \psi \in Q_n(C) = H_n(\mathbb{Z}_2; C \otimes_R C))$$

there exist an $(n-1)$-dimensional quadratic complex (E, θ) over S and n-dimensional quadratic pairs

$$\Gamma_1 \ = \ ((i_1)_! E \longrightarrow D_1, (\delta_1 \theta, (i_1)_! \theta)) \ , \quad \Gamma_2 \ = \ ((i_2)_! E \longrightarrow D_2, (\delta_2 \theta, (i_2)_! \theta))$$

over R_1, R_2 such that the union n-dimensional quadratic complex over R is homotopy equivalent to (C, ψ)

$$(j_1)_! \Gamma_1 \cup (j_2)_! \Gamma_2 \ \simeq \ (C, \psi) \ .$$

The algebraic Poincaré splitting method of Ranicki [48, §§7.5,7.6] gives a Mayer-Vietoris exact sequence in quadratic L-theory

$$\begin{aligned} \dots \longrightarrow L_n^I(S) \oplus \mathrm{UNil}_{n+1} \longrightarrow L_n(R_1) \oplus L_n(R_2) \longrightarrow L_n(R) \\ \longrightarrow L_{n-1}^I(S) \oplus \mathrm{UNil}_n \longrightarrow L_{n-1}(R_1) \oplus L_{n-1}(R_2) \longrightarrow L_{n-1}(R) \longrightarrow \dots \end{aligned}$$

with $L_n(R) \longrightarrow \mathrm{UNil}_n$ split surjections and

$I = \mathrm{im}(K_1(R) \longrightarrow K_0(S)) = \ker(K_0(S) \longrightarrow K_0(R_1) \oplus K_0(R_2)) \subseteq K_0(S)$,

using the algebraic transversality given by (**) to replace the geometric transversality of [48, 7.5.1]. There is a corresponding Mayer-Vietoris exact sequence in symmetric L-theory. This type of algebraic Poincaré transversality was already used in Milgram and Ranicki [32] and Ranicki [50] for the L-theory of Laurent polynomial extensions and the associated lower L-theory.

□

§9. With one bound

The applications of bounded and controlled algebra to splitting theorems in topology and the Novikov conjectures depend on the development of an algebraic theory of transversality: algebraic Poincaré complexes in categories associated to topological spaces are shown to have enough transversality properties of manifolds mapping to the spaces to construct a 'disassembly' map. For the sake of simplicity we shall restrict attention to the bounded algebra of Pedersen and Weibel [43] and Ranicki [50], even though it is the continuously controlled algebra of Anderson, Connolly, Ferry and Pedersen [1] which is actually used by Carlsson and Pedersen [14].

Given a metric space X and a ring A let $\mathbb{C}_X(A)$ be the **X-bounded free A-module** additive category, with objects the direct sum of f.g. free A-modules graded by X

$$M = \sum_{x \in X} M(x)$$

such that $M(K) = \sum_{x \in K} M(x)$ is a f.g. free A-module for every bounded subspace $K \subseteq X$, and with morphisms the A-module morphisms

$$f = \{f(y,x)\} : M = \sum_{x \in X} M(x) \longrightarrow N = \sum_{y \in X} N(y)$$

for which there exists a number $b > 0$ with $f(y,x) = 0 : M(x) \longrightarrow N(y)$ for all $x, y \in X$ with $d(x,y) > b$.

A **proper eventually Lipschitz map** $f : X \longrightarrow Y$ of metric spaces is a function (not necessarily continuous) such that the inverse image of a bounded set is a bounded set, and there exist numbers $r, k > 0$ depending only on f such that for all $s > r$ and all $x, y \in X$ with $d(x,y) < s$ it is the case that $d(f(x), f(y)) < ks$. Such a map induces a functor

$$f_! : \mathbb{C}_X(A) \longrightarrow \mathbb{C}_Y(A) ;$$

$$M = \sum_{x \in X} M(x) \longrightarrow f_! M = \sum_{y \in Y} \Big(\sum_{x \in f^{-1}(y)} M(x) \Big) .$$

If $f : X \longrightarrow Y$ is a homotopy equivalence in the proper eventually Lipschitz category then $f_! : \mathbb{C}_X(A) \longrightarrow \mathbb{C}_Y(A)$ is an equivalence of additive categories, inducing isomorphisms in algebraic K-theory.

Let $\mathbb{P}_X(A)$ be the idempotent completion of $\mathbb{C}_X(A)$, the additive category in which an object (M,p) is an object M in $\mathbb{C}_X(A)$ together with a projection $p = p^2 : M \longrightarrow M$, and a morphism $f : (M,p) \longrightarrow (N,q)$ is a morphism $f : M \longrightarrow N$ in $\mathbb{C}_X(A)$ such that $qfp = f : M \longrightarrow N$. The reduced projective class group of $\mathbb{P}_X(A)$ is defined by

$$\widetilde{K}_0(\mathbb{P}_X(A)) = \operatorname{coker}(K_0(\mathbb{C}_X(A)) \longrightarrow K_0(\mathbb{P}_X(A))) .$$

Example 9.1 A bounded metric space X is contractible in the proper eventually Lipschitz category, so that $\mathbb{C}_X(A)$ is equivalent to the additive category of based f.g. free A-modules, $\mathbb{P}_X(A)$ is equivalent to the additive category of f.g. projective A-modules and

$$\begin{aligned} K_*(\mathbb{C}_X(A)) &= K_*(\mathbb{P}_X(A)) = K_*(A) \quad (* \neq 0) , \\ K_0(\mathbb{C}_X(A)) &= \operatorname{im}(K_0(\mathbb{Z}) \longrightarrow K_0(A)) \ , \ K_0(\mathbb{P}_X(A)) = K_0(A) , \\ \widetilde{K}_0(\mathbb{P}_X(A)) &= \operatorname{coker}(K_0(\mathbb{Z}) \longrightarrow K_0(A)) = \widetilde{K}_0(A) . \end{aligned}$$

□

Suppose given a metric space X with a decomposition

$$X = X^+ \cup X^- .$$

Define for any $b \geq 0$ the subspaces

$$\begin{aligned} X_b^+ &= \{x \in X \mid d(x,y) \leq b \text{ for some } y \in X^+\} , \\ X_b^- &= \{x \in X \mid d(x,z) \leq b \text{ for some } z \in X^-\} , \\ Y_b &= \{x \in X \mid d(x,y) \leq b \text{ and } d(x,z) \leq b \text{ for some } y \in X^+, z \in X^-\} . \end{aligned}$$

The inclusions $X^+ \longrightarrow X_b^+$, $X^- \longrightarrow X_b^-$ are homotopy equivalences in the proper eventually Lipschitz category, so that

$$K_*(\mathbb{C}_{X_b^+}(A)) = K_*(\mathbb{C}_{X^+}(A)) \ , \ K_*(\mathbb{C}_{X_b^-}(A)) = K_*(\mathbb{C}_{X^-}(A)) .$$

Proposition 9.2 (Pedersen and Weibel [43], Carlsson [12]) *For any metric space X and any decomposition $X = X^+ \cup X^-$ there is defined a Mayer-Vietoris exact sequence in bounded K-theory*

$$\begin{aligned} &\dots \longrightarrow K_n(\mathbb{P}_{X^+}(A)) \oplus K_n(\mathbb{P}_{X^-}(A)) \longrightarrow K_n(\mathbb{P}_X(A)) \\ &\xrightarrow{\partial} \varinjlim_b K_{n-1}(\mathbb{P}_{Y_b}(A)) \longrightarrow K_{n-1}(\mathbb{P}_{X^+}(A)) \oplus K_{n-1}(\mathbb{P}_{X^-}(A)) \longrightarrow \dots \end{aligned}$$

with

$$Y_b = \{x \in X \mid d(x,y) \leq b \text{ and } d(x,z) \leq b \text{ for some } y \in X^+, z \in X^-\} .$$

Proof The original proof in [43] (for open cones) and the generalization in [12] use the heavy machinery of the algebraic K-theory spectra. For $n = 1$ there is a direct proof in Ranicki [50], as follows. Every finite chain complex C in $\mathbb{C}_X(A)$ is such that there exist subcomplexes $C^+, C^- \subseteq C$ with $C^\pm$ defined in $\mathbb{C}_{X_b^\pm}(A)$ and $C^+ \cap C^-$ defined in $\mathbb{C}_{Y_b}(A)$ for some $b \geq 0$. Thus C admits a 'Mayer-Vietoris presentation'

$$0 \longrightarrow C^+ \cap C^- \longrightarrow C^+ \oplus C^- \longrightarrow C \longrightarrow 0 .$$

If C is contractible then C^+ and C^- are $\mathbb{P}_{Y_b}(A)$-finitely dominated chain complexes. The reduced version $\widetilde{\partial}$ of the connecting map ∂ in the Mayer-Vietoris exact sequence

$$\begin{array}{l} \ldots \longrightarrow K_1(\mathbb{C}_{X^+}(A)) \oplus K_1(\mathbb{C}_{X^-}(A)) \longrightarrow K_1(\mathbb{C}_X(A)) \\ \xrightarrow{\partial} \varinjlim_b K_0(\mathbb{P}_{Y_b}(A)) \longrightarrow K_0(\mathbb{P}_{X^+}(A)) \oplus K_0(\mathbb{P}_{X^-}(A)) \longrightarrow \ldots \end{array}$$

sends the Whitehead torsion $\tau(C) \in K_1(\mathbb{C}_X(A))$ to the reduced projective class

$$\widetilde{\partial}\tau(C) = [C^+] = -[C^-] \in \varinjlim_b \widetilde{K}_0(\mathbb{P}_{Y_b}(A)) ,$$

which is such that $\widetilde{\partial}\tau(C) = 0$ if and only if there exists a presentation (C^+, C^-) with $C^+, C^-, C^+ \cap C^-$ contractible. See [50] for further details.

□

Example 9.3 For any metric space Y let

$$X = Y \times \mathbb{R} , \quad X^+ = Y \times \mathbb{R}^+ , \quad X^- = Y \times \mathbb{R}^- ,$$

so that

$$X = X^+ \cup X^- , \quad X^+ \cap X^- = Y \times \{0\} .$$

In this case

$$\begin{array}{rcl} K_*(\mathbb{P}_{X^+}(A)) &=& K_*(\mathbb{P}_{X^-}(A)) = 0 \text{ (Eilenberg swindle)} , \\ K_{*+1}(\mathbb{P}_X(A)) &=& \varinjlim_b K_*(\mathbb{P}_{Y_b}(A)) = K_*(\mathbb{P}_Y(A)) . \end{array}$$

The connecting map

$$\partial : K_1(\mathbb{C}_X(A)) = K_1(\mathbb{P}_X(A)) \longrightarrow K_0(\mathbb{P}_Y(A)) ; \tau(C) \longrightarrow [C^+] = -[C^-]$$

is an isomorphism, with $\tau(C)$ the torsion of a contractible finite chain complex C in $\mathbb{C}_X(A)$ and (C^+, C^-) any Mayer-Vietoris presentation of C.

□

A CW complex M is **X-bounded** if it is equipped with a proper map $M \longrightarrow X$ such that the diameters of the images of the cells of M are uniformly bounded in X, so that the cellular chain complex $C(M)$ is defined in $\mathbb{C}_X(\mathbb{Z})$. We shall only be concerned with metric spaces X which are allowable in the sense of Ferry and Pedersen [18], and finite-dimensional X-bounded CW complexes M which are (-1)- and 0-connected in the sense of [18], with a bounded fundamental group π. The cellular chain complex $C(\widetilde{M})$ of the π-cover $\widetilde{M}$ of M is defined in $\mathbb{C}_X(\mathbb{Z}[\pi])$. Similarly for cellular maps, with induced chain maps in $\mathbb{C}_X(\mathbb{Z}[\pi])$.

If $f : M \longrightarrow N$ is an X-bounded homotopy equivalence of X-bounded CW complexes with bounded fundamental group π the X-bounded Whitehead torsion is given by

$$\begin{aligned}\tau(f) &= \tau(\widetilde{f} : C(\widetilde{M}) \longrightarrow C(\widetilde{N})) \\ \in Wh(\mathbb{C}_X(\mathbb{Z}[\pi])) &= \operatorname{coker}(K_1(\mathbb{C}_X(\mathbb{Z})) \oplus \{\pm\pi\} \longrightarrow K_1(\mathbb{C}_X(\mathbb{Z}[\pi])))\end{aligned}$$

with $\widetilde{f} : C(\widetilde{M}) \longrightarrow C(\widetilde{N})$ the induced chain equivalence in $\mathbb{C}_X(\mathbb{Z}[\pi])$. If $X = X^+ \cup X^-$ the algebraic splitting obstruction

$$\partial\tau(f) \in \varinjlim_b K_0(\mathbb{P}_{Y_b}(\mathbb{Z}[\pi]))$$

is such that $\partial\tau(f) = 0$ if and only if f is X-bounded homotopic to an X-bounded homotopy equivalence (also denoted by f) such that the restrictions $f| : f^{-1}(Y) \longrightarrow Y$ are Y-bounded homotopy equivalences, with $Y = X^+, X^-, Y_b$ (for some $b \geq 0$).

The **lower K-groups** $K_{-*}(A)$ of Bass [3, XII] are defined for any ring A to be such that

$$K_1(A[\mathbb{Z}^i]) = \sum_{j=0}^{i} \binom{i}{j} K_{1-j}(A) \oplus \widetilde{\mathrm{Nil}}\text{-groups} .$$

For a group ring $A = \mathbb{Z}[\pi]$

$$Wh(\pi \times \mathbb{Z}^i) = \sum_{j=0}^{i} \binom{i}{j} Wh_{1-j}(\pi) \oplus \widetilde{\mathrm{Nil}}\text{-groups} ,$$

where the lower Whitehead group are defined by

$$Wh_{1-j}(\pi) = \begin{cases} Wh(\pi) & \text{if } j = 0 \\ \widetilde{K}_0(\mathbb{Z}[\pi]) & \text{if } j = 1 \\ K_{1-j}(\mathbb{Z}[\pi]) & \text{if } j \geq 2 . \end{cases}$$

Bass, Heller and Swan [4] proved that $Wh(\mathbb{Z}^i) = 0$ $(i \geq 1)$, so that

$$Wh_{1-*}(\{1\}) = 0 .$$

Example 9.4 (Pedersen [40]) The $\mathbb{R}^i$-bounded K-groups of a ring A are the lower K-groups of A

$$K_*(\mathbb{P}_{\mathbb{R}^i}(A)) = K_{*-i}(A) .$$

The $\mathbb{R}^i$-bounded Whitehead groups of a group π are the lower Whitehead groups

$$Wh(\mathbb{C}_{\mathbb{R}^i}(\pi)) = Wh_{1-i}(\pi) \ (i \geq 1) .$$

□

There is a corresponding development of bounded L-theory.

An involution on the ground ring A induces a duality involution on the X-bounded A-module category

$$* : \mathbb{C}_X(A) \longrightarrow \mathbb{C}_X(A) \ ; \ M = \sum_{x \in X} M(x) \longrightarrow M^* = \sum_{x \in X} M(x)^* ,$$

with $M(x)^* = \mathrm{Hom}_A(M(x), A)$.

Definition 9.5 (Ranicki [49], [50]) The **X-bounded symmetric L-groups** $L^*(\mathbb{C}_X(A))$ are the cobordism groups of symmetric Poincaré complexes in $\mathbb{C}_X(A)$. Similarly for the **X-bounded quadratic L-groups** $L_*(\mathbb{C}_X(A))$.

□

The symmetrization maps $1 + T : L_*(\mathbb{C}_X(A)) \longrightarrow L^*(\mathbb{C}_X(A))$ are isomorphisms modulo 8-torsion. For bounded X $\mathbb{C}_X(A)$ is equivalent to the category of f.g. free A-modules and

$$L^*(\mathbb{C}_X(A)) = L^*(A) \ , \ L_*(\mathbb{C}_X(A)) = L_*(A) .$$

The functor

$$\{\text{metric spaces and proper eventually Lipschitz maps}\}$$
$$\longrightarrow \{\mathbb{Z}\text{-graded abelian groups}\} \ ; \ X \longrightarrow L_*(\mathbb{C}_X(A))$$

was shown in Ranicki [50] to be within a bounded distance (in the non-technical sense) of being a generalized homology theory. The functor is homotopy invariant, and has the following bounded excision property:

Proposition 9.6 (Ranicki [50, 14.2]) *For any metric space X and any decomposition $X = X^+ \cup X^-$ there is defined a Mayer-Vietoris exact sequence in bounded L-theory*

$$\ldots \longrightarrow L_n(\mathbb{C}_{X^+}(A)) \oplus L_n(\mathbb{C}_{X^-}(A)) \longrightarrow L_n(\mathbb{C}_X(A))$$
$$\xrightarrow{\partial} \varinjlim_b L^{J_b}_{n-1}(\mathbb{P}_{Y_b}(A)) \longrightarrow L_{n-1}(\mathbb{C}_{X^+}(A)) \oplus L_{n-1}(\mathbb{C}_{X^-}(A)) \longrightarrow \ldots ,$$

with

$$\begin{aligned} Y_b &= \{x \in X \,|\, d(x,y) \leq b \text{ and } d(x,z) \leq b \text{ for some } y \in X^+, z \in X^-\} , \\ J_b &= \ker(\widetilde{K}_0(\mathbb{P}_{Y_b}(A)) \longrightarrow \widetilde{K}_0(\mathbb{P}_X(A))) \\ &\subseteq \widetilde{K}_0(\mathbb{P}_{Y_b}(A)) \ = \ \mathrm{coker}(K_0(\mathbb{C}_{Y_b}(A)) \longrightarrow K_0(\mathbb{P}_{Y_b}(A))) \ . \end{aligned}$$

The J_b-intermediate quadratic L-groups $L^{J_b}_*(\mathbb{P}_{Y_b}(A))$ are such that there is defined a Rothenberg-type exact sequence

$$\ldots \longrightarrow L_n(\mathbb{C}_{Y_b}(A)) \longrightarrow L^{J_b}_n(\mathbb{P}_{Y_b}(A))$$
$$\longrightarrow \widehat{H}^n(\mathbb{Z}_2; J_b) \longrightarrow L_{n-1}(\mathbb{C}_{Y_b}(A)) \longrightarrow \ldots$$

with $\widehat{H}^*(\mathbb{Z}_2; J_b)$ the Tate $\mathbb{Z}_2$-cohomology groups of the duality involution $* : J_b \longrightarrow J_b$.

□

The **lower L-groups** $L^{\langle -j\rangle}_*(A)$ of Ranicki [45] are defined for any ring with involution A to be such that

$$L_n(A[\mathbb{Z}^i]) \ = \ \sum_{j=0}^{i} \binom{i}{j} L^{\langle 1-j\rangle}_{n-j}(A) \ ,$$

with $L^{\langle 1\rangle}_*(A) = L^h_*(A) = L_*(A)$ the free L-groups and $L^{\langle 0\rangle}_*(A) = L^p_*(A)$ the projective L-groups.

Example 9.7 The $\mathbb{R}^i$-bounded L-groups of a ring with involution A were identified in Ranicki [50] with the lower L-groups of A

$$L_*(\mathbb{C}_{\mathbb{R}^i}(A)) \ = \ L^{\langle 1-i\rangle}_{*-i}(A) \ .$$

□

Definition 9.8 The **X-bounded symmetric signature** of an m-dimensional X-bounded geometric Poincaré complex M with bounded fundamental group π is the cobordism class

$$\sigma^*(M) \ = \ (C(\widetilde{M}), \phi) \in L^m(\mathbb{C}_X(\mathbb{Z}[\pi])) \ ,$$

with ϕ the symmetric structure of the Poincaré duality chain equivalence $[M] \cap - : C(\widetilde{M})^{m-*} \longrightarrow C(\widetilde{M})$.

□

The standard algebraic mapping cylinder argument shows:

Proposition 9.9 *The X-bounded symmetric signature is an X-bounded homotopy invariant of an X-bounded geometric Poincaré complex.*

□

Definition 9.10 Let $(f, b) : (M', \partial M') \longrightarrow (M, \partial M)$ be a normal map from an X-bounded m-dimensional manifold with boundary $(M', \partial M')$ to an X-bounded m-dimensional geometric Poincaré pair $(M, \partial M)$, such that M has bounded fundamental group π, and $\partial f : \partial M' \longrightarrow \partial M$ is an X-bounded homotopy equivalence. The **X-bounded quadratic signature** of (f, b) is the quadratic Poincaré cobordism class

$$\sigma_*(f, b) = (C(f^!), \psi) \in L_m(\mathbb{C}_X(\mathbb{Z}[\pi])) ,$$

with ψ the quadratic structure on the algebraic mapping cone $C(f^!)$ of the Umkehr chain map in $\mathbb{C}_X(\mathbb{Z}[\pi])$

$$f^! : C(\widetilde{M}) \simeq C(\widetilde{M}, \partial\widetilde{M})^{m-*} \xrightarrow{f^*} C(\widetilde{M}', \partial\widetilde{M}')^{m-*} \simeq C(\widetilde{M}') .$$

□

The quadratic Poincaré complex $(C(f^!), \psi)$ in 9.10 can be obtained in two (equivalent) ways: either by the X-bounded version of Wall [58, §§5,6] by first performing geometric surgery below the middle dimension to obtain a quadratic form/formation in $\mathbb{C}_X(\mathbb{Z}[\pi])$ as in Ferry and Pedersen [18], or by the X-bounded version of Ranicki [47], using algebraic Poincaré complexes and the chain bundle theory of Weiss [59].

Proposition 9.11 *The X-bounded quadratic signature is the bounded surgery obstruction of Ferry and Pedersen [18], such that $\sigma_*(f, b) = 0$ if (and for $m \geq 5$) (f, b) is normal bordant to an X-bounded homotopy equivalence.*

□

The symmetrization of the X-bounded quadratic signature is the X-bounded symmetric signature

$$(1 + T)\sigma_*(f, b) = \sigma^*(M' \cup_{\partial f} -M) \in L^m(\mathbb{C}_X(\mathbb{Z}[\pi])) .$$

Let M be an X-bounded CW complex with bounded fundamental group π. See Ranicki [51, Appendix C5] for the construction of the **locally finite assembly maps**

$$A^{lf} : \mathbb{H}^{lf}_\bullet(M; \mathbb{L}_\bullet) \longrightarrow \mathbb{L}_\bullet(\mathbb{C}_X(\mathbb{Z}[\pi])) .$$

The locally finite homology spectrum $\mathbb{H}_{\bullet}^{lf}(M;\mathbb{L}_{\bullet})$ is defined using locally finite sheaves over M of quadratic Poincaré complexes over $\mathbb{Z}$, and the L-spectrum $\mathbb{L}_{\bullet}(\mathbb{C}_X(\mathbb{Z}[\pi]))$ is defined using quadratic Poincaré complexes in $\mathbb{C}_X(\mathbb{Z}[\pi])$. The **$X$-bounded structure groups of** M

$$\mathbb{S}_*^b(M) \;=\; \pi_*(A^{lf} : \mathbb{H}_{\bullet}^{lf}(M;\mathbb{L}_{\bullet}) \longrightarrow \mathbb{L}_{\bullet}(\mathbb{C}_X(\mathbb{Z}[\pi])))$$

are the relative groups in the **X-bounded algebraic surgery exact sequence**

$$\begin{aligned}\ldots \longrightarrow \mathbb{S}_{m+1}^b(M) \longrightarrow H_m^{lf}(M;\mathbb{L}_{\bullet}) &\xrightarrow{A^{lf}} L_m(\mathbb{C}_X(\mathbb{Z}[\pi])) \\ &\longrightarrow \mathbb{S}_m^b(M) \longrightarrow \ldots .\end{aligned}$$

Proposition 9.12 (Ranicki [50], [51])
(i) *An m-dimensional X-bounded manifold M with bounded fundamental group π has an $\mathbb{L}^{\bullet}(\mathbb{Z})$-coefficient fundamental class $[M]_{\mathbb{L}} \in H_m^{lf}(M;\mathbb{L}^{\bullet}(\mathbb{Z}))$ with locally finite assembly the X-bounded symmetric signature*

$$A^{lf}([M]_{\mathbb{L}}) \;=\; \sigma^*(M) \in L^m(\mathbb{C}_X(\mathbb{Z}[\pi])) \, .$$

A normal map $(f,b) : M' \longrightarrow M$ has a **normal invariant**

$$[f,b]_{\mathbb{L}} \in H_m^{lf}(M;\mathbb{L}_{\bullet}) \;=\; H^0(M;\mathbb{L}_{\bullet}) \;=\; [M, G/TOP] \, .$$

The surgery obstruction of (f,b) is the image of the normal invariant under the locally finite assembly map

$$\begin{aligned}\sigma_*(f,b) \;&=\; A^{lf}([f,b]_{\mathbb{L}}) \in \mathrm{im}(A^{lf} : H_m^{lf}(M;\mathbb{L}_{\bullet}) \longrightarrow L_m(\mathbb{C}_X(\mathbb{Z}[\pi]))) \\ &=\; \ker(L_m(\mathbb{C}_X(\mathbb{Z}[\pi])) \longrightarrow \mathbb{S}_m^b(M)) \, .\end{aligned}$$

(ii) *An m-dimensional X-bounded geometric Poincaré complex M has a* **total surgery obstruction**

$$s^b(M) \in \mathbb{S}_m^b(M)$$

such that $s^b(M) = 0$ if (and for $m \geq 5$ only if) M is X-bounded homotopy equivalent to an m-dimensional X-bounded topological manifold. The total surgery obstruction has image $[s^b(M)] = 0 \in H_{m-1}^{lf}(M;\mathbb{L}_{\bullet})$ if and only if the Spivak normal fibration $\nu_M : M \longrightarrow BG$ admits a TOP reduction $\widetilde{\nu}_M : M \longrightarrow BTOP$, in which case $s^b(M) = [\sigma_*(f,b)]$ is the image of the X-bounded surgery obstruction $\sigma_*(f,b) \in L_m(\mathbb{C}_X(\mathbb{Z}[\pi]))$ for any normal map $(f,b) : M' \longrightarrow M$.

(iii) *An X-bounded homotopy equivalence $h : M' \longrightarrow M$ of m-dimensional X-bounded topological manifolds has a* **structure invariant**

$$s^b(h) \in \mathbb{S}^b_{m+1}(M)$$

such that $s^b(h) = 0$ if (and for $m \geq 5$ only if) h is X-bounded homotopic to a homeomorphism. Moreover, for $m \geq 5$ every element $s \in \mathbb{S}^b_{m+1}(M)$ is the structure invariant $s = s^b(h)$ of such an X-bounded homotopy equivalence $h : M' \longrightarrow M$. Thus

$$\mathbb{S}^b_{m+1}(M) = \mathcal{S}^{b,TOP}(M)$$

is the X-bounded topological manifold structure set of M, with a surgery exact sequence

$$\ldots \longrightarrow L_{m+1}(\mathbb{C}_X(\mathbb{Z}[\pi])) \longrightarrow \mathcal{S}^{b,TOP}(M) \longrightarrow [M, G/TOP] \longrightarrow L_m(\mathbb{C}_X(\mathbb{Z}[\pi]))$$

as in Ferry and Pedersen [18, §11].

□

For any subspace $K \subseteq S^N$ define the **open cone** metric space

$$O(K) = \{tx \mid x \in K, t \geq 0\} \subseteq \mathbb{R}^{N+1} ,$$

such that for compact K

$$H^{lf}_{*+1}(O(K); \mathbb{L}_\bullet) = \widetilde{H}_*(K; \mathbb{L}_\bullet) .$$

In particular, $O(S^N) = \mathbb{R}^{N+1}$ and

$$H^{lf}_{*+1}(O(S^N); \mathbb{L}_\bullet) = \widetilde{H}_*(S^N; \mathbb{L}_\bullet) = L_{*-N}(\mathbb{Z}) .$$

Proposition 9.13 (Ranicki [50], [51]) (i) *The locally finite assembly maps*

$$A^{lf} : H^{lf}_*(O(K); \mathbb{L}_\bullet(\mathbb{Z})) \longrightarrow L_*(\mathbb{C}_{O(K)}(\mathbb{Z}))$$

are isomorphisms for any compact polyhedron $K \subseteq S^N$, with $\mathbb{S}^b_*(O(K)) = 0$. Similarly for symmetric L-theory.
(ii) *The symmetric L-theory orientation $[M]_{\mathbb{L}} \in H_m(M; \mathbb{L}^\bullet(\mathbb{Z}))$ of a closed m-dimensional manifold M is a topological invariant.*
Proof (i) For any ring with involution A every quadratic complex (C, ψ) in $\mathbb{C}_{O(K)}(A)$ is cobordant to the assembly $A(\Gamma)$ of a locally finite sheaf Γ over $O(K)$ of quadratic complexes over A. If (C, ψ) is a quadratic Poincaré complex it may not be possible to choose Γ such that each of the stalks is a quadratic Poincaré complex over A — the reduced lower K-groups

$\widetilde{K}_{-*}(A)$ are the potential obstructions to such a quadratic Poincaré disassembly. This is an $O(K)$-bounded algebraic L-theory version of the lower Whitehead torsion obstruction (10.1 below) to codimension 1 splitting of $O(K)$-bounded homotopy equivalences of $O(K)$-bounded open manifolds. For $A = \mathbb{Z}$ the obstruction groups are $\widetilde{K}_{-*}(\mathbb{Z}) = Wh_{1-*}(\{1\}) = 0$ by Bass, Heller and Swan [4]. See [51, Appendix C14] and §10 below for further details.
(ii) Let $M_+ = M \cup \{\text{pt.}\}$. Regard $M \times \mathbb{R}$ as an $(m+1)$-dimensional $O(M_+)$-bounded geometric Poincaré complex via the projection $M \times \mathbb{R} \longrightarrow O(M_+)$, with $O(M_+)$ defined using any embedding $M_+ \subset S^N$ (N large). The symmetric L-theory orientation of M is the $O(M_+)$-bounded symmetric signature of $M \times \mathbb{R}$

$$\begin{aligned} \sigma^*(M \times \mathbb{R}) &= [M]_{\mathbb{L}} \\ \in L^{m+1}(\mathbb{C}_{O(M_+)}(\mathbb{Z})) &= H^{lf}_{m+1}(O(M_+); \mathbb{L}^\bullet(\mathbb{Z})) = H_m(M; \mathbb{L}^\bullet(\mathbb{Z})) . \end{aligned}$$

A homeomorphism $h : M' \longrightarrow M$ determines an $O(M_+)$-bounded homotopy equivalence $h \times 1 : M' \times \mathbb{R} \longrightarrow M \times \mathbb{R}$, so that

$$\begin{aligned} [M]_{\mathbb{L}} &= \sigma^*(M \times \mathbb{R}) = (h \times 1)_* \sigma^*(M' \times \mathbb{R}) = h_*[M']_{\mathbb{L}} \\ &\in H_m(M; \mathbb{L}^\bullet(\mathbb{Z})) = L^{m+1}(\mathbb{C}_{O(M_+)}(\mathbb{Z})) . \end{aligned}$$

See [51, Appendix C16] for further details.

□

Remark 9.14 (i) As in the original proof of the topological invariance of the rational Pontrjagin classes due to Novikov [38] it suffices to prove the topological invariance of signatures of special submanifolds – cf. 2.6. As in the proof of 4.1 suppose given a homeomorphism $h : M'^m \longrightarrow M^m$ of m-dimensional (differentiable) manifolds and a special submanifold $N^{4k} \subset M^m \times \mathbb{R}^j$. Let

$$W = N \times \mathbb{R}^i \subset M \times \mathbb{R}^j \quad (i = m + j - 4k)$$

be a regular neighbourhood of N in $M \times \mathbb{R}^j$, and let

$$W' = (h \times \mathrm{id}_{\mathbb{R}^j})^{-1}(W) \subset M' \times \mathbb{R}^j .$$

Now W' is an $(m+j)$-dimensional $\mathbb{R}^i$-bounded manifold which is $\mathbb{R}^i$-bounded homotopy equivalent to W, so that the $\mathbb{R}^i$-bounded symmetric signatures are such that

$$\sigma^*(W') = \sigma^*(W) = \sigma(N) \in L^{m+j}(\mathbb{C}_{\mathbb{R}^i}(\mathbb{Z})) = L^{4k}(\mathbb{Z}) = \mathbb{Z} .$$

Let $N'^{4k} \subset W'$ be the inverse image submanifold obtained by making the homeomorphism $(h \times \mathrm{id}_{\mathbb{R}^j})| : W' \longrightarrow W$ transverse regular at $N \subset W$, so that N' is the transverse inverse image of $0 \in \mathbb{R}^i$ under $W' \longrightarrow \mathbb{R}^i$. The algebraic isomorphism $L^{m+j}(\mathbb{C}_{\mathbb{R}^i}(\mathbb{Z})) \cong L^{4k}(\mathbb{Z})$ of Ranicki [50] sends $\sigma^*(W')$ to $\sigma(N')$. Thus

$$\sigma(N') = \sigma^*(W') = \sigma^*(W) = \sigma(N) \in \mathbb{Z} ,$$

giving (yet again) the topological invariance of the signatures of special submanifolds.
(ii) The topological invariance of signatures of special submanifolds is a formal consequence of the topological invariance of the symmetric L-theory orientation, as follows. If $N^{4k} \subset M^m \times \mathbb{R}^j$ is a special submanifold there exists a proper map

$$e : M \times \mathbb{R}^j \longrightarrow \mathbb{R}^i \quad (i = m + j - 4k)$$

such that $N = e^{-1}(0)$, and there is defined a commutative diagram

$$\begin{array}{ccc}
H_m(M;\mathbb{L}^\bullet(\mathbb{Z})) = H^{lf}_{m+j}(M \times \mathbb{R}^j;\mathbb{L}^\bullet(\mathbb{Z})) & & \\
\downarrow & \searrow^{e_*} & \\
 & & H^{lf}_{m+j}(\mathbb{R}^i;\mathbb{L}^\bullet(\mathbb{Z})) = L^{4k}(\mathbb{Z}) \\
 & \nearrow_{A} & \\
H^{lf}_{m+j}(M \times \mathbb{R}^j, (M \times \mathbb{R}^j)\backslash N;\mathbb{L}^\bullet(\mathbb{Z})) = H_{4k}(N;\mathbb{L}^\bullet(\mathbb{Z})) & &
\end{array}$$

with A the simply-connected symmetric L-theory assembly map. The symmetric L-theory orientation $[M]_{\mathbb{L}} \in H_m(M;\mathbb{L}^\bullet(\mathbb{Z}))$ has image the signature of N

$$e_*([M]_{\mathbb{L}}) = A([N]_{\mathbb{L}}) = \sigma(N) \in L^{4k}(\mathbb{Z}) = \mathbb{Z} .$$

The topological invariance of the symmetric L-theory orientation $[M]_{\mathbb{L}}$ thus implies the topological invariance of the signatures $\sigma(N)$ of special submanifolds, and hence the topological invariance of the $\mathcal{L}$-genus and the rational Pontrjagin classes $\mathcal{L}(M), p_*(M) \in H^{4*}(M;\mathbb{Q})$ (as in 4.1).

□

§10. Codimension 1 splitting for non-compact manifolds

The obstruction theory for splitting homotopy equivalences of compact manifolds along codimension 1 submanifolds involves both algebraic K- and

L-theory, as recalled in §8. In fact, the approach to the (integral) Novikov conjecture of Carlsson and Pedersen [14] makes use of the obstruction theory for splitting bounded homotopy equivalences of non-compact manifolds along codimension 1 submanifolds, which only requires algebraic K-theory obstructions to be considered.

Bounded Codimension 1 Splitting Theorem 10.1 (Ferry and Pedersen [18, 7.2], Ranicki [50, 7.5]) *Let $h : M'^m \longrightarrow M^m$ be an $X \times \mathbb{R}$-bounded homotopy equivalence of m-dimensional $X \times \mathbb{R}$-bounded manifolds with bounded fundamental group π. Assume the given proper map $\rho : M \longrightarrow X \times \mathbb{R}$ is transverse regular at $X \times \{0\} \subset X \times \mathbb{R}$, so that*

$$N^{m-1} = \rho^{-1}(X \times \{0\}) \subset M^m$$

is a codimension 1 X-bounded submanifold with trivial normal bundle and bounded fundamental group π. The $X \times \mathbb{R}$-bounded Whitehead torsion

$$\tau(h) \in Wh(\mathbb{C}_{X\times\mathbb{R}}(\mathbb{Z}[\pi])) = \widetilde{K}_0(\mathbb{P}_X(\mathbb{Z}[\pi]))$$

is such that $\tau(h) = 0$ if (and for $m \geq 6$ only if) h splits along $N \subset M$.
K-theoretic proof. Make $h : M' \longrightarrow M$ transverse regular at $N \subset M$, and let $N' = h^{-1}(N) \subset M'$, so that as in the K-theoretic proof of 8.1 we have

$$h = h^+ \cup_g h^- : M' = M'^+ \cup_{N'} M'^- \longrightarrow M = M^+ \cup_N M^- .$$

Since h is an $X \times \mathbb{R}$-bounded homotopy equivalence the natural chain map is a chain equivalence in $\mathbb{C}_{X\times\mathbb{R}}(\mathbb{Z}[\pi])$

$$C(\widetilde{N}', \widetilde{N}) \simeq C(\widetilde{M}'^+, \widetilde{M}^+) \oplus C(\widetilde{M}'^-, \widetilde{M}^-) ,$$

and Poincaré duality defines a chain equivalence in $\mathbb{C}_{X\times\mathbb{R}}(\mathbb{Z}[\pi])$

$$C(\widetilde{M}'^+, \widetilde{M}^+)^{m-1-*} \simeq C(\widetilde{M}'^-, \widetilde{M}^-) .$$

The restriction X-bounded normal map

$$(g, c) = h| : N' \longrightarrow N$$

is an X-bounded homotopy equivalence if and only if the chain complex $C(\widetilde{M}'^+, \widetilde{M}^+)$ is chain contractible. The isomorphism given by 9.3

$$\partial : Wh(\mathbb{C}_{X\times\mathbb{R}}(\mathbb{Z}[\pi])) \xrightarrow{\simeq} \widetilde{K}_0(\mathbb{P}_X(\mathbb{Z}[\pi]))$$

sends $\tau(h)$ to the reduced projective class of the $\mathbb{C}_X(\mathbb{Z}[\pi])$-finitely dominated cellular $\mathbb{Z}[\pi]$-module chain complex $C(\widetilde{M}'^+, \widetilde{M}^+)$. For $m \geq 6$ $\tau(h) = 0$ if and only if it is possible to modify N' by X-bounded handle exchanges inside M' until the X-bounded normal map $h| : N' \longrightarrow N$ is a homotopy equivalence, if and only if h splits along $N \subset M$.

L-theoretic proof. The unobstructed case $\tau(h) = 0 \in Wh(\mathbb{C}_{X\times\mathbb{R}}(\mathbb{Z}[\pi]))$ proceeds as in the L-theoretic proof of 10.1 to compute the simple $X \times \mathbb{R}$-bounded topological manifold structure set of M

$$\begin{aligned} \ldots \longrightarrow L^s_{m+1}(\mathbb{C}_{X\times\mathbb{R}}(\mathbb{Z}[\pi])) \longrightarrow \mathbb{S}^{b,s}_{m+1}(M) \\ \longrightarrow H^{lf}_m(M;\mathbb{L}_\bullet) \longrightarrow L^s_m(\mathbb{C}_{X\times\mathbb{R}}(\mathbb{Z}[\pi])) \longrightarrow \ldots \,. \end{aligned}$$

It follows from the algebraic computation of Ranicki [50]

$$L^s_{m+1}(\mathbb{C}_{X\times\mathbb{R}}(\mathbb{Z}[\pi])) \;=\; L_m(\mathbb{C}_X(\mathbb{Z}[\pi]))$$

that there is defined an exact sequence

$$\ldots \longrightarrow \mathbb{S}^b_{m+1}(M\backslash N) \longrightarrow \mathbb{S}^{b,s}_{m+1}(M) \longrightarrow \mathbb{S}^b_m(N) \longrightarrow \mathbb{S}^b_m(M\backslash N) \longrightarrow \ldots \,.$$

The structure set $\mathbb{S}^{b,s}_{m+1}(M)$ of simple $X \times \mathbb{R}$-bounded homotopy equivalences of m-dimensional manifolds $h : M' \longrightarrow M$ is thus identified with the structure set $\mathbb{S}^b_{m+1}(N \longrightarrow M\backslash N)$ of $X \times \mathbb{R}$-bounded homotopy equivalences $h : M' \longrightarrow M$ which split along $N \subset M$

$$\mathbb{S}^{b,s}_{m+1}(M) \;=\; \mathbb{S}^b_{m+1}(N \longrightarrow M\backslash N) \,.$$

□

Example 10.2 For $m \geq 6$ an $X \times \mathbb{R}$-bounded homotopy equivalence of m-dimensional $X \times \mathbb{R}$-bounded manifolds of the type

$$h \;:\; M'^m \longrightarrow M^m \;=\; N^{m-1} \times \mathbb{R}$$

is homotopic to

$$g \times \mathrm{id}_{\mathbb{R}} \;:\; M' \;=\; N' \times \mathbb{R} \longrightarrow M \;=\; N \times \mathbb{R}$$

for an X-bounded homotopy equivalence of $(m-1)$-dimensional X-bounded manifolds $g : N' \longrightarrow N$ if and only if

$$\tau(h) \;=\; 0 \in Wh(\mathbb{C}_{X\times\mathbb{R}}(\mathbb{Z}[\pi])) \;=\; \widetilde{K}_0(\mathbb{P}_X(\mathbb{Z}[\pi])) \,.$$

The algebraic surgery exact sequences for the structure set $\mathbb{S}^{b,s}_{m+1}(N \times \mathbb{R})$ of simple $X \times \mathbb{R}$-bounded homotopy equivalences $h : M' \longrightarrow M$ and the structure set $\mathbb{S}^b_m(N)$ of X-bounded homotopy equivalences $g : N' \longrightarrow N$ are related by an isomorphism

$$\begin{array}{ccccccc}
\dots \longrightarrow & H^{lf}_m(N;\mathbb{L}_\bullet) & \xrightarrow{A^{lf}} & L_m(\mathbb{C}_X(\mathbb{Z}[\pi])) & \longrightarrow & \mathbb{S}^b_m(N) & \longrightarrow \dots \\
& \cong \downarrow & & \cong \downarrow & & \cong \downarrow & \\
\dots \longrightarrow & H^{lf}_{m+1}(N\times\mathbb{R};\mathbb{L}_\bullet) & \xrightarrow{A^{lf}} & L^s_{m+1}(\mathbb{C}_{X\times\mathbb{R}}(\mathbb{Z}[\pi])) & \longrightarrow & \mathbb{S}^{b,s}_{m+1}(N\times\mathbb{R}) & \longrightarrow \dots
\end{array}$$

so that

$$\mathbb{S}^{b,s}_{m+1}(N\times\mathbb{R}) \;=\; \mathbb{S}^b_m(N) \;,$$

and simple $X\times\mathbb{R}$-bounded homotopy equivalences $h : M' \longrightarrow M$ split along $N\times\{0\} \subset M = N\times\mathbb{R}$.

□

Proposition 10.3 *Let N be a compact n-dimensional manifold, and let W be an open $(n+i)$-dimensional $\mathbb{R}^i$-bounded manifold with an $\mathbb{R}^i$-bounded homotopy equivalence $h : W \longrightarrow N\times\mathbb{R}^i$ $(i \geq 1)$. The $\mathbb{R}^i$-bounded Whitehead torsion*

$$\tau(h) \in Wh(\mathbb{C}_{\mathbb{R}^i}(\mathbb{Z}[\pi])) \;=\; Wh_{1-i}(\mathbb{Z}[\pi]) \;\; (\pi = \pi_1(N))$$

is such that $\tau(h) = 0$ if (and for $n \geq 5$ only if) h is $\mathbb{R}^i$-bounded homotopic to

$$g\times \mathrm{id}_{\mathbb{R}^i} \;:\; W \;=\; N'\times\mathbb{R}^i \longrightarrow N\times\mathbb{R}^i$$

for some closed codimension i submanifold $N' \subset W$, with $g : N' \longrightarrow N$ a homotopy equivalence.

Proof See Bryant and Pacheco [8] for a proof based on the geometric twist-glueing technique of Siebenmann [56]. Alternatively, apply 10.2 i times.

□

§11. Splitting the assembly map

This section is an outline of the infinite transfer method used by Carlsson and Pedersen [14] to prove the integral Novikov conjecture by splitting the algebraic L-theory assembly map

$$A \;:\; H_*(B\pi;\mathbb{L}_\bullet(\mathbb{Z})) \longrightarrow L_*(\mathbb{Z}[\pi])$$

for torsion-free groups π with finite classifying space $B\pi$, such that $E\pi$ has a sufficiently nice compactification. The method may be viewed as a particularly well-organized way of avoiding the algebraic K-theory codimension

1 splitting obstructions to deforming homotopy equivalences of manifolds with fundamental group π to homeomorphisms.

The **homotopy fixed set** of a pointed space X with π-action is

$$X^{h\pi} = \text{map}_\pi(E\pi_+, X) ,$$

with $E\pi_+ = E\pi \cup \{\text{pt.}\}$.

Let K be a connected compact polyhedron, regarded as a metric space. The action of the fundamental group $\pi = \pi_1(K)$ on the universal cover $\widetilde{K}$ induces an action of π on the spectrum $\mathbb{L}_\bullet(\mathbb{C}_{\widetilde{K}}(\mathbb{Z}))$, with the fixed point spectrum such that

$$\mathbb{L}_\bullet(\mathbb{C}_{\widetilde{K}}(\mathbb{Z}))^\pi \simeq \mathbb{L}_\bullet(\mathbb{C}_K(\mathbb{Z}[\pi])) \simeq \mathbb{L}_\bullet(\mathbb{Z}[\pi]) .$$

The action of π on the cofibration sequence of spectra

$$\mathbb{H}^{lf}_\bullet(\widetilde{K};\mathbb{L}_\bullet) \xrightarrow{A^{lf}} \mathbb{L}_\bullet(\mathbb{C}_{\widetilde{K}}(\mathbb{Z})) \longrightarrow \mathbb{S}^b(\widetilde{K})$$

determines a cofibration sequence of the homotopy fixed point spectra

$$\mathbb{H}^{lf}_\bullet(\widetilde{K},\mathbb{L}_\bullet)^{h\pi} \xrightarrow{A^{lf}} \mathbb{L}_\bullet(\mathbb{C}_{\widetilde{K}}(\mathbb{Z}))^{h\pi} \longrightarrow \mathbb{S}^b(\widetilde{K})^{h\pi}$$

with a homotopy equivalence

$$\mathbb{H}^{lf}_\bullet(\widetilde{K},\mathbb{L}_\bullet)^{h\pi} \simeq \mathbb{H}_\bullet(K,\mathbb{L}_\bullet) .$$

The infinite transfer maps of Ranicki [51, p. 328]

$$\text{trf} : L_*(\mathbb{Z}[\pi]) = L_*(\mathbb{C}_K(\mathbb{Z}[\pi])) = L_*(\mathbb{C}_{\widetilde{K}}(\mathbb{Z})^\pi) \longrightarrow L_*(\mathbb{C}_{\widetilde{K}}(\mathbb{Z}))$$

extend to define a natural transformation of algebraic surgery exact sequences

$$\begin{array}{ccccccccc}
\dots \to & \mathbb{S}_{m+1}(K) & \longrightarrow & H_m(K;\mathbb{L}_\bullet) & \xrightarrow{A} & L_m(\mathbb{Z}[\pi]) & \longrightarrow & \mathbb{S}_m(K) & \to \dots \\
& \big\downarrow \text{trf} & & \big\downarrow \text{trf} \cong & & \big\downarrow \text{trf} & & \big\downarrow \text{trf} & \\
\dots \to & \mathbb{S}^{b,h\pi}_{m+1}(\widetilde{K}) & \to & H^{lf,h\pi}_m(\widetilde{K};\mathbb{L}_\bullet) & \xrightarrow{A^{lf}} & L_m(\mathbb{C}_{\widetilde{K}}(\mathbb{Z})^{h\pi}) & \to & \mathbb{S}^{b,h\pi}_m(\widetilde{K}) & \to \dots
\end{array}$$

with

$$\mathbb{S}^{b,h\pi}_*(\widetilde{K}) = \pi_*(\mathbb{S}^b(\widetilde{K})^{h\pi}) , \quad L_*(\mathbb{C}_{\widetilde{K}}(\mathbb{Z})^{h\pi}) = \pi_*(\mathbb{L}_\bullet(\mathbb{C}_{\widetilde{K}}(\mathbb{Z}))^{h\pi}) .$$

The composite

$$\mathbb{S}_{m+1}(K) \xrightarrow{\text{trf}} \mathbb{S}^{b,h\pi}_{m+1}(\widetilde{K}) \longrightarrow \mathbb{S}^{b}_{m+1}(\widetilde{K})$$

sends the structure invariant $s(h) \in \mathbb{S}_{m+1}(K)$ of a homotopy equivalence $h : M' \longrightarrow M$ of compact m-dimensional manifolds with a π_1-isomorphism reference map $M \longrightarrow K$ to the $\widetilde{K}$-bounded structure invariant $s^b(\widetilde{h}) \in \mathbb{S}^b_{m+1}(\widetilde{K})$ of the induced $\widetilde{K}$-bounded homotopy equivalence $\widetilde{h} : \widetilde{M'} \longrightarrow \widetilde{M}$ of the universal covers.

The method of infinite transfers first applied by Carlsson [13] to the algebraic K-theory version of the Novikov conjecture has the following application in algebraic L-theory to the integral Novikov conjecture:

Proposition 11.1 *Let π be a group such that the classifying space $B\pi$ has the homotopy type of a finite CW complex, so that π is torsion-free. If the universal cover $E\pi$ of $B\pi$ is realized by a contractible metric space E with a free π-action and such that the locally finite assembly maps are isomorphisms*

$$A^{lf} \ : \ H^{lf}_*(E;\mathbb{L}_\bullet(\mathbb{Z})) \xrightarrow{\simeq} L_*(\mathbb{C}_E(\mathbb{Z}))$$

then the integral Novikov conjecture holds for π, i.e. the assembly maps

$$A \ : \ H_*(B\pi;\mathbb{L}_\bullet(\mathbb{Z})) \longrightarrow L_*(\mathbb{Z}[\pi])$$

are split injections.

Proof The $E\pi$-bounded structure spectrum $\mathbb{S}^b(E\pi)$ is contractible, and hence so is the homotopy fixed point spectrum $\mathbb{S}^b(E\pi)^{h\pi}$. The locally finite assembly map

$$A^{lf}_\pi \ : \ \mathbb{H}^{lf}_\bullet(E\pi;\mathbb{L}_\bullet(\mathbb{Z}))^{h\pi} \longrightarrow \mathbb{L}_\bullet(\mathbb{C}_{E\pi}(\mathbb{Z}))^{h\pi}$$

is a homotopy equivalence, so that there are defined homotopy equivalences

$$\mathbb{H}_\bullet(B\pi;\mathbb{L}_\bullet(\mathbb{Z})) \simeq \mathbb{H}^{lf}_\bullet(E\pi;\mathbb{L}_\bullet(\mathbb{Z}))^{h\pi} \simeq \mathbb{L}_\bullet(\mathbb{C}_{E\pi}(\mathbb{Z}))^{h\pi} .$$

The infinite transfer maps

$$\text{trf} \ : \ \mathbb{L}_\bullet(\mathbb{Z}[\pi]) \simeq \mathbb{L}_\bullet(\mathbb{C}_{E\pi}(\mathbb{Z}))^{\pi} \longrightarrow \mathbb{L}_\bullet(\mathbb{C}_{E\pi}(\mathbb{Z}))^{h\pi} \simeq \mathbb{H}_\bullet(B\pi;\mathbb{L}_\bullet(\mathbb{Z}))$$

induce splitting maps $\text{trf} : L_*(\mathbb{Z}[\pi]) \longrightarrow H_*(B\pi;\mathbb{L}_\bullet(\mathbb{Z}))$ for the assembly maps $A : H_*(B\pi;\mathbb{L}_\bullet(\mathbb{Z})) \longrightarrow L_*(\mathbb{Z}[\pi])$.

□

Example 11.2 Let $\pi = \mathbb{Z}^n$, so that

$$B\pi = T^n \ , \ E = E\pi = \mathbb{R}^n .$$

Compactify E by adding the $(n-1)$-sphere at infinity

$$\overline{E} = \mathbb{R}^n \cup S^{n-1} = D^n ,$$

extending the free $\mathbb{Z}^n$-action on $\mathbb{R}^n$ by the identity on $\partial E = S^{n-1}$. In this case the locally finite assembly isomorphisms

$$\begin{aligned} A^{lf} \ : \ H_*(D^n, S^{n-1}; \mathbb{L}_\bullet(\mathbb{Z})) &= H^{lf}_*(\mathbb{R}^n; \mathbb{L}_\bullet(\mathbb{Z})) \\ &= \widetilde{H}_{*-1}(S^{n-1}; \mathbb{L}_\bullet(\mathbb{Z})) = L_{*-n}(\mathbb{Z}) \\ &\xrightarrow{\simeq} L_*(\mathbb{C}_{\mathbb{R}^n}(\mathbb{Z})) \end{aligned}$$

and the assembly isomorphisms

$$A \ : \ H_*(T^n; \mathbb{L}_\bullet(\mathbb{Z})) \xrightarrow{\simeq} L_*(\mathbb{Z}[\mathbb{Z}^n])$$

were already obtained in Ranicki [45], [50], using the identification of the $\mathbb{R}^n$-bounded L-groups of a ring with involution A with the lower L-groups

$$L_*(\mathbb{C}_{\mathbb{R}^n}(A)) = L^{\langle 1-n \rangle}_{*-n}(A)$$

and the splitting theorem

$$L_*(A[\mathbb{Z}^n]) = \sum_{k=0}^{n} \binom{n}{k} L^{\langle 1-k \rangle}_{*-k}(A) ,$$

with $L^{\langle -* \rangle}_*(\mathbb{Z}) = L_*(\mathbb{Z})$ by virtue of $Wh_{-*}(\{1\}) = 0$.

□

Example 11.3 Let $\pi = \pi_1(M)$ be the fundamental group of a complete closed n-dimensional Riemannian manifold with non-positive sectional curvature M. The universal cover $E = \widetilde{M}$ is a complete simply-connected open Riemannian manifold such that the exponential map at any point $x \in E$ defines a diffeomorphism

$$\exp_x \ : \ \tau_x(E) = \mathbb{R}^n \longrightarrow E$$

by the Hadamard-Cartan theorem, so that $M = B\pi$ is aspherical. The locally finite assembly map

$$A^{lf} \ : \ H^{lf}_*(E; \mathbb{L}_\bullet(\mathbb{Z})) = L_{*-n}(\mathbb{Z}) \longrightarrow L_*(\mathbb{C}_E(\mathbb{Z}))$$

is an isomorphism, so that the integral Novikov conjecture holds for π by 11.1. See Farrell and Hsiang [16] for the original geometric proof, which is generalized by Carlsson and Pedersen [14] (cf. 11.5 below) by abstracting the properties of the π-action on the compactification $\overline{E} = D^n$ near the sphere at ∞ $\partial E = \overline{E} \backslash E = S^{n-1}$.

□

Example 11.4 For any integer $g \geq 1$ let

$$\pi_g = \{a_1, a_2, \ldots, a_{2g} \,|\, [a_1, a_2] \ldots [a_{2g-1}, a_{2g}]\}$$

be the fundamental group of the closed oriented surface M_g of genus g, so that

$$B\pi_g = M_g \ , \ E = E\pi_g = \mathbb{R}^2 \ .$$

For $g = 1$ $M_g = T^2$, as already considered in 11.2. For $g \geq 2$ M_g has a hyperbolic structure, and the free action of π_g on $E = \mathbb{R}^2 = \mathrm{int}(D^2)$ extends to a (non-free) action on $\overline{E} = D^2$, which is the identity on $\partial E = S^1$. The hypotheses of 11.1 are satisfied, so that the assembly maps

$$A \ : \ h_*(B\pi_g) = H_*(B\pi_g; \mathbb{L}_\bullet(\mathbb{Z})) \longrightarrow L_*(\mathbb{Z}[\pi_g])$$

are split injections, and the integral Novikov conjecture holds for π_g. In fact, these assembly maps are isomorphisms, which may be verified by the following argument (for which I am indebted to C.T.C.Wall). By the Freiheitssatz for one-relator groups the normal subgroup $\rho_g \triangleleft \pi_g$ generated by $a_1, a_2, \ldots, a_{2g-1}$ is free, so that π_g is the α-twisted extension of ρ_g by $\mathbb{Z} = \{a_{2g}\}$

$$\{1\} \longrightarrow \rho_g \longrightarrow \pi_g \longrightarrow \mathbb{Z} \longrightarrow \{1\}$$

and

$$\mathbb{Z}[\pi_g] = \mathbb{Z}[\rho_g]_\alpha[z, z^{-1}]$$

is the α-twisted Laurent polynomial extension of $\mathbb{Z}[\rho_g]$, with

$$z = a_{2g} \ , \ \alpha(a_i) = (a_{2g})^{-1} a_i a_{2g} \ (1 \leq i \leq 2g-1) \ .$$

The assembly maps $A : h_*(B\rho_g) \longrightarrow L_*(\mathbb{Z}[\rho_g])$ are isomorphisms by Cappell [9]. A 5-lemma argument applied to the assembly map

$$\begin{array}{ccccccccc}
\ldots \longrightarrow & h_n(B\rho_g) & \xrightarrow{1-\alpha} & h_n(B\rho_g) & \longrightarrow & h_n(B\pi_g) & \longrightarrow & h_{n-1}(B\rho_g) & \longrightarrow \ldots \\
& \downarrow{\scriptstyle A} & & \downarrow{\scriptstyle A} & & \downarrow{\scriptstyle A} & & \downarrow{\scriptstyle A} & \\
\ldots \longrightarrow & L_n(\mathbb{Z}[\rho_g]) & \xrightarrow{1-\alpha} & L_n(\mathbb{Z}[\rho_g]) & \longrightarrow & L_n(\mathbb{Z}[\pi_g]) & \longrightarrow & L_{n-1}(\mathbb{Z}[\rho_g]) & \longrightarrow \ldots
\end{array}$$

from the Wang exact sequence in group homology to the exact sequence of Ranicki [46] for the L-theory of a twisted Laurent polynomial extension (using $Wh(\pi_g) = 0$) shows that the assembly maps

$$A : h_*(B\pi_g) = H_*(B\pi_g; \mathbb{L}_\bullet(\mathbb{Z})) \longrightarrow L_*(\mathbb{Z}[\pi_g])$$

are isomorphisms.

□

Theorem 11.5 (Carlsson and Pedersen [14]) *Let π be a group with finite classifying space $B\pi$ such that the universal cover $E\pi$ is realized by a contractible metric space E with a free π-action, and with a compactification $\overline{E}$ such that:*

(a) *the free π-action on E extends to a π-action on $\overline{E}$ (which need not be free),*
(b) *$\overline{E}$ is contractible,*
(c) *compact subsets of E become small near the boundary $\partial E = \overline{E} \backslash E$, i.e. for every point $y \in \partial E$, every compact subset $K \subseteq E$ and for every neighbourhood U of y in $\overline{E}$, there exists a neighbourhood V of y in $\overline{E}$ so that if $g \in \pi$ and $g(K) \cap V \neq \emptyset$ then $g(K) \subset U$.*

Then the integral Novikov conjecture holds for π.

□

The proof of 11.5 uses infinite transfer maps (as in 11.1), but with the continuously controlled category $\mathbb{B}_{X,Y}(\mathbb{Z})$ of Anderson, Connolly, Ferry and Pedersen [1] replacing the bounded category $\mathbb{C}_E(\mathbb{Z})$ of Pedersen and Weibel [43]. For a compact metrizable space X and a closed subspace $Y \subseteq X$ $\mathbb{B}_{X,Y}(\mathbb{Z})$ is the category with the same objects as $\mathbb{C}_E(\mathbb{Z})$, where $E = X \backslash Y$. A morphism in $\mathbb{B}_{X,Y}(\mathbb{Z})$

$$f = \{f(x', x)\} : A = \sum_{x \in E} A(x) \longrightarrow B = \sum_{x' \in E} B(x')$$

is a $\mathbb{Z}$-module morphism such that for every $y \in Y$ and every neighbourhood $U \subseteq X$ of y there is a neighbourhood $V \subseteq U$ such that

$$f(x', x) = 0 : A(x) \longrightarrow B(x') \ \ (x \in V, x' \in X \backslash U)$$

(or equivalently $f(A(V)) \subseteq B(U)$). If E is dense in X and compact subsets of E become small near the boundary $\partial E = Y$ in $\overline{E} = X$ there is defined a forgetful functor $\mathbb{C}_E(\mathbb{Z}) \longrightarrow \mathbb{B}_{X,Y}(\mathbb{Z})$. This functor induces isomorphisms in K- and L-theory in certain cases with X contractible (e.g. if $E = O(K)$ is the open cone on a compact subcomplex $K \subseteq S^N$ and $X = O(K) \cup K$ is the closed cone, with $Y = K \subset X$), but it is not known if it does so in general. See Pedersen [41] for the relationship between the bounded and continuously controlled categories.

The algebraic transversality of Ranicki [50], [51] is extended in Carlsson and Pedersen [14, 5.4] to prove that the continuously controlled L-theory assembly maps

$$A \ : \ H_*^{lf}(E;\mathbb{L}_\bullet(\mathbb{Z})) \ = \ H_*(X,Y;\mathbb{L}_\bullet(\mathbb{Z})) \longrightarrow L_*(\mathbb{B}_{X,Y}(\mathbb{Z}))$$

are isomorphisms if $E = E\pi$ and $(X,Y) = (\overline{E},\partial E)$ are as in 11.5 – this is the key step in the proof. As in 10.1 there are potential lower Whitehead torsion obstructions to splitting, which are avoided by the computation $Wh_{-*}(\{1\}) = 0$ of Bass, Heller and Swan [4]. The assembly map $A : H_m^{lf,h\pi}(E;\mathbb{L}_\bullet(\mathbb{Z}))\longrightarrow L_m(\mathbb{B}_{\overline{E},\partial E}(\mathbb{Z})^{h\pi})$ in the commutative square

$$\begin{array}{ccc}
H_m(B\pi;\mathbb{L}_\bullet(\mathbb{Z})) & \xrightarrow{\quad A \quad} & L_m(\mathbb{Z}[\pi]) \\
\Big\downarrow{\scriptstyle \mathrm{trf}\ \cong} & & \Big\downarrow{\scriptstyle \mathrm{trf}} \\
H_m^{lf,h\pi}(E;\mathbb{L}_\bullet(\mathbb{Z})) & \xrightarrow{\quad A \quad} & L_m(\mathbb{B}_{\overline{E},\partial E}(\mathbb{Z})^{h\pi})
\end{array}$$

is an isomorphism, giving the splitting of the assembly map

$$A \ : \ H_m(B\pi;\mathbb{L}_\bullet(\mathbb{Z})) \longrightarrow L_m(\mathbb{Z}[\pi]) \ .$$

Example 11.6 As already noted by Carlsson and Pedersen [14], the work of Bestvina and Mess [5] shows that negatively curved groups in the sense of Gromov satisfy the conditions of Theorem 11.5, so that the integral Novikov conjecture holds for these groups. The fundamental groups π of complete Riemannian manifolds (of finite homotopy type) $B\pi$ with non-positive curvature are the main examples of such groups – cf. 11.3.

□

If π is in the class of groups satisfying the conditions of 11.5

$$L_*(\mathbb{Z}[\pi]) \ = \ H_*(B\pi;\mathbb{L}_\bullet) \oplus \mathbb{S}_*(B\pi) \ .$$

It is worth investigating the extent to which $\mathbb{S}_*(B\pi)$ is determined by the Cappell UNil-groups.

References

[1] D. R. Anderson, F. X. Connolly, S. Ferry and E. K. Pedersen, *Algebraic K-theory with continuous control at infinity*, J. Pure and App. Alg. 94, 25–47 (1994)

[2] M. Atiyah, *The signature of fibre bundles*, Papers in the honour of Kodaira, Tokyo Univ. Press, 73–84 (1969)
[3] H. Bass, *Algebraic K-theory*, Benjamin (1969)
[4] ––, A. Heller and R. Swan, *The Whitehead group of a polynomial extension*, Publ. Math. I. H. E. S. 22, 61–80 (1964)
[5] M. Bestvina and G. Mess, *The boundary of negatively curved groups*, Journal of A. M. S. 4, 469–481 (1991)
[6] W. Browder, *Structures on* $M \times \mathbb{R}$, Proc. Camb. Phil. Soc. 61, 337–345 (1965)
[7] ––, *Manifolds with* $\pi_1 = \mathbb{Z}$, Bull. A. M. S. 72, 238–244 (1966)
[8] J. L. Bryant and P. S. Pacheco, K_{-i}*-obstructions to factoring an open manifold*, Topology and its Applications 29, 107–139 (1988)
[9] S. Cappell, *Unitary nilpotent groups and hermitian K-theory I.*, Bull. A. M. S. 80, 1117–1122 (1974)
[10] ––, *A splitting theorem for manifolds*, Inventiones Math. 33, 69–170 (1976)
[11] ––, *On homotopy invariance of higher signatures*, Inventiones Math. 33, 171–179 (1976)
[12] G. Carlsson, *Homotopy fixed points in the algebraic K-theory of certain infinite discrete groups* Proc. James Conf., LMS Lecture Notes 139, 5–10 (1989)
[13] ––, *Bounded K-theory and the assembly map in algebraic K-theory*, in these proceedings
[14] –– and E. K. Pedersen, *Controlled algebra and the Novikov conjectures for K- and L-theory*, Topology (to appear)
[15] F. T. Farrell and W. C. Hsiang, *Manifolds with* $\pi_1 = G \times_\alpha T$, Amer. J. Math. 95, 813–845 (1973)
[16] –– and ––, *On Novikov's conjecture for nonpositively curved manifolds*, Ann. Math. 113, 197–209 (1981)
[17] –– and L. E. Jones, *Rigidity in Geometry and Topology*, Proc. 1990 I. C. M., Kyoto, 653–663 (1991)
[18] S. Ferry and E. K. Pedersen, *Epsilon surgery theory*, in these proceedings
[19] ––, A. A. Ranicki and J. Rosenberg, *A History and Survey of the Novikov Conjecture*, in these proceedings
[20] –– and S. Weinberger, *A coarse approach to the Novikov conjecture*, in these proceedings
[21] M. Gromov, *Geometric reflections on the Novikov conjecture*, in these proceedings
[22] ––, *Positive curvature, macroscopic dimension, spectral gaps and higher signatures*, Functional Analysis on the Eve of the 21st Century (Proc. conf. in honor of I. M. Gelfand's 80th birthday), Progress in Math., Birkhäuser (to appear)
[23] G. Higman, *Units in group rings*, Proc. Lond. Math. Soc. (2) 46, 231–

248 (1940)

[24] F. Hirzebruch, *Topological methods in algebraic geometry*, Grundlehren der mathematischen Wissenschaften 131, Springer (1978)

[25] C. B. Hughes and A. A. Ranicki, *Ends of complexes*, preprint

[26] S. Hutt, *Poincaré sheaves on topological spaces*, preprint

[27] M. Kervaire and J. Milnor, *Groups of homotopy spheres I.*, Ann. Math. 77, 504–537 (1963)

[28] R. Kirby and L. Siebenmann, *Foundational essays on topological manifolds smoothings and triangulations*, Ann. Math. Stud. 88, Princeton (1977)

[29] W. Lück and A. A. Ranicki, *Surgery obstructions of fibre bundles*, J. of Pure and Applied Algebra 81, 139–189 (1992)

[30] G. Lusztig, *Novikov's higher signature and families of elliptic operators*, J. Diff. Geom. 7, 229–256 (1972)

[31] W. Meyer, *Die Signatur von lokalen Koeffizientensystemen und Faserbündeln*, Bonn. Math. Schr. 53 (1972)

[32] J. Milgram and A. A. Ranicki, *The L-theory of Laurent polynomial extensions and genus 0 function fields*, J. f. reine und angew. Math. 406, 121–166 (1990)

[33] J. Milnor, *On manifolds homeomorphic to the 7-sphere*, Ann. of Math. 64, 399–405 (1956)

[34] –– and J. Stasheff, *Characteristic classes*, Ann. Math. Stud. 76, Princeton (1974)

[35] A. S. Mishchenko, *Homotopy invariants of non–simply connected manifolds. III. Higher signatures*, Izv. Akad. Nauk SSSR, ser. mat. 35, 1316–1355 (1971)

[36] S. P. Novikov, *Homotopy equivalent smooth manifolds I.*, Izv. Akad. Nauk SSSR, ser. mat. 28, 365–474 (1965)
English translation: A. M. S. Transl. (2) 48, 271–396 (1965)

[37] ––, *Rational Pontrjagin classes. Homeomorphism and homotopy type of closed manifolds I.*, Izv. Akad. Nauk SSSR, ser. mat. 29, 1373–1388 (1965)
English translation: A. M. S. Transl. (2) 66, 214–230 (1968)

[38] ––, *Manifolds with free abelian fundamental group and applications (Pontrjagin classes, smoothings, high–dimensional knots)*, Izv. Akad. Nauk SSSR, ser. mat. 30, 208–246 (1966)
English translation: A. M. S. Transl. (2) 67, 1–42 (1969)

[39] ––, *The algebraic construction and properties of hermitian analogues of K-theory for rings with involution, from the point of view of the hamiltonian formalism. Some applications to differential topology and the theory of characteristic classes*, Izv. Akad. Nauk SSSR, ser. mat. 34, I. 253–288, II. 478–500 (1970)
English translation: Math. USSR Izv. 4, 257–292, 479–505 (1970)

[40] E. K. Pedersen, *On the K_{-i}-functors*, J. of Algebra 90, 461–475 (1984)

[41] ––, *Bounded and continuous control*, in these proceedings
[42] –– and A. A. Ranicki, *Projective surgery theory*, Topology 19, 239–354 (1980)
[43] –– and C. Weibel, *K-theory homology of spaces*, Proc. 1986 Arcata Algebraic Topology Conf., Springer Lecture Notes 1370, 346–361 (1989)
[44] A. A. Ranicki, *Algebraic L-theory I. Foundations*, Proc. Lond. Math. Soc. 27 (3), 101–125 (1973)
[45] ––, *Algebraic L-theory II. Laurent extensions*, Proc. Lond. Math. Soc. 27 (3), 126–158 (1973)
[46] ––, *Algebraic L-theory III. Twisted Laurent extensions*, Proc. 1972 Battelle Seattle K-theory Conf., Vol. III, Springer Lecture Notes 343, 412–463 (1973)
[47] ––, *The algebraic theory of surgery I., II.*, Proc. Lond. Math. Soc. 40 (3), 87–287 (1980)
[48] ––, *Exact sequences in the algebraic theory of surgery*, Mathematical Notes 26, Princeton (1981)
[49] ––, *Additive L-theory*, K-theory 3, 163–195 (1989)
[50] ––, *Lower K- and L-theory*, London Math. Soc. Lecture Notes 178, Cambridge University Press (1992)
[51] ––, *Algebraic L-theory and topological manifolds*, Cambridge Tracts in Mathematics 102, Cambridge University Press (1992)
[52] –– (ed.), *The Hauptvermutung book*, a collection of papers by Casson, Sullivan, Armstrong, Rourke, Cooke and Ranicki, K-theory Journal book series (to appear)
[53] ––, *Splitting theorems in the algebraic theory of surgery*, (in preparation)
[54] –– and M. Yamasaki, *Controlled K-theory*, Topology and Its Applications 61, 1–59 (1995)
[55] L. Siebenmann, *The obstruction to finding the boundary of an open manifold*, Princeton Ph. D. thesis (1965)
[56] ––, *A total Whitehead torsion obstruction to fibering over the circle*, Comm. Math. Helv. 45, 1–48 (1972)
[57] F. Waldhausen, *Algebraic K-theory of generalized free products I.,II.*, Ann. Math. 108, 135–256 (1978)
[58] C. T. C. Wall, *Surgery on compact manifolds*, Academic Press (1971)
[59] M. Weiss, *Surgery and the generalized Kervaire invariant I.,II.*, Proc. Lond. Math. Soc. 51 (3), 146–230 (1985)

Dept. of Mathematics and Statistics, Edinburgh University, Edinburgh EH9 3JZ, Scotland, UK

email: a.ranicki@edinburgh.ac.uk

Analytic Novikov for topologists

Jonathan Rosenberg

ABSTRACT. We explain for topologists the "dictionary" for understanding the analytic proofs of the Novikov conjecture, and how they relate to the surgery-theoretic proofs. In particular, we try to explain the following points:

(1) Why do the analytic proofs of the Novikov conjecture require the introduction of C^*-algebras?
(2) Why do the analytic proofs of the Novikov conjecture all use K-theory instead of L-theory? Aren't they computing the wrong thing?
(3) How can one show that the index map μ or β studied by operator theorists matches up with the assembly map in surgery theory?
(4) Where does "bounded surgery theory" appear in the analytic proofs? Can one find a correspondence between the sorts of arguments used by analysts and the controlled surgery arguments used by topologists?

The literature on the Novikov conjecture (see [FRR]) consists of several different kinds of papers. Most of these fall into two classes: those based on topological arguments, usually involving surgery theory, and those based on analytic arguments, usually involving index theory. The purpose of this note is to "explain" the second class of papers to those familiar with the first class. I do *not* intend here to give a detailed sketch of the Kasparov KK-approach to the Novikov conjecture (for which the key details appear in [Kas4], [Fac2], and [KS]), since this has already been done in the convenient expository references [Fac1], [Kas2], [Kas3], [Bla], and [Kas5]. Nor do I intend to explain the approach to the Novikov conjecture taken by Mishchenko and Soloviev (found in [Mis1], [Mis2], [MS], [Mis3], and [KS,

---

1991 *Mathematics Subject Classification*. Primary 19J25; Secondary 46L80, 19G24, 19K35, 19K56, 57R67, 55P42, 58G12.

Expanded version of a lecture given at the Oberwolfach meeting on "Novikov conjectures, index theorems and rigidity," 6 September, 1993. I would like to thank my co-organizers Steve Ferry and Andrew Ranicki for encouraging me to write this talk up for publication, and also for useful technical comments. I would also like to thank Erik Pedersen, John Roe, Stephan Stolz, Shmuel Weinberger, and the referee for helpful suggestions about the exposition. Finally, I would like to thank Bruce Williams for noticing a homotopy-theoretic error in the original version of §2.

This work was partially supported by NSF Grant # DMS-92-25063.

Appendix]), using Fredholm representations but not using KK, for which a convenient expository reference is [HsR]. Rather, I intend to concentrate on explaining the "dictionary" for relating the two main classes of papers on the Novikov conjecture, the topological and the analytic, and on trying to find the common ground relating them. I will assume that the reader is already familiar with the reduction of Novikov's original conjecture on homotopy invariance of higher signatures to a statement about the L-theory assembly map

$$A : H_\bullet(X; \mathbb{L}_\bullet(\mathbb{Z})) \to L_\bullet(\mathbb{Z}\pi_1(X)),$$

in the case where the space X is taken to be a $K(\pi, 1)$-space $B\pi$. This aspect of the problem is discussed elsewhere in these proceedings, especially in [Ran3].

§1. Why K-theory of C^*-algebras?

Topologists looking at the analytic literature on the Novikov conjecture often wonder why so much emphasis is placed on the (topological) K-theory of C^*-algebras, when in fact it is known that the original Novikov conjecture has to do with L-theory of group rings, and that certain related problems in the topology of non-simply connected manifolds (concerning Whitehead torsion) have to do with the algebraic K-theory of integral group rings. First let's pin down the objects of study.

1.1. Definition. A **Banach $*$-algebra** is a real or complex Banach algebra, together with an isometric (conjugate-linear) involution $*$. A **$*$-homomorphism** or **$*$-isomorphism** of such algebras means a homomorphism or isomorphism preserving the involutions. A **real or complex C^*-algebra** A is a Banach $*$-algebra which is isometrically $*$-isomorphic to a norm-closed involutive subalgebra of the bounded operators on some Hilbert space (real or complex, as the case may), with involution obtained by sending an operator a to its Hilbert space adjoint a^*, defined by the property that

$$\langle a\xi, \eta\rangle = \langle \xi, a^*\eta\rangle.$$

(In the real case, it is important to remember that real, complex, and quaternionic Hilbert spaces may all be regarded by restriction of scalars as real Hilbert spaces.) If a C^*-algebra A acts on a Hilbert space $\mathcal{H}$, then $M_n(A)$ naturally acts on the Hilbert space $\mathcal{H}^n$, and thus any matrix algebra over a C^*-algebra is also a C^*-algebra.

We also quickly remind the reader of the most crucial special properties of C^*-algebras, which may be deduced either from the Spectral Theorem for self-adjoint operators on a Hilbert space, or else from the algebraic characterization of C^*-algebras as in [Ped, Ch. 1]. First, some basic definitions.

1.2. Definition. An element a of a Banach $*$-algebra A is called **self-adjoint** if $a = a^*$, **positive** (actually "non-negative" would be better, since 0 is not excluded, but the name "positive" has become standard), written $a \geq 0$, if $a = b^*b$ for some other element b. If A has a unit and $a \in A$, the **spectrum** $\operatorname{spec} a$ of a (this has nothing to do with spectra in homotopy theory!!) is the set of $\lambda \in \mathbb{C}$ such that $a - \lambda \cdot 1$ is *not* invertible in A (or in $A_{\mathbb{C}} := A \otimes_{\mathbb{R}} \mathbb{C}$ if A is only an algebra over $\mathbb{R}$). This is always a compact subset of $\mathbb{C}$, and clearly it only depends on the structure of A as an algebra over $\mathbb{C}$ or $\mathbb{R}$, not on the norm. Note that in the real case, since the obvious action of $\operatorname{Gal}(\mathbb{C}/\mathbb{R})$ on $A_{\mathbb{C}}$ fixes $a \in A$, it follows that $\operatorname{spec} a$ is invariant under complex conjugation. If A does not have a unit, then we may always embed A as an ideal of codimension 1 in a C^*-algebra A_+ with unit, obtained by realizing A as an algebra of operators on a Hilbert space $\mathcal{H}$ and considering all operators of the form $a + \lambda \cdot 1$, where λ is a scalar and 1 is the identity operator on $\mathcal{H}$. The spectrum of $a \in A$ may then be defined to be its spectrum in A_+, and in this case, $\operatorname{spec} a$ always contains 0.

Here are the crucial facts we will need.

1.3. Facts. *Let A be a real or complex C^*-algebra, and let $a \in A$.*

(i) *If a is self-adjoint, then $\operatorname{spec} a \subset \mathbb{R}$. Conversely, if a and a^* commute and $\operatorname{spec} a \subset \mathbb{R}$, then $a = a^*$.*

(ii) *If $a \geq 0$, then $\operatorname{spec} a \subset [0, \infty)$. Conversely, if $a = a^*$ and $\operatorname{spec} a \subset [0, \infty)$, then $a \geq 0$.*

(iii) (**Functional Calculus**) *Assume that $a = a^*$, or more generally that a and a^* commute, and that f is a continuous function defined on $\operatorname{spec} a$. (If A is an algebra over $\mathbb{C}$, f can be complex-valued; if A is an algebra over $\mathbb{R}$, then f has to satisfy the condition that $\overline{f(z)} = f(\overline{z})$.) Assume in addition that if A does not have a unit, then $f(0) = 0$. Then there is a unique element $f(a) \in A$ which is contained in the closed subalgebra generated by a, having the property that if f_n is a sequence of polynomials (chosen without constant term, in case A does not have a unit), converging uniformly to f on $\operatorname{spec} a$, then $f_n(a)$ (defined as in any algebra) converges in norm to $f(a)$.*

(iv) *Any $*$-homomorphism φ from A to another C^*-algebra is automatically continuous with closed range, and induces an isometric $*$-isomorphism from $A/\ker\varphi$ onto $\operatorname{range}\varphi$.*

(v) *If A has a unit, any finitely generated projective A-module defined by an idempotent $e = e^2$ in A may also be defined by a self-adjoint idempotent p in A.*

(vi) *The norm on A is determined by the $*$-algebra structure by the* formula $\|a\|^2 = \max \operatorname{spec} a^*a$.

Proof. We omit the proofs of (i)–(iv), which can be found in [Ped, Ch. 1], except for the real case of (iii). Here one first uses the complex case to define $f(a)$ as an element of $A_{\mathbb{C}}$, and then one uses the condition $\overline{f(z)} = f(\bar{z})$ to see that $f(a)$ is invariant under the action of $\mathrm{Gal}(\mathbb{C}/\mathbb{R})$ and thus lies in A. The property (vi) follows from basic facts about operators on Hilbert spaces.

Since (v) is a little less standard (I believe it was first noted by Kaplansky), we include a proof. One can do everything completely algebraically, but to get a better impression of what is going on, suppose A is acting on a Hilbert space $\mathcal{H}$. Then the image of e must be a closed subspace V of $\mathcal{H}$, and with respect to the decomposition $\mathcal{H} = V \oplus V^{\perp}$ of $\mathcal{H}$, e must have a matrix of the form $\begin{pmatrix} 1 & a \\ 0 & 0 \end{pmatrix}$, where $a : V^{\perp} \to V$ is a bounded operator. Then

$$e^* = \begin{pmatrix} 1 & 0 \\ a^* & 0 \end{pmatrix}, \qquad ee^* = \begin{pmatrix} 1 + aa^* & 0 \\ 0 & 0 \end{pmatrix},$$

and in particular, the spectrum of ee^* is contained in $\{0\} \cup [1, \infty)$. Thus if $f(0) = 0$ and $f(t) = 1$ for $t \geq 1$, f is continuous on the spectrum of ee^* and thus (by (iii) above) $p = f(ee^*)$ lies in A and is a self-adjoint projection with the same range as e. We claim $Ae \cong Ap$ as projective modules over A. But in fact $ep = p$ and $pe = e$, so right multiplication by p gives an isomorphism from Ae to Ap, with inverse given by right multiplication by e. □

We should emphasize that these facts are special to C^*-algebras; they fail in some other Banach $*$-algebras.

1.4. To explain the appearance of K-theory of group C^*-algebras, consider a group π with integral group ring $\mathbb{Z}\pi$ and complex group ring $\mathbb{C}\pi$. One can form the (complex) Hilbert space $\ell^2(\pi)$ having π as an orthonormal basis, and by definition, the *reduced group C^*-algebra* $C_r^*(\pi)$ is the completion of $\mathbb{C}\pi$ in the operator norm for its action on $\ell^2(\pi)$ by left multiplication. Most analytic approaches to the Novikov conjecture for the group π, and thus to homotopy invariance of higher signatures for manifolds having π as fundamental group, involve in some way the K-theory of the C^*-algebra $C_r^*(\pi)$. While there are obvious inclusions

$$\mathbb{Z}\pi \hookrightarrow \mathbb{C}\pi \hookrightarrow C_r^*(\pi),$$

it is not in any way expected that these should be close to inducing isomorphisms on K-theory. In fact, in the fundamental example where $\pi \cong \mathbb{Z}^n$ is free abelian, $\tilde{K}_0$ vanishes for both $\mathbb{Z}\pi$ and $\mathbb{C}\pi$, whereas $K_0(C_r^*(\pi)) \cong$

$K^0(T^n)$ has rank 2^{n-1}. Rather, the introduction of $K_0(C_r^*(\pi))$ is designed to facilitate the definition of a "non-simply connected Hirzebruch signature formula," which Novikov recognized from the start to be intimately tied up with the conjecture on higher signatures. In fact, an examination of Novikov's original text in [Nov, §11] shows that he already suggested the introduction of C^*-algebras for this purpose, though only in a special case. (For the original text and an analysis of Novikov's formulation of the higher signature conjecture, see [FRR, §2].)

1.5. To see how C^*-algebras come in, it is worth thinking about how the ordinary signature of a manifold is defined. The Poincaré duality structure (which is the homotopy-theoretic manifestation of the basic geometric property of transversality) gives rise to a symmetric bilinear form on middle-degree cohomology. The signature is extracted from this form by diagonalizing the form (over $\mathbb{R}$) and taking the formal difference of its positive and negative eigenspaces. Thus while it is customary to view the signature as an ordinary integer, it naturally arises *as an element of a group of such formal differences*, that is, as an element of $K_0(\mathbb{R})$ (which happens to be naturally isomorphic to $\mathbb{Z}$). It is thus natural to expect a "generalized signature" also to be a formal difference of positive and negative eigenspaces of a bilinear form, this time arising from Poincaré duality with local coefficients. For this to make sense, it is necessary to work in a ring over which the form is diagonalizable, and this is where C^*-algebras come in. To "diagonalize" a form, one wants to make use of the Spectral Theorem for self-adjoint operators on a Hilbert space, but this requires that our involutive ring be identifiable with a ring of bounded operators on a Hilbert space, in other words, with a (subalgebra of a) C^*-algebra. Then the "positive and negative eigenspaces" of the form will be projective modules over this ring, and thus the signature, their formal difference, will be an element of K_0 of a C^*-algebra completion of the (real or complex) group ring. Signatures are more tractable objects than forms, since (topological) K-theory behaves better than L-theory, because of its useful "rigidity" properties of homotopy invariance and strong excision. Thus the whole machinery of topological K-theory and index theory can be brought to bear on the study of "generalized signatures."

An important question in trying to prove the Novikov conjecture using analytic methods is whether the passage from symmetric forms to signatures loses any essential information. One may view this operation as consisting of two rather different steps: passage from symmetric forms over the group ring to symmetric forms over a *completed* group ring, and passage from symmetric forms to signatures in the context of C^*-algebras. The first of these operations is still not completely understood, though it seems that in many cases it loses information only at the prime 2. (As an example, the natural maps $L_\bullet(\mathbb{Z}) \to L_\bullet(\mathbb{R})$ are not isomorphisms in even degrees, but become isomorphisms after tensoring with $\mathbb{Z}[\frac{1}{2}]$.) However, the second operation, passing from forms to signatures in the context of C^*-algebras, is better understood, and loses almost no information. A key folk theorem in this direction, which I believe may have first been discovered by Gelfand or Mishchenko, is often cited but hard to pin down in the literature,* so we include a proof here. The case where $A = C^{\mathbb{C}}(X)$ is complex and abelian appears in [GM] and as Theorem 4.1 in [Nov].

1.6. Theorem (Folklore). *Let A be a real or complex C^*-algebra with unit, regarded also as ring with involution (in the complex case, note that the involution on scalar multiples of the identity is complex conjugation). If φ is a non-singular hermitian form on a finitely generated projective A-module P, then there is a unique φ-invariant splitting $P = P^+ \oplus P^-$ of P, with respect to which φ is equivalent to the form with matrix*

$$\begin{pmatrix} 1_{P^+} & 0 \\ 0 & -1_{P^-} \end{pmatrix}.$$

Thus isomorphism classes of non-singular hermitian forms over A may be identified with pairs $([P^+], [P^-])$ of isomorphism classes of finitely generated projective modules over A. The "signature" map

$$[P, \varphi] \mapsto [P^+] - [P^-]$$

defines a natural isomorphism of functors

$$\Phi : L_0 \xrightarrow{\cong} K_0$$

(on the category of real or complex C^*-algebras with unit), where $L_0 = L_0^p$ is the usual projective quadratic L-group of A as defined in [Ran1, §1] *(this coincides with the symmetric L-group L^0 since $\frac{1}{2} \in A$).*

Similarly, if φ is a non-singular skew-hermitian form on a finitely generated free A-module $P = A^n$, then φ is equivalent to the form defined by a skew-hermitian element h of $M_n(A)$ with $\operatorname{spec} h \subset \{i, -i\}$. (If A is an algebra over $\mathbb{C}$, then since $i = \sqrt{-1} \in A$,

$$(x, y) \mapsto \varphi(ix, y)$$

*Equivalent statements using slightly different notation appear in [Mis2], [Kar3] and in [KM].

is a non-singular hermitian form, so in the complex case there is no essential difference between hermitian and skew-hermitian forms.)

Proof. Recall that $L^0(A) = L_0(A)$ is the Witt group of non-singular hermitian forms φ defined on finitely generated projective (left) A-modules P, modulo hyperbolic forms. Let (P, φ) be such a form, and let Q be a complement to P, so that $P \oplus Q$ is finitely generated and free, say $P \oplus Q \cong A^n$. We may identify P with $A^n e$, where e is a self-adjoint projection in $M_n(A)$ (which is 0 on Q and the identity on P), and may identify φ with an element $h = h^* \in eM_n(A)e$. (We have used Fact 1.3(v).) Non-singularity of φ means that $h + 1 - e$ is invertible, and thus (using Fact 1.3(i)) its spectrum is contained in

$$(\infty, -\varepsilon] \cup [\varepsilon, \infty)$$

for some $\varepsilon > 0$ (depending on h). Let f be $+1$ on $[\varepsilon, \infty)$ and 0 on $(\infty, -\varepsilon]$. Then (using Fact 1.3(iii)) $f(h + 1 - e)$ is a self-adjoint projection which commutes with h and e; also it is clear that $f(h + 1 - e) \geq 1 - e$ (in other words, if $p = f(h + 1 - e) - (1 - e)$, then $0 \leq p \leq e$). Note that $P = Pp \oplus P(1 - p) = P^+ \oplus P^-$ is exactly the decomposition of P into the "positive" and "negative" eigenspaces of φ, as in the definition of the classical signature in 1.5, in that $hp \geq 0$ and $h(1 - p) \leq 0$. It is obvious from the construction that the submodules P^+ and P^- are φ-invariant. So to show that φ is equivalent to the form with matrix

$$\begin{pmatrix} 1_{P^+} & 0 \\ 0 & -1_{P^-} \end{pmatrix},$$

it is sufficient to replace P by $P^+ \oplus$ (some complement) and h by $hp \oplus 1$, and to show show that if the form φ on a finitely generated free module A^n is given by an invertible positive operator h, then it is equivalent to the standard form

$$\langle a, b \rangle = a_1 b_1^* + \cdots + a_n b_n^*. \quad \dagger$$

(A similar argument then applies to P^-.)

But the fact that φ is given by h means of course that

$$\varphi(a, b) = ahb^*.$$

Since h is positive, it has a square root $h^{\frac{1}{2}}$ (defined using Fact 1.3(iii)), and $h^{\frac{1}{2}}$ is invertible since h is. Thus under the change of basis defined by $h^{-\frac{1}{2}}$, h

†I am taking forms to be A-linear in the first variable and $*$-linear in the second variable, slightly different notation from that used in [Ran1].

is replaced by $h^{-\frac{1}{2}}hh^{-\frac{1}{2}^*} = 1$, as required. On the other hand, any change of basis sends h to khk^* for some k, and thus preserves positivity. So the splitting into positive and negative eigenspaces is unique, and this proves the part of the theorem about classification of forms.

To complete the proof in the hermitian case, observe that we now know that non-singular hermitian forms over A correspond exactly to pairs (P^+, P^-) of finitely generated projective modules. Hyperbolic forms obviously correspond to such pairs with $P^+ \cong P^-$. So the "signature" $[P^+] - [P^-]$ is well-defined on Witt classes and clearly induces an isomorphism from $L^0 = L_0$ to K_0.

As far as the skew-hermitian case is concerned, basically the same proof shows that if $P = A^n e$, where e is a self-adjoint projection in $M_n(A)$, then using the functional calculus we can change variables so that φ is given by (the cut-down to P of) a skew-hermitian element h of $M_n(A)$ with $\operatorname{spec} h \subset \{i, -i\}$ and commuting with the projection e. In the complex case, h is diagonalizable and the classification is the same as in the hermitian case. In the real case, this is not quite enough to classify the skew-hermitian forms. □

1.7. Remark. The functor L_0^h defined using hermitian forms on finitely generated free modules does not have such a simple description, but one can see from the above description that for A as above, $L_0^h(A)$ can be identified with the quotient group

$$\Big\{([P^+],[P^-]) \in K_0(A) \times K_0(A) : [P^+] + [P^-] = 0 \text{ in } \tilde{K}_0(A)\Big\} \Big/ \langle([A],[A])\rangle .$$

This is consistent with the usual Rothenberg exact sequence (derived for example in [Ran1, §1.10])

$$\cdots \to \hat{H}^1(\mathbb{Z}/2; \tilde{K}_0(A)) \to L_0^h(A) \to L_0^p(A) \to \hat{H}^0(\mathbb{Z}/2; \tilde{K}_0(A)) \to \cdots$$

since we see that when $L_0^p(A)$ is identified with $K_0(A)$, the image of $L_0^h(A)$ consists of all

$$[P^+] - [P^-], \quad [P^+] + [P^-] = 0 \text{ in } \tilde{K}_0(A),$$

and thus the cokernel of $L_0^h(A) \to L_0^p(A)$ is just

$$\tilde{K}_0(A)/2\tilde{K}_0(A).$$

Note that this is $\hat{H}^0(\mathbb{Z}/2;\, \tilde{K}_0(A))$ for the trivial action of $\mathbb{Z}/2$ on $\tilde{K}_0(A)$. However, the trivial action is indeed appropriate here, since the action of $\mathbb{Z}/2$ on $\tilde{K}_0(A)$ in the Rothenberg sequence is defined via the map $p \mapsto p^*$ on idempotents. But every finitely generated projective module over a C^*-algebra may be defined by a self-adjoint idempotent in some matrix algebra over A, because of Fact 1.3(v).

Similarly, the kernel of $L_0^h(A) \to L_0^p(A)$ is given by

$$\begin{aligned}\Big\{([P],[P]) : [P] \in K_0(A), \quad 2[P] = 0 \quad \text{in } \tilde{K}_0(A)\Big\} / \langle([A],[A])\rangle \\ \cong \text{2-torsion in } \tilde{K}_0(A) \cong \hat{H}^1(\mathbb{Z}/2;\, \tilde{K}_0(A)).\end{aligned}$$

Thus for C^*-algebras the Rothenberg sequence reduces to

$$0 \to \hat{H}^1(\mathbb{Z}/2;\, \tilde{K}_0(A)) \to L_0^h(A) \to L_0^p(A) \to \hat{H}^0(\mathbb{Z}/2;\, \tilde{K}_0(A)) \to 0,$$

with $\mathbb{Z}/2$ acting trivially. □

While Theorem 1.6 seems to be well-known, the relationship between higher L-groups and higher K-groups for C^*-algebras is more complicated and does not seem to be explained very clearly in the literature. Recall that for a Banach algebra A, possibly without unit, one has topological K-groups $K_\bullet^{\text{top}}(A)$ (which for $\bullet \geq 1$ are just the homotopy groups of $GL(A)$ shifted in degree) which are periodic of period 2 in the complex case and of period 8 in the real case (convenient references are [Kar2] and [Bla]). For algebras of functions, these are the more familiar groups (to topologists—operator algebraists use the groups $K_\bullet^{\text{top}}(A)$ all the time!) $K_\bullet^{\text{top}}(C^{\mathbb{C}}(X)) \cong KU^{-\bullet}(X)$, $K_\bullet^{\text{top}}(C^{\mathbb{R}}(X)) \cong KO^{-\bullet}(X)$. If (X, τ) is a compact Real space in the sense of Atiyah (in other words, a space with an involution τ) and we define

$$C(X, \tau) := \left\{ f \in C^{\mathbb{C}}(X) : \overline{f(x)} = f(\tau(x)) \right\},$$

then $K_\bullet^{\text{top}}(C(X, \tau)) \cong KR^{-\bullet}(X, \tau)$. K_0^{top} coincides with the *algebraic* K-group K_0, but in general, K_n^{top} differs from the algebraic K-group K_n if $n \neq 0$. On the other hand, we also have the surgery groups $L_\bullet^h(A)$ and $L_\bullet^p(A)$ defined forgetting the topology of A and only using the structure of A as a ring with involution. For complex C^*-algebras A, we will see now that the groups $L_\bullet^p(A)$ and $K_\bullet^{\text{top}}(A)$ are naturally isomorphic, whereas for real C^*-algebras, this is only true after inverting 2. The coincidence of $L_\bullet^p(A)$ with $K_\bullet^{\text{top}}(A)$ was first noted for commutative complex C^*-algebras in [Nov, Theorem 4.1], and (for general complex C^*-algebras) is basically equivalent to some of the results in §3 and §4 of [Mis2] (though it's hard at first to see whether that paper is talking about $L^h_\bullet$ or $L^p_\bullet$). A complete proof, using algebraic surgery and rather different from the one that we will give below, was given in [Mil]. That *algebraic* L-theory can recapture *topological* K-theory seems surprising until one remembers Fact 1.3(vi), which asserts that the structure of a C^*-algebra as a topological algebra can be recaptured from its algebraic structure as an involutive algebra over $\mathbb{R}$ or $\mathbb{C}$ (together with the topology of the real or complex numbers).

1.8. Theorem ([Mis2], [Mil]). *Let A be a complex C^*-algebra with unit. Let $L_\bullet(A)$ denote the 4-periodic (projective) quadratic L-groups of A as defined in* [Ran1, §1] *(these also coincide with the symmetric L-groups since $\frac{1}{2} \in A$). Then these groups are 2-periodic, and are naturally isomorphic to the topological K-groups $K^{\text{top}}_\bullet(A)$.*

Proof. The fact that the L-groups are 2-periodic is due to the fact that $\sqrt{-1} \in A$, so that there is a natural bijection between hermitian and skew-hermitian forms over A. Because of Theorem 1.6 and Remark 1.7, we really only need to show that there is a natural isomorphism Φ from $L^h_1(A)$ to $K^{\text{top}}_1(A)$, analogous to the signature map of Theorem 1.6.

Now the group $L^h_1(A)$ is defined** to be the limit (as $n \to \infty$) of the abelianization of what we can call $U(n, n; A)$, the unitary group of the hyperbolic form on $A^n \oplus A^n$ defined by the matrix

$$\sigma_n = \begin{pmatrix} 0 & 1_n \\ 1_n & 0 \end{pmatrix},$$

modulo σ_n itself and modulo the image of $GL(n, A)$ under the map

$$GL(n, A) \to U(n, n; A) : a \mapsto \begin{pmatrix} a & 0 \\ 0 & (a^*)^{-1} \end{pmatrix}.$$

In the notation of [Kar3], the abelianization of $U^{\text{hyp}}(A) := \varinjlim U(n, n; A)$ is denoted ${}_1L_1(A)$. The map $GL(n, A) \to U(n, n; A)$ defines in the limit (after passing to abelianizations) a map $K_1(A) \to {}_1L_1(A)$, whose cokernel in [Kar3] is denoted ${}_1W_1(A)$. However, $GL(n, A)$ and $U(n, n; A)$ are

**This is close to, but not exactly, the form of the definition given in [Ran1, §1.6] using "formations," but is an equivalent formulation obtained by modifying the definition of [what is now called] $L^s_1(A)$ in [Wall, §6]. The class of a unitary $u \in U(n, n; A)$ corresponds to the class of the formation

$$(A^n \oplus A^n, A^n \oplus 0, u(A^n \oplus 0)).$$

topological groups (in fact, Banach Lie groups), and if we divide by the connected components of the identity, $GL(n, A)_0$ and $U(n, n; A)_0$, instead of by the commutator subgroups (which are always contained in the former), then we obtain in place of $K_1(A)$ and ${}_1L_1(A)$ groups which (again in the notation of [Kar3]) are denoted by $K_1^{\mathrm{top}}(A)$ and ${}_1L_1^{\mathrm{top}}(A)$. Théorème 2.3 of [Kar3] shows that ${}_1L_1^{\mathrm{top}}(A)$ is canonically isomorphic to $K_1^{\mathrm{top}}(A) \oplus K_1^{\mathrm{top}}(A)$, with the natural map $K_1^{\mathrm{top}}(A) \to {}_1L_1^{\mathrm{top}}(A)$ corresponding to the diagonal embedding. This means that we have an exact sequence

$$K_1(A) \to U^{\mathrm{hyp}}(A)/U^{\mathrm{hyp}}(A)_0 \to K_1^{\mathrm{top}}(A) \to 0.$$

So to prove the theorem we need to show two things: that σ_n is in $U(n, n; A)_0$, and that the image of $GL(n, A)_0$ in $U(n, n; A)_0$ generates the latter modulo the commutator subgroup.

For our purposes it is useful to make use of the Lie algebra

$$\mathfrak{g} = \left\{ \begin{pmatrix} a & b \\ c & -a^* \end{pmatrix} : b = -b^*,\ c = -c^* \right\}$$

of $U(n, n; A)$, which can be identified with the tangent space to $U(n, n; A)_0$ at the identity. This space has the property that if $X \in \mathfrak{g}$, then $e^{tX} \in U(n, n; A)_0$. Since we are in the complex case,

$$\begin{pmatrix} 0 & i \\ i & 0 \end{pmatrix}, \begin{pmatrix} i & 0 \\ 0 & i \end{pmatrix} \in \mathfrak{g},$$

and exponentiating, we find that

$$\begin{aligned} \exp\left(\tfrac{-\pi}{2}\begin{pmatrix} i & 0 \\ 0 & i \end{pmatrix}\right) \exp\left(\tfrac{\pi}{2}\begin{pmatrix} 0 & i \\ i & 0 \end{pmatrix}\right) &= \begin{pmatrix} -i & 0 \\ 0 & -i \end{pmatrix}\begin{pmatrix} 0 & i \\ i & 0 \end{pmatrix} \\ &= \begin{pmatrix} 0 & 1 \\ 1 & 0 \end{pmatrix} = \sigma_n. \end{aligned}$$

Thus $\sigma_n \in U(n, n; A)_0$. To prove that the image of $GL(n, A)_0$ generates $U(n, n; A)_0$ modulo the commutator subgroup, we apply the (Banach space version of the) Implicit Function Theorem to the map

$$\psi : GL(n, A)_0 \times U(n, n; A)_0 \times U(n, n; A)_0 \to U(n, n; A)_0$$

defined by

$$(h, g_1, g_2) \mapsto \begin{pmatrix} h & 0 \\ 0 & (h^*)^{-1} \end{pmatrix} g_1 g_2 g_1^{-1} g_2^{-1}.$$

If we take (for example)

$$g_1 = \begin{pmatrix} 2 & 0 \\ 0 & \frac{1}{2} \end{pmatrix}, \quad g_2 = \exp \begin{pmatrix} 0 & b \\ c & 0 \end{pmatrix},$$

we obtain

$$g_1 g_2 g_1^{-1} g_2^{-1} = \exp \begin{pmatrix} 0 & 4b \\ \frac{1}{4}c & 0 \end{pmatrix} \exp \begin{pmatrix} 0 & -b \\ -c & 0 \end{pmatrix},$$

and from this it is easy to see that the differential of ψ at the point $(1, g_1, 1)$ is surjective. Thus the image of ψ contains an open neighborhood of the identity in $U(n, n; A)_0$, and thus the image of $GL(n, A)_0$ and the commutator subgroup of $U(n, n; A)_0$ generate all of $U(n, n; A)_0$. This completes the proof. □

If we look carefully at the above proof, we see that the assumption that A is a *complex* (and not just a real) C^*-algebra was only used twice: once to show that the L-groups are 2-periodic, and once to show that $\sigma_n \in U(n, n; A)_0$. If A is a real C^*-algebra, Karoubi's [Kar3, Théorème 2.3] is still valid, and thus the same proof as for Theorem 1.8 yields the following (which seems not to be in the literature).

1.9. Theorem. *If A is a real C^*-algebra, there is a canonical surjection of $K_1^{\text{top}}(A)$ onto $L_1^h(A)$, with kernel of order at most 2 (generated by the class of σ_n).*

1.10. Remark. Note that one cannot do any better than this, since $K_1^{\text{top}}(\mathbb{R}) \cong \mathbb{Z}/2$ and L^p_{odd} vanishes for any semisimple ring [Ran1, Proposition 1.2.3(iii)], in particular for a field, so that $L_1^p(\mathbb{R}) = L_1^h(\mathbb{R}) = 0$.

Now the Rothenberg exact sequence quoted in Remark 1.7 continues to the left as

$$\begin{aligned} L_2^h(A) &\to L_2^p(A) \to \hat{H}^0(\mathbb{Z}/2; \tilde{K}_0(A)) \\ &\to L_1^h(A) \to L_1^p(A) \to \hat{H}^1(\mathbb{Z}/2; \tilde{K}_0(A)) \to L_0^h(A), \end{aligned}$$

and since the map $\hat{H}^1(\mathbb{Z}/2; \tilde{K}_0(A)) \to L_0^h(A)$ is injective, the map $L_1^h(A) \to L_1^p(A)$ is always surjective. In the *complex* case where the map $L_2^h(A) \to L_2^p(A)$ can be identified with the map $L_0^h(A) \to L_0^p(A)$, it follows from Remark 1.7 that $L_1^h(A) \to L_1^p(A)$ is always an isomorphism. However in the real case, the map $L_1^h(A) \to L_1^p(A)$ need *not* be an isomorphism. For if

$$A = \mathbb{R} \oplus \cdots \oplus \mathbb{R} \quad (n \text{ summands}),$$

A is still semisimple, so $L_1^p(A) = 0$. And $L_2^p(A) = 0$ since any symplectic form over A is a hyperbolic form on a projective module. So from the Rothenberg exact sequence,

$$L_1^h(A) \cong \hat{H}^2(\mathbb{Z}/2; \tilde{K}_0(A)) \cong (\mathbb{Z}/2)^{n-1}.$$

This is consistent with Theorem 1.9, since $K_1^{\text{top}}(A) \cong (\mathbb{Z}/2)^n$.

One may note as well that the projective L-groups L_*^p have the good categorical properties

$$L_*^p(A_1 \times A_2) \cong L_*^p(A_1) \oplus L_*^p(A_2) \quad \text{(additivity)},$$

$$L_*^p(M_n(A)) \cong L_*^p(A) \quad \text{(Morita invariance)}$$

[Ran2, §22] in analogy with K-theory. The free L-groups L_*^h don't have such good properties, because the reduced projective class group $\widetilde{K}_0$ doesn't have them. □

If we are willing to invert 2, then the results of Karoubi in [Kar1] and [Kar3] yield the following in the real case.

1.11. Theorem (essentially due to Karoubi). *If A is a real C^*-algebra, then the groups $K_\bullet^{\text{top}}(A)\left[\frac{1}{2}\right]$, $L_\bullet^h(A)\left[\frac{1}{2}\right]$, and $L_\bullet^p(A)\left[\frac{1}{2}\right]$ are canonically isomorphic, and periodic with period 4.*

Proof. Because the groups $\hat{H}^\bullet(\mathbb{Z}/2; \_)$ are always 2-torsion, the Rothenberg sequence shows that (for any ring) the natural map $L_\bullet^h(A)\left[\frac{1}{2}\right] \to L_\bullet^p(A)\left[\frac{1}{2}\right]$ is an isomorphism. Also it is well-known that the 8-periodicity of $K_\bullet^{\text{top}}(A)$ becomes a 4-periodicity after inverting 2 (see for example [Kar2, Theorem III.2.11]). Since Theorems 1.6 and 1.9 already handle the cases $\bullet \cong 0, 1 \mod 4$, we need only explain how to deal with the cases $\bullet \cong 2, 3 \mod 4$. Here we need to apply Théorème 0.1 and Théorème 0.4 of [Kar1], for which parts of the proofs appear in [Kar3] and [Kar4]. As in the case above, [Kar3, Théorème 2.3] shows that there is a natural isomorphism from ${}_1L_\bullet^{\text{top}}(A)$ to $K_\bullet^{\text{top}}(A) \oplus K_\bullet^{\text{top}}(A)$, and thus from ${}_1L_\bullet^{\text{top}}(A)/K_\bullet^{\text{top}}(A) = {}_1W_\bullet^{\text{top}}(A)$ (Karoubi's notation) to $K_\bullet^{\text{top}}(A)$. Now after inverting 2, ${}_1W_\bullet^{\text{top}}(A)$ becomes what Karoubi calls ${}_1W_\bullet(A)\left[\frac{1}{2}\right]$ [Kar1, Théorème 0.1 and Théorème 0.4]. Applying periodicity, this is ${}_{-1}W_{\bullet-2}(A)\left[\frac{1}{2}\right]$, which in turn coincides with $L_{\bullet-2}^h(A, -1) \cong L_\bullet^h(A)\left[\frac{1}{2}\right]$. (Karoubi's W-group is the same as Witt group, and thus the same as the Wall group, in degree 0, and differs from the usual L-group in degree 1 by at most a $\mathbb{Z}/2$ coming from the class of σ.) □

1.12. It is worth pointing out (this fact is well-known to specialists in C^*-algebras) that the complex group ring $\mathbb{C}\pi$ of any group π has a *unique maximal* C^*-algebra-completion, denoted $C^*_{\max}(\pi)$. It has the universal property that any involution-preserving homomorphism from $\mathbb{C}\pi$ into the algebra of bounded operators on a Hilbert space must factor through the canonical inclusion $\mathbb{C}\pi \hookrightarrow C^*_{\max}(\pi)$. Thus the study of generalized signatures naturally leads to a study of $K_0(C^*_{\max}(\pi))$. Since the algebra $C^*_{\max}(\pi)$ is fairly intangible, however, most analytic proofs of the Novikov conjecture make do with the more concrete algebra $C^*_r(\pi)$ defined above, which is necessarily a quotient of $C^*_{\max}(\pi)$. The canonical map $C^*_{\max}(\pi) \twoheadrightarrow C^*_r(\pi)$ is an isomorphism if and only if π is amenable. More generally, the group π is called K-*amenable* when this map induces an isomorphism on topological K-groups. This condition is now known to be satisfied for discrete subgroups of $SO(n, 1)$ and $SU(n, 1)$, as well as for amenable groups. However it fails when π has Kazhdan's property T, for example, if π is a lattice subgroup in $Sp(n, 1)$ $(n \geq 1)$ or in a simple Lie group of $\mathbb{R}$-rank ≥ 2.

For purposes of studying which closed manifolds admit Riemannian metrics of positive scalar curvature, I was forced in [Ros1] and [Ros3] to deal with C^*-algebra completions of the real group ring $\mathbb{R}\pi$ rather than of the complex group ring $\mathbb{C}\pi$. This does not make an essential difference in the theory: one still has a universal C^*-algebra completion $C^*_{\max}(\pi)$ and a reduced C^*-algebra $C^*_r(\pi)$ (the completion of $\mathbb{R}\pi$ for its action on $\ell^2(\pi)$ by left multiplication). When it is necessary to distinguish this from the complex C^*-algebra, one can write $C^*_{r,\mathbb{R}}(\pi)$.

It seems that to the extent that there is a good C^*-algebraic analogue of the Borel conjecture (which involves surjectivity of an assembly map and not just injectivity), this analogue (of which the leading candidate is called the Baum-Connes conjecture) must involve $C^*_r(\pi)$ and not $C^*_{\max}(\pi)$; the K-theory of the latter is in many cases simply "too large". Of course, for K-amenable groups, the question of which C^*-algebra completion of $\mathbb{C}\pi$ to use makes no difference.

§2. Identity of the assembly and index maps

Now that we have seen how the K-theory of C^*-algebras is naturally related to higher signatures, we come to the issue of proofs of the Novikov conjecture itself. Analytic proofs for a group π tend to be arguments for injectivity of a certain index map, usually denoted μ or β, which roughly speaking maps the KU- or KO-homology of $B\pi$ to the K-groups of some C^*-algebra completion of the real or complex group ring.

To relate this to the surgery-theoretic assembly map as defined, say, in

[Ran2] or in [Q2], we need first of all a version of Theorems 1.8 and 1.11 *on the level of spectra* and not just on the level of individual groups. While it is in some sense known to the experts that such a version should exist, it does not seem to be written down anywhere, so we provide some details. But even in the complex case, just as in Theorem 1.11, it is necessary to invert 2. The reason is basically that L-spectra when localized at 2 are always Eilenberg-MacLane spectra [TaW, Theorem A], whereas the spectra for topological K-theory behave quite differently at the prime 2.

2.1. Theorem. *Let A be a C^*-algebra, either real or complex, and let $\mathbb{L}_\bullet(A)$ be the L-theory spectrum of A defined as in [RanL] and [Ran2]. (The p decoration would be more natural here, but it really doesn't matter since we need to invert 2 anyway. $\mathbb{L}^p$ and $\mathbb{L}^h$ become equivalent after inverting 2.) Let $\mathbb{K}^{\text{top}}_\bullet(A)$ denote the spectrum for topological K-theory of A, with homotopy groups $K^{\text{top}}_\bullet(A)$. (This may be realized as an Ω-spectrum with 0-th space $K_0(A) \times BGL(A)$. In the complex case, each even-numbered space may be taken to be $K_0(A) \times BGL(A)$ and each odd-numbered space may be taken to be $GL(A)$. In the real case, every 8-th space may be taken to be $K_0(A) \times BGL(A)$. Each space in the spectrum has the homotopy type of a CW-complex, by [Pal].) We localize the spectra as in [Ad, §3.1].*

Then there is a natural equivalence of spectra

$$\mathbb{K}^{\text{top}}_\bullet(A)\left[\tfrac{1}{2}\right] \to \mathbb{L}_\bullet(A)\left[\tfrac{1}{2}\right]$$

which on homotopy groups induces the isomorphisms of Theorems 1.8 and 1.11.

Proof. The proof that follows is rather nonconstructive. It would be desirable to have a more constructive argument based on directly constructing a map of (unlocalized) infinite loop spaces which induces an equivalence on homotopy groups after localization, but we have encountered various difficulties in doing this.

First we observe that the theorem is true for $A = \mathbb{R}$. To see this, note that $\mathbb{L}_\bullet(\mathbb{Z})$ is a periodic delooping of the infinite loop space

$$\mathbb{L}_0(\mathbb{Z}) \simeq \mathbb{Z} \times G/\text{Top},$$

[Ran2, p. 136], which after inverting 2 becomes equivalent to the 0th space $\mathbb{Z} \times BO$ in the spectrum of real K-theory $\mathbb{K}^{\text{top}}_\bullet(\mathbb{R})$ [MaM, Theorem 4.28]. The equivalence

$$\mathbb{Z} \times G/\text{Top}\left[\tfrac{1}{2}\right] \simeq \mathbb{Z} \times BO\left[\tfrac{1}{2}\right]$$

can be chosen to respect the natural H-space structures (coming on the left from the Cartesian product of a surgery problem with a manifold, and coming on the right from the tensor product of vector bundles) [MaM, Corollary 4.31]. On the other hand, the equivalence is also an infinite loop map [MaM, beginning of Ch. 6]. The equivalence

$$\mathbb{L}_0(\mathbb{Z}) \simeq \mathbb{Z} \times G/\mathrm{Top}$$

also respects the infinite loop space structures. The spectrum $\mathbb{L}_\bullet(\mathbb{Z})$ is a $\mathbb{L}^\bullet(\mathbb{Z})$-module spectrum [Ran2, Appendix B], in essence via the tensor product of a quadratic form with a symmetric form, and the multiplication

$$\mathbb{L}^\bullet(\mathbb{Z}) \wedge \mathbb{L}_\bullet(\mathbb{Z}) \to \mathbb{L}_\bullet(\mathbb{Z})$$

obviously corresponds to the "Cartesian product" H-space structure on $\mathbb{Z} \times G/\mathrm{Top}\left[\frac{1}{2}\right]$. So putting everything together and taking the periodic deloopings, we obtain an *equivalence of spectra*

$$\mathbb{L}_\bullet(\mathbb{Z})\left[\tfrac{1}{2}\right] \simeq \mathbb{K}_\bullet^{\mathrm{top}}(\mathbb{R})\left[\tfrac{1}{2}\right].$$

Furthermore, the inclusion $\mathbb{Z} \hookrightarrow \mathbb{R}$ induces isomorphisms on L-groups after inverting 2, and the symmetrization map $\mathbb{L}_\bullet(\mathbb{Z}) \to \mathbb{L}^\bullet(\mathbb{Z})$ respects $\mathbb{L}^\bullet(\mathbb{Z})$-module structures and becomes an equivalence after inverting 2, so that we obtain *equivalences of ring spectra*

$$\mathbb{K}_\bullet^{\mathrm{top}}(\mathbb{R})\left[\tfrac{1}{2}\right] \simeq \mathbb{L}^\bullet(\mathbb{Z})\left[\tfrac{1}{2}\right] \xrightarrow{\simeq} \mathbb{L}^\bullet(\mathbb{R})\left[\tfrac{1}{2}\right].$$

The fact this equivalence is compatible with products amounts to saying that it matches up the respective periodicity operators (which shift degree by 4).

Now consider a general C^*-algebra A. Then $\mathbb{K}_\bullet^{\mathrm{top}}(A)$ is naturally a module spectrum over $\mathbb{K}_\bullet^{\mathrm{top}}(\mathbb{R})$, while $\mathbb{L}_\bullet(A)$ is naturally a module spectrum over $\mathbb{L}^\bullet(\mathbb{R})$, in each case via the tensor product. And we have seen that the equivalence

$$\mathbb{K}_\bullet^{\mathrm{top}}(\mathbb{R})\left[\tfrac{1}{2}\right] \simeq \mathbb{L}^\bullet(\mathbb{R})\left[\tfrac{1}{2}\right]$$

is compatible with products. So because of Theorems 1.8 and 1.11, it's enough to know:

2.2. Theorem (Bousfield). *Two $\mathbb{K}_\bullet^{\mathrm{top}}(\mathbb{R})\left[\frac{1}{2}\right]$-module spectra with the same homotopy groups are homotopy-equivalent. (More precisely, any isomorphism of the homotopy groups of such spectra, compatible with the module action of the ring*

$$\pi_\bullet\left(\mathbb{K}_\bullet^{\mathrm{top}}(\mathbb{R})\left[\tfrac{1}{2}\right]\right) = K_\bullet^{\mathrm{top}}(\mathbb{R})\left[\tfrac{1}{2}\right] \cong \mathbb{Z}\left[\tfrac{1}{2}\right][t, t^{-1}],$$

where t is in degree 4, is realized by an equivalence of spectra.)

Proof. This follows immediately from the Universal Coefficient Theorem for $\mathbb{K}_\bullet^{\mathrm{top}}(\mathbb{R})$-module spectra of Bousfield [Bou, Theorems 4.6 and 9.6].‡ □

We want to use Theorem 2.1 to relate the analytic literature on the Novikov conjecture to the topological literature. For this we need to recall the basic construction in the analytic literature, the Mishchenko/Kasparov/Baum-Connes index map.

2.3. Definition. Let π be a group, and let $C_r^*(\pi)$ denote its real or complex reduced C^*-algebra. (The construction works equally well with real and complex coefficients.) The index map β (also sometimes denoted μ) from $K_\bullet(B\pi)$ (meaning $KU_\bullet(B\pi)$ in the complex case, $KO_\bullet(B\pi)$ in the real case) to $K_\bullet^{\mathrm{top}}(C_r^*(\pi))$, first studied in this generality in [Kas2], is defined as follows. First define the universal bundle $\mathcal{V}$ over $B\pi$ with fibers which are finitely generated projective (right) $C_r^*(\pi)$-modules, by letting $\mathcal{V} = E\pi \times_\pi C_r^*(\pi)$. This defines a class $[\mathcal{V}]$ in the Grothendieck group of such bundles, $K^0(B\pi; C_r^*(\pi))$, first studied for this purpose in [MS]. Then β is basically the operation of "slant product" with this class. To make this rigorous, think of $K_\bullet(B\pi)$ as $\varinjlim K_\bullet(X)$, where X runs over the finite subcomplexes of $B\pi$ indexed by inclusion. Then $[\mathcal{V}]$ pulls back to a class $[\mathcal{V}]_X$ over any such X, which we can think of as belonging to

$$K^0(X; C_r^*(\pi)) \cong K_0(C(X) \otimes C_r^*(\pi)),$$

with $C(X) \otimes C_r^*(\pi)$ here denoting the (completed) tensor product of C^*-algebras, in other words, $C(X, C_r^*(\pi))$. (Restriction to a finite subcomplex makes X compact, and thus makes $C(X)$ a commutative C^*-algebra with unit.) The "slant product" to be used here is the Kasparov intersection product of [Kas1]

$$K_\bullet(X) \cong KK_\bullet(C(X), \mathbb{C}) \xrightarrow{[\mathcal{V}]_X \otimes_X} KK_\bullet(\mathbb{C}, C_r^*(\pi)) \cong K_\bullet^{\mathrm{top}}(C_r^*(\pi)), \dagger\dagger$$

for which convenient expository references are [Fac1] and [Bla, §18].‡‡

‡I am indebted to Stephan Stolz for pointing out to me the relevance of Bousfield's work.

††In the real case, replace $\mathbb{C}$ everywhere by $\mathbb{R}$.

‡‡Since we are not using quite the most general case of the Kasparov product, there are various ways of simplifying the definition of the pairing here, for example by using the Connes-Higson notion of "asymptotic morphism" in [CoH].

Calling β the "index map" is justified by the following interpretation, which explains precisely how it arises in the analytic proofs. Namely, let M be a closed manifold equipped with a map $f : M \to B\pi$, and suppose D is an elliptic operator on M, for example the signature operator on an oriented manifold, or the Dirac operator on a spin manifold. The definition of the Kasparov groups is such that D immediately defines a class $[D] \in K_\bullet(M)$, which we can push forward to a class $f_*([D]) \in K_\bullet(B\pi)$. In fact, by a fairly easy theorem of Conner-Floyd type, all classes in $K_\bullet(B\pi)$ arise this way (a much more precise result along these lines has been given by Baum, and is announced in [BD], though the complete proof never appeared). Then $\beta(f_*([D])) \in K_\bullet^{\mathrm{top}}(C_r^*(\pi))$ is the index in the sense of Mishchenko-Fomenko [MF] of $D_{\mathcal{V}}$, meaning D with "twisted coefficients" in the flat bundle $\mathcal{V}$. This follows immediately from the interpretation of the Kasparov product in terms of index theory, explained in [Bla, §24] or in [Ros4].

2.4. As we will see in the next section, there is another way of defining β, which comes from going up from $B\pi$ to its universal cover $E\pi$ and doing everything equivariantly with respect to π. From this point of view, β is the map in *equivariant* Kasparov K-homology defined by the π-equivariant "collapse map" $E\pi \to *$ to a point, followed by the isomorphism $K_\bullet^\pi(*) \cong K_\bullet^{\mathrm{top}}(C_{\max}^*(\pi))$ [Kas2, §4] and the map on topological K-groups induced by the projection $C_{\max}^*(\pi) \twoheadrightarrow C_r^*(\pi)$ mentioned above in 1.12. Here we note that $K_\bullet^\pi(E\pi) \cong K_\bullet(B\pi)$ since π acts freely on $E\pi$, so that we indeed get a map $K_\bullet(B\pi) \to K_\bullet^{\mathrm{top}}(C_r^*(\pi))$. Chasing the definitions of the various Kasparov products involved shows that in this case we get back to the previous construction of β.

We can justify the name "index map" for β from this point of view as well. If D is an elliptic operator on a closed manifold M, defining a class $[D] \in K_0(M)$, then its analytic index (in the usual sense) is exactly the image of $[D]$ in $K_0(*) \cong \mathbb{Z}$ under the "collapse map" $M \to *$. If D commutes with an action of a compact group G on M, then D defines an equivariant class in equivariant K-homology $K_0^G(M)$, and its equivariant index is the image of this class in $K_0^G(*) \cong R(G)$ under the (G-equivariant) "collapse map" $M \to *$. (See [Bla, §24] and [Ros4] for an explanation.) What we are doing here is exactly analogous, except that we are replacing G by π and doing π-equivariant index theory on the covering space of M defined by a map $M \to B\pi$.

2.5. Remark. The map which we have just called β is closely related to a similar map, often denoted μ, which Baum and Connes have conjectured to be an isomorphism (see [BCH]). The relationship is the following.

One can define a π-space $\underline{E}\pi$ (see [BCH]) which is universal for *proper* actions of π, just as $E\pi$ is universal for free actions. (When π is finite, $\underline{E}\pi$ is just the one-point π-space $*$, while $E\pi$ is of course infinite-dimensional.) Then the Baum-Connes map $\mu : K_\bullet^\pi(\underline{E}\pi) \to K_\bullet^{\text{top}}(C_r^*(\pi))$ can be defined similarly, as the map in equivariant Kasparov K-homology defined by the π-equivariant "collapse map" $\underline{E}\pi \to *$, followed by the isomorphism $K_\bullet^\pi(*) \cong K_\bullet^{\text{top}}(C_{\max}^*(\pi))$ and the map on topological K-groups induced by the projection $C_{\max}^*(\pi) \twoheadrightarrow C_r^*(\pi)$ mentioned above in 1.12. Since $\underline{E}\pi = E\pi$ if π is torsion-free, in this case μ reduces to β. For groups with torsion, $\underline{E}\pi \times E\pi$ is a free contractible π-space, thus another model for $E\pi$, and thus projection onto the first factor gives (up to equivariant homotopy) a π-map $E\pi \to \underline{E}\pi$ through which the collapse map $E\pi \to *$ factors, and in this way β factors through μ.

2.6. Now we address the question of how one gets from information about β or μ to information about the Novikov conjecture. If one is only interested in homotopy invariance of higher signatures, it is not necessary to invoke spectra, and two arguments are available, one due to Kasparov, described in [Kas2, §9] and in [Kas5], and the other due to Kaminker-Miller [KM]. These show directly (without any reference to L-theory assembly) that if D is the signature operator on an oriented closed manifold M^{2n} equipped with a map $f : M \to B\pi$, then $\beta(f_*([D])) \in K_\bullet^{\text{top}}(C_r^*(\pi))$ is an oriented homotopy invariant. Thus if β is rationally injective, $\text{Ch}\, f_*([D]) = 2^n f_*([M] \cap \mathcal{L}(M)) \in H_\bullet(B\pi; \mathbb{Q})$ is an oriented homotopy invariant, $\mathcal{L}$ being the Atiyah-Singer renormalization of the total Hirzebruch L-class. This conclusion is obviously equivalent to the usual formulation of the Novikov conjecture.

However, it is more interesting to ask how *integral* statements about β are related to *integral* versions of the Novikov conjecture. Here Theorem 2.1 turns out to be quite relevant.

2.7. Theorem. *Let A^h and A^p denote the assembly maps in the sense of* [Ran2] *for (periodic) L^h- and L^p-theory, respectively. Then for any con-*

nected CW-complex X, the following diagram commutes:

$$\begin{array}{ccc}
H_\bullet(X;\mathbb{L}^h_\bullet(\mathbb{Z})) & \xrightarrow{A^h} & L^h_\bullet(\mathbb{Z}\pi_1(X)) \\
\cong\downarrow & & \downarrow \\
H_\bullet(X;\mathbb{L}^p_\bullet(\mathbb{Z})) & \xrightarrow{A^p} & L^p_\bullet(\mathbb{Z}\pi_1(X)) \\
\iota_*\downarrow & & \downarrow\iota_* \\
H_\bullet(X;\mathbb{L}^p_\bullet(\mathbb{R})) & \xrightarrow{A^p} & L^p_\bullet(\mathbb{R}\pi_1(X)) \\
\cong\downarrow\text{invert 2} & & j\downarrow\text{invert 2} \\
H_\bullet(X;\mathbb{K}^{\text{top}}_\bullet(\mathbb{R})) & \xrightarrow{\beta} & K^{\text{top}}_\bullet(C^*_r(\pi_1(X))).
\end{array}$$

Here the maps ι_* are induced by the inclusions $\mathbb{Z}\hookrightarrow\mathbb{R}$, $\mathbb{Z}\pi_1(X)\hookrightarrow\mathbb{C}\pi_1(X)$, and the map j is map on L^p-groups induced by the inclusion $\mathbb{C}\pi_1(X)\hookrightarrow C^*_r(\pi_1(X))$, followed by the isomorphism of Theorem 1.11. (One can also replace $\mathbb{R}$ by $\mathbb{C}$ in the diagram.)

Reduction of the problem. The commutativity of the upper two squares is obvious from the general machinery of [Ran2], so we need to prove commutativity of the lower square. Furthermore, the assembly maps are really the induced maps on homotopy groups of certain maps of spectra (which with minor abuse of notation we'll denote by the same letters), so we will do slightly better and show that that the diagram of spectra

$$\begin{array}{ccc}
\mathbb{H}_\bullet(X;\mathbb{L}^p_\bullet(\mathbb{R})\left[\tfrac{1}{2}\right]) & \xrightarrow{A^p} & \mathbb{L}^p_\bullet(\mathbb{R}\pi_1(X))\left[\tfrac{1}{2}\right] \\
\cong\downarrow & & \downarrow j \\
\mathbb{H}_\bullet(X;\mathbb{K}^{\text{top}}_\bullet(\mathbb{R})\left[\tfrac{1}{2}\right]) & \xrightarrow{\beta} & \mathbb{K}^{\text{top}}_\bullet(C^*_r(\pi_1(X)))\left[\tfrac{1}{2}\right]
\end{array}$$

is homotopy-commutative. We sketch two different arguments for this.

First Proof. The first proof depends on the description of assembly for L-theory given in [Ran2, Appendix B], together with the homotopy-theoretic description of β in [Ros3, Theorem 2.2] along the lines of the description of assembly in [Lod]. Let $f : X \to B\pi_1(X)$ be the classifying map for the universal cover of X. The references just cited give parallel descriptions of the maps A^p and β as composites

$$\begin{aligned}
\mathbb{H}_\bullet(X;\mathbb{L}^p_\bullet(\mathbb{R})) = X_+\wedge\mathbb{L}^p_\bullet(\mathbb{R}) &\xrightarrow{f_*\wedge id} B\pi_1(X)_+\wedge\mathbb{L}^p_\bullet(\mathbb{R}) \\
&\xrightarrow{\sigma\wedge id} \mathbb{L}^p_\bullet(\mathbb{R}\pi_1(X))\wedge\mathbb{L}^p_\bullet(\mathbb{R}) \xrightarrow{\mu} \mathbb{L}^p_\bullet(\mathbb{R}\pi_1(X)),
\end{aligned}$$

$$\begin{aligned}\mathbb{H}_\bullet(X; \mathbb{K}^{\mathrm{top}}_\bullet(\mathbb{R})) = X_+ \wedge \mathbb{K}^{\mathrm{top}}_\bullet(\mathbb{R}) &\xrightarrow{f_*\wedge id} B\pi_1(X)_+ \wedge \mathbb{K}^{\mathrm{top}}_\bullet(\mathbb{R}) \\ &\xrightarrow{B\iota\wedge id} \mathbb{K}^{\mathrm{top}}_\bullet(\mathbb{R}\pi_1(X)) \wedge \mathbb{K}^{\mathrm{top}}_\bullet(\mathbb{R}) \xrightarrow{\mu} \mathbb{K}^{\mathrm{top}}_\bullet(C^*_r(\pi_1(X))).\end{aligned}$$

Here σ is the "preassembly" map of [Ran2, Appendix B] (we are using the fact that symmetric and quadratic L-theory agree for algebras over a field of characteristic 0) and ι is the inclusion of $\pi_1(X)$ into $GL(C^*_r(\pi_1(X)))$, $B\iota$ is the induced map of classifying spaces (recall the "identity component" of $\mathbb{K}^{\mathrm{top}}_0(C^*_r(\pi_1(X)))$ is just $BGL(C^*_r(\pi_1(X)))$), and μ in each case is an appropriate "multiplication" map. Now as shown in [Ran2, Appendix B], preassembly is compatible with Loday-type assembly; in other words, the following diagram commutes:

$$\begin{array}{ccc} B\pi_1(X)_+ \wedge \mathbb{K}^{\mathrm{top}}_\bullet(\mathbb{R})\left[\tfrac12\right] & \xrightarrow{\mu\circ(\sigma\wedge id)} & \mathbb{L}^p_\bullet(\mathbb{R}\pi_1(X))\left[\tfrac12\right] \\ \cong\downarrow & & \downarrow j \\ B\pi_1(X)_+ \wedge \mathbb{K}^{\mathrm{top}}_\bullet(\mathbb{R})\left[\tfrac12\right] & \xrightarrow{\mu\circ(B\iota\wedge id)} & \mathbb{K}^{\mathrm{top}}_\bullet(C^*_r(\pi_1(X)))\left[\tfrac12\right]. \end{array}$$

This clearly yields the desired result. □

Second Proof. The second proof uses the abstract characterization of assembly maps found in [WW1, §2]. (The idea that one could use the Weiss-Williams results for this purpose I learned from reading [CarP].) The main result from [WW1, §2] that we need is the following; see also [WW2] for an easy independent treatment.

2.8. Theorem [WW1, §2], [WW2]. *Given a homotopy-invariant functor F from spaces (without basepoint) to spectra, such that $F(\emptyset)$ is contractible, there is a unique strongly excisive homotopy-invariant functor (in other words, a generalized homology theory) $F^{\%}$ and a natural transformation $F^{\%} \to F$ (called assembly) which is the "best approximation to F from the left by a homology theory," in the sense that given any strongly excisive homotopy-invariant functor E with a natural transformation $\varphi: E \to F$, E factors through $F^{\%}$. (Here "strongly excisive" means $F^{\%}$ takes homotopy pushout squares of spaces to homotopy pushout squares of spectra, and preserves arbitrary coproducts, up to homotopy equivalence.)*

2.9. Corollary (implicit in [WW1, §2], **also stated in** [WW2, §1]**).** *In the context of Theorem 2.8, there is a natural equivalence of functors*

$$F^{\%}(X) \simeq |X|_+ \wedge F(*),$$

and if E is a strongly excisive homotopy-invariant functor with a natural transformation $\varphi : E \to F$, then φ may be identified with the assembly for F provided that

$$\varphi(*) : E(*) \to F(*)$$

is an equivalence of spectra.

Proof. The first statement comes up in the proof of Theorem 2.8, and comes from the characterization of generalized homology theories in terms of spectra. Then if E is a strongly excisive homotopy-invariant functor with a natural transformation $\varphi : E \to F$, we also have a natural equivalence

$$E(X) \simeq |X|_+ \wedge E(*),$$

and φ factors through the assembly natural transformation $F^{\%} \to F$. This factorization $\varphi^{\%}$ gives an equivalence of spectra

$$\varphi^{\%}(X) : E(X) \simeq |X|_+ \wedge E(*) \to |X|_+ \wedge F(*) \simeq F^{\%}(X)$$

for all X provided that

$$\varphi(*) : E(*) \to F(*)$$

is an equivalence of spectra. □

In the context of [WW1], "spaces" means CW-complexes, but as explained there, the theorem is valid on other categories of spaces, for example simplicial complexes. We want to apply the Weiss-Williams theorem to the functor

$$F : X \mapsto \mathbb{K}_\bullet^{\mathrm{top}}(C_r^*(\pi_1(X))),$$

but of course one encounters the difficulty that $\pi_1(X)$ depends on a choice of basepoint, and thus $X \mapsto \pi_1(X)$ is not well-defined on unbased spaces. To get around this, we use instead the C^*-algebra of the fundamental *groupoid* of a space, $\pi(X)$, which was also used in [RosW] for basically the same reason of functoriality. Here $\pi(X)$ is the topological groupoid with X as its space of objects, whose morphisms from $x \in X$ to $y \in X$ are the homotopy classes (rel $\{0, 1\}$) of paths $\gamma : [0, 1] \to X$ with $\gamma(0) = x$, $\gamma(1) = y$. If X is a connected space which is nice enough for covering space theory to apply, e.g., a CW-complex, then $\pi(X)$ can be identified with $\tilde{X} \times_\Gamma \tilde{X}$, where $\tilde{X} \to X$ is a universal covering space of X and Γ is the associated group of covering transformations, and thus carries a natural topology. If in addition X is locally compact and carries a canonical measure class of full support (for instance, if X is the geometric realization of a *locally finite* simplicial complex—we can take the class represented by Lebesgue measure on each simplex), then $\pi(X)$ is a locally compact groupoid admitting a Haar system (the measure class is needed to define the Haar system), and $C_r^*(\pi(X))$ is well-defined in the sense of [Ren]. More simply (and this is what we'll do here), if X is a simplicial complex or simplicial set (and now one can dispense with the local finiteness and the measure), one can interpret $\pi(X)$ to mean the simplicial fundamental groupoid of X, which has as its set of objects the vertices of X (with the discrete topology), and whose morphisms from a vertex x to a vertex y are the homotopy classes (rel $\{0, 1\}$) of simplicial paths γ from some subdivision of $[0, 1]$ into X, with $\gamma(0) = x$, $\gamma(1) = y$. While this is not literally the same groupoid as before, it is equivalent to it, and has the advantages that $C_r^*(\pi(X))$ is always well-defined, and that a simplicial map gives rise to a $*$-homomorphism of the associated C^*-algebras. Furthermore, $C_r^*(\pi(X))$ is strongly Morita equivalent to the (C^*-algebraic) direct sum of the $C_r^*(\pi_1(|X|, x_j))$, where we choose one basepoint x_j in each component of X. (This results from a trivial modification of [RosW, Theorem 2.5] to cover the disconnected case.) A simplicial map of polyhedra gives a map of the associated fundamental groupoids that only depends of the homotopy class of the map, so

$$F : X \mapsto \mathbb{K}_\bullet^{\text{top}}(C_r^*(\pi(X)))$$

is a homotopy functor on the category of simplicial sets. (In fact, though we don't need to know it, the homotopy type of the spectrum $\mathbb{K}_\bullet^{\text{top}}(A)$ depends only very weakly on the C^*-algebra A; as we mentioned once before, since $\mathbb{K}_\bullet^{\text{top}}(A)$ is a $\mathbb{K}_\bullet^{\text{top}}(\mathbb{R})$-module spectrum, its homotopy type as a spectrum, after localizing away from 2, only depends on the topological K-groups of A as groups. This follows immediately from Theorem 2.2.) As such, F comes with an assembly natural transformation

$$F^{\%}(X) \simeq |X|_+ \wedge F(*) = |X|_+ \wedge \mathbb{K}_\bullet^{\text{top}}(\mathbb{R}) \to \mathbb{K}_\bullet^{\text{top}}(C_r^*(\pi(X))).$$

The universality statement in Theorem 2.8, together with Corollary 2.9, now proves that the assembly $F^{\%} \to F$ coincides with β; in fact, it gives an easy proof that the two definitions of β in 2.3 and 2.4 coincide.*** And similarly,

$$G : X \mapsto \mathbb{L}_\bullet^p(\mathbb{R}\pi(X))$$

***In order to apply the argument, one needs to know that β (using either the definition in 2.3 or the one in 2.4) may be viewed not just as a map of groups but also as a natural transformation of homotopy functors from spaces to spectra. We explain how to do this from the perspective of 2.3; the case of 2.4 is similar. Recall that we defined $K_\bullet(B\pi)$ as $\varinjlim K_\bullet(X)$, where X runs over the finite subcomplexes of $B\pi$ indexed by inclusion, and β was defined using the Kasparov intersection product (over $C(X)$) with the class $[\mathcal{V}]_X \in K^0(X; C_r^*(\pi)) \cong K_0(C(X) \otimes C_r^*(\pi))$. Regard $K_\bullet(X)$ as the coefficient is a homotopy functor on the category of simplicial sets, and $G^{\%}$ coincides with A^p. Since j is a natural transformation $G\left[\frac{1}{2}\right] \to F\left[\frac{1}{2}\right]$, it yields a homotopy-commutative square

$$\begin{array}{ccc} G^{\%}\left[\frac{1}{2}\right] & \longrightarrow & G\left[\frac{1}{2}\right] \\ {\scriptstyle j_*}\downarrow & & \downarrow{\scriptstyle j} \\ F^{\%}\left[\frac{1}{2}\right] & \longrightarrow & F\left[\frac{1}{2}\right], \end{array}$$

which is exactly what we needed. □

This now yields an application to the integral Novikov conjecture.

2.10. Corollary. *If, for some group π, the index map*

$$H_\bullet(B\pi; \mathbb{K}_\bullet^{\mathrm{top}}(\mathbb{R})) \xrightarrow{\beta} K_\bullet^{\mathrm{top}}(C_r^*(\pi))$$

is injective (after inverting 2), then so is the L-theory assembly map

$$H_\bullet(X; \mathbb{L}_\bullet^p(\mathbb{R})) \xrightarrow{A^p} L_\bullet^p(\mathbb{R}\pi_1(X))$$

(after inverting 2). Similarly, if $\beta\left[\frac{1}{2}\right]$ is split injective, then so is $A^p\left[\frac{1}{2}\right]$.

Proof. This is immediate from the commutative diagram in Theorem 2.7. □

Of course, we can also work with complex instead of real C^*-algebras.

§3. Bounded and controlled surgery

In this final section, we will try to relate the topologists' notions of bounded and controlled surgery to the analytic work on the Novikov conjecture. While bounded and controlled surgery deal with non-compact surgery

groups of a homology theory

$$Y \mapsto K_\bullet(X \times Y) = KK_\bullet(C(X) \otimes C(Y), \mathbb{R}),$$

and $K_\bullet^{\mathrm{top}}(C_r^*(\pi))$ as the coefficient groups of the homology theory

$$Y \mapsto KK_\bullet(C(Y), C_r^*(\pi)).$$

Then the more general form of the Kasparov product (with "coefficients") gives a map of homology theories, and it's easy to check naturality.

problems, with measurement of distances by means of reference maps to some metric space, they can be applied to problems about compact manifolds merely by passage to the universal cover. As such they are naturally related to the classical Novikov conjecture, as well of course to certain "coarse" analogues that may not be related to compact problems. The idea of passage from surgery on compact manifolds, related to the Novikov conjecture, to bounded or controlled surgery on the non-compact universal cover, makes an appearance in [Ran2, Appendix C], in [CarP], in [Gr], in [FW], in [Ran3], and in many other references which I will not attempt to enumerate here. Similarly, there is starting to be a considerable parallel literature on the analytic side, typified by [Roe1], [Roe2], and [HR2]. I will not attempt a "grand synthesis" of all of this work, but rather something much more modest. Namely, one sees in [CarP] and in [Ran2, Appendix C, §C4] the basic point of view that assembly maps can be viewed as "partial forget control" maps, from continuous control to bounded control.††† The secret to splitting assembly maps, in other words to proving versions of the Novikov conjecture, thus seems to be *to find a way to go back, from bounded control to continuous control*. This in fact is the main theme in the analytic literature as well.

We begin with a review of some of the relevant algebraic and topological concepts.

3.1. Definition. Let (X, d) be a metric space and let R be a ring. Let $\mathcal{C}_X(R)$ be the **X-bounded projective R-module** additive category of Pedersen and Weibel [PW]. This is defined as follows: an object in $\mathcal{C}_X(R)$ is a direct sum of finitely generated projective R-modules graded by X

$$A = \bigoplus_{x \in X} A(x)$$

such that for every subspace $K \subseteq X$ of finite diameter, only finitely many of the modules $A(x)$, $x \in K$, are non-zero. A morphism $A \to B$ in $\mathcal{C}_X(R)$ is roughly speaking a matrix of finite bandwidth (called by some authors finite

†††In fact, this point of view first appeared in [Q1], though the treatment there of what later came to be known as assembly is somewhat different, and harder to match up with what we are doing here.

To be honest, there are several competing notions of control that are relevant here, and the kind of "continuous control" used in [CarP] is different from what we call "continuous control" below, which is also known as "asymptotic control." Our approach to assembly here is closer to that in [Car1, §III] than to that in [CarP]. However, the basic philosophy is pretty much the same.

propagation distance), in other words, an R-module morphism defined by "matrix entries"

$$f = \{f(y, x) : A(x) \to B(y)\}$$

such that there exists a number $b > 0$ with $f(y, x) = 0 : A(x) \to B(y)$ for all $x, y \in X$ with $d(x, y) > b$. For X compact, the "finite bandwidth" condition is always trivially satisfied, so $\mathcal{C}_X(R)$ is just equivalent to the category of finitely generated projective R-modules. If R is a ring with involution $*$, then there is an induced involution on $\mathcal{C}_X(R)$ defined by applying $*$ "pointwise" in X, and so the L-theory of $\mathcal{C}_X(R)$ is defined.

3.2. The Pedersen-Weibel category $\mathcal{C}_X(R)$ was originally introduced in [PW] for the purpose of giving a good realization of K-homology, since it turns out by the main theorem of [PW] that if $X = O(Y) \subseteq \mathbb{R}^{N+1}$ is the open cone on a compact subset $Y \subseteq S^N$, then

$$K_\bullet(\mathcal{C}_X(R)) \cong H_\bullet^{lf}(O(Y); \mathbb{K}_\bullet(R)) \cong \tilde{H}_{\bullet-1}(Y; \mathbb{K}_\bullet(R)).$$

Here $\mathbb{K}_\bullet(R)$ is the (nonconnective) algebraic K-theory spectrum of the ring R. When $R = \mathbb{R}$, the algebraic K-theory spectrum $\mathbb{K}_\bullet(\mathbb{R})$ maps to the topological K-theory spectrum $\mathbb{K}_\bullet^{\mathrm{top}}(\mathbb{R})$, and so $K_\bullet(\mathcal{C}_X(R))$ maps to the topological K-homology of Y (with a degree shift). A direct construction of this map in terms of the Kasparov realization of topological K-homology was attempted in [Ros2, Theorem 3.4], but as John Roe has kindly pointed out to me, the proof given there is incorrect. (In the notation of the proof of [Ros2, Theorem 3.4], the error is that there is no reason why the action ψ of $C(X)$ on the algebraic direct sum of the $A_{t\cdot x}$ should be a $*$-homomorphism with respect to the inner product $\langle\cdot, \cdot\rangle_0$.) However, we still ought to be able to match up "bounded topology" with the analytic literature based on KK-theory, and indeed this is done in [PRW], though only by non-constructive methods and in the case where X is the open cone on a compact polyhedron. Matching the "bounded topology" Novikov conjecture with the analytic literature on the Novikov conjecture involves the "simply connected assembly map" A^{lf} of [Ran2, Appendix C] for the L-theory of $\mathcal{C}_{\tilde{X}}(\mathbb{Z})$, where $\tilde{X}$ is the universal cover of a space X, and relating it by transfer to the usual assembly map

$$A^h : H_\bullet(X; \mathbb{L}_\bullet^h(\mathbb{Z})) \to L_\bullet^h(\mathbb{Z}\pi_1(X)).$$

However, we have to take the action of $\pi_1(X)$ into account and look at equivariant theories.

Recall that if a group π acts on a (based) space X, the *homotopy fixed set* $X^{h\pi}$ is defined to be the set

$$X^{h\pi} := \mathrm{Map}(E\pi_+, X)^\pi$$

of based equivariant maps from the universal π-space $E\pi_+$ to X (we've added a disjoint basepoint to $E\pi$ in order to stay in the based category). The obvious (based) map from the actual fixed set X^π to the homotopy fixed set $X^{h\pi}$ may or may not be an equivalence.

We will need the observation that if $\tilde{X}$ is the universal covering of a connected compact polyhedron with fundamental group π and if R is a ring, then π operates on the category $\mathcal{C}_{\tilde{X}}(R)$. If R is a PID (so that all projective R-modules are free and $\mathbb{L}^p_\bullet(R) = \mathbb{L}^h_\bullet(R) := \mathbb{L}_\bullet(R)$), then in the definition of the category $\mathcal{C}_{\tilde{X}}(R)$ we may as well replace the word "projective" by the word "free," and the subcategory of invariants for this action may be identified with $\mathcal{C}^{\text{free}}_X(R\pi)$. This is defined the same way as $\mathcal{C}_X(R\pi)$, but again with the word "projective" replaced by the word "free," since $R\pi$ is exactly the free R-module with π as basis, and since $R\pi$-module morphisms of finitely generated free $R\pi$-modules have finite bandwidth. The category $\mathcal{C}^{\text{free}}_X(R\pi)$ in turn is equivalent to the category of finitely generated free $R\pi$-modules, since X is compact. Thus

$$\mathbb{L}_\bullet(\mathcal{C}_{\tilde{X}}(R))^\pi \simeq \mathbb{L}^h_\bullet(R\pi)$$

(this observation was first made in [CarP]), while

$$\mathbb{H}(X;\, \mathbb{L}_\bullet(R)) \simeq \mathbb{H}^{lf}(\tilde{X};\, \mathbb{L}_\bullet(R))^\pi \simeq \mathbb{H}^{lf}(\tilde{X};\, \mathbb{L}_\bullet(R))^{h\pi}$$

if π is torsion-free (this latter fact is proved in [Car1, Corollary II.5], [Car2, §II] and in [CarP], in fact for general spectra in place of $\mathbb{L}_\bullet(R)$).

3.3. Theorem [CarP], [Ran3, §8]. *If X is a connected compact polyhedron with torsion-free fundamental group π and universal covering $\tilde{X}$, and R is a PID (we are interested in the cases $\mathbb{Z}$, $\mathbb{R}$, and $\mathbb{C}$), then there is a natural homotopy-commutative diagram*

$$\begin{array}{ccc} \mathbb{H}(X;\, \mathbb{L}_\bullet(R)) & \xrightarrow{A^h} & \mathbb{L}^h_\bullet(R\pi) = \mathbb{L}_\bullet(\mathcal{C}_{\tilde{X}}(R))^\pi \\ {\scriptstyle \mathrm{trf}}\downarrow{\scriptstyle \cong} & & {\scriptstyle \mathrm{trf}}\downarrow \\ \mathbb{H}^{lf}(\tilde{X};\, \mathbb{L}_\bullet(R))^{h\pi} & \xrightarrow{A^{lf,\,h\pi}} & \mathbb{L}_\bullet(\mathcal{C}_{\tilde{X}}(R))^{h\pi}. \end{array}$$

The vertical arrows may be identified with the natural maps from fixed sets to homotopy fixed sets.

In particular, if $\mathbb{H}^{lf}(\tilde{X};\, \mathbb{L}_\bullet(R)) \xrightarrow{A^{lf}} \mathbb{L}_\bullet(\mathcal{C}_{\tilde{X}}(R))$ is an equivalence, then (since this implies $A^{lf,\,h\pi}$ is also an equivalence) A^h is split injective.

3.4. The rough idea of the methods of [CarP] (this is not exactly what they do, but only for technical reasons) is first to identify $H^{lf}(\tilde{X};\, \mathbb{L}_\bullet(R))$ with the L-theory of a continuously controlled (and not just bounded) category of R-modules over $\tilde{X}$, and to identify the locally finite assembly map

$$\mathbb{H}^{lf}(\tilde{X}; \mathbb{L}_\bullet(R)) \xrightarrow{A^{lf}} \mathbb{L}_\bullet(\mathcal{C}_{\tilde{X}}(R))$$

with a "partial forget control" map. Then if there is a model for $E\pi$ with a contractible metrizable compactification Y on which π acts, so that compact subsets of $E\pi$ become small when translated toward the boundary, one tries to use the geometry of the compactification to show that

$$\mathbb{H}^{lf}(E\pi; \mathbb{L}_\bullet(R)) \xrightarrow{A^{lf}} \mathbb{L}_\bullet(\mathcal{C}_{E\pi}(R))$$

is a weak homotopy equivalence. The idea (which is also basic to [FW] and to many of the other references we have cited) is that "pushing to the boundary" converts bounded control to continuous control, and thus provides a splitting to the forgetful map. If one can show this, then one can apply Theorem 3.3 to finish the argument.

We will see that all of this has a counterpart in the analytic literature as well. One could trace this through almost all of the analytic papers on the Novikov conjecture, especially in [Mis3], [HsR], [Gr], and [Hu], but to the limit the discussion I will concentrate here on the programs of Kasparov and Roe as expressed, say, in [Kas4], [Kas5], [Roe2], and [HR2].

Now let's specialize to the case where the coefficient ring R is $\mathbb{R}$ or $\mathbb{C}$. We'll write everything out for the case of $\mathbb{R}$, but $\mathbb{C}$ works exactly the same way. Following the principles which we have discussed earlier in this paper, the analytic counterpart to $\mathbb{L}_\bullet(\mathcal{C}_{\tilde{X}}(\mathbb{R}))$ should be the topological K-theory spectrum of a certain "C^*-completion" of the category $\mathcal{C}_{\tilde{X}}(\mathbb{R})$. There are a number of technical problems in defining this completion (which also came up in [Ros2, §3]), which are due to the fact that morphisms in the category $\mathcal{C}_{\tilde{X}}(\mathbb{R})$ are not necessarily bounded for any obvious norm, though they are "locally bounded" over compact subsets of $\tilde{X}$. To see exactly what is going on it is convenient first to simplify the category slightly by using the observation that $\tilde{X}$, being the universal cover of a compact metrizable space with fundamental group π, is "coarsely equivalent" to $|\pi|$, the group π viewed as a metric space with regard to the word length metric d defined by some set of generators. Thus the category $\mathcal{C}_{\tilde{X}}(\mathbb{R})$ is equivalent to the category $\mathcal{C}_{|\pi|}(\mathbb{R})$, for which the underlying metric space $|\pi|$ is countable (as a set). Then we may form the separable Hilbert space $\mathcal{H} \cong \ell^2(\pi) \otimes \ell^2$ (completed tensor product) which is the Hilbert space direct sum of one copy ℓ^2_g of ℓ^2 (the standard infinite-dimensional separable Hilbert space) for each element g of π. Each morphism in $\mathcal{C}_{|\pi|}(\mathbb{R})$ may be identified with an unbounded operator on $\mathcal{H}$, more specifically, by the operator defined by a matrix with rows and columns indexed by π, where the (g, h) matrix entry is a finite-rank operator from ℓ^2_h to ℓ^2_g, and where this entry vanishes as soon as $d(g, h)$ is sufficiently large (the "finite bandwidth" condition). The set of these operators is a $\mathbb{R}$-algebra $A(|\pi|)$ with an involution $*$, which has the convenient property of "stability," i.e., $M_n(A(|\pi|))$ is naturally isomorphic to $A(|\pi|)$ itself. However, $A(|\pi|)$ is a ring without unit, since the identity operator on ℓ^2_g is not of finite rank. For most purposes this is not a problem, since $A(|\pi|)$ has an "approximate identity" consisting of distinguished self-adjoint idempotents p_α, where α ranges over all maps $\pi \to \mathbb{N}$, in the sense that $A(|\pi|) = \varinjlim_\alpha p_\alpha A(|\pi|)p_\alpha$. (Cf. [HR1, Lemma 3.8].) The idempotent p_α is given by the diagonal matrix whose (g, g)-matrix entry is the orthogonal projection from ℓ^2_g onto the span of the first $\alpha(g)$ basis vectors. There is a naturally partial ordering on the maps $\alpha : \pi \to \mathbb{N}$, which corresponds to the usual partial order on idempotents as applied to the p_α, coming from the usual order $\leq$ on $\mathbb{N}$. Note also that the objects of the category $\mathcal{C}_{|\pi|}(\mathbb{R})$ are in natural bijection with the p_α's, so that for example an algebraic Poincaré complex in $\mathcal{C}_{|\pi|}(\mathbb{R})$ corresponds to an algebraic Poincaré complex of projective modules over some $p_\alpha A(|\pi|)p_\alpha$. Thus one can identify the L-theory spectrum of the category $\mathcal{C}_{\tilde{X}}(\mathbb{R})$ with the L^p-theory spectrum of the nonunital ring $A(|\pi|)$, by which we mean $\operatorname{hocolim}_\alpha \mathbb{L}^p(p_\alpha A(|\pi|)p_\alpha)$. (Here we are implicitly using the fact that $\mathbb{L}^p$, though not $\mathbb{L}^h$, is functorial for ring $*$-homomorphisms which are non-unital, but which map the unit of the first ring to a proper idempotent in the second.)

Now let $A^{\mathrm{bdd}}(|\pi|)$ be the subring of $A(|\pi|)$ consisting of matrices which act as *bounded* operators on $\mathcal{H}$. This is a $*$-closed algebra of Hilbert-space operators, so it has a C^*-algebra completion $C^*(|\pi|)$. It is trivial to see that this is the same as the algebra of the same name introduced in [Roe2] and [HR1, Definition 3.5]. Now we can give an analytic counterpart to Theorem 3.3.

3.5. Theorem. *In the situation of Theorem 3.3, the following diagram is homotopy-commutative. Here the map i_* is induced by the inclusion i : $A^{\mathrm{bdd}}(|\pi|) \hookrightarrow A(|\pi|)$ and (to save space) we have omitted explicit mention* *of the fact that everything is to be localized away from 2.*

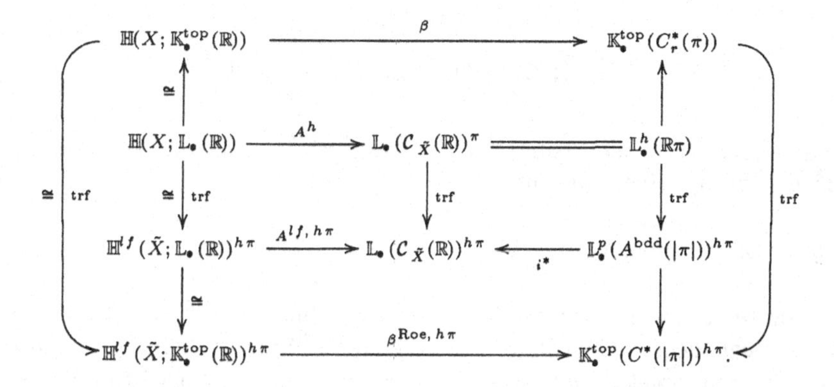

The horizontal map on the bottom is the analogue of β in "coarse Baum-Connes" theory, as explained in [HR2]. *In fact, the outer square in this diagram is the "spacified" version of a diagram appearing at the end of* §6 *of* [HR2].

Proof. Most of this follows immediately from Theorem 2.7 and Theorem 3.3. The fact that the middle and outer vertical arrows on the left are equivalences is, as we mentioned before, proved in [Car1, Corollary II.5], [Car2, §II] and in [CarP]. We only need to explain the arrows and the commutativity on the far right and very bottom of the diagram. The map $\mathbb{K}^{\text{top}}_\bullet(C^*_r(\pi)) \xrightarrow{\text{trf}} \mathbb{K}^{\text{top}}_\bullet(C^*(|\pi|))^{h\pi}$, as explained at the end of §6 of [HR2], arises from the fact that π acts naturally on $C^*(|\pi|)$, with fixed-point algebra Morita-equivalent (in the C^*-algebra sense) to $C^*_r(\pi)$. Thus the vertical arrow on the far right is induced by the inclusion $C^*(|\pi|)^\pi \hookrightarrow C^*(|\pi|)$. Similarly, π acts naturally on $A^{\text{bdd}}(|\pi|)$, with fixed-point algebra (algebraically) Morita-equivalent to $\mathbb{R}\pi$, so one has a similar vertical arrow trf : $\mathbb{L}^h_\bullet(\mathbb{R}\pi) \to \mathbb{L}^p_\bullet(A^{\text{bdd}}(|\pi|))^{h\pi}$ induced by the inclusion $A^{\text{bdd}}(|\pi|)^\pi \hookrightarrow A^{\text{bdd}}(|\pi|)$, and this arrow and trf : $\mathbb{K}^{\text{top}}_\bullet(C^*_r(\pi)) \to \mathbb{K}^{\text{top}}_\bullet(C^*(|\pi|))$ fit together in a homotopy-commutative square. The horizontal arrow at the bottom of the diagram is result of passing to homotopy fixed sets in the "spacified" version of the composite

$$K^{lf}_\bullet(\tilde{X}) \to KX_\bullet(\tilde{X}) \cong KX_\bullet(|\pi|) \cong K^{lf}_\bullet(\underline{E}\pi) \xrightarrow{\mu} K^{\text{top}}_\bullet(C^*(|\pi|))$$

discussed in [HR2], where the commutativity of the outer square is also proved. □

3.6. Corollary (basically discovered independently by several authors). *If, in the situation of Theorem 3.3,*

$$\mathbb{H}^{lf}(\tilde{X}; \mathbb{K}_{\bullet}^{\mathrm{top}}(\mathbb{R})) \xrightarrow{\beta^{\mathrm{Roe}}} \mathbb{K}_{\bullet}^{\mathrm{top}}(C^*(|\pi|))$$

is an equivalence, then (since this implies $\beta^{\mathrm{Roe}, h\pi}$ is also an equivalence) A^h is split injective (after inverting 2).

Proof. Chase the diagram. □

Now we can see in the analytic literature almost exact parallels to the topological theory. The coarse assembly map β^{Roe} may again be regarded as a "partial forget control" map from continuously controlled to bounded topology. If there is a model for $E\pi$ with a contractible metrizable compactification Y on which π acts, so that compact subsets of $E\pi$ become small when translated toward the boundary, then "pushing to the boundary" converts bounded control to continuous control, and thus provides a splitting to the coarse assembly map. Corollary 3.6 then gives a form of the integral Novikov conjecture.

Furthermore, one can do away with the assumption that π be torsion-free. When π has torsion, the natural conjecture is that of Baum and Connes (see 2.5 above) with $\underline{E}\pi$ replacing $E\pi$. This has a counterpart for L-theory as well. Work in progress of Carlsson, Pedersen, and Roe (among others) gives an analogue of Corollary 3.6 in this context.

REFERENCES

[Ad] J. F. Adams, *Infinite Loop Spaces*, Annals of Math. Studies, vol. 90, Princeton Univ. Press, Princeton, 1978.

[BCH] P. Baum, A. Connes, and N. Higson, *Classifying space for proper actions and K-theory of group C*-algebras*, C*-algebras: A 50-Year Celebration (R. S. Doran, ed.), Contemp. Math., vol. 167, Amer. Math. Soc., Providence, RI, 1994, pp. 241–291.

[BD] P. Baum and R. G. Douglas, *K-homology and index theory*, Operator Algebras and Applications (R. V. Kadison, ed.), Proc. Symp. Pure Math., vol. 38 (Part 1), Amer. Math. Soc., Providence, RI, 1982, pp. 117–173.

[Bla] B. Blackadar, *K-Theory for Operator Algebras*, Math. Sciences Research Inst. Publications, vol. 5, Springer-Verlag, New York, Berlin, Heidelberg, London, Paris, Tokyo, 1986.

[Bou] A. K. Bousfield, *A classification of K-local spectra*, J. Pure Appl. Algebra **66** (1990), 121–163.

[Carl] G. Carlsson, *Proper homotopy theory and transfers for infinite groups*, Algebraic Topology and its Applications (G. Carlsson, R. Cohen, W.-C. Hsiang, and J. D.

S. Jones, eds.), M. S. R. I. Publications, vol. 27, Springer-Verlag, New York, 1994, pp. 1–14.

[Car2] G. Carlsson, *Bounded K-theory and the assembly map in algebraic K-theory*, these proceedings.

[CarP] G. Carlsson and E. Pedersen, *Controlled algebra and the Novikov conjectures for K- and L-theory*, Topology (to appear).

[CoH] A. Connes and N. Higson, *Déformations, morphismes asymptotiques et K-théorie bivariante*, C. R. Acad. Sci. Paris Sér. I **311** (1990), 101–106.

[Fac1] T. Fack, *K-théorie bivariante de Kasparov*, Séminaire Bourbaki, 1982–83, exposé no. 605, Astérisque **105–106** (1983), 149–166.

[Fac2] T. Fack, *Sur la Conjecture de Novikov*, Index theory of Elliptic Operators, Foliations, and Operator Algebras (J. Kaminker, K. C. Millet, and C. Schochet, eds.), Contemp. Math., vol. 70, Amer. Math. Soc., Providence, RI, 1988, pp. 43-102.

[FRR] S. Ferry, A. Ranicki, and J. Rosenberg, *A history and survey of the Novikov conjecture*, these proceedings.

[FW] S. Ferry and S. Weinberger, *A coarse approach to the Novikov conjecture*, these proceedings.

[GM] I. M. Gelfand and A. S. Mishchenko, *Quadratic forms over commutative group rings, and K-theory*, Funct. Anal. and its Appl. **3** (1969), 277–281.

[Gr] M. Gromov, *Geometric reflections on the Novikov conjecture*, these proceedings.

[HR1] N. Higson and J. Roe, *A homotopy invariance theorem in coarse cohomology and K-theory*, Trans. Amer. Math. Soc. **345** (1994), 347–365.

[HR2] N. Higson and J. Roe, *On the coarse Baum-Connes conjecture*, these proceedings.

[HsR] W.-C. Hsiang and H. D. Rees, *Mishchenko's work on Novikov's conjecture*, Operator algebras and K-theory (San Francisco, 1981) (R. G Douglas and C. Schochet, eds.), Contemp. Math., vol. 10, Amer. Math. Soc., Providence, RI, 1982, pp. 77–98.

[Hu] S. Hurder, *Exotic index theory and the Novikov conjecture*, these proceedings.

[KM] J. Kaminker and J. Miller, *Homotopy invariance of the analytic index of signature operators over C^*-algebras*, J. Operator Theory **14** (1985), 113–127.

[Kar1] M. Karoubi, *Périodicité de la K-théorie hermitienne*, Hermitian K-Theory and Geometric Applications (H. Bass, ed.), Lecture Notes in Math., vol. 343, Springer-Verlag, Berlin, Heidelberg, and New York, 1973, pp. 301–411.

[Kar2] M. Karoubi, *K-Theory: An Introduction*, Grundlehren der math. Wissenschaften, vol. 226, Springer-Verlag, Berlin, Heidelberg, New York, 1978.

[Kar3] M. Karoubi, *Théorie de Quillen et homologie du groupe orthogonal*, Ann. of Math. **112** (1980), 201–257.

[Kar4] M. Karoubi, *Le théorème fondamental de la K-théorie hermitienne*, Ann. of Math. **112** (1980), 259–282.

[Kas1] G. G. Kasparov, *The operator K-functor and extensions of C^*-algebras*, Izv. Akad. Nauk SSSR, Ser. Mat. **44** (1980), 571–636; English translation, Math. USSR–Izv. **16** (1981), 513–572.

[Kas2] G. G. Kasparov, *K-theory, group C^*-algebras, and higher signatures: Conspec-*

tus, I, II, preprint, Inst. for Chemical Physics of the Soviet Acad. of Sci., Chernogolovka; reprinted with annotations, these proceedings.

[Kas3] G. G. Kasparov, *Operator K-theory and its applications: elliptic operators, group representations, higher signatures, C^*-extensions*, Proc. International Congress of Mathematicians, Warsaw, 1983, vol. 2, Polish Scientific Publishers, Warsaw, 1984, pp. 987–1000.

[Kas4] G. G. Kasparov, *Equivariant KK-theory and the Novikov conjecture*, Invent. Math. **91** (1988), 147–201.

[Kas5] G. G. Kasparov, *Novikov's conjecture on higher signatures: the operator K-theory approach*, Representation Theory of Groups and Algebras (J. Adams, R. Herb, S. Kudla, J.-S. Li, R. Lipsman, and J. Rosenberg, eds.), Contemporary Math., vol. 145, Amer. Math. Soc., Providence, 1993, pp. 79–100.

[KS] G. G. Kasparov and G. Skandalis, *Groups acting on buildings, operator K-theory, and Novikov's conjecture*, K-Theory **4**, Special issue in honor of A. Grothendieck, part 4 (1991), 303–337.

[Lod] J.-L. Loday, *K-théorie algébrique et représentations de groupes*, Ann. scient. Éc. Norm. Sup. **9** (1976), 309–377.

[MaM] I. Madsen and J. Milgram, *The Classifying Spaces for Surgery and Cobordism of Manifolds*, Ann. of Math. Studies, vol. 92, Princeton Univ. Press, Princeton, NJ, 1979.

[Mil] J. Miller, *Signature operators and surgery groups over C^*-algebras*, Preprint, 1992.

[Mis1] A. S. Mishchenko, *Infinite dimensional representations of discrete groups and higher signatures*, Math. U.S.S.R.-Izv. **8** (1974), 85–111.

[Mis2] A. S. Mishchenko, *C^*-algebras and K-theory*, Algebraic Topology, Aarhus 1978 (J. L. Dupont and I. Madsen, eds.), Lecture Notes in Math., vol. 763, Springer-Verlag, Berlin, Heidelberg, and New York, 1979, pp. 262–274.

[Mis3] A. S. Mishchenko, *Controlled Fredholm representations*, these proceedings.

[MF] A. S. Mishchenko and A. T. Fomenko, *The index of elliptic operators over C^*-algebras*, Izv. Akad. Nauk SSSR, Ser. Mat. **43** (1979), 831–859; English translation, Math. USSR–Izv. **15** (1980), 87–112.

[MS] A. S. Mishchenko and Yu. P. Soloviev, *Infinite-dimensional representations of fundamental groups, and formulas of Hirzebruch type*, Dokl. Akad. Nauk SSSR **234** (1977), 761–764; English translation, Soviet Math. Dokl. **18** (1977), 767–771.

[Nov] S. P. Novikov, *Algebraic construction and properties of Hermitian analogs of K-theory over rings with involution from the viewpoint of the Hamiltonian formalism. Applications to differential topology and the theory of characteristic classes*, Izv. Akad. Nauk SSSR, Ser. Mat. **34** (1970), 253–288, 475–500; English translation, Math. USSR–Izv. **4** (1970), 257–292, 479–505.

[Pal] R. S. Palais, *Homotopy theory of infinite dimensional manifolds*, Topology **5** (1966), 1–16.

[PRW] E. K. Pedersen, J. Roe, and S. Weinberger, *On the homotopy invariance of the boundedly controlled signature of a manifold over an open cone*, these proceedings.

[PW] E. K. Pedersen and C. Weibel, *K-theory homology of spaces*, Algebraic Topology, Proc., 1986 (G. Carlsson R. L. Cohen, H. R. Miller, and D. C. Ravenel, eds.), Lecture Notes in Math., vol. 1370, Springer-Verlag, Berlin, Heidelberg, and New York, 1989, pp. 346–361.

[Ped] G. K. Pedersen, *C^*-algebras and their Automorphism Groups*, London Math. Soc. Monographs, vol. 14, Academic Press, London and New York, 1979.

[Q1] F. Quinn, *Ends of maps, II*, Invent. Math. **68** (1982), 353–424.

[Q2] F. Quinn, *Assembly maps in bordism-type theories*, these proceedings.

[Ran1] A. Ranicki, *Exact Sequences in the Algebraic Theory of Surgery*, Princeton Math. Notes, vol. 26, Princeton Univ. Press, Princeton, 1981.

[Ran2] A. Ranicki, *Algebraic L-Theory and Topological Manifolds*, Cambridge Tracts in Math., vol. 102, Cambridge Univ. Press, Cambridge, 1992.

[Ran3] A. Ranicki, *On the Novikov conjecture*, these proceedings.

[RanL] A. Ranicki and N. Levitt, *Intrinsic transversality structures*, Pacific J. Math. **129** (1987), 85–144.

[Ren] J. Renault, *A groupoid approach to C^*-algebras*, Lecture Notes in Math., vol. 793, Springer-Verlag, Berlin, Heidelberg, and New York, 1980.

[Roe1] J. Roe, *Hyperbolic metric spaces and the exotic cohomology Novikov conjecture*, K-Theory **4** (1990), 501–512; *Erratum*, ibid. **5** (1991), 189.

[Roe2] J. Roe, *Coarse cohomology and index theory on complete Riemannian manifolds*, Mem. Amer. Math. Soc. no. 497 (1993).

[Ros1] J. Rosenberg, *C^*-algebras, positive scalar curvature, and the Novikov conjecture, III*, Topology **25** (1986), 319–336.

[Ros2] J. Rosenberg, *K and KK: topology and operator algebras*, Operator Theory / Operator Algebras and Applications (W. B. Arveson and R. G. Douglas, eds.), Proc. Symp. Pure Math., vol. 51 (Part 1), Amer. Math. Soc., Providence, RI, 1990, pp. 445–480.

[Ros3] J. Rosenberg, *The KO-assembly map and positive scalar curvature*, Algebraic Topology, Poznań 1989 (S. Jackowski, B. Oliver, and K. Pawałowski, eds.), Lecture Notes in Math., vol. 1474, Springer-Verlag, Berlin, Heidelberg, and New York, 1991, pp. 170–182.

[Ros4] J. Rosenberg, *Review of* **Elements of KK-theory** *by K. K. Jensen and K. Thomsen*, Bull. Amer. Math. Soc. **28** (1993), 342–347.

[RosW] J. Rosenberg and S. Weinberger, *An equivariant Novikov conjecture*, with an appendix by J. P. May, K-Theory **4**, Special issue in honor of A. Grothendieck, part 1 (1990), 29–53.

[TaW] L. Taylor and B. Williams, *Surgery spaces: formulae and structure*, Algebraic topology, Waterloo, 1978, Lecture Notes in Math., vol. 741, Springer-Verlag, Berlin, Heidelberg, New York, 1979, pp. 170–195.

[Wall] C. T. C. Wall, *Surgery on Compact Manifolds*, London Math. Soc. Monographs, vol. 1, Academic Press, London and New York, 1970.

[WW1] M. S. Weiss and B. Williams, *Automorphisms of manifolds and algebraic K-theory: Finale*, preprint, Univ. of Notre Dame.

[WW2] M. S. Weiss and B. Williams, *Assembly*, these proceedings.

DEPARTMENT OF MATHEMATICS, UNIVERSITY OF MARYLAND, COLLEGE PARK, MD 20742, U.S.A.

email: **jmr@math.umd.edu**

Printed in the United States
By Bookmasters